Cereals

HANDBOOK OF PLANT BREEDING

Editors-in-Chief:

JAIME PROHENS, *Universidad Politecnica de Valencia, Valencia, Spain*
FERNANDO NUEZ, *Universidad Politecnica de Valencia, Valencia, Spain*
MARCELO J. CARENA, *North Dakota State University, Fargo, ND, USA*

Volume 1
Vegetables I: Asteraceae, Brassicaceae, Chenopodicaceae, and Cucurbitaceae
Edited by Jaime Prohens and Fernando Nuez

Volume 2
Vegetables II: Fabaceae, Liliaceae, Solanaceae and Umbelliferae
Edited by Jaime Prohens and Fernando Nuez

Volume 3
Cereals
Edited by Marcelo J. Carena

Marcelo J. Carena
Editor

Cereals

Springer

Editor

Prof. Dr. Marcelo J. Carena
North Dakota State University
Corn Breeding & Genetics
Dept. of Plant Sciences
Dept #7670
374D Loftsgard Hall
Fargo ND 58108-6050
USA
marcelo.carena@ndsu.edu

ISBN 978-0-387-72294-8 e-ISBN 978-0-387-72297-9
DOI: 10.1007/978-0-387-72297-9

Library of Congress Control Number: PCN Applied for

Printed on acid-free paper.

9 8 7 6 5 4 3 2 1

springer.com

Preface

Plant breeding is a discipline that has evolved with the development of human societies. Similar to the rapid changes in other disciplines during the twentieth century, plant breeding has changed from selection based on the phenotype of individuals to selection based on the information derived at the deoxyribonucleic acid (DNA) level in molecular genetic laboratories and data from replicated field experiments. The initial beginnings of plant breeding occurred when humans made the transition from a nomadic hunter–gatherer lifestyle to the development of communities, colonies, tribes, and civilizations. The more sedentary lifestyle required that adequate food supplies (both plant and animal) were available within the immediate surrounding areas. The plants available within the immediate areas became very important to sustain the food, fuel, fiber, and feed needs of the local settlements. Hence, the greater the grain and forage yields of the native plants, the greater the sustainability of the needs of the local settlements. They recognized the relative importance of some plant species that could meet the needs of the settlements and practiced selection of individual plants that had greater grain and/or forage yields. Seed was saved from desirable plants to perpetuate the plants in the next growing season. By present-day standards, the methods of selection would seem simplistic because selection was based only on the phenotype of individual plants. But the selection methods were effective to develop landrace cultivars that provided substance for the local settlements to prosper and expand into regional civilizations. The landrace cultivars also were the germplasm resources for future generations of plant breeding. The original plant breeders, therefore, provided the plant resources for the development of human societies and the germplasm resources to sustain modern human societies. The major contributions of the early plant breeders were to develop domesticated crop species, dependent on humans (in some instances for survival) from their wild progenitors.

Domestication of our major crop species from their wild progenitors occurred over broad areas and time frames. The extent and rapidity of the distribution of the different domesticated crops depended on human movements within and among different areas of the world. It is estimated, for example, that maize (*Zea mays* L.) was domesticated 7,000–10,000 years ago in southern Mexico and Guatemala. Maize, however, was unknown outside the Western Hemisphere until Columbus

(1493) brought maize seed upon his return to Europe. The potential of maize was recognized and spread rapidly throughout the world. Similar patterns occurred for the other domesticated crop species. Because of the different needs of the different societies and the different environments inhabited, the next stage of plant breeding occurred. The selection techniques of the domesticators were used to develop cultivars adapted to their specific environments. Within the domesticated crop species, different landraces were developed that had the desired traits for the local needs and customs and environmental conditions. By 1900, it was reported, for example, that more than 800 distinctive open-pollinated cultivars were available in the United States. Until 1900, the plant breeding selection methods emphasized selection of individual phenotypes, but modifications were being made to improve selection effectiveness, such as the progeny test suggested by Vilmorin in 1858. Although the early plant breeders did not have a knowledge of Mendelian genetics (and his predecessors, they did observe that progeny tended to resemble their parents) and scientific methods to separate genetic and environmental effects (i.e., heritability) in trait expression, the early plant breeders were effective in domestication of wild, weedy plants for human use and the development of improved strains and cultivars that provided the germplasm resources for twentieth century plant breeders.

Plant breeding is often described as the art and science of developing superior cultivars. Art is defined as the skill in performance acquired by experience, study, or observation, which were certainly strong traits of the early plant breeders, whereas science is defined as the knowledge attained through study or practice. The distinctions between art and science are not always clear because even with experimental field and molecular data, subjective decisions are often necessary in choices of parents, progenies to consider for further testing, choices of testers, stage of testing, etc. But the relative importance of the art and science of plant breeding was reversed during the nineteenth and twentieth centuries with the emphasis on science (data driven) replacing emphasis on art (phenotypic appearance). The scientific basis of plant breeding was enhanced in the early part of the twentieth century by several developments, including the rediscovery of Mendel's laws of inheritance; a greater understanding of Darwin's theory of evolution based on Mendelian genetics; development of field experimental methods (randomization, replication, and repetition) to make valid comparisons among cultivars; theoretical basis for the inheritance of complex traits designated as quantitative traits; integration of the concepts of evolution, Mendelian genetics, and quantitative genetics to provide a basis to understand (and predict) response to selection; the importance of recycling of germplasm (both via pedigree selection within crosses of related lines and genetically broad-based populations) to enhance consistent genetic advance; and the advances made during the latter part of the twentieth century in molecular genetics on qualitative trait loci. Each of the developments impacted plant breeding methods in different ways, but collectively, all have been important to provide a firm and valid genetic basis for developing superior cultivars for the producers. Each of the advances was made to give greater emphasis to selection based on genotypic differences. During the past 100 years, plant breeding has changed from selection based on individual phenotypes to selection at the DNA level for selection for primarily genetic differences. This trend will continue in the future with greater emphasis at the DNA, gene, and phenotypic levels.

This volume is a summary and an update on the breeding methods that have evolved for our major cereal crop species, especially those based on breeding experience, often not presented in books. Similar to other research disciplines, rapid changes occur annually for the scientific basis of plant breeding. Although the basic genetic information and techniques of plant breeding continue to evolve, the basic concepts of plant breeding to develop superior cultivars remain the same; integrate all the available information to enhance the effectiveness and efficiency of our choice of parental materials, genetic enhancement of germplasm resources, estimate breeding values of progenies with greater levels of precision, and develop genetically diverse cultivars with greater tolerances to pest and environmental stresses as well as greater quality for a healthier diet. There is documented evidence that significant genetic improvements for greater yields have been made in cultivated crop species during the twentieth century. Similar genetic improvements are needed to meet human needs (e.g., biofuels) during the twenty-first century. Genetic information at the DNA level will continue to provide basic scientific information and will, hopefully, have a greater role in the future. Similar to other scientific disciplines, the science of plant breeding will continue to evolve for development of superior cultivars with the necessary traits to continue to provide adequate nutritional food supplies to sustain continued population expansions in a world of finite dimensions. Plant breeders have and will continue to develop cultivars. Plant breeding has and will continue to have important roles to ensure the future health of the world's human societies.

Fargo, ND Marcelo J. Carena
Ames, IA Arnel R. Hallauer

Contents

Section II Adding Value to Breeding

Contributors

J.A. Anderson
Department of Agronomy and Plant Genetics, University of Minnesota, St. Paul, MN 55108, USA

S. Baezinger
Department of Agronomy and Horticulture, University of Nebraska-Lincoln, Lincoln, NE 68588, USA

Y. Barrière
Unité de Génétique et d'Amélioration des Plantes Fourragères, INRA, Route de Saintes, BP6, F-86600 Lusignan, France

W. Berzonsky
North Dakota State University, Department of Plant Sciences, NDSU Dept. 7670, Po Box 6050, Fougo, ND 58108-6050

M.J. Carena
North Dakota State University, Department of Plant Sciences, NDSU Dept. 7670, Po Box 6050, Fougo, ND 58108-6050

S. Ceccarelli
The International Center for Agricultural Research in the Dry Areas (ICARDA), Aleppo, Syria

K.V. Cooper
P.O. Box 689, Stirling, SA 5152, Australia

F.A. van Eeuwijk
Wageningen University, Applied Statistics, 6700 AC Wageningen, the Netherlands

E. Elias
North Dakota State University, Department of Plant Sciences, NDSU Dept. 7670, Po Box 6050, Fougo, ND 58108-6050

J.C. Emile
Unité Expérimentale Fourrages et Environnement, INRA, Les Verrines, F-86600 Lusignan, France

J. Franckowiak
Department of Primary Industries and Fisheries, Hermitage Research Station, Warwick, Queensland, Australia

H.H. Geiger
University of Hohenheim, Institute of Plant Breeding, Seed Science, and Population Genetics, D-70593 Stuttgart, Germany

S. Grando
The International Center for Agricultural Research in the Dry Areas (ICARDA), Aleppo, Syria

R. Graybosch
USDA-ARS and Department of Agronomy and Horticulture, University of Nebraska-Lincoln, Lincoln, NE 68588, USA

S. Guillaumie
Unité de Génétique et d'Amélioration des Plantes Fourragères, INRA, Route de Saintes, BP6, F-86600 Lusignan, France

E.P. Guimaraes
Food and Agriculture Organization of the United Nations (FAO), Viale delle Termi di Caracalla, Crop and Grassland Service (AGPC), 00153 Rome, Italy

A.R. Hallauer
Department of Agronomy, Iowa State University, Ames, IA 50011, USA

R.G. Henzell
Department of Primary Industries, University of Queensland, Queensland, Australia

R. Horsley
North Dakota State University, Department of Plant Sciences, NDSU Dept. 7670, Po Box 6050, Fougo, ND 58108-6050

D.R. Jordan
Department of Primary Industries, University of Queensland, Queensland, Australia

H. Gómez Macpherson
Instituto de Agricultura Sostenible, CSIC, 14071 Córdoba, Spain

F.A. Manthey
North Dakota State University, Department of Plant Sciences, NDSU Dept. 7670, Po Box 6050, Fougo, ND 58108-6050

T. Medianer
University of Hobenbeim, State Plant Breeding Institute, D-70593 Stuttgalt, Germany

M. Mergoum
North Dakota State University, Department of Plant Sciences, NDSU Dept. 7670, Po Box 6050, Fougo, ND 58108-6050

L. Munck
Department of Food Science, Quality and Technology, Spectroscopy and Chemometrics Group, University of Copenhagen, Frederiksberg, Denmark

R.J. Peña
Wheat Program, International Maize and Wheat Improvement Center (CIMMYT), Mexico DF 06600, Mexico

M. Pichon
UMR5546, Pôle de Biotechnologie Végétale, 24 chemin de Borde Rouge, BP17, F-31326 Castanet-Tolosan, France

J.K. Ransom
North Dakota State University, Department of Plant Sciences, NDSU Dept. 7670, Po Box 6050, Fougo, ND 58108-6050

A.J. Lozano del Rio
UAAAN, Dept. de Fitomejoramiento, Buenavista, Saltillo, Coahuila, Mexico, CP 25315

I. Romagosa
Centre UdL-IRTA, University of Lleida, Lleida, Spain

C. Royo
Institute for Food and Agricultural Research and Technology, Generalitat de Catalunya, Cereal Breeding, Lleida, Spain

D.F. Salmon
Field Crop Development Centre, Alberta Agriculture and Food, 5030-50th Street, Lacombe, AB, T4L 1W9, Canada

D. van Sanford
Department of Plant and Soils Sciences, University of Kentucky, Lexington, KY 40546, USA

P.B. Schwarz
North Dakota State University, Department of Plant Sciences, NDSU Dept. 7670, Po Box 6050, Fougo, ND 58108-6050

P.K. Singh
North Dakota State University, Department of Plant Sciences, NDSU Dept. 7670, Po Box 6050, Fougo, ND 58108-6050

R.P. Singh
Wheat Program, International Maize and Wheat Improvement Center (CIMMYT), Mexico DF 06600, Mexico

W.T.B. Thomas
Scottish Crops Research Institute, Invergowrie, Dundee, UK

S.S. Xu
USDA-ARS, Northern Crop Science Laboratory, Fargo, ND 58108-6050, USA

Section I
Cereal Crop Breeding

Maize Breeding

Arnel R. Hallauer and Marcelo J. Carena

Abstract Maize (*Zea mays* L.) originated from teosinte (*Zea mays* L. spp Mexicana) in the Western Hemisphere about 7,000 to 10,000 years ago. Maize was widely grown by Native Americans (e.g. it was the first crop in North Dakota) in the U.S. during the 1600s and 1700s. The practical value of hybrid vigor or heterosis traces back to the controlled hybridization of U.S. southern Dents and northern Flints by farmers in the 1800s. Inbreeding and hybridization studies in the public sector dramatically change maize breeding. The Long Island (led by Shull) and Connecticut (led by East) public research groups created the inbred-hybrid concept (hybrid maize) which allowed industry to exploit the practical and economical value of heterosis. The hybrid maize technology was rapidly adopted by U.S. farmers and generated genetic gains for grain yield at a rate of 1.81 kg ha^{-1} $year^{-1}$. However, emphasis on the exploiting the inbred-hybrid concept detracted from further improvements on open-pollinated cultivars and their cultivar crosses.

Maize breeding is the art and science of compromise. Multi-trait selection, multi-stage testing, and multi-progeny evaluation are common for discarding thousands of lines and hybrids. Maize breeding has unique features that are different from any extensively cultivated self-pollinated crop. Breeding techniques from both self and cross-pollinated crops are utilized in maize. The fundamentals of maize breeding remain the same: germplasm improvement (e.g. recurrent selection), development of pure-lines by self-pollination, production of crosses between derived lines, identification of hybrids having consistent and reliable performance across an extensive number of environments, and production of the best hybrid for use by the farmer. Each successful hybrid has its own unique combination of genetic effects and allelic frequencies often limiting sample sizes for QTL experiments relative to classical quantitative genetic studies. The main limitation of traditional methods of maize breeding is to determine the genetic worth of lines in hybrid combinations. Most of the economically important traits in maize breeding are inherited quantitatively. Their importance is recognized by molecular geneticists through their emphasis in QTL experiments, molecular markers, mark-

M.J. Carena(✉)
North Dakota State University, Department of Plant Sciences, NDSU Dept. 7670, PO Box 6050, Fargo, ND 58108–6050, e-mail: marcelo.carena@ndsu.edu

M.J. Carena (ed.), *Cereals*,
DOI: 10.1007/978-0-387-72297-9, © Springer Science + Business Media, LLC 2009

er-assisted selection to predict early and late generation combining abilities, and/or ultimately gene-assisted selection through specific genome selection (e.g. metaQTL analyses) and/or association mapping. Information in maize genetics has significantly expanded in the past 50 years until the unraveling of the genome sequence in 2008. However, the limiting factor for genetic improvement remains the same: good choice of germplasm. The most sophisticated breeding methods and/or technologies carrying all of the genetic information available will have limited success if poor choices of germplasm are made. Biotechnology continues to be an important addition to the breeding process for single-gene traits while conventional breeding continues to be the key for improving economically important traits of quantitative inheritance. This chapter starts with a general introduction followed by pre-breeding and the incorporation of exotic germplasm, currently led by the USDA-GEM network. The integration of recurrent selection methods with inbred line development programs follows with the classical example of B73, the public line derived from BSSS that generated billions of dollars to the hybrid industry. The chapter continues with the inheritance of quantitative traits, and methods of line development and hybrids. Finally, the concepts of heterotic groups, heterotic patterns, and inbred line recycling are detailed for exploiting heterosis and hybrid stability including multi-trait selection utilizing indices. A summary is included at the end of the chapter.

1 Introduction

The evolution of maize (*Zea mays* L.) breeding methods is similar to other major cultivated crop species. Plant breeding started when humans made the transition from hunter–gatherers to living in more concentrated and organized societies. To meet the needs of the concentrated societies, the human needs for food, feed, fiber, and fuel, the plants within the surrounding native vegetation were observed and selected to meet their needs. The plants were highly adapted to the particular settlements and survived without human intervention. The choice of the plant species selected depended on the prevalence of the available plants and the needs of the settlements. The choice of plants selected was different in different areas of the world where the original settlements were being established.

Maize is one of the few major cultivated crop species that originated in the Western Hemisphere. Information suggests that maize arose in the highlands of southern Mexico and Guatemala about 7,000 to 10,000 years ago. Similar to other crop species, maize arose from a wild, weedy species native to the area. Collective information during the past 60 years suggests that teosinte (*Zea mays* L.: ssp. Mexicana) was the putative parent of modern-day maize (Wilkes, 2004). From the initial settlements to the highly developed societies of the native populations, selection of the more productive plants was conducted to meet the needs of the societies. Hence, maize arose from the wild, weedy type teosintes to produce types that became dependent on humans for survival. By the time European explorers arrived in the Western Hemisphere, maize was an important component of the societies throughout the Western Hemisphere. Columbus brought maize seeds to Europe after his first voyage in 1492, and maize became widely distributed upon its introduction to Europe (Mangelsdorf, 1974). Cortés, when he invaded Mexico in 1618, and DeSota, when he explored the area that is present-day southeastern United States in 1636, both found maize widely grown by the native populations throughout the respective areas (Marks, 1993; Hudson, 1994). Maize also was an important crop for the early European settlements established in the seventeenth and eighteenth centuries. Selection procedures similar to the methods of the native populations were used by the Europeans to further the development of more productive strains of maize; that is, phenotypic selection of individual plants and ear traits that were desired for their culture needs and environments. Galinat (1988), Goodman and Brown (1988), and Wilkes (2004) have summarized the information on the origin and on development of maize in the Western Hemisphere.

Although the transition from a wild species to a modern cultivated species was similar to other crops in many aspects, maize, however, has had some different properties, other than its origin in the Western Hemisphere. Maize is a cross-pollinated species with unique and separate male (tassel) and female (ear) organs. Maize breeding has unique features that are different from the other extensively cultivated grain species, such as rice (*Oryza sativa* L.), wheat (*Triticum vulgare* L.), soybeans (*Glycine max* Merr.), oats (*Avena sativa* L.), and barley (*Hordeum vulgare* L.), which are primarily self pollinated. Techniques from both self- and cross-pollinated crops are utilized in maize. To ensure control of parentage, hand pollinations are necessary where pollen (male gametes) collected from the tassel are either applied to the silks (female gametes) of the same plant (self-pollination) or to silks of different plants (cross-pollination). Controlled pollinations in maize breeding are conducted daily when plants are shedding pollen and have receptive silks. Techniques, however, have been developed that are used by nearly all maize breeders to produce good seed set by hand pollinations (Russell and Hallauer, 1980; Hallauer, 1994).

Because maize had become a very important source of feed for livestock, there was an interest in developing greater yielding maize cultivars. Data on US average maize yields had not changed significantly from 1865 to 1935 (Fig. 1). Beal (1880) reported on controlled crosses of open-pollinated cultivars and their potential for increasing maize yields. Other studies on cultivar crosses were reported, but varietal crosses were not extensively used. Parental control may have been a factor for the inconsistent results. Richey (1922) summarized data for 244 cultivar crosses and reported that the superiority of the cultivar crosses over the greatest yielding parent cultivar was not great enough to attract growers to the use of cultivar crosses. However, the economic potential of population hybrids through the population–hybrid concept utilizing extensively improved populations needs further consideration (East and Hayes, 1911; Hayes, 1956; Darrah and Penny, 1975; Carena, 2005a). Inbreeding and hybridization studies by Shamel (1905), East (1908), Shull (1908, 1909, 1910), and Jones (1918) dramatically changed maize breeding. The suggestions of Shull (1910) and Jones (1918) stimulated greater interest in the possibilities of hybrids produced from pure lines. The suggestions of the inbred–hybrid concept created greater interests that the public concept could impact maize yields. In 1922, a comprehensive effort was

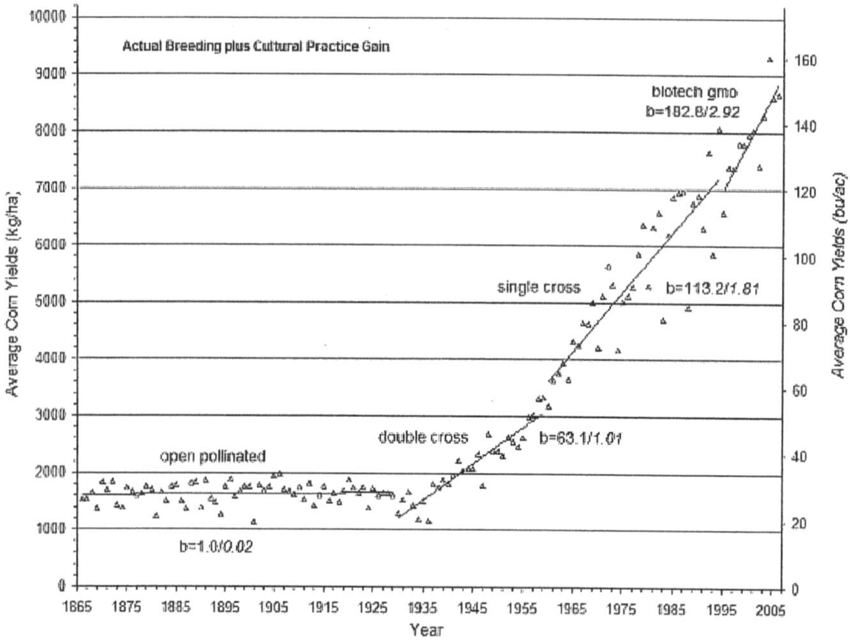

Fig. 1 Average US maize yields from 1865 to 2006 for different types of cultivars grown and regression (*b*) value for the different eras of different types of cultivars (USDA-NASS 2005 data prepared by F. Troyer and E. Wellin)

made by the US Department of Agriculture (USDA) and the state agricultural experiment stations (SAES) to test the new concept as a method to increase US maize yields. Extensive inbreeding studies to develop inbred lines and testing in hybrid combinations were conducted. The land-race cultivars (open-pollinated cultivars) were the initial germplasm sources for developing inbred lines. A few hybrids were tested in 1924, but it was 1935 before double-cross hybrids were generally available to the growers (Hallauer, 1999a). During the 70 years (1865 to 1935) average US maize yields had shown no improvement (or about 18.8 q ha^{-1}). The superiority of the double-cross hybrids compared with the open-pollinated cultivars convinced maize producers to use hybrids. By 1950 nearly 100% of the US Corn Belt growers were using double-cross hybrids. Average US maize yields gradually increased (1.01 kg ha^{-1} year^{-1}; Troyer, 2006) from 1935 to 1960. Because of intensive breeding and testing, the grain yields and agronomic traits of the newer inbreds were improved. Based on breeding results and the theory of genetic variability among different types of hybrids (Cockerham, 1961), conditions for the large-scale production of single-cross were available in the 1960s. The replacement of double-cross hybrids by single-cross hybrids resulted in greater yield increases (1.81 kg ha^{-1} year^{-1}; Troyer, 2006). Currently, single-cross hybrids are used on nearly 100% of the US maize area and in other temperate areas of the world. Because of economic conditions and environmental stresses, more complex hybrids and/or improved land-race cultivars are used in other areas of the world.

Where possible, however, the newer technologies to identify superior hybrids are emphasized for all major world maize producing areas. In lesser developed areas, mass selection methods are used to improve the currently grown land-race cultivars, sometimes referred to as farmer breeders (Dowswell et al., 1996).

2 General

The basic feature of all plant improvement programs is to increase the frequency of favorable allelic combinations. In maize breeding, this feature is common to all aspects related to maize improvement: introduction and adaptation of exotic germplasm, improvement of germplasm resources, pedigree selection to develop improved inbred lines, backcrossing to incorporate alleles and/or allelic combinations into otherwise desirable inbred lines, and conversion programs to improve and/or change the chemical composition of either the grain or the stover. The principles of recycling in maize breeding has been used since the Native Americans selected within teosinte to develop modern maize to the present-time when maize breeders make good-by-good crosses of inbred lines to initiate pedigree selection for developing inbred lines (Gepts, 2004). The inbred lines derived from the F_2 generation from crosses of inbred lines are usually referred to as recycled lines because they would include germplasm from the parental lines.

A common theme throughout the history of maize breeding has been selection of the superior individuals in a population, intermating the superior individuals, and selection of the superior individuals in the reconstituted population; this repetition of selection and intermating is continued during the lifetime of the breeding program. Until the development of inbred–hybrid concept in the twentieth century, phenotypic (or mass) selection of the superior individuals within the open-pollinated cultivars was the more common form of selection. Mass selection was effective in developing cultivars adapted to specific environments, cultivars with distinctive plant and ear traits, and cultivars with different maturities. The Native Americans developed distinctive cultivars distributed throughout the Western Hemisphere before the arrival of the European explorers. Similar methods were used by the European colonists on the cultivars developed by the natives. Sturtevant (1899) reported that there were nearly 800 unique open-pollinated cultivars in the United States. Although the mass selection methods were effective in developing identifiable open-pollinated cultivars, the methods were not effective in developing greater yielding cultivars (Fig. 1). Lack of parental control (poor isolation) and low heritability of the complex trait yield probably were the primary factors that limited the effect of mass selection for this particular trait. Better methods were needed to determine the genetic difference among phenotypes.

Rediscovery of Mendelism in 1900 stimulated research in the genetics and breeding of maize. Inbreeding studies by Shamel (1905), East (1908), and Shull (1908), the use of pure lines by Shull (1909, 1910) and Jones (1918), and the exploitation of the inbred–hybrid concept by the seed industry subsequently changed the landscape of maize breeding during the twentieth century. The open-pollinated cultivars developed by the Native Americans and the European colonists main role was as sources of germplasm to initiate development of inbred lines for use in hybrids.

There are several distinct phases in comprehensive maize breeding programs: prebreeding to evaluate and develop germplasm resources; genetic improvement of germplasm; and development and testing of inbred lines for use in hybrids. In most instances, equal weights are not given to each phase in individual breeding programs. Each phase does not directly contribute to developing inbred lines, but each phase can either directly or indirectly contribute to inbred line development.

3 PreBreeding

Prebreeding includes the introduction, adaptation, evaluation, and improvement of germplasm resources for use in breeding programs. Prebreeding usually does not provide directly new cultivars for the growers. It rather develops germplasm resources that are either directly or indirectly used to develop new cultivars. Prebreeding is not a recent concept and has been an important component in the development of present-day single-cross hybrids. Several stages of prebreeding have preceded the inbred–hybrid concept of maize breeding and continue today. The transition from a wild, weedy species to a species dependent on humans for its survival was the initial stage of prebreeding for modern maize, followed by the selection of open-pollinated cultivars adapted to environmental niches in nearly all maize growing regions of the world. The methods used to develop the open-pollinated cultivars were not systematic but the open-pollinated cultivars provided the germplasm for developing the first-cycle inbred lines that were the parents of the double-cross hybrids grown in the 1930s and 1940s. The development of the open-pollinated cultivars provided a wealth of germplasm for the twentieth century maize breeders.

Further development of germplasm resources was very limited during the period of 1910–1950 because extensive effort was given to developing breeding methods for effective and efficient development of inbred lines as parents of hybrids. Although Brown (1953) and Wellhausen (1956) emphasized that ~98% of the world's maize germplasm was being ignored, prebreeding efforts were either very limited or ignored. Prebreeding requires long-term goals which are not popular in either, the public or the private sector breeding programs and/or granting agencies. Immediate, short-term results are often difficult to measure and/or do not lead to development of commercial products or unique research. In most instances, researchers in both sectors need to provide evidence that progress is being attained, which may be difficult in the short term. Hence, prebreeding is not a popular research topic for young scientists who are under pressure for promotion and tenure in the public sector and to develop products that generate income in the private and/or public sectors. Funding has been a restraint, either being absent or inadequate, to support the long-term goals of prebreeding usually concentrating scientist efforts on research based on funding as opposed to research based on needs.

Prebreeding during the past 50 years has been more of an effort by individuals who appreciate the possible contributions that exotic germplasm can contribute to modern-day maize breeding programs. It has only been recently that a consortium of public and private individuals and organizations was formed to identify, improve, and develop a broad array of germplasm for present-day maize breeding programs (Pollak, 2003).

The development of the open-pollinated cultivars was an important contribution to the ultimate success of the inbred–hybrid concept. Although the open-pollinated cultivars were developed before the rediscovery of Mendelism in 1900, different individuals had different objectives in mind for the different geographical areas and anticipated uses. Consequently, the different open-pollinated cultivars often had distinctive plant and ear traits. Allele frequencies for different traits would differ among cultivars either because of intentional selection by humans or because of the environmental effects. The methods and materials also varied. DeKruif (1928), Wallace and Brown (1988), and Troyer (2001, 2006), for example, described the methods and materials used to develop landraces Leaming, Reid Yellow Dent, Lancaster Sure Crop, Krug, Minnesota 13, etc., all of which contributed useful inbred parents of the early double-cross hybrids. The only common theme used in developing these cultivars was that the originators desired to develop cultivars that, in their judgments, met the needs of the growers for their specific environments. For the more widely used cultivars (e.g., Reid Yellow Dent and Lancaster Sure Crop), additional strains were developed, such as Troyer Reid, Black's Reid Yellow Dent, Iodent, McCullock's Reid Yellow Dent, Osterland's Reid Yellow Dent, Richey Lancaster, etc. The wealth of available open-pollinated cultivars provided maize breeders choices for use in early breeding programs. Some cultivars were more useful sources of individual lines than others. The geographic areas of developing the open-pollinated cultivars (e.g., Lancaster Sure Crop in southeast Pennsylvania and Reid Yellow Dent in Delaven County, Illinois) led to the widely acclaimed heterotic groups of Reid Yellow Dent and Lancaster Sure Crop, which was to have a significant role in developing greater yielding hybrids in the US Corn Belt. Crosses between known genotypes (heterotic groups) that express a higher level of heterosis caused heterotic patterns to become established (Carena and Hallauer, 2001b). The development of the open-pollinated cultivars was an important prebreeding activity. Maize breeders (1920–1950) extensively sampled the better open-pollinated cultivars to develop inbred lines that were used extensively until 1950; for example, WF9, L317, I205, C103, 38-11, Hy, 187-2, Tr, 461, etc. (Crabb, 1947). After the initial samplings of the open-pollinated cultivars, further samplings were not successful in developing inbred lines that were superior to the initial sampling, which would be expected if the original samplings were adequate. Emphasis on developing the inbred–hybrid concept detracted from further improvements of the open-pollinated cultivars. Prebreeding essentially ceased in the early 1900s within the open-pollinated cultivars. Performance of crosses between open-pollinated cultivars were first reported by Beal (1880) and continued until about 1920.

Because of experimental methods and perhaps some relationships among cultivars (choice of germplasm), superiority of cultivar crosses was not consistent, and interest in cultivar crosses was not widespread (Richey, 1922). Greater interest and emphasis given to the potentials of the inbred–hybrid concept created less interest in further selection within the open-pollinated cultivars.

Interest in prebreeding was revived on a limited scale with concerns of the limited sources of germplasm included in US maize breeding programs during the 1950s and 1960s (Brown, 1953, 1975; Wellhausen, 1956, 1965). Greater impetus to this concern occurred with the southern corn leaf blight (*Bipolaris maydis* [Nisik.] Shoem.) epidemic in 1970 (Tatum, 1971). Although the southern corn leaf blight epidemic of 1970 occurred because of the extensive use of the Texas male-sterile cytoplasm source in production of the hybrid seed, concerns also were expressed of the genetic vulnerability of the major cultivated crop species (Anonymous, 1972). In most instances, the pedigrees of the germplasm used to develop the more important cultivars could be traced to a very limited number of ancestors. Isolated studies were conducted by interested individuals on the possible uses of germplasm that was normally not an important component of US breeding programs. Griffing and Lindstrom (1954), Kramer and Ullstrup (1959), Goodman (1965), Thompson (1968), Nelson (1972), and Moll et al. (1962, 1965) are examples where specific objectives were tested, but, in all instances, no major comprehensive long-term research programs were developed to follow up on the issues addressed. Griffing and Lindstrom (1954) crossed nine inbred lines (three adapted, three exotic, and three with 25–50% exotic germplasm) in diallel crosses. They found that inbred lines with 25–50% non-Corn Belt germplasm had combining abilities for grain yield greater than the 100% Corn Belt inbred lines; Goodman (1965) reported greater genetic variability in an exotic population compared with an adapted population; Thompson (1968) found that exotic germplasm had greater tonnage, but a lower quality silage, compared with adapted germplasm; and Moll et al. (1962, 1965) found there was a limit to genetic divergence and the expression of heterosis in crosses between adapted and exotic cultivars. In most instances, exotic germplasm infers the germplasm was acquired from some geographical area and was not adapted to the area for intended use. A more general usage of exotic germplasm includes all germplasm (adapted and nonadapted) that has not had selection and evaluation for direct use in applied breeding programs (Lonnquist, 1974). The specific studies did not resolve concerns about the limited genetic diversity in applied breeding programs but useful information was gleaned from the research for possible future use.

Interest in the potential of exotic germplasm in maize breeding was researched for different goals and interests. Because of sites of origin of maize in tropical and subtropical areas, it seemed that accessions from these areas would possess greater resistance and/or tolerance to major pests of maize because of year around exposure to the major pests of maize. Evidence suggests exotic germplasm does possess greater resistance to some of the major pests of maize. Kim et al. (1988) evaluated nine inbred lines, including six of tropical and subtropical origin, in a diallel mating design for resistance to feeding by the second generation European corn borer

(*Ostrinia nubilalis*, Hűbner). They reported the exotic inbred lines had greater resistance to second generation European corn borers and would be good sources of resistance if photoperiod sensitivity does not impede inbred line development. Holley et al. (1989) reported that tropical hybrids crossed with US Corn Belt testers had better resistance to kernel ear rot (*Fusarium moniliforme*). Tropical populations crossed with US Corn Belt populations suggested that the tropical populations possessed unique alleles for resistance to common rust (*Puccinia sorghi* Schw.), gray leaf spot (*Cercospera zea-maydis* Tehon and Daniels), and southern corn leaf blight (*Helminthosporium maydis* Nisik. and Miyake) that were not present in a widely used hybrid (Kraja et al., 2000). Holley and Goodman (1988a) also reported a greater level of resistance to southern corn leaf blight among 100% tropical inbred lines. Temperate adapted, 100% tropical inbred lines, evaluated per se and in hybrids exhibited greater resistance to *Diplodia maydis* (Berk.) Sacc., than did indigenous inbred lines (Holley and Goodman, 1988b). In addition to pest resistance, exotic sources of germplasm have been screened to determine if unique alleles can be identified that affect kernel quality traits. Campbell et al. (1995a) reported highly significant genetic variation for starch properties among 26 exotic inbred lines and suggested that screening for desirable starch property values would be useful. Evaluation of two heterozygous populations containing 50% exotic germplasm, but homozygous for the sugary (su2) locus, had increased genetic variation for starch thermal properties compared with inbred lines fixed at the su2 locus, suggesting the presence of modifiers that could be used to modify normal su2 starch (Campbell et al., 1995b).

Studies have been conducted evaluating the potential of exotic populations and their crosses and crosses between exotic and adapted population to determine their relative performance for grain yield and other important agronomic traits that are essential in modern maize production. The diallel mating design and testcrosses of exotic materials and adapted testers have been the more common methods for evaluating exotic sources. Evaluations have been made in both tropical and temperate regions. Crossa et al. (1990) evaluated diallel crosses of 25 recognized Mexican races at three elevations in Mexico and reported heterosis was expressed in several race crosses. In a 10-parent population diallel evaluated in the US Corn Belt, Mongoma and Pollak (1988) reported that BSSS(R)C10, an adapted population of primarily Reid Yellow Dent germplasm, had the best general combining ability (GCA), particularly with a Mexican dent population. Crossa et al. (1987) reported that populations with lower estimates of variety heterosis were among the better populations for mean cross performance, based on a 13-parent diallel of maize populations. They suggested that the relations between populations and their heterotic patterns would be needed for the correct choice of populations to include in reciprocal recurrent selection (RRS) programs. Diallel crosses between seven exotic populations and two US Corn Belt populations had greater grain yields among adapted by exotic crosses (50% adapted germplasm) compared with crosses having 100% adapted germplasm (Michelini and Hallauer, 1993). Echandi and Hallauer (1996) evaluated a diallel of eight populations including four 100% tropical populations, previously adapted to Iowa, and four US Corn Belt populations and the populations themselves for grain yield and seven agronomic traits in Iowa. The two greatest yielding crosses were BSSS(R)C12 × BSCB1(R)C12 (12 cycles of RRS in Iowa completed) and BS10(FR)C10 (10 cycles of RRS in Iowa completed) × BS29 (an adapted strain of Suwan-1), suggesting BS29 has potential in the Lancaster Sure Crop heterotic group (Menz and Hallauer, 1997).

Evaluation of exotic or exotic derived germplasm has been accomplished via use of testcrosses with either adapted single crosses or elite representatives of US Corn Belt heterotic groups rather than use of diallel crosses. Lonnquist (1974) compared both methods and reported the use of one or two elite testers from each heterotic group permitted more consistent assignment of exotic populations into the US Corn Belt heterotic groups. Mishra (1977) and Stuber (1978) reported good agreement between diallel and testcross information, but Christensen (1984) reported poor agreement between the two methods. Stuber (1978) crossed 285 exotic collections to three adapted single-cross testers that were evaluated in North Carolina. The best testcrosses were further evaluated 2 years in regional trials and the best four testcrosses had yields greater than 90% as much as the best commercial hybrids. Gutierrez-Gaitan et al. (1986) testcrossed 24 Mexican populations, developed primarily by CIMMYT, to two populations representing the primary heterotic group of the US Corn Belt; BS13(S)C3, a representative of Reid Yellow Dent and BS26, a representative of Lancaster Sure Crop. Testcrosses, testers, and populations themselves were evaluated in Mexico and the US Corn Belt. Grain yields of testcrosses did not differ significantly from the adapted tester populations in the US Corn Belt, and the US Corn Belt materials performed better than expected when evaluated in the Mexican environments.

Tallury and Goodman (1999) included all possible single, three-way, and double-cross hybrids among three primarily temperate and three adapted inbred lines in yield trials. Single-cross hybrids with 50–60% adapted germplasm produced grain yields equal to the commercial checks. Elite inbred lines (B73 and Mo17) representing Iowa Stiff Stalk Synthetic (BSSS) and non-BSSS heterotic groups were crossed to seven tropical populations and hybrids were evaluated for gray leaf spot, southern corn leaf blight, and common rust (Kraja et al., 2000). The exotic sources had favorable dominant alleles for each of the leaf diseases. Kraja et al. (2000) recommended that testcrosses to a series of tropical populations be made using the same inbred tester(s).

Holley and Goodman (1988b) evaluated the yield potential of tropical maize derivatives derived from diallel crosses of nine tropical hybrids. Selection was initiated within each cross during eight generations of inbreeding for acceptable maturity and other desirable agronomic traits. After eight generations of inbreeding and selection, 34 inbred lines were crossed to two US Corn Belt maturity testers with the testcrosses evaluated for 2 years at three North Carolina locations. They derived inbred lines from 100% tropical germplasm that had testcrosses that were adapted for agronomic traits to the southern United States, matured 1 week later than B73, had plant heights and grain moisture levels of testcrosses within the range of commercial hybrids used in the area, and about 25% of the testcrosses had grain yields similar to the commercial checks. Holley and Goodman (1988b) also found that the derived inbred lines were relatively insensitive to photoperiod effects, which has been a major concern with attempting to integrate tropical germplasm into temperate area breeding programs. They credited the insensitivity to photoperiod as a result of integrating complementary genetic systems from different tropical germplasm sources.

The use of tropical germplasm to broaden the genetic base of temperate area as sources of abiotic and biotic sources of pest resistance and for new traits is not without difficulty. Lack of adaptation is the primary limitation to determine which sources may have greatest potential to contribute useful genes and combinations of genes for temperate areas. Judicious selection of germplasm and careful selection, however, can overcome the handicaps of photoperiod effects (Holley and Goodman, 1988b). Lack of adaptation and lower mean yields of tropical materials compared with adapted temperate materials are, however, major difficulties for their immediate use, necessitating in most instances, longer-term breeding programs. Lack of adaptation is the primary reason two to three backcrosses are recommended when integrating germplasm from tropical sources into temperate materials. Holland and Goodman (1995) were able to develop photoperiod insensitive versions of the 40 original exotic accessions by a combination of crossing four plants of each exotic accession to an adapted inbred line and then intermating the earliest plants among the four full-sib families. This method was used for two additional generations to produce the photoperiod insensitive versions of the original exotic accessions. Hainzelin (1998) used a combination of mass selection and backcrossing of exotic materials to adapted germplasm to reduce the effects of photoperiod, which is similar to the method used by Holland and Goodman (1995). Photoperiod effects can also be reduced by crossing to a very early source followed by selection for adaptation (Gerrish, 1983; Holley and Goodman, 1988b) or by crossing improved unadapted sources followed by selection or by identifying photoperiod insensitive exotic sources (Oyervides-Garcia et al., 1985).

Another approach to adapt tropical materials to temperate areas includes selection for earlier maturity and desirable plant types during inbreeding in segregating populations. These populations included primarily backcross populations derived either from biparental crosses or 100% tropical hybrids. Eagles and Hardacre (1990) derived S_1 progenies from the backcross of an elite US Corn Belt population to a Mexican highland population to develop materials for the cool, temperate climate of New Zealand. S_2 progenies were derived from the S_1 progenies and S_2 testcrosses were evaluated. Grain yields of the selected S_2 testcrosses were similar to the S_2 testcrosses of the US Corn Belt recurrent parent, the S_2 testcrosses from the backcrosses had greater root lodging, but acceptable grain moisture levels. Caton (1999) for subtropical materials and Whitehead (2002) for tropical materials evaluated backcrosses derived from crosses between US Corn Belt and CIMMYT populations. Heterotic alignments of the respective areas were used in producing the population crosses: BSSS populations were crossed to primarily Tuxpeno sources and non-BSSS populations crossed to primarily non-Tuxpeno materials. All populations used in the crosses were derived from recurrent selection programs in Iowa and at CIMMYT. The crosses and backcrosses were produced in Mexico with the evaluations of backcross progenies and testcrosses of selected backcross progenies conducted in Iowa (Whitehead et al., 2006). There were 684 subtropical and 891 tropical backcross progenies evaluated 1 year at Ames, IA. Based on data for maturity, grain yield, root and stalk strength, and ear droppage, 142 subtropical and 181 tropical backcrosses were crossed to elite US Corn Belt testers and evaluated at five and seven US Corn Belt locations. Evaluation of backcrosses to temperate recurrent parents (25% tropical) and testcrosses of superior backcrosses (12.5% tropical) with elite temperate testers had flowering dates and harvest moisture levels that were, in most instances, not significantly greater than the recurrent parents and adapted checks. The objective of the research was to integrate elite exotic materials into elite temperate materials to combine favorable alleles for grain yield and other agronomic traits into germplasm pools that were adapted to temperate environments. The two-stage selection and testing program with multiple-trait selection was used to identify the superior backcrosses progenies that were intermated to form four germplasm pools (BS35, BS36, BS37, and BS38). Results suggested 25% elite exotic germplasm can be incorporated in the important US heterotic groups without disrupting the combining ability for grain yield expressed in the BSSS and non-BSSS crosses.

One concern when attempting to adapt exotic materials to temperate areas is the optimum proportion of exotic germplasm needed to include in adapted materials before initiating selection. Crossa and Gardner (1987) stated that the primary disadvantage regarding selection within populations backcrossed to adapted populations was that useful alleles present at a lower frequency in the nonrecurrent exotic population would have a greater chance of being lost with backcrossing to the adapted parent. Conversely, alleles from the adapted parent would have less chance of being lost in backcross populations than in populations with only 50% adapted germplasm. Albrecht and Dudley (1987) assessed the relative breeding value of four populations with different proportions of exotic germplasm. Random sets of 80–100 S_1 families were evaluated from populations that included 0, 25, 50, and 100% exotic germplasm. The set of S_1 progenies derived from the backcross population with 75% adapted germplasm had the greatest predicted genetic gain for grain yield itself and would be the more favorable population to initiate selection. Hameed et al. (1994) included 18 exotic inbred lines and their F_2 and backcross populations that were evaluated in testcrosses to B73 and Mo17. Grain yield increased in the backcross population versus the F_2 populations suggesting that backcrossing to the superior parent was the better method. Majaya and Lambert (1992) crossed five diverse Brazilian inbreds to two Lancaster Sure Crop inbred lines and then backcrossed to the two adapted lines. Selected backcross families were backcrossed again to the adapted lines to form the second backcrosses. Selection of the backcross families was based on multiple leaf and stalk rot pathogens, earlier maturity, and phenotypes similar to the recurrent parents. The best 26 backcross families from either the first or second backcrosses were evaluated as FRB73 testcrosses in Illinois. Generally, the families from the first backcross had better testcross grain yields than the check hybrids. Hofbeck et al. (1995) investigated the effects of backcrossing and intermating in an adapted ×

adapted tropical population by evaluating 100 unselected lines derived for each combination of three generations (50%, 75%, and 87.5%) of backcrossing and three cycles (0, 3, and 5) of intermating. Backcrossing shifted the means, reduced genetic variation, and developed earlier maturity levels, whereas intermating had no significant effects on the population. Hofbeck et al. (1995) concluded that backcrossing was more useful for the incorporating of exotic germplasm into temperate germplasm than intermating.

Mass selection (phenotypic recurrent selection) methods have been used effectively to adapt 100% tropical and tropical × US Corn Belt populations to temperate environments. Also, stratified mass selection has been successfully utilized to adapt late maturing temperate populations into the US north central region (Carena et al., 2008). Genter (1976) suggested that any system of cyclical selection that improves adaptation would decrease the frequency of the less desirable individuals and increase grain production. He conducted ten cycles of mass selection within 25 Mexican populations for erect, disease free plants with mature grain at harvest time. The selections from the 25 Mexican populations were intermated to form a single population having increased grain yield, fewer days to flowering, drier grain at harvest, greater stalk lodging, and no changes for root lodging. Six cycles of mass selection for earlier flowering within seven late flowering synthetic populations were completed in Minnesota at 65,000–87,000 plants per hectare (Troyer and Brown, 1976). The mass selection procedure included intermating the earliest 5% for flowering via use of bulk pollen. Cycle comparisons showed significant linear associations between cycle number and number of days to flowering, pollen shed-silking interval, grain moisture, and plant and ear heights. Troyer and Brown (1976) concluded that mass selection for earlier flowering at greater plant densities was effective for adapting later flowering synthetic population crosses for earlier maturity zones. Carena et al. (2008) and Eno (in press) showed similar results of successful adaptation of BS11(FR)C13 and BSK(HI)C11 improved populations after three cycles of stratified mass selection utilizing 22,500 seeds and selecting the 400 plants with earliest silk emergence and evaluation across nine environments in 2005, 2006, and 2007.

Hallauer (1999b) summarized the results of mass selection for earlier flowering in four tropical cultivars to reduce to photoperiod effects for possible use as germplasm sources for US Corn Belt breeding programs. For each cycle of mass selection, 10,000–15,000 seeds were planted in an isolated field and the 250 earliest flowering plants were marked for selection. Selection was based on silk emergence with no selection for pollen shed. Response to selection was similar for each tropical cultivar (Table 1). Average linear response for earlier flowering was -3.3 days cycle^{-1} of selection. Correlated responses to selection for earlier flowering included reduced ear height and increased grain yield. Grain yields increased because of greater adaptation to temperate environments. Other correlated responses included reduced tassel size, reduced root and stalk lodging, reduced plant height, reduced infection by *Ustilago maydis* (DC.) Cda. Narro (1990) evaluated Compuesto Selection Precoz after 15 cycles of half-sib recurrent selection for earliness. Compuesto Selection Precoz was formed by intermating

Table 1 Response to mass selection for earlier flowering in Eto Composite, Antigua Composite, Tuxpeno Composite, and Suwan-1 maize cultivars and correlated responses for ear height and grain yield

Cycle of selection	Eto Composite (BS16)		Antigua Composite (BS27)			Tuxpeno Composite (BS28)			Suwan-1 (BS29)		
	Days to silk (no.)[a]	Ear height (cm)	Days to silk (no.)	Ear height (cm)	Grain yield (q ha^{-1})	Days to silk (no.)	Ear height (cm)	Grain yield (q ha^{-1})	Days to silk (no.)	Ear height (cm)	Grain yield (q ha^{-1})
C_0	116	212	91	143	7.0	95	131	43.3	105	141	41.0
C_1	112	192	91	146	12.5	90	101	56.0	99	131	56.3
C_2	110	182	82	137	37.2	86	93	58.0	96	124	61.7
C_3	106	178	79	133	46.1	81	86	56.0	93	120	54.5
C_4	100	146	76	121	50.9	79	81	50.0	90	114	67.8
C_5	–	–	74	117	50.4	79	81	58.0	92	121	62.0
C_6	–	–	74	124	50.9	–	–	–	–	–	–
B[b]	−3.8	−15	−3.2	−5	19.3	−3.3	−9	3.0	−2.6	−4	5.9

[a]Number of days from planting to 50% silk emergence
[b]Estimates of linear response over cycles of mass seletion

15 high-yielding tropical materials. The goal of the selection program was to develop an earlier flowering, high yield cultivar for use in tropical areas. Selection was practiced at two locations in Mexico. After 15 cycles of selection for earliness, evaluations were conducted at 12 locations (nine tropical and three temperate) to determine responses (direct and indirect) to selection for earlier flowering. Time from planting to flowering decreased 0.46 days cycle^{-1} ($b = -10.46$), which was less than that reported by Troyer and Brown (1972, 1976) and Hallauer (1999b). For the one temperate location (Ames, IA), direct response was −1.30 day cycle^{-1}, which was similar to the data reported by Troyer and Brown (1972, 1976). Indirect response included reductions in grain yield, grain moisture, plant and ear heights, and leaf area. Mass selection is a very cost effective method for adapting exotic sources to temperate environment, but the adapted exotic sources require greater breeding efforts within breeding programs. This is because mass selection does not include any intentional inbreeding to reduce the genetic load of deleterious recessive alleles and no testcrossing with adapted materials is involved to determine the combining ability of exotic materials with adapted materials. For the tropical cultivars that Hallauer (1999b) adapted to temperate environments, the adapted tropical cultivars, however, had good performance when compared as cultivars themselves and in crosses with previously selected Corn Belt synthetic cultivars (Hallauer, 2003). Suwan-1 (BS29) performance itself and in crosses was similar to US Corn Belt synthetic cultivars that had undergone 10 or more cycles of RRS. Suwan-1 (BS29) and Tuxpeno Composite (BS28) are currently undergoing reciprocal half-sib recurrent selection and have flowering dates and harvest grain moisture levels similar to US Corn Belt populations (Menz and Hallauer, 1997; Hallauer, 2002).

Tropical cultivars grown in temperate environments are characterized as having tall stature, larger leaves, larger tassels, longer growing season because of photoperiodism, greater susceptibility to *Ustilago maydis*, lower grain yield, and consequently, a poor grain-to-stover ratio (<0.40). Thompson (1968) reported, for example, that the tropical cultivars provided greater tonnage for silage but had lower quality silage because of reduced grain production. Hallauer (1999b) found that grain yields increased with selection for earlier flowering in Antigua Composite, Tuxpeno Composite, and Suwan-1. On the average, days from planting to flowering decreased 3.3 days cycle^{-1} of selection and grain yields increased 9.4 q ha^{-1} cycle^{-1} because the tropical cultivars became more adapted to the temperate environments. Carena et al. (2008) and Eno (2008) found, on average across populations, that planting to flowering decreased 2.1 days year^{-1} of selection and grain yields increased 3.5 q ha^{-1} year^{-1} when adapting temperate materials to North Dakota.

The effects of photoperiod have been the major constraint in evaluating the relative potential of tropical materials for temperate areas (Goodman, 1985). It does not seem, however, that the use of tropical materials in temperate area breeding programs is limited by photoperiodism. Research by Gerrish (1983), Holley and Goodman (1988b), Holland and Goodman (1995), and Hallauer (1999b) suggests there is genetic variation for photoperiodism within tropical materials and that selection is effective for reducing the effects of photoperiodism. It seems a few major genes control photoperiodism. Mass selection by Troyer and Brown (1972, 1976) and Hallauer (1999b) was effective in developing earlier flowering strains of tropical cultivars that had grain yields similar to adapted US Corn Belt cultivars. Just because tropical materials are adapted to temperate environments, it does not infer that they are of immediate use to modern corn breeding programs. Similar to other germplasm sources considered for breeding programs, the adapted tropical materials need to meet the current standards for grain yield, root and stalk strength, pest tolerance, and maturity that are present in adapted materials used in current breeding programs. Additional selection will be needed. Hallauer (1999b), for example, initiated reciprocal half-sib recurrent selection (RRS) with BS28 (adapted strain of Tuxpeno Composite) and BS29 (adapted strain of Suwan-1) to improve grain yield and other agronomic traits. Days-to-flowering and harvest grain moisture of the BS28 and BS29 half-sib families are similar to those for BSSS(R) and BSCB1, two adapted populations that have been under half-sib RRS since 1949 in Iowa (Keeratinijakal and Lamkey, 1993).

Photoperiodism limits making direct comparisons between tropical and adapted cultivars in temperate environments. Hence, crosses of tropical cultivars with earlier maturity materials, backcrosses to adapted recurrent parents, and testcrosses of tropical cultivars to adapted testers often have been used to determine the relative potential of tropical cultivars in temperate area breeding programs. The proportion of the tropical germplasm in the materials evaluated can range from 12.5% to 50% with 25% to 50% the more common ranges for the central US Corn Belt. The lesser the amount of tropical germplasm included in the evaluation trials, the greater the opportunity that useful germplasm from the tropical cultivars may be eliminated. This, of course, would detract from the original goals of introducing tropical materials to increase genetic diversity and introduce useful alleles from the tropical materials in temperate area breeding goals. Dudley (1984a, b) suggested a method where a series of crosses are made between adapted and exotic materials to identify which exotic sources would contribute useful alleles that not currently present in the adapted sources. Crossa (1989) has discussed theoretically the choice of selecting the appropriate populations and the ideal percentages of exotic germplasm to integrate the more useful alleles of the two sources. The suggestions of Dudley (1984a, b) and Crossa (1989) have had limited empirical testing but the concepts could have greater application in studies in marker-assisted selection (MAS) and/or gene assisted selection (GAS) if adequate molecular markers or, ideally genes, become available in the tropical materials. It seems more likely that MAS will have greater applications in transferring desirable chromosome segments from elite tropical lines into elite adapted lines, in addition to aiding backcrossing on GMO single-gene traits.

Long-term programs for the introduction, evaluation, and adaptation of tropical materials for temperate environments are limited. The program at North Carolina State University led by M. M. Goodman has had the greatest impact in the United States (Goodman, 1999a, b). His approach has been to introduce and evaluate tropical inbred lines and hybrids. Crosses, backcrosses, and testcrosses have been used to identify selections whose performances have, in some instances, been either equal to or similar to the adapted hybrid checks (Goodman, 1999a, b; Holley and Goodman, 1988b; Uhr and Goodman, 1995; Holland and Goodman, 1995; Hawbaker et al., 1997). In some instances, inbred lines included 100% tropical germplasm. The program has used conventional maize breeding methods of inbreeding and testcross evaluation to identify inbred lines that have been released. In addition to grain yield and other agronomic traits, lines during development were also screened for most of the leaf diseases common to southeastern United States, but also present in the US Corn Belt. Goodman has also been an active leader and contributor of the Latin American Maize Program (LAMP) and the Germplasm Enhancement of Maize (GEM) program. The North Dakota State University (NDSU) maize breeding program used a similar approach for adaptation of released GEM lines to the northern US Corn Belt. Since the year 1999, early maturing (<95 RM) GEM derived lines have become available to increase the on-farm genetic diversity of early maturing maize. It is the first effort to incorporate tropical and late temperate genetically diverse early maturing (<100RM) materials in North America. Carena (2008) the Iowa State University maize breeding program used a different approach. This program used mass selection to adapt tropical cultivars that are important in tropical areas around the world for Iowa (Hallauer, 1999b). The tropical cultivars subjected to mass selection included Eto Composite from Colombia (during 1960s), Antigua Composite from Antigua (during 1970s), Tuxpeno Composite from Mexico (during 1980s), Suwan-1 from Thailand (during 1980s), and Tuson Composite from Cuba (during 1990s). Tuxpeno Composite was formed from different strains of Tuxpeno developed by Elmer Johnson at CIMMYT (Johnson et al., 1986), whereas Tuson Composite was formed by intermating five strains of Tuson provided by M. M. Goodman. It required 6 to 8 cycles (years) of mass selection to adapt most tropical cultivars to central Iowa. The adapted strains, designated as BS16 (Eto Composite), BS27 (Antigua Composite), BS28 (Tuxpeno Composite), and BS29 (Suwan-1) have flowering dates, harvest grain moisture levels, and plant and ear heights similar to the Iowa populations included in long-term recurrent selection programs. Tuson Composite has not been evaluated, but after 10 cycles of mass selection flowering dates and plant and ear heights are similar to adapted populations. No inbreeding and testcrosses of lines have been developed from any of the adapted tropical populations. Three cycles of S_1–S_2 recurrent selection have been completed in BS16 for grain yield, agronomic traits, resistance to first-generation European corn borer, and one cycle for tolerance to viruses. Three cycles of half-sib RRS have been completed for BS28 and BS29 (Hallauer, 2002). No inbred lines have been developed and released from any of the adapted tropical cultivars. Further improvements are needed for general agronomic performance, particularly greater root strength. From 1995 to 2005, a conversion program involving populations derived from long-term recurrent selection programs in Iowa and at CIMMYT was conducted. Crosses and backcrosses were produced in Mexico. Alignments of heterotic groups for the respective areas were retained; that is, BSSS populations crossed to primarily Tuxpeno materials, and non-BSSS populations crossed to non-Tuxpeno materials (Whitehead et al., 2006). Backcross progenies were evaluated for grain yield and moisture, root and stalk strength, and days to flower; superior backcross progenies crossed to adapted testers (LH185 for BSSS and LH198 for non-BSSS) and evaluated at 5 to 7 Iowa locations. Remnant seed of the superior backcross progenies per se and in testcrosses was intermated to form BS35 and BS36 (subtropical) and BS37 and BS38 (tropical) and released. These four populations include 25% subtropical and tropical germplasm. No inbred lines have been developed from the backcrosses.

Two other long-term programs for evaluating germplasm have a greater scope than those previously discussed; one, CIMMYT, has an international mandate to evaluate and develop materials for tropical areas worldwide and the other, GEM, primarily emphasizes the introduction and evaluation of exotic germplasm within the United States. The comprehensive program conducted by CIMMYT includes different stages for the development and evaluation of germplasm pools and populations for the subtropical, tropical, and tropical highland areas of the world (Vasal et al., 1982). The original goal was to provide improved cultivars for the smaller, subsistent farmers, but emphasis also has been given to developing inbred lines and hybrids during the past 20 years. Initially, 27 gene pools were formed that had specified adaptation, maturity, grain color, and texture. The gene pools represented broad reservoirs of genes formed by intermating diverse cultivars, cultivar crosses, and hybrids that possessed the specified traits of the 27 pools. It seems at this time no attention was given to heterotic groups. Major emphasis was given to pool and population development rather than hybrids. The emphasis for development of hybrids occurred later, and then greater concern was given to heterotic groups in tropical areas (Vasal et al., 1999). Modified ear-to-row selection was used to select for desirable progenies and plants within each pool. If other promising materials were identified that could enhance the pools, the new introductions were integrated into the respective pools. The more promising families and selections from the pools are entered in the advanced unit for more extensive testing, including international trials. Within the 27 populations included in the advanced unit, full-sib family selection is practiced. Based on extensive testing and selection, selected families are entered in elite experimental variety trials that are conducted by all interested national programs. Data from the international trials identified superior cultivars that were seed increased and distributed where requested. The comprehensive program met the goal of providing improved seeds to subsistent farmers, and germplasm from the CIMMYT program has become widely distributed and used in the lesser developed areas of the world (Dowswell et al., 1996). Although the use of hybrids has increased in the areas emphasized by CIMMYT's population improvement program, the products of the selection within the pools and advanced units provided the germplasm for the development of inbred lines and hybrids. The formation and selection within the genetic pools and advanced units provided the germplasm for developing inbred lines similar to the selected open-pollinated cultivars for the US maize breeding programs.

The GEM project in the United States is based on the information derived from LAMP. The goals of LAMP were to characterize and regenerate the maize accessions held in the Latin American germplasm banks. There were 12,113 accessions evaluated, and after five stages of evaluation and selection 268 elite accessions were identified (Salhuana et al., 1997). After the completion of LAMP, it was realized that additional selection was needed if the accessions were to contribute useful germplasm to present-day breeding programs. It was also realized that the task was too great for any one research unit to conduct the comprehensive program needed to genetically improve the accessions because of limited funding and staffing. An enhancement proposal was prepared and the US Congress appropriated $500,000 annual funding for the GEM project. The ultimate objective of GEM is to improve and broaden the germplasm base of hybrids grown by American farmers (Pollak, 2003). The GEM project is a cooperative effort including personnel and facilities of the USDA-ARS, universities, private industry, international, and non-governmental organizations to collaboratively broaden the maize germplasm base by improving germplasm from exotic sources. During 2007, GEM cooperators included 24 private companies, 20 public institutions, one nongovernment agency, and 11 international cooperators. The GEM breeding protocol is a modified pedigree method originating from crosses with adapted proprietary inbred lines provided by GEM cooperator companies. Early generation testing starts at the S_2 generation (first-year trials) followed by further testing at the S_3 generation (second-year trials). The testers in all stages are elite proprietary inbred lines. In the US Corn Belt, the focus is breeding of 25% tropical breeding crosses and 25% and 50% temperate breeding crosses, whereas at Raleigh, NC emphasizes 50% breeding crosses; that is, an additional backcross is used in the US Corn Belt to evaluate the exotic accession. GEM is an active, mature breeding program adequately funded to sustain the breeding efforts for the long term. GEM inbred lines that seem to offer useful genes for maize breeding programs are made available for public use. Materials from GEM will broaden the genetic base of US breeding programs. Pollak and Salhuana (1999) used restriction fragment length polymorphisms to determine relations among six Caribbean LAMP accessions and the adapted US inbred lines B73 and Mo17. They found that the LAMP accessions were different from B73 and Mo17 and that the Caribbean accessions also were very diverse among themselves. The limited sample studied by Pollak and Salhuana (1999) suggests even greater diversity can be expected from the broadly based exotic accessions being studied by GEM. GEM materials have been incorporated into US public programs to develop new sources of pest resistance (University of Illinois, University of Delaware, Texas AM University, and Cornell University)), grain quality (Truman State University, North Dakota State University, and Iowa State University), early maturity and moving GEM germplasm northward (North Dakota State University), drought tolerance (North Dakota State University and Texas AM University), and silage products (University of Wisconsin).

Suggestions of the potential of exotic germplasm to contribute useful alleles to temperate area breeding programs have been made for 50 years. It is estimated that temperate area breeding programs have used less than 5% of the available maize germplasm. Although specific research studies have been reported during the past 50 years, sustained long-term programs involving exotic germplasm were not continued because they were either not financially supported and discontinued or supported at levels not conducive for making significant contributions to applied breeding programs. The GEM program may be able to overcome past failures and provide useful germplasm with exotic components in the future. Incorporating exotic germplasm into adapted germplasm will require the same evolutionary methods that occurred in the formation of new races; that is, the US Corn Belt Dent race developed by merging the favorable alleles from the Northern Dents and Southern Dents followed by selection. The goal is difficult because intense selection during the past 100 years has developed an elite pool of genotypes that have favorable combinations of alleles. Any interruption of the favorable gene complexes by the introduction of exotic germplasm will cause changes in performance that may not be desirable. At present, the contributions of exotic germplasm to current inbred line development are limited but promising. Goodman (1999b) reported that exotic germplasm included in the inbred parents of US hybrids had increased from 1% in 1984 to 3% in 1996. And the largest proportion (1.99%) came from use of exotic temperate germplasm 41.2504B, a selection from Maize Amargo used to develop B64 and B68. Mikel and Dudley (2006) summarized information of the proprietary lines included to produce US single-cross hybrids. They found that the majority of current germplasm used in developing inbred lines originated from seven lines and that 63% of the lines had germplasm that traced to BSSS. BSSS was developed in the 1930s by intermating 16 inbred lines that had primarily Reid Yellow Dent germplasm at the Iowa State University corn breeding program. It is difficult to visualize how with the limited germplasm base of BSSS that BSSS germplasm could still have such an important role in current US hybrids. The recycling methods used in BSSS itself and in line development since 1939 have created a finely-tuned, unique complexes of genes (e.g., epistatic combinations) to produce inbred lines that in hybrids have consistently high levels of performance. If exotic germplasm was introduced in BSSS, the exotic germplasm would need to be carefully integrated to not disrupt the gene complexes developed during the past 60 years of selection and intermating. This will be a challenge where there has been a long history of selection within adapted germplasm.

4 Recurrent Selection

Systematic genetic improvement of maize populations and inbred lines requires the use of some type of cyclical continuous selection. Cyclical selection (phenotypic) was used to make the transition from a wild, weedy species to a cultivated species and in developing the different races, cultivars, and strains of the open-pollinated varieties. The originators of the popular open-pollinated cultivars Leaming, Reid Yellow Dent, and Lancaster Sure Crop used cyclical selection (mainly phenotypic) methods to develop cultivars that conformed to their concepts for developing superior cultivars for their environments. In some instances, germplasm from other sources were introduced to modify certain traits that they wanted to enhance. This process was continued by other individuals to develop substrains of the more popular open-pollinated cultivars to develop Osterland Reid, Black's Reid, Iodent,

Krug, Richey Lancaster, etc. This process did not use structured selection methods, rather the ideotypes selected depended on the individual concepts to develop better performing cultivars and, in earlier years, how the selected ears conformed to specified score card standards. Success depended on the patience and perseverance of the interested seedsmen. Interest in open-pollinated cultivars decreased with the suggestions of Shull (1910) and Jones (1918) for developing hybrids and the industry exploitation of the inbred–hybrid concept. Public efforts were complemented by the private sector being responsible for the practical success of hybrid maize (Carena and Wicks III, 2006). Simple selection methods were used effectively to develop unique open-pollinated cultivars (Sturtevant, 1899). However, average US maize grain yields did not improve from 1865 to 1935 (Troyer, 2006; Fig. 1). Failure to improve grain yields of the open-pollinated cultivars with the recycling methods used probably was because selection was based on individual plants (low heritability), selection on plant and ear traits that did not impact grain yield (low correlations), inadequate control of male gametes (poor or no isolation), and poor plot techniques (poor control of local environmental effects). Modifications for reducing some of the limitations for effectiveness of mass selection were proposed by Gardner (1961) and the extensive use of population improvement schemes and choice of elite maize germplasm were good suggestions but too late to influence the seedsmen in the 1800s and early 1900s.

The inbred–hybrid concept of Shull (1910) and Jones (1918) dominated US maize breeding and research from 1922 to 1950. Emphasis was given to developing inbred lines, testing of hybrids, and developing breeding methods to enhance the effectiveness of identifying inbred lines and their hybrids. The more popular open-pollinated cultivars were the primary source germplasm for developing inbred lines. Recycling methods for germplasm improvement and inbred lines were minimal. During the 1950s, there was greater interest in recycling to enhance genetically broad-based populations and improve the vigor and productiveness of inbred lines. Although average US maize yields increased 63.1 kg ha^{-1} from 1935 to 1965, it seemed that grain yields of double-cross hybrids had a plateau. Hence, there was a reexamination of breeding methods, germplasm sources, and inbred lines to determine how further progress can be attained.

The primary germplasm sources for developing the first-cycle inbred lines used in double-cross hybrids were the open-pollinated cultivars and no attention had been given to improve the original germplasm sources. Resampling of the open-pollinated cultivars had not been fruitful, which would be expected if the initial samplings had been adequate to capture the genetic variation within the open-pollinated cultivars. Heterosis had been exploited in developing the double-cross hybrids, but determining the genetic basis of heterosis has been elusive. Because the use of hybrids had become established and demanded by the producers, the suggestions of procedures to enhance genetically broad-based populations were influenced by the genetic effects considered of greatest importance in the expression of heterosis in hybrids. Hence, selection methods were suggested that emphasized selection for genetic effects that the originators considered of greater importance in the expression of heterosis in hybrids.

The selection schemes suggested for the genetic improvement of populations are designated as recurrent selection. Although recurrent selection is usually associated with population improvement and has methods of selection with different objectives when compared to pedigree selection, the concepts of recurrent selection also are applicable to pedigree selection methods for recycling of inbred lines (Hallauer, 1985). Effective recurrent selection methods contribute selections that can be used in pedigree selection (Hallauer and Miranda, 1988). In all instances the goal of recurrent selection schemes is to increase the frequency of the favorable alleles for the trait(s) under selection. The traits considered in recurrent selection are complex traits that are not amenable to Mendelian analyses, designated as quantitatively inherited traits. To genetically improve the quantitative traits, recycling methods, or recurrent selection methods are used to incrementally increase the frequency of the favorable alleles. Gradual genetic improvements are the norm for traits under recurrent selection. To accomplish the goals of recurrent selection, the following breeding procedures are used after the choice of populations to include in recurrent selection has been determined: develop progenies (or families) that adequately sample the genetic variation of the population; evaluate the progenies in replicated trials to determine their relative breeding values within the target environment(s); intermate the progenies with the superior breeding values to form a population for the next cycle of selection; and repeat the three stages for continued selection. The more important questions asked for each stage is: how many? How many progenies should be developed for evaluation? How many replications of data are needed to differentiate breeding values among the progenies? And, how many progenies should be intermated to form the next cycle population for continued selection? Specific answers to each question depend on trait under selection, relative heritability (h^2) of the trait under selection, types of progenies being evaluated, environmental effects and experimental techniques in determining the precision of the breeding values of the progenies, methods and facilities to use for intermating, and funds, facilities (e.g., winter nurseries), and personnel available to complete each of the three phases. Specific guides cannot be offered, but empirical data from the recurrent selection programs do provide some general guides. It seems 100 to 200 families should be sampled, based on empirical data reported by Marquez-Sanchez and Hallauer (1970a, b) for the standard errors of the additive genetic and the nonadditive genetic components of variance. The number of replications depends on the types of progenies (relative h^2s) evaluated, the environmental variation ($G \times E$) within target environments, and the expected experimental precision of the evaluation trials (LSD's). In Iowa, the h^2s for grain yield can vary from 0.45 for half-sib family selection to 0.90 for S_2 progeny selection when tested at four locations with two replications per location with similar error mean squares. In North Dakota, the h^2s for grain yield can range from 0.39 to 0.87 for the same methods when tested at three locations with two replications per location with similar error mean squares. In addition, reciprocal full-sib recurrent selection showed h^2s for grain yield at 0.78 when tested at six locations with two replications per location in North Dakota. Depending on differences among locations and the frequency of severe storms (locations lost), choices can be made for the distributions of replications (more or fewer) among locations; one replication at eight locations (more expensive) or eight replications at one location (less expensive but greater risk). The combination of replications and locations one uses is based on previous experiences. Except for mass selection, the number of progenies intermated to generate genetic variability for future cycles of selection has usually ranged from 10 to 20 selected progenies. Initially, $n = 10$ was the more common number intermated primarily because the diallel mating design (45 crosses) was used for intermating. If a greater number of progenies is included for intermating by the diallel mating design, the number of crosses increases rapidly (if $n = 20$, 190 crosses). Harrison (1967) introduced the incomplete diallel mating scheme which rapidly reduces the space requirements for intermating when n is large. If $n = 20$, 380 nursery rows are needed for diallel mating design, but only 40 nursery rows are needed with the Harrison mating design (Hallauer, 1985). Rawlings (1970), based on certain conditions, suggested that the number of parents intermating should be 16 to permit both short- and long-term response to recurrent selection. The number of progenies sampled, evaluated, and intermating ultimately has to be determined by the researcher based on goals of recurrent selection, trait(s) considered in selection resources available, and past experiences of the breeding program. Johnson (1982) developed some guidelines for conducting recurrent selection for GCA based on a two-locus theory. His theoretical studies suggested that the linkage disequilibrium in the initial population used to initiate recurrent selection has a permanent effect upon future selection response; that one or more generations of intermating before each cycle would reduce the effects of linkage disequilibrium in future cycles of selection; and that additional intermating the population(s) before recurrent selection is initiated would be more efficient in reducing the effects of linkage disequilibrium than intermating between cycles of selection. Hanson (1959) determined that 3 to 5 generations of intermating would be adequate to break up initial linkage blocks. Hence, if linkage disequilibrium is a concern in the population(s) considered for a long-term recurrent selection program, it would be prudent to include at least two to three generations of intermating before selection is initiated. One would not want to restrict future progress if linkages are a concern. Concerns for linkage disequilibrium would be greater in populations formed by intermating a restricted set of inbred lines, such as developing synthetic cultivars for base populations. It seems more positive long-term response would be realized by conducting more cycles of selection than more intermating between cycles of selection (i.e., keep years small as possible). An extensive review of the responses attained in long-term selection studies and how genetic effects and population sizes affect responses to selection were given by Janick (2004).

Maize breeders have several options available as to the recurrent selection method they consider appropriate for their situations. Because sufficient quantities of seed for evaluation can be produced for evaluation trials by either self- or cross-pollination, maize breeders have more options than for most other crop species. Most options include the original suggestions of Jenkins (1940), Hull (1945), Comstock et al. (1949), and Gardner (1961) and modifications thereof. Jenkins (1940) judged that additive genetic effects were more important in grain yield of maize populations and that selection for additive effects should be emphasized. Jenkins (1940) main interest was to increase the frequency of favorable alleles in cultivars that may be more suitable for use in stress environments rather than double-cross hybrids. The selection method proposed by Jenkins (1940) was designated as selection for GCA (Sprague and Tatum, 1942). Hull (1945), however, was of the opinion that nonadditive effects (dominance and/or epistasis) were of greater importance for the expression of heterosis and that selection should be emphasized for specific combining ability (SCA) (Sprague and Tatum, 1942). Hull (1945) suggested that experiments should be conducted within the same populations that included inbred (S_1 or S_2) progeny selection and half-sib family selection with use of either an inbred line or single-cross as the tester. He believed that the responses from the parallel selection studies would provide definitive evidence of the relative importance of additive versus nonadditive genetic effects in maize. Tanner and Smith (1987) and Horner et al. (1989) reported data for the study suggested by Hull (1945). In both studies, inbred progeny selection was initially superior to half-sib family selection, but in later cycles the responses to half-sib family selection was greater than for inbred progeny selection. Horner et al. (1989) attributed the greater response via half-sib family selection was because of the overdominant genetic effects expressed in the half-sib family selection program, but Tanner and Smith (1987) interpreted their results were mainly due to additive genetic effects. Comstock et al. (1949) proposed RRS because they showed theoretically that RRS would be equally effective to selection programs that emphasized selection for either GCA (additive genetic effects) or SCA (nonadditive genetic effects). If both additive and nonadditive effects were important in trait expression, then RRS would be more effective than the methods suggested by Jenkins (1940) and Hull (1945). Gardner (1961) suggested modifications to increase the effectiveness of mass selection. He imposed a grid system within isolated fields to reduce the effects of the microenvironments within the isolation field. Most of the modifications suggested were variations of the basic methods proposed by Jenkins (1940), Hull (1945), and Comstock et al. (1949) to improve effectiveness of selection, adapt to conditions for specific areas, and to make changes based on information from previous studies.

The choice of recurrent selection method to use is an important decision. Incremental genetic improvements of quantitative traits usually require long-term commitments to realize significant genetic improvements. Each stage of recurrent selection requires making good decisions for the populations included for selection including sample sizes and the extent of data required to differentiate the breeding values of the tested progenies. Changes and adoption of newer technology that enhances efficiency and effectiveness of selection can be made during the course of recurrent selection programs. But the critical component is the choice of population (s) that one believes can contribute useful germplasm to breeding programs for developing inbred lines and hybrids.

Recurrent selection programs should not be considered separate aspects of applied breeding programs. The usefulness of recurrent selection methods can only be fully realized if they are integrated with the applied breeding programs to develop superior cultivars (Hallauer, 1985). There are no time limits of recurrent selection programs unless it seems the genetic variability has been exhausted or the goals of selection (selection limits) have been attained. Effectiveness of recurrent selection is similar to other breeding methods. They are effective screens to determine genetic differences (breeding values) among progenies. For most pests of maize, effective screens have been developed to ensure uniform presence of pests to reduce the incidence of escapes. Laboratory techniques are available to rapidly discriminate among progenies for kernel and plant chemical compositions. For grain yield, evaluation trials may be adjusted with different allocations of replications and locations, use of more efficient experimental field designs to reduce experimental errors, and use of statistical analyses that exploit the information available in the field data; that is to improve techniques to permit us to identify the superior progenies.

The choice of recurrent selection methods depends on the traits considered in selection, genetic variability of the trait, preferred breeding method, and facilities available. The choices vary from the relatively simple method of mass selection to the relatively more complex RRS methods with modifications made for particular situations or to increase effectiveness of selection (Table 2). The parameters listed in Table 2 for each selection method affect genetic response to selection. If accurate estimates of genetic variability are available for the population(s) considered in selection, genetic responses (Δ_G) to selection can be determined for which method will have the greatest response to selection as a prediction. Eberhart (1970) developed a general formula that can be used to compare the relative genetic responses expected for each selection method. Eberhart (1970) suggested that $\Delta_G = (ck\,\sigma^2_{g'})/(y\,\sigma_P)$ be used to include the parameters that affect selection: c is the parental control, where $c = 1$ if the same materials evaluated are also used to intermate the selected progenies; k is a function of the intensity of selection (Table 6.1, Hallauer and Miranda, 1988); $\sigma^2_{g'}$ is the genetic variability among progenies or families; y is the number of years required to complete each cycle of selection; and σ_P is the square root of the phenotypic variance. The expectations of c, y, and $(\sigma^2_{g'})$ are listed in Table 2 for each of the listed methods for both intrapopulation and interpopulation selection methods. An example of the relative Δ_G for intrapopulation selection is listed in Table 3, based on the estimates of the genetic variability for BSSS maize population (Silva and Hallauer, 1975). Genetic gain estimates are ultimately expressed on a per year basis, where per year gain expresses the gain comparable to the selection methods that require 1 year for completion. There are greater Δ_G for selection methods that require more years because parental control and relative heritability are greater than for the more simple methods where one cycle can often be completed annually. The estimates of Δ_G included in Table 3 are conservative. Unlike most other maize populations, the estimates of σ^2_A (169) and σ^2_D (193) are similar and smaller than for other maize populations (Table 5.2, Hallauer and Miranda, 1988). Summaries of estimates of σ^2_A and σ^2_D by Hallauer and Miranda (1988) showed that estimates of σ^2_A were 2 to 4 times greater than the estimates of σ^2_D of other maize populations (Table 5.1, Hallauer and Miranda, 1988). Options for recurrent selection does include whether intrapopulation or interpopulation

Table 2 A summary of the parameters used to predict genetic gain (Δ_G) for different methods of recurrent selection

Methods of selection	Seasons per cycle	Parental control	σ^2_G [a]	
	y	C	σ^2_A	σ^2_D
Mass				
Original	1	$\leq$0.5[b]	1.00	1.00
Selection after flowering (Gardner, 1961)	1	0.5	1.00	1.00
Selection before flowering	1	1.0	1.00	1.00
Ear-to-row				
Original (Hopkins, 1899)	1	$\leq$0.5[b]	0.25	0.00
Modified (Lonnquist, 1964)	1	0.5	0.25[c]	0.00
Modified–modified (Compton and Comstock, 1976)	2	1.0	0.25[c]	0.00
Half-sib families				
Population as tester (Jenkins, 1940)				
Remnant half-sib seed	2	1.0	0.25	0.00
Self-pollinated seed	2	2.0	0.25+	0.00
Poor inbred of population	2	2.0	0.25+	0.00
Full-sib families				
Plant-to-plant crosses	2	1.0	0.50	0.25
Self-pollinated seed	3	2.0	0.50	0.25
Selfed progeny				
S_1 (Hull, 1945)	2	1.0	1.00	0.25
S_2	3	1.0	1.50	0.18
S_n	$n + 1$	1.0	~2.00[d]	~0.00[d]
S_1 modified[e] (Dhillon and Khehra, 1989)	1	0.5	1.00	0.25
Reciprocal recurrent selection				
Half-sib (Comstock et al., 1949)	2	1.0	0.25	0.00
Modified-1 (Paterniani and Vencovsky, 1977)	3	0.25	0.25	0.00
Modified-2 (Paterniani and Vencovsky, 1977)	2	0.5	0.25	0.00
Modified-3 (Russell and Eberhart, 1975)	2	2.0	0.25	0.00
Full-sib (Hallauer and Eberhart, 1970)	2	1.0	0.50	0.25
Modified (Marquez-Sanchez, 1982)	3	0.5	0.50	0.25

The parameters include number of seasons per cycle (areas with one crop per season but off-season nurseries can be used to either produce progenies or intermate selected progenies), parental control, and the expectations of the additive genetic (σ^2_A) and dominance (σ^2_D) components of variance of the components of variance among progenies (σ^2_G). Parents used to produce the progenies included for selection were noninbred

[a] Predicted gains were based on the formula suggested by Eberhart (1970) as $\Delta_G = (ck\,\sigma^2_{g'})/(y\,\sigma_P)$

[b] Parental control equals 0.5 if adequate isolation was used

[c] If within plot plant selection was used then 0.375 σ^2_A would be added to predicted gain

[d] Depending on the level of inbreeding at which progenies are evaluated but coefficients approach 2 and 0 as F approaches one for single-seed descent with no selection

[e] The modified S_1 scheme also includes testcross phase

Table 3 Predicted genetic gain (Δ_G) for different intrapopulation recurrent selection methods that were calculated for areas with one growing season (but off-season nurseries are used for intermating) based on components of variance estimated in Iowa Stiff Stalk Synthetic (Silva and Hallauer, 1975)

Methods of recurrent selection	Years per cycle (no.)	Parental control (c)	Predicted gain (Δ_G)	
			Per cycle (gm plant^{-1})	Per year (gm plant^{-1})
Mass selection, original	1	0.5	3.87	3.87
Mass selection, modified[a]	1	0.5	4.07	4.07
Ear-to-row, original	1	0.5	2.23	2.23
Ear-to-row, modified[a]	1	0.5	4.65	4.65
Ear-to-row, modified–modified[a]	2	1.0	9.31	4.65
Half-sib family selection				
Remnant half-sib seed intermated	2	1.0	9.31	4.65
S_1 seed intermated	2	2.0	18.62	9.31
Full-sib family selection	2	1.0	12.68	6.34
Inbred progeny selection				
S_1	2	1.0	20.64	10.32
S_1 modified[a]	1	0.5	12.52	12.52
S_2	3	1.0	27.25	9.08
S_n^b	4	1.0	29.39	7.33

The estimates of components of variance for estimation of Δ_G included $\sigma_A^2 = 169$, $\sigma_{AE}^2 = 92$, $\sigma_D^2 = 193$, $\sigma_{DE}^2 = 75$, $\sigma^2 = 185$, and $\sigma_W^2 = 1301$, based on the prediction formula given by Eberhart (1970), $\Delta_G = (ck\,\sigma_{g'}^2)/(y\,\sigma_P)$. Units of measure are gm plant^{-1}. Selection intensity was 10% for a k value of 1.755

[a]Modifications were suggested for mass, ear-to-row, and S_1 selection by Gardner (1961), Lonnquist (1964), Compton and Comstock (1976), and Dhillon and Khehra (1989)

[b]S_n progenies derived by single-seed descent with use of off-season nurseries to develop progenies where $F = \sim 1.0$

methods are used (Table 2). Although it is perceived that interpopulation recurrent selection methods are more complex than intrapopulation recurrent selection, it is not really the case. For half-sib RRS compared with half-sib intrapopulation (Table 3), the methods are similar except two parallel half-sib programs are conducted for half-sib RRS. The Δ_G expected for half-sib would include the sum of the Δ_G for each of the two populations (Table 12.13, Hallauer and Miranda, 1988). For full-sib RRS, only one set of full-sib families is evaluated and the Δ_G formula is similar to Δ_G for intrapopulation recurrent selection.

Mass selection has been effective for improving traits of maize with relatively high heritability, such as disease resistance (Jenkins et al., 1954; Genter, 1976), ear and plant heights (Acosta and Crane, 1972; Genter, 1976), ear length (Hallauer et al., 2004a), ear moisture content (Cross et al., 1987), earlier flowering (Troyer and Brown, 1972, 1976; Hallauer, 1999b; Carena et al., 2008 and Eno, 2008), insect resistance (Zuber et al., 1971), pericarp thickness (Ito and Brewbaker, 1981), seed size (Odhiambo and Compton, 1987), prolificacy (Lonnquist, 1967; Torregroza, 1973), and grain yield (Gardner, 1976). Gardner (1976) reported 3% average gain in grain yield after 15 cycles of mass selection in the Hayes Golden cultivar. After 15 cycles, there were no consistent increases in grain yield which he attributed to a decrease in the additive genetic variance. Also because of urban sprawl, the original field for selection was lost, and the selection study was transferred to another area, where the environmental effects were different to cause a delay in response to selection for cycles 16–30 (Hallauer, 1992). But Hallauer and Sears (1969) reported no significant improvements in grain yield after six cycles of mass selection in two open-pollinated varieties, Krug and Iowa Ideal. Mass selection is effective, at least initially. It seems that after the initial increases in frequency of major alleles have been attained that response to selection either decreases or reaches a plateau (Gardner, 1976). Correlated responses for increased grain yield were not realized with mass selection for components of yield even though significant responses were realized for the yield components (Odhiambo and Compton, 1987; Hallauer et al., 2004) except in isolated cases (Lonnquist, 1967; Mareck and Gardner, 1979). Mass selection is the oldest method used in plant selection, and the method continues to play a role in maize breeding, particularly in developing countries where improved open-pollinated cultivars are used (Dowswell et al., 1996).

Lonnquist (1964) indicated that the ear-to-row selection procedure is a method for between and within family selection. Except for mass selection, forms of ear-to-row selection have been used longer than the other methods in Table 2. The longest, continuous selection study in maize is the long-term selection experiment conducted at the University of Illinois in the open-pollinated cultivar Burr's White. The experiment was initiated in 1896 (Hopkins, 1899). The goal of the experiment was to determine if the chemical composition of the maize kernel could be altered by selection. Dudley and Lambert (2004) summarized 100 generations of selection for divergent protein and oil content; selection was effective in all instances. Dudley and Lambert (2004) presented a detailed history and analyses of the study that used a form of ear-to-row selection. Vasal et al. (1982) reported the use of modified ear-to-row selection scheme for improvement of five gene pools at CIMMYT after four to nine cycles of selection of grain yield, flowering time, and ear height. Average response per cycle across the five gene pools was 479.8 kg ha^{-1} (12.5%) for grain yield, 2.4 days (3.1%) earlier flowering, and 13.4 cm (12.7%) shorter ear height relative to the original (C_0) gene pools.

Half-sib family selection probably has greater use than any selection method in maize improvement. Half-sib family selection usually implicates the use of a tester to develop the half-sib families. Both of the original suggestions of recurrent selection used half-sib families. Jenkins (1940) used the source population as tester (GCA) whereas Hull (1945) suggested use of either an inbred or a single-cross as tester (SCA). Hence, the primary difference between the proposals of Jenkins (1940) and Hull's (1945) is the tester's genetic base. Half-sib family selection for GCA was initiated in BSSS with IA13, a double-cross hybrid, as the tester, designating the population as BS13 (Hallauer, 1992). Half-sib family selection was continued in BS13 until 1970, when changed to S_1–S_2 progeny selection. The half-sib family phase of recurrent selection in BS13 was effective in identifying inbred lines B14, B37, B73, and B84, which have been widely used in hybrids and as germplasm in pedigree breeding to develop recycled inbred lines (Mikel and Dudley, 2006).

Full-sib family recurrent selection has not been used to the same extent as half-sib recurrent selection. For intrapopulation improvement for both additive and dominance genetic effects, it seems full-sib family selection should have received greater interest. Moll and Hanson (1984), CIMMYT (Vasal et al., 1982), and NDSU (Carena, 2005a) have reported use of full-sib family selection. After 8 to 10 cycles of full-sib family selection, Moll and Hanson (1984) reported grain yield increases of 17.1 q ha^{-1} (26.2%) for Jarvis, 5.3 q ha^{-1} (6.5%) for Indian Chief, and 17.4 q ha^{-1} (20.6%) for the Jarvis × Indian Chief population cross. Jarvis and Indian Chief are open-pollinated cultivars adapted to North Carolina. The CIMMYT maize breeding program has made greater use of full-sib family recurrent selection than others. One example is the selection for grain yield, days-to-silk, and plant height for eight tropical cultivars (CIMMYT, 1984). After four to five cycles of full-sib selection, average responses per cycle of selection were 12.7 q ha^{-1} (5.9%) for grain yield, −0.6 days-to-silking (−2.6%), and −1.0 cm (−4.6%) for reduced plant height. CIMMYT has also used full-sib family selection for reduced ear height, grain quality, and pest resistance that included within family selection with among family selection. NDSU has currently eight full-sib selection programs evaluating grain yield, grain moisture at harvest, lodging resistance, test weight, and grain quality traits.

Use of progenies that are obtained by inbreeding has been used for several traits, particularly for pest resistance and grain quality. The level of inbreeding used is arbitrary, but usually either S_1 (F = 0.5) or S_2 (F = 0.75) progenies are used to reduce the length of each cycle of selection. In some instances, an unselected set inbred lines (F = ~1.0) derived by single-seed descent are used because all of variation among progenies are due to additive genetic effects. Inbred progeny recurrent selection (S_1) was suggested by Hull (1945) to compare with half-sib recurrent selection. Multiple inbred generations (say S_1 and S_2) have been used to reduce the higher costs associated with grain yield trials. S_1 progenies are developed by self-pollination of hundreds of plants within the population. The S_1 progenies may be screened in single replicate plots for diseases and insects common to the area. Progenies are discarded based on pest ratings and a less number of progenies retained for testing for grain yield and agronomic traits in replicated yield trials. One can usually expect significant response to inbred progeny recurrent selection for traits considered in selection. The additive genetic variance among inbred progenies are usually two to four times greater than among noninbred progenies (Table 2), which increases the h^2 of the traits if effective screens are used to measure genetic differences among progenies. Concern has been expressed that the experimental errors would be greater (e.g., grain yield) with inbred progeny field trials compared with noninbred trials but larger experimental errors are not necessarily linked with inbred progeny trials. Some have compared the coefficients of variation ($CV = \sigma/\overline{X} \times 100$), which are generally larger for inbred progeny trials because the mean ($\overline{X}$) of inbred progenies are 40–70% (depending on level of inbreeding) less than grain yield of noninbred progenies.

Inbred progeny selection for grain yield, however, has not been sustained for the long term (Tanner and Smith, 1987; Hallauer and Miranda, 1988; Horner et al., 1989). It seems there is a rapid fixation of the major alleles contributing to grain yield after two to four cycles of inbred progeny selection. Significant improvements for grain yield are realized from inbred progeny selection for the first two to four cycles of selection and level off in later cycles. Inbred progeny selection also was not effective in BS13(HI)C7 after seven cycles of half-sib family recurrent selection. It seems maize populations are more responsive in the long term if crosses (half-sib or full-sib families) are made and evaluated. Maize is nearly 100% cross pollinated and does not seem amenable to inbred-progeny selection for the long term. Comparison of S_1 progeny and half-sib family selection have been conducted to verify the suggestion of Hull (1945) of the types of genetic effects of greater importance that affect response for grain yield. Tanner and Smith (1987) evaluated the response to S_1–S_2 progeny and half-sib family selection after eight cycles of recurrent selection in a substrain (BSKC0) of the Krug open-pollinated cultivar. Response to inbred progeny selection was greater at the C_4 compared with the C_4 of half-sib family selection. At the C_8, they found greater response to selection via half-sib family selection; greatest grain yield was attained by the C_4 via inbred progeny selection with no further gains with four additional cycles of selection. Crosses were made between populations developed from the two methods of recurrent selection with 7.1% midparent heterosis for the $C_4 \times C_4$ cross and 14.1% for the $C_8 \times C_8$ cross, suggesting different alleles were selected by the two methods. Inbreeding depression (ID) estimates showed 7.9% less ID in the C_8 population from inbred-progeny selection versus the C_8 population from half-sib family selection. Horner et al. (1989) reported similar results comparing inbred-progeny and half-sib family recurrent selection in the Fla.767 maize population. They interpreted the different responses to selection to overdominant effects which were expressed in the half-sib families when crossed with the inbred tester, F6, which verified Hull's (1945) original suggestion that such comparisons of the two recurrent selection methods would provide evidence of the relative importance of the types of genetic effects important in maize crosses.

The combination of S_1 or S_2 selection on a progeny basis and evaluation of S_1 or S_2 testcross is a common feature of applied maize breeding programs. Inbred progenies that have superior GCA and SCA are retained in the breeding nursery for further inbreeding and selection and additional evaluation in crosses. Elite inbred lines can be used in pedigree breeding to develop recycled lines. This protocol is continued to enhance the combination of alleles important in hybrids. Hence, testing identifies superior inbred lines that can be recurrently selected to develop better inbred lines (see Fig. 3, Troyer, 1999). Although selecting among inbred progenies, based on testcrossing, has been extremely effective in developing genetically improved hybrids, formal recurrent selection experiments to compare relative effectiveness of inbred-progeny versus half-sib selection are limited to those reported by Tanner and Smith (1987) and Horner et al. (1989).

Interpopulation recurrent selection (RRS) studies have not been used as frequently as intrapopulation selection. Interpopulation recurrent selection is considered for those crop species that develop hybrids for the producers. Grain yield is the trait that has been subjected to most interpopulation selection studies. Comstock et al. (1949) showed that RRS is equally effective selecting for additive (GCA) and nonadditive (SCA) genetic effects. Interpopulation selection includes two initial populations, and direct response to selection is measured in the population cross. Responses in the two parent populations themselves would be indirect responses to selection, which is contrary to intrapopulation selection where direct response would be either in the population itself or testcrosses. Falconer (1960) showed that the expression of heterosis (H) can be formulated as $H = \Sigma y^2 d$, where y is difference(s) in allele frequencies, d is the level of dominance, and summed over loci. If $d = 0$, there is no expression of heterosis, and y will determine the magnitude of heterosis when d is greater than zero. With RRS we desire to increase y (i.e., increase the difference in allele frequencies), which would increase the rate of direct response in the cross.

A summary of RRS studies reported in the literature is given in Table 4. Average direct response for the 20 published programs was 4.8% cycle^{-1}, ranging from a low of 0.8% (Moll and Robinson, 1966) to 7.5% (Eyherabide and Hallauer, 1991). For specific RRS programs conducted for greater number of cycles, the average direct responses were generally greater and more consistent, which it would be expected to smooth out the variations of responses among cycles. If RRS is effective in selection of complementary alleles that affect heterosis (i.e., increase y, the difference in allele frequencies), the expression of heterosis would be expected to increase from crosses ($C_0 \times C_0$) of the original populations (C_0) to crosses ($C_n \times C_n$) of the advanced selected populations (C_n). Extensive data are not available, but information from five RRS programs suggests RRS is effective in selection of complementary sets of alleles (or combinations of alleles) that enhance heterosis (Table 5). Average midparent heterosis increased from 7.3% for the $C_0 \times C_0$ crosses to 37.4% for the $C_n \times C_n$ crosses, a 5.1-fold increase. Average direct

Table 4 Summary of reported direct responses of reciprocal recurrent selection for grain yield in maize populations

References	Cycles of selection (no.)	Direct response (% cycle^{-1})
Douglas et al. (1961)	3	5.8
Moll and Robinson (1966)	3	0.8
Moll and Stuber (1971)	6	3.5
Gevers (1975)	3	5.8 (S_0 plants)
Gevers (1975)	3	3.3 (S_1 plants)
Moll et al. (1977)	6	2.1 (Single crosses)
Paterniani and Vencovsky (1977)	1	7.5
Paterniani and Vencovsky (1978)	3	3.5
Darrah et al. (1978)	3	7.0
Smith (1983)	8	4.3
Hallauer et al. (1983)	8	3.6
Lambert (1984)	2	5.3
Moll and Hanson (1984)	10	2.7
Darrah (1986)	5	5.5
Eyherabide and Hallauer (1991)	8	7.5
Keeratinijakal and Lamkey (1993)	11	7.0
Schnicker and Lamkey (1993)	11	6.5
Betran and Hallauer (1996)	9	6.1 (Single crosses)
Menz et al. (1999)	6	4.4 (Population tester)
Menz et al. (1999)	6	1.6 (Inbred tester)
Average		4.8

response was 4.6% cycle^{-1} in the population crosses versus indirect responses of 0.3% cycle^{-1} in the populations themselves. The direct and indirect rates of responses are expected because greater emphasis is given to selection based on testcrosses with the opposing populations used as the respective testers. The rates of direct and indirect responses are similar for full-sib and half-sib RRS. Most of the RRS programs have been discontinued but additional four reciprocal full-sib recurrent selection programs were added based on alternative heterotic patterns found by Carena (2005b) for early maturing US regions. In this case, S_1 progenies are produced and then grown in pair-crosses in order to increase the production of seed, the goal being an increased number of testing locations to reduce $G \times E$ interactions during index selection of best progenies closing the gap between predicted and observed genetic gain.

On the basis of the data derived from diallel crosses between US open-pollinated cultivars, Kauffmann et al. (1982) suggested that Leaming (originating in Ohio) and Midland (originating in southeast Kansas) have the potential to be an alternative to the Reid Yellow Dent and Lancaster Sure Crop heterotic groups because of the heterosis (21%) expressed in the Leaming × Midland cultivar cross. Carena and Hallauer (2001a) evaluated the Leaming and Midland cultivars after three cycles of S_1 and S_2 recurrent selection that emphasized grain yield, root and stalk strength, European corn borer resistance (Leaming) and maturity (Midland) for the cultivars themselves and cultivar crosses (Table 5). Direct responses for grain yield in

Table 5 Realized direct and indirect responses and estimates of midparent heterosis for reciprocal recurrent selection programs conducted in temperate area maize populations that emphasized selection for increased grain yield

Populations	References	Cycles of selection (no.)	Response cycle^{-1}(%)		Midparent heterosis (%)	
			Direct	Indirect	$C_0 \times C_0$	$C_n \times C_n$
Jarvis and Indian Chief	Moll and Hanson (1984)	10	2.7	3.1 −10.7	6.6	28.9
BS10 and BS11	Eyherabide and Hallauer (1991)	8	7.5	3.0 1.6	2.5	39.6
BSSS and BSCB1	Keeratinijakal and Lamkey (1993)	11	7.0	2.0 0.0	25.4	76.0
BS21 and BS22	Menz et al. (1999)[a]	6	4.4	−0.2 0.5	1.0	25.4
BS21 and BS22	Menz et al. (1999)[b]	6	1.6	−15.9 −0.5	1.0	17.2
Average			4.6	0.3	7.3	37.4
Leaming and Midland	Carena and Hallauer (2001a)[c]	3	9.4 11.4	11.4	4.9	17.7

[a]Populations used as testers

[b]Inbred lines used as testers: A632 for BS21 and H99 for BS22

[c]Inbred progeny (S_1–S_2) recurrent selection used within populations for the heterotic pattern suggested by Kauffmann et al. (1982)

Leaming and Midland and in the cultivar crosses were similar. For the Leaming and Midland strains used in recurrent selection, midparent heterosis increased from 4.9% to 17.7% although only intrapopulation selection was emphasized. Neither Leaming (early maturity) nor Midland (late maturity) populations are adapted to central Iowa and the traits included in selection were for adaptation to central Iowa. It seems selection was effective for improving cross performance (indirect effects) with intrapopulation selection which agrees with other populations reported in Carena and Wicks III (2006). But the current status of the Leaming–Midland heterotic groups would not be competitive with the Reid Yellow Dent–Lancaster Sure Crop heterotic groups because of the emphasis given to the latter for the past 60 years. The use of newer heterotic groups would require the same intensive selection similar to the established heterotic groups.

In situations where producers demand hybrids, RRS would be the appropriate recurrent selection method. Heterotic groups have been identified for applied hybrid breeding programs, and populations of the heterotic groups would be the logical populations for use in RRS (e.g., Kitale II and Ecuador 573). Other populations used in RRS are synthetic cultivars developed by intermating selected inbred lines that are representative of the heterotic groups, for instance, BSSS and BSCB1,

BS21 and BS22, Leaming and BS22, BS21 and CGSS, BS21 and CGL, and NDSAB and BS21 (Keeratinijakal and Lamkey, 1993; Menz and Hallauer, 1997; Carena, 2005a; Melani and Carena, 2005; Carena and Wicks III, 2006; Jumbo and Carena, 2007). Inbred lines developed from RRS would represent the heterotic groups in testing and producing single-cross hybrids. In areas where hybrids are not an option, intrapopulation recurrent selection seems the better choice to improve cultivars themselves. However, the population–hybrid concept might be an alternative (Carena, 2005a).

Responses to direct and indirect recurrent selection programs are determined either by including the same checks in each of the cycles of evaluation trials as a base of reference or by conducting replicated trials of the different cycle populations and/or cycle population crosses and testcrosses where appropriate to determine direct response to selection. Both methods have been used. Better information can be obtained if all cycle populations and/or cycle population crosses are included in replicated trials to permit direct comparisons with the same error terms. For long-term selection studies, the inclusion of all cycles for evaluation usually requires seed reproduction to ensure all entries have the same seed quality. Because we are evaluating populations and/or their crosses, adequate plant and seed samples are needed to adequately represent the genetic variation and means of the populations. Sample sizes similar to those used to develop progenies (n = 100 to 200) seem appropriate. Hence, adequate size populations and/or population crosses are reproduced (usually by hand pollinations in breeding nurseries) by pollinating all plants, harvest all pollinated ears, and form balanced bulks (equal number seeds from each ear) to be used in evaluations to determine responses to selection. After preliminary analyses to estimate means of cycle populations and/or population crosses from the replicated evaluation trials, estimates of response to selection can be determined (Table 6). Often, the difference between the advance cycle population (Cn) and the original population (C_0) divided by number (n) of cycles has been used as measure of selection responses per cycle. If it is desired to present on a per year (y) basis, the difference can be divided by y, if more than 1 year cycle^{-1} has been utilized or, $(C_n - 1\ C_0)/C_0 \times 100$ = total percentage gain, divide by n = percentage gain per cycle, and finally divide percentage gain per cycle by years = percentage gain per year. All forms have been used. Both the linear regression coefficient and Smith's (1983) model to estimate response to selection have been used more frequently. For the three methods, the estimates of responses tend to be greater with use of $(C_n - C_0)/n$ than with the linear regression and the Smith model estimates. Estimates of responses of selection are similar for linear regression and the Smith model, but the R^2 values were greater with use of Smith model. The interpretations of the data and the conclusions would not change regardless of which method was used to estimate response to selection. It seems the $(C_n - C_0)/n$ estimates would tend to be biased upward and the Smith model provides a better fit to model because of consistently greater R^2 values and its inclusion of inbreeding and genetic drift effects.

More extensive reviews of the selection methods available and experimental data are given by Hallauer et al. (1988), Hallauer and Miranda (1988), Hallauer

Table 6 Estimates of response to 15 cycles of mass selection for earlier flowering in Compuesto Selecion Precoz population (Narro, 1990)

Trait	Estimates of responses to selection				
	$C_n - C_0/n$	Linear regression (*b*)		Smith model (1983)	
	$C_{15} - C_0/15$	*b*	R^2	2(ALI + DLI)[a]	R^2
Days-to-flower (no.)	−0.55	−0.46	0.89	−0.42	0.89
Grain yield (q ha^{-1})	−1.03	−0.96	0.81	−0.86	0.99
Grain moisture (%)	−0.06	−0.09	0.63	−0.10	0.99
Plant height, cm	−2.50	−2.22	0.27	−2.29	0.99
Ear height (cm)	−2.26	−2.00	0.51	−2.21	0.99
Leaf area (cm^2)	−202.93	−162.54	0.71	−150.16	0.99
Root lodging, %	−0.07	−0.12	0.30	0.07	0.91
Stalk lodging (%)	−0.10	−0.05	0.46	0.12	0.94

Direct response to selection is for days-to-flower and changes in other traits are correlated responses. Responses to selection were determined as $(C_n - C_0)/n$, linear regression coefficient (*b*), and linear function of ALI and DLI parameters (Smith, 1983)

[a]ALI is one-half the change in mean with selection due to effects of homozygous loci, and DLI is one-half the change in mean due to effects of heterozygous loci (Smith, 1983)

(1992), and Pandey and Gardner (1992). Evidence indicates that recurrent selection methods from mass selection to RRS are effective. One instance of no response to selection was reported by Williams and Davis (1983) who were selecting for greater resistance to stalk tunneling by southwestern corn borer (*Diatraea grandiosella* [Dyar]). Response was not realized because of low frequency of alleles for resistance and/or screening methods were not adequate to discern differences among genotypes. To be more productive and contribute to future genetic advances, recurrent selection should be conducted on a continuous basis to permit incremental genetic improvements that are cumulative over time (Eberhart et al., 1967).

Advances in molecular genetics have identified genomic segments that affect quantitative traits, designated as quantitative trait loci (QTL) (Tanksley, 1993). Dekker and Settar (2004) examined the effectiveness for use of molecular markers for improvement of quantitative traits for different genetic and selection methods. Selection on the basis of phenotype was compared with MAS. Relative comparisons were made via computation simulation with an additive genetic model assumed in all instances. Some general conclusions included (1) Use of molecular information that maximizes short-term response was expected to reduce long-term response compared with selection on phenotype alone; (2) For the situations that included 10 and 50 polygenes, a slightly greater portion of polygenes was lost in early generations (five to ten generations) with MAS than with phenotypic selection because reduced emphasis on polygenes; (3) Selection strategies for MAS that maximized long-term selection response were not recommended if greater emphasis placed on short-term selection response but they also state that decisions made to maximize short-term response can affect future response because of selection intensity and inbreeding; (4) QTL effects have no impact on selection strategies if the goal is to maximize long-term response but QTL estimates must be used to maximize short-term selection response; (5) Dominance effects were not expected to affect selection strategies for optimal MAS that maximize long-term response. For most comparisons made by Dekker and Settar (2004), the selection responses for long-term recurrent selection programs were similar for MAS and phenotypic selection. In genetically broad-based maize populations, different schemes of recurrent selection can be used to improve grain yield. Additive genetic effects usually are of greatest importance but nonadditive effects (dominance, overdominance, and epistasis) become important in long-term recurrent selection to enhance the heterotic groups for inbred line and hybrid development. Even within commercial breeding programs, long-term goals are used with continued selection within specific groups of inbred lines (Troyer, 1999; Duvick, 2004). Recycling, or recurrent selection, is long-term if continued incremental genetic gains are desired for the future, which was more than 60 years for the example presented by Troyer (see Fig. 3, Troyer, 1999). It seems an optimal balance has to be used with MAS (for greater short-term response) and what is expected 100 years in future, long-term response. Maize breeders emphasize breeding and selection strategies to improve current materials, which have been developed by previous long-term selection to enhance genetic complexes that have been identified in elite hybrids. Use of half-sib and full-sib family RRS methods would require molecular markers for two populations and every selection cycle, and how they would combine with the opposing population. The maize genome sequence is scheduled to be unraveled in 2008. Once we know the genes (not the markers) and, unlike the prediction of Bernardo (2001), selection efficiency of phenotypes could be complemented with gene information in order to enhance selection for quantitative traits in crops (Hammond and Carena, 2008). Genotypic information is not the bottleneck anymore compared to accurate phenotyping. Private–public interactions, especially between molecular biologists and breeders, should be able to solve the current limitations to significantly increase genetic gains in the future with wise use of technology (e.g. association mapping, metaQTL, and classical quantitative genetics on large samples until genes are known) and most efficient conventional breeding techniques.

5 Inheritance of Quantitative Traits

Maize breeders emphasize selection for a matrix of traits that are not amenable to Mendelian analyses. Effectiveness of selection for the different traits depends on developing effective screens for the traits individually and collectively. Although we assume the traits are quantitatively inherited, more consistent differences in maturities (less complex trait) are more easily measured than determining the relative consistent differences for grain yield (more complex trait). There are some traits that are usually considered to be quantitative but for which a major gene(s) has been found to significantly affect their expression. Jenkins et al. (1954) used recurrent selection to increase the frequency of alleles for resistance to northern corn leaf blight (*Helminthosporium turcicum* Pass.). Three major genes that contributed to resistance to northern corn leaf blight were found that reduced losses from the leaf disease (Smith and White, 1988). The long-term selection experiment conducted at Illinois used the ear-to-row method to modify protein and oil content of the maize kernel (Dudley and Lambert, 2004), but Mertz et al. (1964) reported that the opaque mutant gene (o_2) drastically changes grain protein composition and increased lysine content of the maize endosperm. Coors and Lauer (2001) reported that the brown midrib (bm) mutant alleles, the leafy allele (Lfy), and mutant waxy (wx) allele had been studied to increase silage quality. Although the individual genes had a direct impact for silage quality traits, they individually did not have long-term impact for the trait modifications. For the o_2 gene, there were nearly 30 years of dedicated research to integrate the correct combination of modifier alleles to develop cultivars that had acceptable levels of kernel and plant traits for use by the projected clientele (Vasal, 2001). The resistance conferred by the genes for northern corn leaf blight were soon overcome by mutant forms of the causal fungus and the resistant genes became an integral part of the package to select for general horizontal resistance. Coors and Lauer (2001) indicated that the genes projected to improve silage quality where not as successful as originally projected because effects of the alleles on other plant and ear traits were undesirable, similar to the effects of the o_2 allele. One major gene used effectively was the fertility restoring gene (Rf_1) that was used to restore fertility in Texas type cytoplasmic male sterility (cms-T), used in the production of hybrid seed. However, the cms-T genetic system was used effectively to produce hybrid seed corn until 1970 when the cms-T system was found to be highly susceptible to a strain of southern corn leaf blight [Bipolaris maydis (Nisik.) Shoem.]. Until 1970, nearly 90% of hybrid seed was produced using the cms-T system but by 1972, nearly all the seed production reverted to use of normal cytoplasm. Except for conversion of inbred lines to strains of cms-T (female) and Rf_1 (males) to use as parents in seed production, the genetic system was not used in breeding programs in an effort to improve quality and quantity of specific plant and ear traits (e.g., bm_2 or o_2).

The impacts of single major genes have been minor (e.g. except for business targeted GMO traits) compared with the overall usage of maize for food, feed, fuel (ethanol), and fiber, but the individual mutant alleles have contributed to the development of specialty crops that are used for specialized products (Hallauer, 2001, 2004). In each instance, however, the values of the mutant alleles were enhanced by modifier alleles and extensive breeding efforts to develop cultivars that had acceptable levels for agronomic traits to enable production of the specialty corns (Hallauer, 2001). The major factor that affected their use was that the mutant alleles tended to reduce yields 10–20%. All of the agronomic traits that were needed to develop useful cultivars also were quantitatively inherited.

Extensive theoretical and empirical studies have been conducted in maize (Hallauer and Miranda, 1988). Maize received greater attention earlier because the inbred–hybrid concept of Shull (1910) had become a reality by 1940. Hybrid maize was rapidly accepted by the US Corn Belt producers and was essentially grown on 100% of the US Corn Belt hectarage by 1950. The inbred–hybrid concept was developed empirically. The expression of heterosis in crosses of inbred lines was observed, but the genetic basis of heterosis was not resolved (Gowen, 1952). Definitive evidence on the theories of heterosis was not convincing and studies were conducted to determine the genetic effects that affected the quantitative traits. The accumulation of favorable dominant alleles and overdominance were the two primary theories suggested for expression of heterosis. The suggestions of Jenkins (1940), Hull (1945), and Comstock et al. (1949) for the genetic improvement of populations also were based on what were the primary genetics important for selection within maize populations. Two different situations (hybrids vs populations) were considered, but the relative importance of additive and nonadditive genetic effects was of interest.

Mather (1949), Comstock and Robinson (1952), and Cockerham (1956a, 1963) suggested mating designs for the estimation of genetic components of variance and average levels of dominance of genes affecting the inheritance of quantitative traits. Hallauer and Miranda (1988) summarized studies conducted in maize. Grain yield was of greatest interest with 99 studies reported, but estimates also were reported for 18 other traits (Table 5.1, Hallauer and Miranda, 1988). For grain yield, the average of the estimates of the additive genetic component of variance (σ^2_A) was 1.6 times greater than the average of the estimates of the components of variance due to dominance effects (σ^2_D). Often the estimates of σ^2_A were 2 to 4 times greater than the estimates of σ^2_D for grain yield. Average level of dominance was 0.94, or nearly complete dominance of alleles affecting grain yield. For the other traits, σ^2_A explained more of the total genetic variance (σ^2_G). Because it seemed that additively of alleles with partial to complete dominance effects were of greater importance. Gardner (1961) and Lonnquist (1964) suggested modifications of mass and ear-to-row selection to emphasize selection for additive gene effects. Different types of populations were studied, and the estimates of levels of dominance within F_2 populations created by selfing the cross (F_1) from two inbred lines were often overdominant, which supported the theory that heterosis was conditioned by alleles with overdominant effects. The studies were repeated by intermating plants within the F_2 populations 4–15 generations (Table 7). When the intermated populations were evaluated, the estimates of levels of dominance were partial to complete dominance. The greater the number of intermating generations, the greater the decrease in the levels of dominance was obtained. The estimates of σ^2_A, σ^2_D, and levels of dominance were influenced by linkage effects, as shown by Mather (1949). Mather showed that linkage effects bias the estimates of σ^2_A and σ^2_D. With coupling phase linkages, estimates of σ^2_A and σ^2_D have positive biases. If repulsion phase linkages, estimates of σ^2_A have negative biases, whereas estimates of σ^2_D have positive biases. Hence, the estimates of levels of dominance were overestimated because of repulsion phase linkages, with the linkage biases in the estimates of σ^2_A and σ^2_D reduced with intermating. The estimates of levels of dominance in F_2 populations were explained by pseudooverdominance because of linkage effects. From the studies conducted in maize populations, the evidence suggests that σ^2_A was

Table 7 Estimates of levels of dominance of genes for F_2 and intermated F_2 populations

Generation[a]	F_2 populations				
	$C_{121} \times NC_7$[b]	M14 × 187-2[b]	B73 × B84[c]	B73 × Mo17[c]	B73 × Mo17[d]
F_2	1.68	1.98*	1.53*	1.28	1.17
F_4	–	1.04	–	–	–
F_8	1.24	0.72	0.62*	0.95	–
F_{10}	–	–	–	–	0.80
F_{13}	1.09	–	–	–	–
F_{15}	–	0.62*	–	–	–

*Significantly different from 1.0
[a]Number of generations F_2 populations were intermated
[b]Gardner (1963)
[c]Han and Hallauer (1989)
[d]Cook (1998)

the more important component of the genetic variability for all traits, with dominance effects of greater importance for yield than for the other plant and ear traits.

The effects of epistasis were assumed absent for purposes of estimating σ^2_A and σ^2_D. Cockerham (1954, 1956b), however, derived the covariances of relatives that included epistasis and effects of linkage. But more complex mating designs were needed to estimate epistatic components of variances (Cockerham, 1956a). Attempts to estimate epistatic components have not been successful (Hallauer, 2006). Silva and Hallauer (1975), for example, used the combination of mating designs suggested by Cockerham (1956a). They were unable to obtain realistic estimates of the epistatic components of variance (most were negative) with estimates of σ^2_A and σ^2_D accounting for more than 90% of σ^2_G for all traits within the BSSS population. Wolf and Hallauer (1997) and Wolf et al. (2000) used different mating designs and they also were not able to quantify epistatic variance in the F_2 population of B73 × Mo17. The assumption of no epistasis in the estimation of σ^2_A and σ^2_D was necessary for estimation purposes, but it was acknowledged that epistatic effects were probably important in the inheritance of quantitative traits. Although epistatic components of variance have not been estimable, mean comparisons among different generations and hybrids have detected significant epistatic effects. Gamble (1962), using the generation-mean analysis proposed by Hayman (1958), Russell (1971), and Russell and Eberhart (1970) using factorial analyses to estimate the effects of individual mutant alleles in their phenotypes, and Bauman (1959) and Moreno-Gonzalez and Dudley (1981) comparing hybrids with different genetic expectations are examples of studies where significant epistatic effects were detected for grain yield and other plant and ear traits. The difficulty for the estimation of epistatic components of variance occurs because the coefficients of the epistatic components of variances in the expectations for the covariances of relatives are correlated with the coefficients of the σ^2_A and σ^2_D components of variance. This situation is not unique for the estimation of epistatic components of variances. Although the estimation of genetic effects by use of generation means is simpler and more precise compared with the estimation of genetic components of variance, Hayman (1960) emphasized that similar problems for estimation of epistatic effects do occur. If only additive and dominance effects are significant, they are unique estimates. But if significant epistatic effects are detected then the estimates of additive and dominance effects are not unique. For the estimation of components of genetic variances, the genetic components of σ_A^2 and σ_D^2 are initially estimated. If epistatic components of variance are sequentially estimated after σ_A^2 and σ_D^2, the estimates change dramatically for σ_A^2 and σ_D^2 and estimates of the epistatic components of variance often are negative and exceed their standard errors (Silva and Hallauer, 1975).

The information derived from the quantitative analyses of the relative importance of different genetic effects within maize populations did not advance our knowledge of the genetic effects important in the expression of heterosis. Generation mean analyses detected significant additive, dominance, and epistatic variation; estimates of components of σ_A^2 and σ_D^2 for F_2 populations suggested that levels of dominance were in the overdominant range, but these estimates were reduced to partial-to-complete dominance after intermating; estimates of components of σ_A^2 and σ_D^2 within genetically broad based populations suggested greater portion of σ_G^2 was due to σ_A^2; and estimation of epistatic components of variance was generally futile. A vast body of information on the inheritance of quantitative traits in maize has been reported (Hallauer and Miranda, 1988). Estimates of components of variance within maize populations suggest that σ_A^2 accounts for the greatest portion of σ_G^2, which seems reasonable because positive responses to selection within and between populations have been realized in all instances. The different types of generation mean analyses include the crosses and their derivatives produced from inbred lines. Significant estimates of additive, dominant, and epistatic effects were detected in all instances. The expression of heterosis in maize hybrids is determined by the complimentary interactions of alleles at each locus (dominance) and the interactions of alleles between loci (epistasis) of the two inbred parents. Hence, the estimates of genetic effects via generation mean analyses seem valid. Trying to make translations from estimates of components of variance within populations to the genetic effects operative in specific crosses of inbred lines does not seem reasonable. Estimates are specific for the specific materials used in analyses. The estimates of σ_A^2, σ_D^2, and σ_G^2 for populations were determined from a representative sample of genotypes (F_2s) for the population, whereas the estimates of genetic effects via generation mean analyses are for the complex of alleles in one F_1 formed by crossing inbred lines. If components of epistatic variances were estimable this would assist in determining with greater precision the genetic effects within populations.

Dudley and Moll (1969), Moll and Stuber (1974), and Dudley (1997) have discussed how the information on the inheritance of quantitative traits can be applied in planning selection and breeding strategies for cultivar development. The estimates of the genetic components of variance did not contribute directly to the explaining of the genetic basis of heterosis, but the estimates did provide information on the relative heritabilities (h^2) of plant and ear traits and relations (r_s) between traits. The estimates of h^2 were useful in planning selection and breeding strategies for improvement of populations and development of inbred lines and hybrids. Lush (1945) suggested the terms of broad-sense (σ^2_G / σ^2_P) and narrow-sense (σ^2_A / σ^2_P) h^2s, where σ^2_P is the estimate of phenotypic variance. Maize breeders have several options for the types of progenies used in selection and the extent of testing progenies in their target environments. Heritability estimates also are needed to predict Δ_G in recurrent selection schemes. In some instances h^2 estimates can be determined from the evaluation trials as $h^2 = (h^2 = (\sigma^2_{g'})/(\sigma^2/re + \sigma^2_{g'e}/e + \sigma^2_{g'}))$, where $\sigma^2_{g'}$ is the additive genetic component of variance for progenies, σ^2 is the experimental error, $\sigma^2_{g'e}$ is the genotype $\times$ environment interaction, r is the number of replications, and e is the number of environments. These types of estimates are on a progeny-mean basis. In some instances, a new selection program is planned and estimates of components of variance are not available. For these instances, mating designs are imposed on the population(s) of interest to estimate the genetic components, which are used to predict and compare the expected genetic gains for different selection strategies. Table 8 illustrates how the estimates of components of genetic variances can impact different selection methods, particularly with modifications made for the different methods (Table 3). Heritability estimates are specific for populations and testing situations, and can change over time if significant changes in allele frequencies

Table 8 Estimates of heritability (h^2) for individuals and progenies that can be considered in different selection schemes with the expected genetic components of variance (σ^2_A and σ^2_D) included in numerator and the phenotypic variance for the number of replications and environments used

Unit of selection	Expected heritabilities of the units selected[a]
Mass: Original	$\sigma^2_A/(\sigma^2_W + \sigma^2 + \sigma^2_{ge} + \sigma^2_g)$
Modified (Gardner, 1961)	$\sigma^2_A/(\sigma^2_W + \sigma^2_{ge} + \sigma^2_g)$
Ear-to-row: Hopkins (1899)	$(1/4)\,\sigma^2_A/(\sigma^2 + \sigma^2_{ge} + \sigma^2_g)$
Modified-1 (Lonnquist, 1964)	$(1/4)\,\sigma^2_A/[\sigma^2/r + (\sigma^2_{ge} + \sigma^2_g)]$
Modified-2 (Compton and Comstock, 1976)	$(1/4)\,\sigma^2_A/[\sigma^2/r + (\sigma^2_{ge} + \sigma^2_g)]$
Modified-3 (Marquez-Sanchez and Gomez-Montrel, 1988)	$(1/4)\,\sigma^2_A/[\sigma^2/r + (\sigma^2_{ge} + \sigma^2_g)]$
Half-sib	$(1/4)\,\sigma^2_A/(\sigma^2/re + \sigma^2_{ge}/e + \sigma^2_g)]$
Full-sib	$[(1/2)\,\sigma^2_A + (1/4)\,\sigma^2_D]/(\sigma^2/re + \sigma^2_{ge}/e + \sigma^2_g)$
Inbred	
S_1	$[\sigma^2_A + (1/4)\,\sigma^2_D]/(\sigma^2/re + \sigma^2_{ge}/e + \sigma^2_g)$
Modified (Dhillon and Khehra, 1989)	$[\sigma^2_A + (1/4)\,\sigma^2_D]/[\sigma^2/r + (\sigma^2_{ge} + \sigma^2_g)]$
S_2	$[(3/2)\,\sigma^2_A + (3/16)\,\sigma^2_D]/(\sigma^2/re + \sigma^2_{ge}/e + \sigma^2_g)$
S_7	$\tilde{2}\,\sigma^2_A/(\sigma^2/re + \sigma^2_{ge}/r + \sigma^2_g)$

[a]The variance components include additive genetic (σ^2_A), dominance (σ^2_D), total genetic (σ^2_g), within plot (σ^2W), experimental error (σ^2), and genetic $\times$ environment interaction (σ^2_{ge}). r and e are the number of replications and environments, respectively. If within plot selection is conducted, there is an additional component that can be included in Δ_G (e.g., Compton and Comstock, 1976)

occur with selection (Hanson, 1963). The component $\sigma^2_{g'}$ was used by Eberhart (1970) for the general Δ_G formula. In Table 2 the expected components of variance (σ^2_A and σ^2_D) for the different progenies are included in the numerator to represent $\sigma^2_{g'}$. The estimates of h^2 will vary with the type of materials (individuals or progenies) under study, but the estimates of h^2 also will vary among populations for the same materials because allele frequencies affect the estimates of the genetic components of variance and will vary if different combinations of replications (r) and environments are used to determine the progeny means. Estimates of h^2 for mass selection are restricted to one environment. The estimates of h^2 are relative for each trait and vary among traits depending on how many genetic factors affect trait expression. For mass selection and ear-to-row selection estimates of components of variance will need to be obtained for the population under study to determine h^2 in order to predict Δ_G using mating designs that permit estimation of σ^2_A. For the other materials, h^2 estimates can be obtained from ANOV if progenies are evaluated in replicated trials (Smith et al., 1981). But estimates from mating designs also can be used to estimate the h^2 used for predicting Δ_G for progeny selection, as illustrated in Table 3. One general problem with individual plant selection is to separate the genetic and environmental effects for variation among plants. Inbred lines are homozygous and homogeneous and F_1s between inbred lines are homogeneous; variation among plants of inbred lines and F_1s would be due to microenvironment effects, provided the inbred lines and hybrids did not include off-type plants because of contamination. F_2 and backcross populations are heterozygous and heterogeneous and would be expected to include genetic and microenvironmental effects. Burton (1951), Mahmud and Kramer (1951), Warner (1952), and Weber and Morthy (1952) have offered methods for estimation of individual plant h^2s. The main concern were that estimates of environmental effects among inbred plants may be overestimated because of their lower vigor (Mahmud and Kramer, 1951) and that estimates among F_1 plants may be underestimated because of their greater vigor (Burton, 1951). Warner (1952) and Weber and Morthy (1952) attempted to reduce these concerns by using F_2 and backcross populations, which are heterozygous and heterogeneous. Schmidt and Hallauer (1995) tested the four suggested methods and found large differences among the h^2 estimates for the four methods for the same F_2 populations as well as among F_2 populations. An example is the cross of B14A $\times$ L317: estimates of h^2 for grain yield were 1.8, 75.4, 61.0, and 36.0% for the methods suggested by Burton (1951), Mahmud and Kramer (1951), Weber and Morthy (1952), and Warner (1952), respectively. There were a range of differences among methods among crosses as well as the differences among methods within the same cross. The differences were caused primarily by the combination of generations used to determine the environmental effects (σ^2_W) among plants.

Estimates of h^2s for the different plant and ear traits and phenotypic and genetic correlations between traits provided information on effectiveness of selection for different traits and/or combinations of traits. Grain yield was of greatest interest but the estimates of h^2 for grain yield were usually smaller than for traits considered as components of yield (e.g., ear length, ear diameter, number of kernel rows, kernel depth, number of ears, number of kernels, etc.). Because it is more time consuming and expensive to evaluate progenies for grain yield, it seemed reasonable that selection based on yield components that are positively correlated with yield, have higher h^2s, and are easier to measure would have correlated response for greater grain yields. Odhiambo and Compton (1987), for kernel depth, Carena et al. (1998), for prolificacy, and Hallauer et al. (2004), for ear length, were not successful in using indirect selection based on yield components, to realize a correlated response of increased grain yield. If the predicted genetic gain for trait 1 is expressed as $\Delta_{G1} = k_1 h_1^2 \sigma_{P1}^2$ or $k_1 h_1 \sigma_{A1}$ and trait 2 is expressed as $\Delta_{G2} = k_2 h_2 \sigma_{A2}$, we will predict the expected direct response for traits 1 and 2, respectively. Assume k_1 and k_2 and r and y are the same for both traits and that traits 1 and 2 are positively correlated. If trait 2 is genetically correlated with trait 1, the mean of trait 2 will change with selection applied for trait 1. The predicted change in trait 2 with direct selection applied for trait 1 can be expressed as $\Delta_{G2.1} = r_{G12} h_1 h_2 \sigma_{P2} k_1$. The question to be answered is whether it may be more efficient to select for trait 1 when a change in trait 2 is desired rather than direct selection for trait 2 per se. If we assume that the genetic correlation between trait 1 (ear length) and trait 2 (grain yield) is positive and that direct selection for ear length is easier and less expensive than direct selection for grain yield, what are the opportunities for the indirect improvement of grain yield with direct selection for ear length? Indirect selection is more effective than direct when h^2s of the trait selected (ear length, Table 9) is greater than the trait desired to improve by correlated response (grain yield, Table 9) and/or the correlation between the two traits is very high. The $\Delta_{G2.1}$ in Table 9 indicates correlated response is affected more by the levels of the genetic correlations between the two traits than the levels of heritabilities. Neither Odhiambo and Compton (1987) for seed size nor Hallauer et al. (2004) for ear length were successful for increasing grain yield selecting for greater seed size or greater ear length. In both instances, however, selection for decreased

Table 9 Estimates of direct responses for ear length (Δ_{G1}) and grain yield (Δ_{G2}) and correlated response for grain yield (g plant^{-1}) with selection for ear length (cm) ($\Delta_{G2.1}$)

Predicted response	Heritabilities		σ_A^2		Genetic correlation (r_{G12})	Response to		Relative efficiency
	Ear length	Grain yield	Ear length	Grain yield		Direct	Indirect	
Δ_{G1}	0.68	–	138	–	–	1.71 cm	–	–
Δ_{G2}	–	0.39	–	169	–	14.33 g	–	–
$\Delta_{G2.1}$	0.68	0.39	138	169	0.66	–	12.41	0.87
$\Delta_{G2.1}$	0.68	0.39	138	169	0.38	–	7.17	0.51
$\Delta_{G2.1}$	0.68	0.39	138	169	1.00	–	18.74	1.31
$\Delta_{G2.1}$	0.92	0.39	138	169	0.90	–	19.58	1.37
$\Delta_{G2.1}$	0.92	0.39	138	169	0.66	–	14.40	1.00
$\Delta_{G2.1}$	0.92	0.39	138	169	0.38	–	8.33	0.58

Phenotypic variance (σ_P^2) was 2.02 for ear length and 427 for grain yield. Selection differential (k) was 1.76 in all instances seed size and shorter ear length decreased grain yield. Past selection for grain yield seemingly had increased the proper combinations of alleles for the various components of yield and the heritabilities and genetic correlations were not large enough to affect grain yield improvement based only on one component of yield.

Theoretical, computer simulation, and empirical studies on the inheritance of quantitative traits of maize have provided information that has impacted all phases of maize breeding. The studies did not directly impact selection and breeding methods for cultivar development, but the information provided a genetic basis for developing and implementing breeding strategies. Cockerham (1961) examined theoretically the genetic differences among single-cross, three-way cross, and double-cross hybrids. He found that if only additive genetic effects were operative in hybrid performance, that selection among single-cross hybrids would be twice as effective as selection among double-cross hybrids; the advantage would be even greater if dominance and epistatic effects were important. Indirect evidence suggests that interaction effects are important in hybrids and have probably become more important in the recycling of elite inbred lines. Use of single-cross hybrids is common practice in most concentrated maize production areas of the world. In the United States, average maize yields have increased 110.4 kg ha^{-1} since the introduction of single-cross hybrids, the greatest sustained rate of gain since 1965 (Fig. 1, Troyer, 1999). The empirical breeding methods supported the theoretical study. Eberhart (1970) developed a very useful method for comparing rates of predicted genetic gains for different selection methods because the variables included were needed to make direct comparisons among methods. Eberhart (1964) and Smith (1979, 1988) developed models for evaluating response to recurrent selection. Smith's (1983) model included parameters that provided information on the contributions of additive and dominance effects to the response to selection and provided estimates of the effects of genetic drift because of small population sizes. The methods suggested by Eberhart (1964) and Smith (1979, 1983) have become standard procedures in determining the rates of response to selection and the genetic basis for the rates of response. Cress (1967), Jones et al. (1971), and Peiris and Hallauer (2005) studied RRS methods algebraically and via computer simulation to determine the most effective methods. Cress (1967) concluded the two initial populations should be intermated before initiating selection. The intermated population would include the alleles of both original populations. RRS would be imposed on the intermated population to develop two subpopulations with complementary alleles. Jones et al. (1971) compared half-sib and full-sib RRS. They concluded that full-sib RRS was superior to half-sib RRS at less intense selection and when environmental effects were large relative to the genetic variance. Using empirical estimates of variation among half-sib and full-sib families, Jones et al. (1971) suggested that the selection differential should be 1.2 times greater for full-sib RRS than for half-sib RRS to give similar response. Peiris and Hallauer (2005) concluded simulation studies comparing 20 cycles of half-sib and full-sib RRS for genetic models that included epistasis. There were 22 genetic models considered. Genetic response to selection of full-sib RRS was similar to half-sib RRS for 21 of the 22 initial genetic models with S_1 progenies used as the recombination units intermated for each selection cycle. The linear response of half-sib RRS was 1.7 times greater than full-sib RRS for the genetic model that included complete dominance and dominance × dominance epistasis with S_1 progenies used for intermating. There were no significant differences between half-sib and full-sib RRS with S_2 progenies used for intermating, but use of S_2 progenies increased selection response for both half-sib and full-sib RRS. Compared with half-sib RRS, full-sib RRS requires 50% fewer testcrosses, and full-sib RRS has the same selection responses for 21 of 22 genetic models examined. Empirical data for full-sib RRS with BS10 and BS11 and half-sib RRS with BSSS and BSCB1 indicate the two types of families evaluated are equally effective for improvement of the population crosses (Table 5).

Rawlings and Thompson (1962), Smith (1986), and Bernardo (1991, 1992) examined two important aspects of maize breeding: the more effective tester for discriminating differences among new lines for their relative combining abilities and the appropriate generation of inbreeding for evaluating the combining ability of new lines. Derived theory and computer simulations were used in the studies. Jenkins (1935), Sprague (1946a), Jensen et al. (1983), Rodriguez and Hallauer (1991), Lile and Hallauer (1994), and Castellanos et al. (1998) reported empirical data that early testing was effective in identifying new lines that had above average combining ability. Rawlings and Thompson (1962) examined models that included different allele frequencies in testers for different levels of dominance. Genetic variation among testcrosses was greatest with testers that had a low frequency of favorable alleles at all levels of dominance. Smith (1986) reported similar results for relative value of testers and there was a poor relation between inbred line and testcross performance. Bernardo (1991, 1992) determined that the relation between early and later generation testcrosses was good (most r_s greater than 0.9) and found that effective early generation testing was limited primarily by nongenetic effects (i.e., low h^2s of testcrosses). The theoretical information supported the contention of Jenkins (1935) and Sprague (1946a) that the combining ability of lines was determined in the early generations of inbreeding and does not change significantly with continued inbreeding.

6 Inbred Line Development

Development of superior inbred lines is the goal of every maize breeder. Compared with other major cultivated crop species, the development of inbred lines (or pure lines) is not the ultimate cultivar for the producers, which is the case for obligate self-pollinating crop species. Development of inbred lines is only the first stage in maize breeding. The breeding system, designated as pedigree selection, is the method used in maize breeding, as well as the one used in other crop species. Pedigree selection method includes keeping accurate records to maintain genetic identity of progenies (pedigrees) during inbreeding, selection, and evaluation. Pedigree selection (record keeping) is used any time inbreeding and selection are initiated within any type of population (open-pollinated, synthetic, and composite cultivars and F_2 and backcross populations), but pedigree selection usually is associated with inbreeding and selection within F_2 populations developed from crosses of homozygous elite parents. Pedigree selection, however, can be initiated within any type of population where accurate records are maintained to trace the genetic history (genealogy) of lines developed via inbreeding and selection among progenies and individuals within progenies. With the use of improved germplasm sources, the development of inbred lines is a relatively easy task; the major task is determining the combination of inbred lines that can be used as parents of superior hybrids. Initially, when the inbred–hybrid concept was proposed, the germplasm sources for developing inbred lines were (what was considered at the time) the more productive open-pollinated cultivars (Jenkins, 1936). The first-cycle inbreds were the parents of the first double-cross hybrids. Because of the poor vigor and poor grain yields of the inbred lines, these were the factors that prompted use of double-cross hybrids because the more productive single-cross hybrids were used as the parents. One of the main factors that permitted use of single-cross hybrids was the more vigorous and more productive inbred lines developed by recycling of better inbred lines (e.g., Idt, Troyer, 1999) and from improved genetically broad-based populations with recurrent selection (e.g., B14, B37, B73, B84, B104 from BSSS). Discussions on the logic and experimental basis of maize breeding during the double-cross era were given by Jenkins (1936), Sprague (1946b, 1955), and Richey (1950).

Selection was effective for developing distinct open-pollinated cultivars for the different environments within the United States, but the selection methods had no effect on average grain yields (Fig. 1). Because of the increasing needs for the rapidly expanding livestock industry in the US Midwest, different methods were needed to improve grain yields. Decisions were made to test the suggestions of Shull (1910) and Jones (1918) to determine if the use of hybrids was an effective method to increase maize grain yields. In 1922, the USDA and the state experiment stations initiated programs to develop inbred lines to form hybrids. The locally adapted open-pollinated cultivars were the source germplasm for inbreeding plans to develop inbred lines. Depending on the location and source germplasm, some locations were more effective than others for developing inbred lines that were used extensively in double-cross hybrids (Hallauer, 1990). The group of first cycle inbred lines was developed from numerous sources (Jenkins, 1936). Thousands of inbred lines were developed by only a few had major roles as parents in the production of single crosses. US13 (WF9 × 38-11)(Hy × L317) was a widely grown double-cross hybrid and the parental inbreds were developed by breeders located in different states (Crabb, 1947). Ia939 (Os420 × Os426)(L289 × I205) was another widely grown double-cross hybrid, and the four parental inbred lines were developed by M. T. Jenkins from his initial 1922 breeding nursery. Lindstrom (1939) estimated that of the 27,641 lines that had been self-pollinated one to three generations only 2.4% were presumably useful inbred lines. Although most of the first cycle inbred lines developed primarily from open-pollinated cultivars were not used directly as parents of hybrids, germplasms of some first-cycle inbreds have been very persistent in pedigrees of the next cycle(s) inbred lines: for example,

C103, I205 (Idt), Oh43, Oh07, 187-2, B164, etc. (Troyer, 1999). Other inbred lines developed initially did not have great impact on production of double-cross hybrids, but they were to have important roles in future breeding programs, such as Mt.42 and ND203 (Rinke and Sentz, 1962). The open-pollinated cultivars provided the germplasm for the initial cycle of inbreds, which provided the germplasm for future inbred line development by pedigree selection and the formation of synthetic cultivars such as BSSS and BSCB1. The use of open-pollinated cultivars as sources of breeding germplasm essentially ceased after 1940: greater emphasis was given to recycling the elite inbred lines. Bauman (1981) reported that the use of open-pollinated, synthetic, and composite cultivars had essentially been replaced with use of elite line crosses as the major germplasm sources for initiating line development. Historical perspectives on the initial maize breeding programs for source germplasm for line development, individuals involved in the early breeding programs, and pedigrees and descriptions of the first-cycle inbreds were provided by Jenkins (1936), Crabb (1947), Gerdes et al. (1994), and Troyer (1999, 2004a, b, 2006). Mikel and Dudley (2006) presented information on the breeding history of inbred lines developed in recent years that also emphasizes the limited use of genetically broad-based germplasm for line development in modern, commercial breeding programs. Lines developed from different cycles of recurrent selection in BSSS, however, was a major contributor (63%) of germplasm of the newer inbred lines.

Judicious choice of germplasm is the key element in successful breeding programs. If unfortunate choices are made and the frequency of favorable alleles or allelic combinations are low, or absent, the breeder will have limited success even with the best technology or tool available. In modern breeding programs, elite inbred lines are used that enhanced the favorable complexes of genes that have been selected in past cycles of breeding and intermating, and the genetic variation is similar to other types of populations (Fountain and Hallauer, 1996). Consistent incremental genetic gains are realized in the hybrids with each cycle of selection (Fig. 1; Troyer, 2006; Mikel, 2006; Mikel and Dudley, 2006).

Development of maize inbred lines is easier within recycled germplasm sources. Compared with the first-cycle inbred lines developed from open-pollinated cultivars, modern inbred lines are more vigorous, more productive, have better agronomic traits and consequently easier to maintain. Because maize has separate male (tassel) and female (ear shoots) inflorescences, maize is essentially 100% cross pollinated. To produce pure self-pollinated seed, it is essential to cover the tassels and ear shoots before pollen shed and manually transfer pollen (male gametes) from the tassel and distribute on the silks (female gametes) of the ear shoots. Effectively techniques of making pollinations in maize have evolved and relatively little training is required to have capable pollinators (Russell and Hallauer, 1980; Hallauer, 1987, 1994). Self-pollination is the usual process used to develop inbred lines. Some, however, believed that self-pollination was too severe because of rapid fixation of alleles, which would reduce effective selection among plants and progenies during inbreeding. Collins (1909) and Stringfield (1974) suggested the broad-line development of lines by sib matings of plants rather than by self-pollination. Because maize is essentially cross-pollinated, inbreeding depression (ID) occurs when inbreeding occurs. Falconer (1960) has shown that ID can be expressed as $\Sigma 2pqdF$, where p and q are allele frequencies, d is level of dominance, F is level of inbreeding, and summed over loci. Genetic expectation per 1% increase in F can be expressed as mean of noninbred ($\overline{X}_0$) minus mean of inbred ($\overline{X}_F$) divided by F: that is, $(\overline{X}_0 - \overline{X}_F)/F = 2pqd$. Hence, ID is determined by three factors and can be different for different populations and different traits because of allele frequencies and levels of dominance. Good and Hallauer (1977) compared rates of inbreeding for two groups of inbred derived by single-seed descent from BSSS: that is, 247 lines derived by selfing 250 S_0 plants, and 243 lines derived by sib mating. Rates of ID per 1% increase in F were −10.4653 q ha^{-1} by self pollination and −0.4511 q ha^{-1} by full-sib matings. Although the rate of fixation of alleles was slower by sib mating than self-pollination, the rates of ID were similar. Average grain yields at 50% homozygosity were 45.2 q ha^{-1} by selfing one generation versus 45.9 q ha^{-1} by three generations of sib mating and 30.3 q ha^{-1} after three generations of selfing versus 31.2 q ha^{-1} after nine generations of sib mating. At similar levels of inbreeding, the differences between means were not significantly different, which is expected. Stringfield (1974) hypothesized that further selection could be effective but the broad-line concept has not been widely used. Hallauer and Miranda (1988) presented summaries of the reported estimates of ID for different traits within different populations. Generally, there was a decrease for the mean of traits with increased levels of inbreeding with grain yield tending to have the greatest rate. The only trait that had positive increases with increased levels of inbreeding was days-to-flower, which indicates the lines had delayed flowering with increased levels of inbreeding. The differences between grain yield (−ID) versus days-to-flower (+ID) reflect the estimates of dominance: dominant favorable alleles for grain yield and dominance effects for earlier days-to-flower.

Allele frequencies also affect ID. Rodriguez and Hallauer (1991) estimated percentages of ID in populations that have been under recurrent selection (Table 10). Estimates of ID for grain yield decreased in all populations except for BSCB1(R)C10. For BSSSC0 the average yield of the S_1 generation was 36% lower than the noninbred C_0 population. After ten cycles of half-sib RRS, average yield of the S_1 generation was 15% lower than the noninbred C_0 population. BS13(S)C4 also was developed from the same BSSSC0 population as BSSS(R)C10; average grain yield of the S_1 generation of BS13(S)C4 was 14% greater than the noninbred C_0 population. The comparison of the two strains of BSKC0 also showed less ID with use of inbred progeny selection recurrent selection versus half-sib recurrent selection. The advanced cycles of BSCB1C0 per se have also shown decreased grain yield and vigor (Keeratinijakal and Lamkey, 1993). It seems random effects of genetic drift and/or inbreeding have affected BSCB1(R)C0 grain yield but this does not seem to explain the increase in ID. Dominance effects seem to have greater importance in BSSS (tester population for BSCB1) and perhaps the effects of RRS have selected complementary alleles in BSCB1 that tend to be recessive. Indirect evidence from the lines used for intermating between recent cycles of selection support the data of Table 10; most lines tend to be more vigorous and have better grain yields and agronomic traits compared with earlier cycles. ID varies among traits, but the estimates of ID indicate that for populations we can expect 40%, 10%,

Table 10 Estimates of inbreeding depression (*ID*) for grain yield in the original (C_0) maize populations and after several cycles of inter- and intrapopulation recurrent selection

Populations	Interpopulation (RRS)		Populations	Intrapopulation	
	$S_1/C_0 \times 100$[a]	$S_1/C_i \times 100$[b]		$S_1/C_0 \times 100$[a]	$S_1/C_i \times 100$[b]
BSSSC0	64	–	BS13(S)C4[c]	114	87
BSSS(R)C10	85	75		–	–
BSCB1C0	63	–	BSKC0	48	–
			BSK(S)C8	102	74
BSCB1(R)C10	53	63	BSK(HI)C8	93	70
BS10C0	67	–	BS12C0	44	–
BS10(FR)C10	72	63	BS12(HT)C7	134	70
BS11C0	59	–	BS2C0	50	–
BS11(FR)C10	80	76	BS2(S)C4	88	70
Inter: $\overline{X}$: $C_0 = 63$ q ha^{-1}			BS16C0	67	–
$\overline{X}$: $C_i = 81$ q ha^{-1}			BS16(S)C3	60	55
Intra: $\overline{X}$: $C_0 = 53$ q ha^{-1}			BSTLC0	62	–
$\overline{X}$: $C_i = 95$ q ha^{-11}			BSTL(S)C3	94	75

Bulks of ~100 S_1 lines were used for each population

[a] S_1 generations of all populations compared with the noninbred C_0 populations

[b] S_1 generations of C_i populations compared with the noninbred C_i populations

[c] A strain of BSSS that has undergone seven cycles half-sib and four cycles of S_1-S_2 recurrent selection. The estimate of *ID* of S_1/C_0 was made using BSSSC0

10% decreases in grain yield, plant height, and ear length, respectively, after one generation of self-pollination. After seven generations of self pollination, the expected decreases are 70%, 25%, and 25% for grain yield, plant height, and ear length, respectively.

Greatest emphasis is given to the F_2 populations of either hybrids or planned crosses of elite lines as sources for inbred line development (Bauman, 1981). Bailey and Comstock (1976) and Bailey (1977) conducted computer simulation studies for different linkage and heritability levels and different frequencies of favorable alleles in the parents of the F_1 crosses to determine the frequency of the favorable alleles in progenies derived from the F_2 populations. The foundation population (F_2) was formed from the cross of two homozygous lines and the F_2 obtained by selfing the F_1. Some of their conclusions included: probability of homozygosity of favorable alleles was higher for selected 10% of lines versus unselected lines; higher with higher heritabilities; and higher with fewer number of loci included in selection. The frequencies of favorable alleles in complete coupling were higher than in absence of linkage in 30 of 32 possible comparisons, and effects were greater for tighter linkages, higher heritabilities, and with selection. With repulsion linkages, the favorable allele frequencies were lower compared with independent assortment (no linkage) in 31 of 32 possible comparisons. The results seem reasonable because with favorable alleles in coupling phase linkages in the F_1 they would tend to be together during meiosis and favorable alleles would be more frequent than with no linkage. Bailey (1977) discussed the situations where the genetic value of the parents used to produce the F_1 has either equal or unequal genetic values. With parents having equal genetic values one could be confident of recovering a line from the F_2 population superior to either of the parent lines. But if one parent has 70% of the favorable alleles, there will be more coupling linkages and the probability of getting a line as good as or better than the better parent was about 7%. These estimates assumed h^2 value of 0.5. If heritability was less than 0.5 and with limited selection, the probabilities of obtaining genotypes superior to the parents is reduced. Bailey (1977) also concluded that if 60 loci affect the trait of interest that whatever the genetic value of the parents there is virtually no chance of recovering a genotype containing the favorable allele for the 60 loci. The simulation studies of Bailey and Comstock (1976) support the experiences of maize breeders having success with selection in F_2 populations of elite line crosses. There are an unknown number of loci affecting grain yield, but continued incremental gains by recycling crosses of elite lines certainly provides complexes of favorable alleles that can be gradually improved. The poor × elite line crosses are not an option because the favorable gene combinations of the elite line are diluted by the, perhaps, rare recombinations that occur during meiosis. However, they seem to be valuable crosses for increasing the basic knowledge of chromosome segments controlling quantitative or mostly qualitative traits (e.g., QTL experiments). Examples are available for the crosses that insert a mutant allele (e.g., o_2, bm, fl_2, etc.) into widely used elite line in the desire to have, say, an opaque-2 version of the elite line. It seldom, if ever, happens. Within F_2 populations, maximum linkage disequilibrium would be expected. In some respects breeders desire to retain linkage blocks (combinations of genes) that have been effective in productive hybrids. But one may also desire to maximize genetic variation to capture the more desirable alleles of both parents.

Covarrubias-Prieto et al. (1989) examined the genetic variation present in the F_2 populations of B73 × Mo17 and B73 × B84 and after five generations of intermating 250 plants within the F_2 populations. The two single crosses represent two types of crosses that may be considered for line development. B73 × Mo17 was one of the great single-cross hybrids grown in the US Corn Belt and the two parents represent the familiar heterotic groups of the US Corn Belt. The B73 × B84 single cross includes two lines derived from different cycles of recurrent selection in BSSS and would be more closely related than the B73 × Mo17 hybrid. Data were collected for 14 plant and ear traits. The B73 × B84 hybrid had 28.3% less grain yield than B73 × Mo17 (Table 11). Comparisons of the F_2 generations show that the (B73 × Mo17)F_2 had 47.9% less grain yield than the F_1. Comparisons of the F_1 and F_2 generations illustrate the differences for the allele frequencies and/or levels of dominance for the two crosses. The differences in Table 11 would be expected

Table 11 Average grain yield (q ha^{-1}) and grain moisture (%) of the F_1, F_2, and F_2 intermated generations for B73 × Mo17 and B73 × B84

Generation	Grain yield (q ha^{-11})		Grain moisture (%)		Stalk lodging (%)	
	B73 × Mo17	B73 × B84	B73 × Mo17	B73 × B84	B73 × Mo17	B73 × B84
F_1	104.8	75.1	21.8	22.7	4.6	7.1
F_2	54.6	61.0	22.2	22.9	7.5	9.5
F_2 Syn 1	57.8	56.2	20.9	21.9	9.6	11.4
F_2 Syn 2	55.6	58.3	20.8	22.6	8.6	8.1
F_2 Syn 3	57.8	54.0	20.8	22.5	11.2	10.1
F_2 Syn 4	56.2	58.3	21.2	21.9	9.1	8.8
F_2 Syn 5	57.8	58.9	20.5	22.0	6.0	12.4
F_2 Syn 6	58.9	60.5	20.6	21.8	7.4	11.1
$\bar{X}$ (Syn i)	57.4	57.7	20.8	22.2	8.5	10.2
b_1[a]	0.05	−0.28	−0.19*	−0.13	−0.28	0.27

There were ~250 plants intermated in the F_2 and subsequent intermated generations, designated as F_2 Syn i (adapted from Covarrubias-Prieto et al., 1989)
*Significant at $P \leq 0.05$
[a]Linear regression for F_2 and F_2 Syn i generations

Table 12 Estimation of components of genetic variance (σ_G^2) among 100 S_1 progenies for grain yield, plant height, ear length, and days to flower developed from the F_2 and F_2 Syn 5 generations for the B73 × Mo17 and B73 × B84 single crosses (adapted from Covarrubias-Prieto et al., 1989)

Traits	B73 × Mo17		B73 × B84	
	F_2	F_2 Syn 5	F_2	F_2 Syn 5
Grain yield (q ha^{-1})	64.4	52.6	22.6	34.2
Plant height (cm)	170.0	168.6	87.1	86.8
Ear length (mm × 10)	7.9	6.9	6.5	9.6
Days-to-flower (no.)	5.6	3.9	2.2	2.8

because of the genetic background of the parents included in the crosses. One hundred S_1 progenies were evaluated in replicated trials to estimate the genetic variability (σ_G^2) in the F_2 populations and after five generations (F_2 Syn. 5) of intermating (Table 12). Except for grain yield, the estimates of σ_G^2 were similar, and the effects of five generations of intermating were minimal. The greatest difference was the estimates for grain yield for the F_2 populations: 64.4 for B73 × Mo17 versus 22.7 for B73 × B84. The estimates for the unrelated line F_2(B73 × Mo17) was nearly three times greater than for the related line F_2(B73 × B84). If repulsion phase linkages were predominant (which they probably were), the estimates of σ_G^2 would be biased upwards. All estimates of σ_G^2 were greater in the unrelated line cross compared with the related line cross. Intermating had relatively small effect for increasing the estimates of σ_G^2. Getschman and Hallauer (1991), Han and Hallauer (1989), and Covarrubias-Prieto et al. (1989) used different mating systems to estimate σ_G^2 in the two crosses. The ratios of the estimates of σ_G^2 for the unrelated/related line crosses were 1.5 (Getschman and Hallauer, 1991), 3.2 (Han and Hallauer, 1989), and 2.8 (Covarrubias-Prieto et al., 1989) for an average ratio of 2.5 with (B73 × Mo17)F_2 population having 2.5 times more σ_G^2 than the (B73 × B84)F_2 population. Although unrelated line F_2 populations would be expected to have greater σ_G^2, compared with related line F_2 populations, their advantages for developing new inbred lines may compromise the coupling phase linkages past selection have developed for specific heterotic groups. Bailey and Comstock (1976) showed that it would be more difficult to develop lines when extremes in genetic value of parents than when genetic values of the two lines are more similar. However, certain hybrids might be good exploring as an alternative. A more common practice is to select within F_2 populations developed from crosses of elite lines within heterotic groups, for example, B73 × B84.

In lesser developed maize growing areas where the economic conditions may not support use of F_1 seed on an annual basis, the data in Table 11 illustrate that if F_2 seed of hybrids is used there is a reduction in yield but yields remain relatively constant thereafter. More productive hybrids should be used and farmer-breeders could practice selection to sustain the yields. The average yields of F_2 and subsequent intermated generations were very similar for the wide cross (57.0 q ha^{-1}) and the related line cross (58.1 q ha^{-1}). Either type of hybrid could be used as seed source when F_2 and subsequent generations are used by local farmers, which may assist in alleviating the poorer yields in areas where improved seed is not used (Dowswell et al., 1996). A similar concept of exploiting heterosis in developing countries is the population–hybrid concept where seed is produced relatively cheap in isolated fields with detasseling female populations (Carena, 2005a; Carena and Wicks III, 2006).

Self-pollination is the more common practice to develop homozygous inbred lines. Although Collins (1909) and Stringfield (1974) and others considered self-pollination a too severe form of inbreeding to permit effective selection during inbreeding, sib-matings are slower approach to homozygosity than by self-pollination. For example, expected homozygosity for 10 generations of half-sib matings

(0.654) is less than two generations of selfing (0.750) and 10 generations of full-sib matings (0.859) equals three generations of selfing (0.875) (see Table 9.1, Hallauer and Miranda, 1988). Time is of the essence to obtain and test homozygous lines. Hence, self-pollination is the usual method. Selection within and among inbred lines is certainly practiced during inbreeding. Bauman (1981) summarized responses of how effective maize breeders considered visual selection for nine plant traits. The nine traits were rated for relative importance (1 more to 4 less important) and effectiveness of visual selection (1 more to 4 less effective). The rankings ranged from 1.2 for root strength and grain yield to 3.0 for erect leaf habit for importance of traits and ranged from 1.3 for flowering date to 3.2 for grain yield for effectiveness of visual selection. The trend was that effectiveness of visual selection was inversely related to the importance of the trait. This trend was confirmed by the simple product–moment correlation (r) estimated between importance of trait and effectiveness of visual selection; $r = -0.89 \pm 0.60$. This relation for the nine traits does not infer that visual selection among and with progenies during inbreeding should not be considered. Screening for pest tolerance, observations on reactions to heat and drought stress, kernel type, ear type, and seed set, etc. are traits that may not be directly reflected in crosses with other lines but they certainly do not detract from their expression in crosses. Genetic variation among lines ($2F\sigma_G^2$) and within lines $(1 - F)\ \sigma_G^2$ will affect relative effectiveness of selection (Table 13). The σ_G^2 rapidly increases among lines and decreases within lines. Individual plant heritabilities are inherently lower than among progenies, so the opportunities for selection within progenies decreases rapidly with the corresponding decrease in σ_G^2, which was Stringfield's (1974) concern. Progenies acquire their primary phenotype by the S_3 (F_5) generation and the differences are obvious in the breeding nursery. Major selections are made among progenies with attention given to the plant (at pollination) and ears (at harvest) traits during the first two to four generations of self-pollination. The number of self-pollinations made within rows (three to five) during the first two to four generations of selfing depends on the observable phenotypic variation within rows. Because of the rapid turnaround in data analyses, final selections among progenies often are not made until the testcross data are available

Table 13 Distribution of genetic variance (σ_G^2) among and within progenies under continuous selfing for additive (σ_A^2) and dominance (σ_D^2) components of variance assuming $p = q = 0.5$

Generation of selfing (S)	Among lines ($2F\ \sigma_G^2$)		Within lines [$(1 - F)\ \sigma_G^2$]	
	σ_A^2	σ_D^2	σ_A^2	σ_D^2
S_1	1.0000	0.2500	0.5000	0.5000
S_2	1.5000	0.1875	0.2500	0.2500
S_3	1.7500	0.1094	0.1250	0.1250
S_4	1.8750	0.0586	0.0625	0.0625
S_5	1.9375	0.0303	0.0313	0.0313
S_6	1.9688	0.0154	0.0156	0.0156
S_n	~2.0000	~0.0000	~0.0000	~0.0000

to determine which progenies have better combining abilities, which ultimately determines the genetic worth of new lines.

The tendency for plant tissue culture for generating genetic variation has been reported. Larkin and Scowcroft (1981) suggested that the use of tissue culture as a model source to generate genetic variation could be useful in plant breeding programs. Lee et al. (1988) evaluated the relationship between culture age and somaclonal variation for several agronomic traits from a maize single cross. Lee et al. (1988) evaluated 305 tissue culture-derived lines and 48 control lines as lines per se and in a testcross in six trials. Tissue culture-derived lines and their testcrosses generally had lower grain yield and moisture than the control lines. Although the tissue culture-derived lines were generally inferior, the greatest yielding line per se in three of six trials and the best line in five of six trials were derived from tissue culture. However, the odds of identifying the best line would favor tissue culture-derived methods because 257 more tissue culture-derived lines were included. They also calculated the phenotypic correlations between S_2 and testcross performance for grain yield and moisture and stalk lodging. Grain yield of S_2 lines was not significantly correlated with grain yield of their testcrosses in any trial, which is similar to previous correlations reported for the traditional methods of developing lines (e.g., Jensen et al., 1983). S_2-testcross correlations for grain moisture and stalk lodging were significantly positive in most trials but their magnitude was generally low. Lee et al. (1988) concluded that tissue culture may generate variation for agronomic traits. Because of the tendency to generate a large proportion of inferior lines via tissue culture, the method may require screening larger populations of lines to identify superior lines. Genetic variation from use of tissue culture did not seem promising for increasing grain yield, but the trend towards earlier maturity could be useful.

Genetic variation in maize breeding has not been a limiting factor in maize breeding. Maize breeders can make elite line × elite line crosses to create the genetic variation they desire. The regeneration of plants via tissue culture may be a limiting factor for many elite lines. Except for some specific traits that are more simply inherited than grain yield and most traits considered important for modern inbred lines, it seems tissue culture-derived genetic variation has limited use. The genetic variation within elite line crosses may be less than genetic variation generated by mutation breeding and tissue culture breeding, but the genetic variation is often useful and permits effective selection.

It was previously stated that the ultimate success of a breeding program depends on the proper choices of parents to form segregating populations to initiate line development. In most instances, the choices of parents may be dictated by the relative importance of lines in producing commercial hybrids, and the desire is to continually improve the workhorse lines (e.g., see Fig. 3, Troyer, 1999). The inventory of inbred lines within an organization can be searched to identify other elite lines within the same heterotic group that would be logical choices in making crosses. The inbred lines selected may be very important as parents of hybrids in different areas (better pest resistance) or different maturity zones. B14 (BSSS) and Oh43 (W8 × Oh40B), for example, were very important parents of hybrids grown in the central US Corn Belt. B14 (crossed with earlier maturity lines Mt42 and ND203) and Oh43 (crossed with earlier maturity line A171) were crossed, backcrossed, and selected to have earlier maturity strains (A632, A641, and A619) of B14 and Oh43 to produce hybrids adaptable to earlier maturity zones (Rinke and Sentz, 1962). In some instances, obvious choices of parents to use in crosses with elite lines may not be available. Dudley (1982, 1984a, b) developed a method that identifies a parent(s) that have a class of alleles that are not present in the parents of a superior hybrid. The goal is to identify a parent that has alleles not present in either parent and can be incorporated in at least one parent to improve the parent(s) of an otherwise superior hybrid. The goal was to determine which one of the original inbred lines used in the hybrid could be improved to improve the performance of the hybrid. Dudley (1982) identified eight classes (A through G) of loci that have the favorable alleles (plus alleles) either in one of the original parents (P_1 and P_2) and another parent (P_W) that may be used to improve either P_1 or P_2. The G class of loci included favorable alleles that were present in P_W but not present in P_1 or P_2. Dudley (1982, 1984a, b) derived a series of equations that represented the genotypic values of parents (P_1, P_2, and P_W) and crosses of the parents ($P_1 \times P_2$, $P_1 \times P_W$, and $P_2 \times P_W$). He derived a series of equations that permitted solutions for six classes of alleles (classes A and H were deleted because P_1, P_3, and P_W either had all plus or all minus alleles). Two examples were discussed. The hybrid C103 (P_1) × B37 (P_2) was tested by crossing each line to B73, N28, Mo17, and Oh43 (P_Ws). The nine hybrids and parents were evaluated and estimates of G were determined for B73 (12.2), N28 (24.0), Mo17 (−4.5), and Oh43 (6.5). The greatest improvement would be contributions of favorable alleles from N28 to improve B37. Based on the genetic backgrounds of each P_W, this would be the logical choice. N28 also derived from BSSS but more distantly related than B37 and B73. The hybrid B73 × Mo17 was considered and each parent crossed to B37, N28, C103, and Oh43. For this hybrid none of the P_W lines would contribute favorable alleles to improve the line because μ G estimates for B37 (−5.0), N28 (−7.0), C103 (−22.8), and Oh43 (−18.5) indicated they would not contribute favorable alleles to improve B73 × Mo17. This result also seems logical because B73 × Mo17 was an excellent hybrid and the P_W lines were older lines. The method proposed by Dudley (1982, 1984) is relatively easy to conduct and relatively precise because means are used to estimate the parameters. The main problem, or concern, in the evaluation trials is evaluating the parents and hybrids in the same trials because of vigor differences. Field plot techniques could be modified to either use larger plots to reduce competition effects, split-plot designs with parents and hybrids in whole plots, or separate experiments which may have different experimental errors. Inbred trials tend to have greater error mean squares, but adjustments can be made for making comparisons among the inbred and hybrid entries. Modifications of the original method suggested by Dudley (1982, 1984a, b) have been made by Dudley (1987), Gerloff and Smith (1988), and Bernardo (1990).

Maize breeders use all the information available to assist them in the selection of genotypes with superior breeding values. Johnson (2004) discussed the use of molecular markers to assist the breeders in identifying lines that had superior breeding values. He used the method commonly used in maize breeding by crossing two elite inbred lines and deriving the F_2 population in which to initiate selection. Johnson (2004) used F_2 populations because linkage disequilibrium is at a maximum and they are considered ideal for the application of molecular markers in maize breeding. The goal of the study was to determine the relative combining abilities of S_1 and S_3 direct descendants in testcrosses. The testers were homozygous inbred lines and the genetic variation among testcrosses would conform to an additive genetic model. Each of the S_1 and S_3 lines was genotyped with RFLP markers covering the genome that were spaced 20 cM apart. Two field experiments were conducted to determine the efficiency of marker assisted selection for measuring the combining abilities of lines derived from F_2 populations. The results from both experiments clearly indicated that marker scores were heritable and significantly contributed to the prediction of advanced generation combining ability. The results of Johnson's (2004) study agreed with those of Eathington et al. (1997) who also concluded that molecular markers increased the accuracy of determining the relative combining abilities of superior advanced generation lines. Data generated from genotyping have reduced their limitations with SSR markers and lately with SNP markers at a point in which currently the limiting factor is accurate phenotyping.

One of the common decisions maize breeders frequently have to make is how many S_0 plants of an F_2 population is an adequate sample. Breeders desire to have an adequate sample that represents the σ^2_G of the F_2 population. Bauman (1981) asked the question from a survey of 130 US corn breeders of how many S_0 plants are sampled from F_2 populations to initiate inbred line development. A broad range of responses were obtained, but the modal response was about 500 S_0 plants per F_2 population. It is the breeders' judgment based on previous experience, potency of parental lines used in the crosses, nursery space available, and the specific goals of line development for each cross relative sample size. There is no specific answer for each breeder. But samples of 500 S_0 plants per F_2 population seem large. If 10,000 nursery rows are available, the breeders could sample 500, 200, or 100 individual S_0 plants from 20, 50, and 100 F_2 populations, respectively. Based on indirect, empirical data from quantitative genetic studies conducted to estimate components of genetic variance and selection responses from recurrent selection studies, it seems sample sizes of 150 to 200 should be adequate (Hallauer and Miranda, 1988). Marquez-Sanchez and Hallauer (1970a, b) estimated standard errors of σ^2_A and σ^2_D estimates, and the point where further decreases in standard errors of the estimates of components of σ^2_A and σ^2_D was attained was at sample sizes of 160 to 180 individuals. Breeders would need to make judgments whether crosses are between inbred lines that have different origins or pedigrees or crosses are between more closely related inbred lines; pedigree and molecular marker information would indicate genetic divergence of the two parent inbred lines of the F_2 populations (Williams and Hallauer, 2000). If more σ^2_G expected, larger S_0 plant samples may be desirable to include the range of σ^2_G in the F_2 population, whereas σ^2_G would be less in more closely related lines (e.g., Table 12). Experience by maize breeders who have had a long career in recycling elite inbred lines in crosses with other elite lines for a specific target environment will have good judgment of the appropriate sample sizes needed to develop inbred lines that are superior to the parental inbred lines, whether 50, 100, 200, or 500 individuals. Because significant genetic improvements in inbred lines and hybrids have been realized during the past 50 years with a relatively few elite inbred lines, greater precision will be needed to determine the genetic worth of newer inbred lines because σ^2_G may be reduced (Hallauer, 2002). More precise information will be needed, such as use of molecular markers to assist in choice of parents to cross and in selection among progenies derived from the cross (Knapp, 1998; Johnson, 2004). Mathematical models could also help in the probability of finding better parents once genes are known from the maize genome sequence (Hammond and Carena, 2008).

7 Doubled Haploids

The potential use of doubled haploids has intrigued maize breeders because it enables breeders to develop homozygous genotypes from heterozygous parents in a single generation. It requires seven generations of self-pollination to develop inbred lines from a heterogeneous F_2 population with an expected inbreeding coefficient of 0.992. Depending on resources available, such as off-season nurseries, it may require 3–6 years to develop S_7 lines with an inbreeding coefficient of 0.992. For mature, long-term breeding programs, the time-frame for developing inbred lines is usually not a limiting factor. Selections within 50–100 F_2 generations are initiated each season, providing lines at different generations of inbreeding each season. Effective selection for pest tolerance, heat and drought stress, maturity, ear and plant type, grain yield, etc. can be accomplished during inbreeding. The more general problem with the traditional methods of maize breeding is to determine the genetic worth of the derived lines in hybrids. For new breeding programs, where time is an important factor, haploid breeding methods are a possible alternative to jump-start the program.

Chase (1949) reported the frequencies of haploids in a double-cross hybrid, its two parental single-cross hybrids, and the four parental inbreds of the two single-cross hybrids. The tester stocks carried the dominant allele for purple plumule (Pu), whereas all the seed parents carried the recessive allele for this gene. The purple marker was used to identify the haploids, which occurred spontaneously. Chase (1949) classified 38,684 seedlings and 43 haploids were identified, an average frequency across the seven seed parents of about 1:900 or 0.1%. Chase (1952) continued his studies and suggested the use of haploid breeding as an alternative method for developing maize inbred lines. Thompson (1954) compared the relative combining abilities of lines developed from doubled haploids with lines developed by the traditional methods of self-pollination. Both sets of lines were derived from the same source (BSSS). The doubled haploid lines were a random sample of lines with respect to combining ability. There were no differences between the means for the two sets of testcrosses and the frequency distributions were similar. B67 and

B69 were inbred lines derived from haploids that were released for public use, but it seems neither inbred line was used in hybrids. Goodsell (1961) and Kermicle (1969) also suggested methods that identified haploids. Goodsell's (1961) method identified paternal haploids. Kermicle (1969) reported on use of the indeterminate gametophyte (*ig*) factor that produced unusual effects in the embryo sack. The method devised by Kermicle (1969) was more attractive because of a greater frequency of haploids. Spitko et al. (2006) conducted extensive studies to determine the aptitude of different source materials and methods to induce the frequency and regeneration of haploids to develop new doubled haploid lines. From their studies, they suggested that haploid regeneration was primarily an inherited trait in their materials with a heterosis effect in the F_1 progeny. Chase (1949) also reported that the incidence of haploidy in a single-cross hybrid could be predicted by the behavior of its component inbred lines, which also suggests that haploidy is a heritable trait.

Snape (1989) discussed the possibilities of use of doubled haploids in plant breeding. He indicated the technology has developed to the point that doubled haploid lines can be produced in sufficient numbers to contribute directly to breeding programs. Snape (1989) suggested the greatest advantage of the doubled haploid systems is increasing the efficiency of selection for quantitative traits. The efficiency of selection among doubled haploids occurs because the additive genetic variance (σ^2_A) among lines homozygous lines is twice the σ^2_A among F_2 plants, that is, $2\sigma^2_A$ (Table 13). Snape (1989) also suggested that doubled haploids can contribute to the efficiency of recurrent selection schemes used in maize breeding. The doubled haploids would increase genetic variability among progenies because additive genetic variance is twice the additive genetic variance of the random mating population, and nonadditive genetic effects would not be a hindrance in selection. But in maize breeding, nonadditive effects are important if the germplasm resources are to contribute lines for use in hybrids (Table 5). For those programs that are developing genetically broad-based populations themselves for use by the producers the method may not have merit. Snape (1989) does concede that a generation of haploidization and chromosome doubling between cycles of selection would increase time for each cycle, which would reduce genetic gain per season (Tables 2 and 3). Doubled haploid breeding does have one great advantage over traditional methods of self-pollination of lines to homozygosity: homozygous lines could be developed in one generation if the technology for identifying haploids and doubling chromosome number permits having an adequate number of inbred lines for evaluation in hybrids. The same methods would be necessary in each heterotic group. The decision to conduct doubled haploid breeding would be difficult for each breeding program. Similar to the comments for genetic variation generated from use of tissue culture, the development of inbred lines by the traditional methods of self-pollination is usually not a limiting factor. The time spent in developing inbred lines by self-pollination is not wasted effort. Selection among and within lines for traits, other than grain yield and root and stalk strength, can be effective for more highly heritable traits that are required in modern hybrids. As Thompson (1954) has shown, the doubled haploid lines are a random sample for combining ability. The use of molecular markers might assist the breeder in identifying doubled haploids that are superior for quantitative traits. Otherwise, the doubled haploids would undergo initial testcross evaluation as homozygous lines.

8 Hybrids

The only reason maize breeders isolate inbred lines is to develop parental inbred lines for the production of hybrids. Inbred lines have been developed by single-seed descent for genetic studies, but the genetic studies were related to obtaining estimates of ID (Hallauer and Sears, 1973; Good and Hallauer, 1977), estimation of genetic variability (Obilana and Hallauer, 1974), and what are designated as recombinant inbred lines (RIs) for molecular genetic studies (Lee, 1995). It soon became obvious to the maize breeders that it was easier to develop inbred lines than to determine their relative worth in hybrids. Initially, breeders self-pollinated S_0 plants of open-pollinated cultivars and continued selfing with selection until lines approximated homozygosity. At this time, crosses between inbred lines were made and evaluated as double-cross hybrids. The main problem was that the possible number of double-cross hybrids among a group of inbred lines became extremely large. The number of possible double-cross hybrids among n lines is $[n(n-1)(n-2)(n-3)]/8$. If $n = 10$, there are 630 possible double-cross hybrids. It was found that the inbred lines required testing in hybrids because the relations between inbred lines and the hybrids were too poor to be of predictive value, both for the first-cycle inbred lines (Jenkins, 1928) and for inbred lines derived from improved sources (Gama and Hallauer, 1977). Jenkins and Brunson (1932) suggested use of the testcross method to make an initial screening of inbred lines for their relative combining with use of a common tester(s). Testcrosses were evaluated and the inbred lines with above average combining ability were then tested as double crosses. But the number of remaining inbred lines was often too great for testing as double crosses. Jenkins (1934) tested a scheme where single-cross and testcross data could be used to predict the better performing double crosses. Because of the number of possible lines developed in breeding nurseries and the resources required to self-pollinate until relative homozygous, Jenkins (1935) and Sprague (1946a) proposed testing the lines in earlier generations of inbreeding to eliminate lines that seem to have poorer combining ability. Jenkins (1935) concluded that the combining ability of a line was established early in the inbreeding process and remained relatively stable in succeeding generations of inbreeding. Sprague and Tatum (1942) partitioned the combining ability of inbred lines into general (GCA) and specific (SCA) combining effects, with GCA as the average performance of lines, due primarily additive genetic effects, and SCA the performance of specific crosses, primarily due to nonadditive genetic effects. There were, however, some who questioned the effectiveness of early testing (e.g., Richey, 1944). Because of the genetic variability among and within lines during successive generations of inbreeding, it was considered that early testing may eliminate lines that may have different and better combining abilities as homozygous lines. This could happen. But Jenkins (1935) and Sprague (1946a) never stated that a perfect correlation between testcrosses made at earlier and more advanced inbreeding levels would exist; they proposed only selection of those lines that exhibited above average combining ability. Subsequent empirical data have supported the feasibility of early testing. Recurrent selection methods use early testing in all instances (usually at the S_0 or S_1 generation), and positive responses have been realized in all instances for improvement of populations.

Protocols for the development of inbred lines and double-cross hybrids were essentially standardized by 1945. There were two issues, however, that generated vigorous debates among maize breeders: (1) efficiency of early testing to determine the relative combining ability of lines; and (2) the best tester(s) to use for evaluation of the combining ability of new lines. Two issues relative to early testing were that the effects of selection during inbreeding may be negated and the large number of experimental field plots that would be required to evaluate the early inbred-generation lines. Selection during inbreeding process could be continued but only among those lines that had above average combining ability. Breeders can continue to practice visual selection among and within lines for desired plant types, seed set, maturity, etc. that have relatively higher heritabilities. However, Bauman's (1981) survey indicated that visual selection was inversely related to the relative importance of traits, so that visual selection should not affect those lines that had above average combining ability. Plot numbers required to evaluate a large number of early generation testcrosses is a major concern. Bauman's (1981) survey showed the following results for testing at the different generations of inbreeding: S_1 (0%), S_2 (18%), S_3 (33%), S_4 (27%), S_5 (9%), and other (13%). Only 18% of the maize breeders surveyed tested new lines earlier than S_3 generations, 22% tested at the S_5 or later generations, and 60% tested new lines at the S_3 and S_5 inbreeding generations, a compromise between early and later generation testing. Bauman (1981) also reported there was at least a 50% discard of new lines from the S_1 to S_2, S_2 to S_3, and S_3 to S_4 inbreeding generations. Maize breeders were obviously practicing intense selection among the early inbreeding generations before testcrossing (a 92% discard from S_1 to S_4 from Bauman's survey). Bauman's (1981) survey of 130 maize breeders was conducted before refinements were made in developing equipment to plant and harvest experimental field trials. Rapid progress was made in developing data recorders on harvesters that could be downloaded for rapid turn around of data analyses. Plots could be harvested and data recorded in 25 seconds or less versus 10–15 min for hand-harvested plots. Modern plot equipment has significantly reduced the concerns of plot numbers and has favored early generation testing.

Theoretical and applied studies have been conducted to determine the relations between inbred lines per se and their testcrosses and between inbred lines tested in earlier inbreeding generations and their derived lines tested after further inbreeding. A summary of studies that compared relation between inbred lines and their crosses indicated that the average correlation for grain yield was 0.22 (Table 8.9, Hallauer and Miranda, 1988). Jensen et al. (1983) conducted a study to simulate a commercial breeding program with elite germplasm. Jensen et al. (1983) collected data for S_2 progenies per se, S_2 progeny testcrosses, and S_5 testcrosses of lines derived from the S_2 progenies and estimated the correlations between S_2 progenies per se and S_5 testcrosses and between S_2 and S_5 testcrosses. For grain yield, the correlation between S_2 progenies per se and S_5 testcrosses was 0.14 compared with correlation of 0.67 between S_2 and S_5 testcrosses. They concluded that the S_2 testcrosses were a better predictor of S_5 testcrosses than S_2 progenies per se. Smith (1986) conducted a computer simulation study to compare lines per se and their testcross performance when crossed to what was considered good, average, and unrelated testers. For the genetic model used, Smith (1986) reported the correlations between lines per se and their testcrosses were 0.22 (good tester), 0.28 (unrelated tester) and 0.34 (average tester). His conclusions were similar to those of Jensen et al. (1983) that progeny per se performance was not a good predictor of testcross performance. Obaidi et al. (1998) also conducted a simulation study to determine selection response that emphasized lines per se but selection was based on testcrosses. They suggested that testing of lines as early as the S_0 generation was important. Seminal studies by Bernardo (1991, 1992) examined theoretically the correlations between early-generation testcrosses versus later-generation testcrosses. He derived the correlations between generation testcrosses based on the relation developed by Rawlings and Thompson (1962) for the variability among testcrosses: $\text{Var}(\text{TC}_n) = (1 + F_n)\, 0.5\, pq[a + (s - r)d]^2$. In the absence of selection, the testcross means S_n and $S_{n'}$ individuals or lines are identical. The genetic covariance (Cov) between S_n and $S_{n'}$ testcrosses (TC) becomes $\text{Cov}(\text{TC}_n, \text{TC}_{n'}) = \text{VarTC}_n$. The covariance between TC_n and $\text{TC}_{n'}$ is equal to the genetic variance among S_n testcrosses. Thus, the genetic correlation ($r_{GnGn'}$) between S_n and $S_{n'}$ testcrosses is $r_{GnGn'} = \text{Cov}(\text{TC}_n, \text{TC}_{n'})/[\text{VarTC}_n)(\text{VarTC}_n]^{0.5}$, which becomes $r_{GnGn'} = [(1 + F_n)/(1 + F_{n'})]^{0.5}$. Thus, the genetic correlation between S_n and $S_{n'}$ testcrosses is equal to the square root of the genetic variances and is a function of the inbreeding

Table 14 Expected genetic correlations ($r_{GnGn'}$) between testcrosses of S_n and $S_{n'}$ maize lines and correlations between testcross phenotypic value at generation n and true genetic value (r_{PnGx}) at homozygosity

S_n line	Inbreeding coefficient (F_n[a])	$S_{n'}$ line						h^2_o		
		S_2	S_3	S_4	S_5	S_6	S_x	0.25	0.50	0.75
				$r_{GnGn'}$					r_{PnGx}	
S_1	0.0	0.82	0.76	0.73	0.72	0.71	0.71	0.35	0.50	0.61
S_2	0.5		0.93	0.89	0.88	0.87	0.87	0.50	0.67	0.78
S_3	0.75			0.97	0.95	0.94	0.94	0.57	0.75	0.86
S_4	0.878				0.98	0.98	0.97	0.60	0.78	0.89
S_5	0.9375					0.99	0.98	0.62	0.80	0.91
S_6	0.96875						0.99	0.62	0.81	0.92

The r_{PnGx} values are for three heritabilities of S_0 testcrosses (adapted from Bernardo, 1992)
[a]Assuming several S_0 plants are crossed to tester.

coefficients of the two selfing generations (Table 14). For example, if $n = S_1$ (F = 0.5) and $n' = S_4$ (F = 0.938), the correlation, $r_{S1,S4} = [(1.5)/1.935]^{0.5}$ or 0.8804. The phenotypic correlation between S_n and $S_{n'}$ testcrosses is $r_{PnPn'} = r_{GnGn'}h_n h_{n'}$, where h_1 and h_2 are the square roots of the heritabilities for S_n and $S_{n'}$ testcrosses. Bernardo (1991) concluded that the effectiveness of early testing was limited mainly by nongenetic effects (i.e., low testcross heritability values). Bernardo (1992) also studied the probabilities of retaining genetically superior lines during early-generation testing. The loss of potentially superior inbred lines with early testing depends on the relative heritabilities of the testcross trials. If heritabilities are consistently low, larger proportions of lines need to be retained at earlier rather than later generation testing if concern for loss of genetically superior lines from earlier generation testing. To reduce the odds of losing superior lines via early generation testing, one could alter testing methods (replications and environments) to increase the h^2 values on a progeny mean basis. Heritabilities of testcrosses are usually around 0.5 for two replications at four environments. One will never know how many potentially genetically superior lines have been discarded because of early-generation testing. Plant breeding is a game of odds because it involves a series of fortuitous choices to identify elite inbred lines. But it seems earlier-generation testcrosses are becoming more common than at the time of Bauman's (1981) survey. Some breeders have crossed F_2 S_0 plants to testers, which is similar to the methods used in some recurrent selection programs. It seems that testcrosses for the first (S_1) and second (S_2) generations of inbreeding are more appropriate because selection among S_1 and S_2 progenies can be made at the same time the testcrosses are produced. Effective selection among S_1 and S_2 progenies can be made for pest tolerance, drought tolerance, and general plant and ear traits that are necessary to be considered as parents of hybrids.

The choice of testers to discriminate the relative combining of lines has always been considered important. In contrast to earlier years, when unique heterotic groups were not defined, the choice of testers is usually elite inbred lines from the opposite heterotic group. Rawlings and Thompson (1962) theoretically examined the genetic variance among testcrosses for different testers for different levels of dominance. Genetic variation among testcrosses with a tester having $p = q = 0.5$ was the same for all levels of dominance, which would simulate using the parent population as the tester. If the tester is nearly homozygous for the favorable alleles that are also favorable alleles for the progenies being tested, the genetic variation among testcrosses was less and equal to zero if complete dominance of favorable alleles. The opposite situation occurred if allele frequencies were either zero or very low; greatest genetic variation among testcrosses occurred for all levels of dominance with a low frequency of favorable alleles in the tester. With the formation of heterotic groups and with selection and testing of lines to enhance the expression of heterosis, favorable complementary groups of alleles are increased in the opposite heterotic groups. Betrán and Hallauer (1996), for example, found that hybrid performance increased in hybrids produced between lines of BSSS and BSCB1 after nine cycles of half-sib RRS. Hallauer and Lopez-Perez (1979) evaluated 50 unselected S_1 progenies and 50 S_8 progenies (derived by single-seed descent from the same S_1 progenies) in testcrosses with five testers selected for their expected differences in allele frequencies for grain yield. The 50 S_1 lines and four of five testers were derived from BSSS. The five testers included BSSS (source population of lines), BS13(S)C1 (derived from BSSS after seven cycles of half-sib recurrent selection and one cycle of S_1 recurrent selection), BSSS-222 (an S_8 line derived by single-seed descent from BSSS and was one of the poorest yielding lines per se), B73 (an elite line developed from BS13(HT)C5 after five cycles of half-sib recurrent selection), and Mo17 (developed from cross of C103 × 187-2 and not related to BSSS). Based on Rawlings and Thompson's (1962) results, it was assumed that the testcross σ^2_G estimates would be greater for BSSS-222 (poor yielding line) and Mo17 (unrelated line) and lesser for B73 (an elite line). The empirical data agreed with expectations (Table 8.3, Hallauer and Miranda, 1988). The estimate of testcross σ^2_G was greater for BSSS-222 and Mo17 testcrosses and smallest for B73 testcrosses; σ^2_G for B73 testcrosses was not significantly different from zero. BSSS and BS13(S)C1 had intermediate estimates of σ^2_G for their testcrosses which were expected because allele frequencies were probably at intermediate levels rather than fixed for the three inbred lines. The average estimates of σ^2_G for the S_8 testcrosses were 2.1 times greater than the average estimates of S_1 testcrosses. At the S_8 generation, the differences in estimates of σ^2_G were not as great as at the S_1 generation but B73 testcrosses had the smallest estimate of σ^2_G for its testcrosses. The data generally agree with the genetic expectations that a poor tester (lower frequencies of favorable alleles) would have the greater variation among testcrosses. BSSS-222 (poor tester) and Mo17 (unrelated tester) had similar estimates of σ^2_G. In maize breeding programs that consider the BSSS and non-BSSS heterotic groups, Mo17 would be the appropriate choice of tester to determine combining abilities of BSSS lines. Genetic correlations were calculated between S_1 and S_8 testcrosses for the five testers, the result was only 0.34 for the unselected 50 lines of BSSS. Although the average genetic correlations were lower than expected, when the relative yields of S_1 and S_8 testcrosses are graphically plotted, the trend is for the greater yielding S_1 testcrosses to predict the greater yielding S_8 testcrosses; for example, genetic correlation between S_1 and S_8 generation testcrosses for the poor yielding line, BSSS-222, was $r_G = 0.42$, but the S_1 generation testcrosses correctly predicted 34 of 50 testcrosses at the S_8 generation. Genetic correlations between S_7 lines per se and S_8 testcrosses was $r = 0.04$ which agrees with the studies of Jensen et al. (1983) and Smith (1986). Correlations reported for lines developed from elite germplasm that had undergone selection during inbreeding had greater correlations between early and later generation testcrosses. Jensen et al. (1983) for S_2 versus S_5 testcrosses ($r = 0.67$) and Lile and Hallauer (1994) for two sets of selection lines ($r = 0.97$ and 0.86) reported greater correlations than with use of unselected lines (Hallauer and Lopez-Perez, 1979). The use of marker-assisted selection during the development of lines and their evaluation was found to increase significantly the prediction of advanced generation combining ability (Johnson, 2004; Eathington et al., 1997). For programs that have the resources to use molecular markers within their programs, the use of molecular markers will have an impact on increasing the genetic information of parents and their testcrosses, which will hopefully aid in making continued genetic advances for the newer hybrids.

The choice of tester(s) is an easier decision than before heterotic groups were recognized. The final products of maize breeding programs are hybrids. Hybrids are tested that include crossing the elite lines from their respective heterotic groups. Hence, the initial and advanced testing of new inbred lines will always include testers that are inbred lines that represent elite germplasm of the breeding program. Priority should be given to testers from the opposite group over poor testers for the same amount of genetic variability. Public breeding programs are encouraged to use testers derived from commercial programs. Breeding goals may also be to improve either the male or the female parent of a successful hybrid (Dudley, 1982). If it is considered that improvement of the male parent would improve the hybrid, the female parent of the hybrid may be the logical choice of tester. In other instances, an entirely new hybrid may be the goal and the inbred lines from the respective heterotic groups may be tested to two to five inbred lines from the opposite heterotic group.

9 Types of Hybrids

There are several types of hybrids that have been made and tested in maize. Beal (1880) suggested that use of crosses between open-pollinated cultivars was a method that could be considered to improve maize yields. During the period from 1880 to 1920, crosses of open-pollinated cultivars were produced and tested, but cultivar crosses never had an impact on the US maize production (Richey, 1922; Table 10.1, Hallauer and Miranda, 1988). Because of poor pollen control and experimental techniques as well as choice of cultivars to cross, a consistent advantage of cultivar crosses relative to their parents was not evident. The suggestions of Shull (1910) and Jones (1918) were followed up and had a major impact on the future maize production in the United States and after WWII in other areas of the world. Double-cross hybrids had essentially replaced open-pollinated cultivars in the United States by 1950 and were the principle cultivars used until about 1965. During the 1950s, it seemed further genetic advances from use of double-cross hybrids had attained a plateau and that refinements and modifications were needed to continue the advances made with the use of double-cross hybrids (1.01 kg ha^{-1}, Troyer, 2006). Several changes had occurred during the 1950s to consider the use of single-cross hybrids: agronomic practices that included use of synthetic fertilizers and herbicides, increased plant densities, rapid improvements of field equipment, rapid expansion of the commercial hybrid maize industry, and improved inbreds developed by recycling of previously used inbreds or extracted from synthetic populations (e.g., B14 and B37). The conditions were appropriate to consider using simpler-type hybrids. Cockerham (1961) also determined theoretically that selection among single-cross hybrids would be more effective than among three-way and double-cross hybrids.

The transition from use of double-cross hybrids to simpler-type hybrids was immediate and permanent in most developed temperate areas. The methodology for developing single-cross hybrids was simpler and reduced the requirements for seed production because production of double-cross hybrids required maintenance of the four parental inbred lines, seed production fields to produce the two single-cross hybrid parents, and seed production fields to produce the double-cross hybrid (see Fig. 8.1, Dowswell et al., 1996). Use of single crosses reduced the number of production fields from seven to three. However, there was still the challenge to have highly vigorous female lines for seed production.

Cockerham (1961) demonstrated theoretically that selection among single crosses was two times greater than among double crosses, considering only σ_A^2 among the two types of crosses. But does the greater selection advantages among the different types of hybrids translate to greater yields of single crosses versus either double crosses or three-way crosses? If epistatic effects contribute to the heterosis expressed for grain yield in single crosses, single crosses would be expected to have greater grain yields than either three-way or double crosses because the uniquely, favorable linkage blocks of genes or epistatic effects could be disrupted in meiosis because of genetic recombination in the single-cross parents used to produce three-way and double crosses. Data that compared grain yields for the three types of hybrids were summarized by Hallauer and Miranda (1988, Table 9.8). Grain yields averaged across all studies did not show a clear advantage of single crosses relative to three-way crosses, but both types of crosses, on average, were superior to double crosses. Weatherspoon (1970) summarized the comparisons for the three types of hybrids, and the single crosses were either equal to or superior to three-way crosses, and three-way crosses were superior to double crosses. Schnell (1975) reexamined Weatherspoon's (1970) data (Table 15). The data showed the same trend as previous comparisons, and the expected greatest yields followed the same trend as the observed data. Weatherspoon (1970) produced the three balanced sets of crosses from nine unrelated inbred lines. The more homogeneous crosses had greater standard deviations than the double crosses and greater interactions with environments, which agreed with previously reported data.

Table 15 Comparisons for grain yield (q ha^{-1}) for 36 single-cross hybrids between nine unrelated inbred lines and a balanced set of 36 three-way and double crosses from the same nine inbred lines

Crosses	Average grain yield	Standard deviation	Extremes		Expected greatest yield
			Least	Greatest	
Single crosses	65.1	8.8	43.6	81.5	83.7
Three-way crosses	62.0	6.2	47.7	72.9	75.1
Double crosses	60.3	3.8	54.0	67.7	68.3
Predicted –					
Three-way crosses	65.1	6.4	47.4	80.1	–
Double crosses	65.1	4.8	52.5	79.1	–

Analysis was by Schnell (1975) based on data reported by Weatherspoon (1970)

Although the single-cross hybrids tended to have greater interactions across environments, the stability of some single-cross hybrids was as good as the more heterogeneous hybrids. Weatherspoon (1970) concluded that stability was under genetic control and that more extensive testing would be required to identify those single-cross hybrids that have good, stable performance across environments. Average observed yields were 65.1, 62.0, and 60.3 for single, three-way, and double crosses, respectively (Table 15). Based on estimates of standard deviations and range for each cross, Schnell (1975) determined that the greatest expected yield were 83.7 q ha^{-1} for a single cross, 75.1 q ha^{-1} for a three-way cross, and 68.3 q ha^{-1} for a double cross. His data clearly suggested that one could expect to identify single crosses that had greater yields than double crosses. Weatherspoon (1970) and Schnell (1975) suggested that the greater yield of single-cross and three-way hybrids was because of greater use of dominance and epistatic effects. Although costs of seed production for single-cross hybrids may be three to five times (or more) greater than production of double-cross hybrids, seed costs have not restricted the rapid acceptance and use of single-cross hybrids in major temperate maize producing areas. Greater uniformity, improved standability, and improved yields of single-cross hybrids negated the extra seed costs (Fig. 1). In other maize growing areas of the world, seed costs are a concern and hybrids that can be produced at lower costs are more popular (Table 8.2, Dowswell et al., 1996; Carena and Wicks III, 2006).

10 Heterotic Groups

The recognition of heterotic groups simplified decisions relative to choices of testers and crosses to test between newer inbred lines. The concept of heterotic groups is different from the one for heterotic patterns. Heterotic patterns are crosses between known genotypes that express a high level of heterosis (Carena and Hallauer, 2001b). They became established by relating the heterosis of crosses with the origin of the parents included in the crosses (Hallauer et al., 1988). Heterotic groups were determined empirically; they were not identified by either theoretical or computer simulation studies. Heterotic groups include germplasm sources that when crossed with each other produce consistently better crosses than when crosses are made within heterotic groups. How did the heterotic groups evolve? There is no specific answer because each heterotic group evolved over time from the parental germplasm in its formation, selection goals for introduction of new germplasm, and, finally, how productive was the germplasm within heterotic groups. Presently, the heterotic groups of BSSS and non-BSSS are frequently used as a base of reference (Mikel and Dudley, 2006). But BSSS was not developed initially as a heterotic group from non-BSSS materials, but to develop source materials with greater stalk strength (Hallauer et al., 1983). The frequently cited heterotic groups of the US Corn Belt include Reid Yellow Dent and Lancaster Sure Crop. Reid Yellow Dent was developed in central Illinois whereas Lancaster Sure

Crop was developed in southeastern Pennsylvania (Wallace and Brown, 1988). Both were open-pollinated cultivars that were synthesized from different germplasm, the originators had different goals in developing them, environments were different where they originated, and both became widely known cultivars. Because of their origins, differences in frequencies of alleles for many loci would be expected. Falconer (1960) demonstrated that the expression of heterosis was dependent on nonadditive genetic effects, and that the magnitude of heterosis was dependent on the differences in allele frequencies. Hence, it seems reasonable to expect greater differences in allele frequencies between Lancaster Sure Crop and Reid Yellow Dent than differences between local cultivars in their respective areas. The same comments are equally valid to other proposed heterotic patterns, such as Leaming × Midland (Kauffmann et al., 1982; Carena and Hallauer, 2001b) as well as BS21 × CGSS, BS21 × CGL, BS21 × NDSAB, and BS22 × Leaming (Carena, 2005a, b; Melani and Carena, 2005). Geographical isolation developed cultivars with different allele frequencies. In addition, selection has caused cultivars to differ in allelic frequencies as well at a point where intra-population recurrent selection methods have increased heterosis in as much as RRS programs.

Extensive testing of inbred lines from different source populations showed trends how inbred lines from different populations reacted in crosses. G. F. Sprague observed the Reid Yellow Dent and Lancaster Sure Crop heterotic pattern while preparing the 1939 and 1940 annual reports, based on data from A. A. Bryan (p. 537, Troyer, 2006). Double-cross hybrids that were produced from single crosses whose parents were from the same source tended to have better yields. Eckhardt and Bryan (1940) reported that the greatest yielding double crosses where two parental sources (A and B) were involved were those in which the two lines from cultivar A was used as one parent in the double-cross hybrid and two lines from cultivar B used as the second parent; that is, (A × A) × (B × B) was greater yielding than (A × B) × (A × B). They also found that if early (E) and late (L) lines were used as parents that (E × E) × (L × L) double crosses were more uniform for traits associated with maturity than (E × L) × (E × L) double crosses.

The concept of heterotic groups for breeding purposes was recognized by the 9th Corn Improvement Conference of the North Central Region of the United States. Stringfield (1947) suggested that grouping of inbred lines should be emphasized for the improvement of lines (i.e., recycling of lines) within lines of the same heterotic group. Initially, the inbred lines were more or less assigned randomly to the two breeding groups. Later, assignment of inbred lines to the breeding groups was based on the origin of the lines, probably because of the data collected by A. A. Bryan and the report of Eckhardt and Bryan (1940). The grouping of inbred lines was primarily based on whether they were of Reid Yellow Dent origin and non-Reid Yellow Dent origin. Subsequent breeding plans emphasized recycling of inbred lines within the two heterotic groups. Heterotic groups are common features within breeding programs for genetic improvement of elite lines because breeders desire to maintain favorable genetic combinations that enhance heterosis. Intensive breeding efforts have continued, but the US Corn Belt heterotic groups have evolved to be designated as Iowa Stiff Stalk Synthetic (primarily Reid Yellow Dent in origin) and non-Iowa Stiff Stalk Synthestic (Mikel and Dudley, 2006). Assignment or classification of inbred lines to heterotic groups has been based primarily on the pedigrees of the inbred lines. Molecular marker data also have proven to be very effective for determining the genotypes of inbred lines and their relative relations to other lines within heterotic groups (e.g., Smith, 1988; Smith et al., 1985; Smith and Smith, 1991; Melchinger et al., 1991; Gethi et al., 2002; and many others). The use of molecular markers also can differentiate the genetic distances among inbred lines within heterotic groups more precisely than pedigree information, which is useful for making planned crosses for breeding purposes. Attemps for predicting hybrid performance with molecular marker, testcross, and diallel data are still under way (Barata and Carena, 2006). However, more research is needed across genotypes and environments.

Heterotic groups were identified and developed by maize breeders (Hallauer et al., 1988). There are heterotic groups broadly defined, but more specific heterotic groups (e.g. subgroups) are available within large breeding programs for specific families of inbred lines. They would be more narrowly defined genetically, but they involve inbred lines that have been very important as parental inbred lines of successful, widely grown hybrids. More extensive discussion of heterotic groups was given by Hallauer et al. (1988) and Coors and Pandey (1999) for temperate regions, and Dowswell et al. (1996) and Goodman (1985) for tropical and subtropical regions and in developing countries.

11 Heterosis

The term heterosis is more widely associated with maize than any other important crop species. Because of the success in implementing the concepts of Shull (1910) and Jones (1918) developed in the public sector, hybrid maize has been recognized worldwide. The genetic basis of heterosis has remained elusive with theories advanced to explain the phenomenon, including interaction of dominant favorable alleles, intra-allelic interactions (overdominance), inter-allelic interactions (epistasis), complementary action of linked genes (pseudo-overdominance), and the nongenetic physiologic stimulation (Hallauer and Miranda, 1988; Richey, 1950). The significance of heterosis and its genetic basis have received extensive discussion at two symposia (Gowen, 1952; Coors and Pandey, 1999). The expression of heterosis in crosses has been observed for more than 200 years, but the early hybridizers were interested how parental traits were expressed in crosses and if the traits were recovered in the offspring of the crosses (Goldman, 1999). Their interests were more scientific and not to develop cultivars that exploited heterosis for commercial use. When the techniques and technology to produce adequate seed of double-cross hybrids for use by the producers were available, hybrid maize became a reality and was considered one of the greatest plant breeding achievements of the twentieth century. Heterosis was exploited to the fullest extent and a very competitive commercial hybrid seed industry developed to meet the demand.

The success of hybrid maize encouraged research in other fields as well as other horticultural, and forestry crops (Coors and Pandey, 1999).

Heterosis has been exploited to the fullest extent possible, but definitive evidence for the genetic basis of heterosis has been elusive. It is agreed that allelic interactions are operative; intra-allelic, inter-allelic, and probably both are equally important. The first heterosis conference emphasized maize with the main topic of dominance of favorable alleles versus overdominance (Gowen, 1952). The second heterosis conference of the 1990s was more extensive and discussed a greater array of topics, including the heterosis observed in different crop species, molecular techniques to study expression of heterosis, and the possible importance of epistasis. Duvick (1999) presented a summary of data for maize hybrids grown from 1930 to 1980. He concluded that heterosis has an important role in maize hybrid yields, heterosis has increased in absolute amounts but percentage of heterosis decreased because inbred line yields have increased (results of recycling inbred lines), and that heterosis will increase in smaller increments in the future because of greater rates of improvement for inbred lines. The experimental data confirms field observations. First cycle inbred lines are difficult to maintain because of poor seed yields relative to twenty-first century inbred lines (Notes: In 2004, a few inbred lines yielded more than 62.5 q ha^{-1} in a replicated trial at Ames, IA; in 2005, the top two NDSU lines yielded 65.0 and 62.0 q ha^{-1} in a replicated trial at Fargo and Casselton, ND with an average yield of 33.0 q ha^{-1} for the whole experiment including five commercial lines that were lower yielding than these NDSU lines). The trends will continue. Inbred lines will continue to be improved for biotic and abiotic stresses by both traditional breeding methods and information from molecular genetics.

An explicit genetic basis of the expression of heterosis for each hybrid is probably not realistic. Each single-cross hybrid is a unique cross between two elite inbred lines, and the cross represents the additive genetic effects of each parent, additive × additive epistatic effects of each parent, interaction effects of the alleles of both parents as well as the epistatic effects that include dominance, which suggests a very complex genetic system. And the relative importance of each type of genetic effects also will be different for each hybrid. Widely grown single-cross hybrids, such as WF9 × C103, A632 × A619, and B73 × Mo17, included different parental inbreds that were derived from different source germplasm by different breeding methods. Each of the parental inbreds was also a progenitor of recycled lines, suggesting that each line contributed important additive genetic effects to their derived progenies. Each of the three single crosses had one parent from the Reid Yellow Dent heterotic group and one parent primarily from the Lancaster Sure Crop heterotic group. Hence, in addition to the genetic effects, the gene frequencies of favorable alleles would be different to magnify the expression of heterosis. The level of expression is not predictable because of allele interactions, similar to the estimates of SCA. Extensive testing is conducted to identify those crosses that have the unique combination of genetic effects and allele frequencies. Molecular markers have been effective to profile inbred lines and often but not always assign to appropriate heterotic groups, but, at this time, molecular genetics is not able to predict the final hybrid. The ultimate selections depend on replicated field trials to determine which cross has the combination of genetic effects that have consistent, high performance under the environmental effects where tested. Discussions on topics related to heterosis were given by Hallauer and Miranda (1988), Hallauer et al. (1988), Coors and Pandey (1999), and Troyer (1999, 2006).

12 Stability of Cultivars

Consistent and reliable performance across locations and years (environments) is highly desirable for an organization that releases, recommends, and promotes the sale of cultivars for use by the producers. The survival of an organization that relies on seed sales needs to have a satisfied clientele. Genotype by environment interactions are common occurrences in plant breeding. At some point, plant breeders will need to evaluate inbred lines to determine their breeding values; that is, determine the relative proportion of the measured phenotypic (P) expression that is due to genetic (G) and environmental (E) effects. If the genetic (G) and environmental (E) effects are additive (P = G + E) we can separate the two sources of variability that affect the phenotypic expression of traits. Cultivars are usually developed for specified target environments, which may be either narrowly defined geographically (e.g., State of Iowa or State of North Dakota), by soil type (Clarion-Webster soil association or Fargo Vertic-Haplaquolls), by maturity (e.g., 700–800 maturity group or 100–300 maturity group), by latitude (e.g., 37–41° north latitude or 45–49° north latitude), or more broadly defined environments such as the northern US Corn Belt, central US Corn Belt, southeastern United States, tropical areas with acid soils, etc. A common occurrence is that the phenotypic measurements of the same cultivars vary among environments either in rank or relative magnitude; that is, they interact among environments (GE) such that P = G + E + (GE). If GE is serious, breeders need to investigate causes to determine possible factors that cause GE. It could be because of different soil types, variation within experimental areas, rainfall patterns, irrigated versus nonirrigated, etc. Choices or changes in experimental designs and analyses may be appropriate when large number of entries is included and breeders desire to make head-to-head comparisons. With correct experimental designs and analyses, the variation among the measured phenotypes (σ_{P}^2) can be partitioned as $\sigma_{\mathrm{P}}^2 = \sigma_{\mathrm{G}}^2 + \sigma_{\mathrm{E}}^2 + \sigma_{\mathrm{GE}}^2$. This additive expression can be used to estimate either heritability (h^2) or repeatability (r^2) of cultivars on a progeny mean as h^2 (or r^2) $= \sigma_{\mathrm{G}}^2 / [\sigma^2 / re + \sigma_{\mathrm{GE}}^2 / e + \sigma_{\mathrm{G}}^2]$. The h^2 or r^2 depends on the source of the cultivars being tested. Usually, a group of selected testcrosses or hybrids are evaluated and r^2 would be the appropriate parameter. For a recurrent selection program where all the progenies are an unselected sample of a specific population, estimates of h^2 would be appropriate. If the $G \times E$ tests significant in the ANOVA, the cultivars interacted with environments, but the significance of $G \times E$ does not indicate which cultivar interacted or the relative size of the interactions among cultivars. It is of interest to determine cultivar stability across environments, or how individual cultivars perform across environments without knowing the specific environmental effects causing $G \times E$. Methods have been suggested to estimate parameters for individual cultivar response to a series of environments.

Stability of genotypes was examined relative to heterozygosity and heterogeneity of genotypes in terms of homeostasis and magnitude of interactions with environments (Adams and Shank, 1959; Shank and Adams, 1960). Generally, it was found that the more heterozygous and/or heterogeneous genotypes had greater homeostasis and less interactions with environments (Sprague and Federer, 1951). Plaisted and Peterson (1959) partitioned the $G \times E$ source of variation to provide an estimate of $G \times E$ for each genotype. The genotype with the lowest value was considered to have the best stability. They considered environments random effects and cultivars as fixed effects so relative estimates of $G \times E$ were for the specific set of cultivars (Model I). For this method there were $n(n-1)/2$ ANOVAs and an estimate of σ^2_{GE} for each cultivar.

Findlay and Wilkinson (1963) developed an analysis to determine the relative adaptation of a group of cultivars to include in plant breeding programs. The analysis was based on a regression of cultivars (dependent variable) on environment means (independent variable), where the environment means were determined by the mean of all cultivars grown at each environment. Environments were arranged from the poorest to the best for the regression analysis. The estimated regression value (b) of each cultivar on the environment means was used as a measure of stability. A regression value of $b = 0$ indicates that cultivar had stable performance over the series of environments the cultivars were tested. A stable cultivar with $b = 0$ would have consistent, say yield, across all environments, but the cultivars could have either a low mean or high mean across environments. The cultivar with the highest mean and $b = 0$ would be desirable. Eberhart and Russell (1966) independently developed a similar regression analysis to determine the stability in performance of cultivars across a series of environments. The regression model was the same as the one used by Findlay and Wilkinson (1963), but they extended their model to include estimates of deviations from regression (s^2_{d}) for each cultivar as well as estimates of regression coefficients. With calculations of the deviations about regression for each cultivar, Eberhart and Russell (1966) used the following interpretations of the two estimated parameters for each cultivar: b is a measure of response of cultivars to changes (poorest to the best) in environments; and s^2_d is a measure of stability of cultivars across environments. With Eberhart and Russell (1966) analysis, three parameters are available to characterize each cultivar: mean ($\overline{\mathrm{X}}$) across environments, response to changes in environment (b), and s^2_d a measure of stability.

Freeman and Perkins (1971) had some questions relative to the validity of the regression analyses suggested by Findlay and Wilkinson (1963) and Eberhart and Russell (1966). The concern was because cultivar means regressed on environments were correlated with the environmental index (independent variable); that is, the cultivar mean was included in the environmental index with the cultivar regressed on environmental index. They suggested the cultivar mean should be excluded from the environmental index before conducting regression analysis for the cultivar. An alternative was to include a set of check cultivars in each environment and use the check means to construct the environmental index. Practical considerations would need to be considered. Would the inclusion of 10 checks provide better estimates of environment means than the 100 cultivars being tested? In maize breeding 50–100 (and more) cultivars are commonly included in experiments. Use of an adequate number of checks to estimate environment means would reduce the number of cultivars included. Compromises would be needed. It seems that if 50–100 experimental cultivars are included, that the environmental means would be estimated with greater precision and the mean of each cultivar will have minimum impact on the overall environment means. Greater number of environments probably is more important to determine responses of each cultivar for different environments. Wricke (1962) suggested the ecovalence analysis to determine stability of genotypes, which is similar to the method suggested by Weatherspoon (1970) where $X_{ij} - X_i + X_j + \overline{X}$ estimated for each cultivar (X_{ij}) which would be an effect from the marginal means for cultivars (X_i) and environments (X_j). Wricke (1962) squared the effects, whereas Weatherspoon (1970) did not. Weatherspoon concluded that cultivar and heterogeneity contributed to stability of performance. Although there are acknowledged statistical deficiencies of the regression analyses, the Eberhart and Russell (1966) analysis has been found useful because of its simplicity and interpretations of the regression coefficients and deviations from regression parameters.

13 Selection Indices

New cultivars have been screened with multistage and multitrait selection of economically important traits under certain breeding goals across numerous environments. Plant breeding always includes selection for multiple traits either during inbreeding and selection or during evaluation trials. The art of plant breeding for traits that are highly heritable can be successful but usually there are multiple traits that need attention. The mental matrix may not be consistent from day-to-day or year-to-year to give proper weights to different traits. Objective selection becomes more difficult as the number of traits increases and/or the traits have a more complex inheritance. Maize breeders need more objective means to be more consistent for evaluating differences among cultivars. Objective analyses also reduce possible biases individuals may inadvertently use for selecting among different groups of materials. Different selection indices have been suggested to assist maize breeders for the simultaneous consideration of all traits that were considered important in developing cultivars. Smith (1936) proposed use of selection indices in plant breeding, and several modifications and alternative types of selection indices have been proposed since the original suggestion by Smith (1936). The basic feature of all selection indices is a linear function of phenotypic values for the different traits. Observed values for each trait are weighted in some manner to provide a composite evaluation of the genotypic values for a series of cultivars. The genotypic value of a particular cultivar is the average value when the cultivar is tested in a large number of environments. The general features of a selection index (I) can be described as $I = b_1P_1 + b_2P_2 + \ldots b_nP_n$, where P_1 is the observed phenotypic value of the ith trait, and b_i is the weight assigned to the ith trait. Construction of the Smith (1936) and similar indices is rather complex because estimates of variances and covariances and economic weights are required. It is not feasible in many instances to determine the variances and covariances of traits if several populations are under selection and different populations are under selection in succeeding generations. Also, because estimates of variances and covariances are dependent on allele frequencies, estimates for one population may not be appropriate for other populations and not appropriate if selection significantly changes allele frequencies within the same population. Baker (1986) has provided a comprehensive discussion on the theory, the purpose, construction of different selection indices, and their relative efficiencies.

Selection indices provide an objective method for determining the relative merits of a series of cultivars. To be practical, but yet useful, maize breeders need simpler selection indices that use data acquired from their evaluation trials to assist them when a large number of cultivars are evaluated. Independent culling levels can be used, but if 7–17 traits are measured and considered important for the success of a cultivar, then other methods are considered. Simplicity is useful provided the indices can provide an objective basis for differentiating among the cultivars being tested.

Elston (1963) proposed the multiplicative index. The multiplicative index is sometimes referred to as a weight-free index because it does not require the use of index weights or economic weights. General form of the multiplicative index is $I = (\overline{X}_1 - K_1)(\overline{X}_2 - K_2)\ldots(\overline{X}_n - K_n)$, where the K_i is the minimum specified value of the ith trait. The K_i values are chosen by the breeder and could vary among experiments depending on the quality of data and the relative importance of traits for different types of cultivars under test. In addition to being weight-free index, the multiplicative index does not require estimates of genetic and phenotypic variances and covariances. Because the multiplicative index is a curvilinear index, it is not available to predict genetic gains. Baker (1974), however, reported that the multiplicative index can be approximated for use as a linear index, where the weights are the reciprocals of the phenotypic standard deviations (σ_P) of the traits included in the index. The choice of the K_i values will depend on the emphasis the breeder gives to each trait. If greater emphasis is to be given to yield, the K value may be set to two standard errors above the mean for all cultivars, where K values for the other traits may be the overall mean. K values can be adjusted after initial cycles of selection if changes in trait expression occur because of past selection (later maturity, reduced stalk lodging, disease pressure, etc.)

Mulamba and Mock (1978) suggested the use of the rank-summation index to make selections among cultivars. The rank-summation index basically involves ranking each of the cultivars (1 to n) for each of the traits and calculating the index by summing the trait ranks for each cultivar: $I = \sum_{n-1}^{n}$ rank (X_i), where we have n traits with each cultivar ranked for the n traits. Cultivars having the lowest index values are chosen. The primary advantages of the rank index are simplicity, estimates of genetic and phenotypic parameters are not required, data can be transformed so that variances are similar for each trait, and does not require any specification of economic weights, although economic weights can be used. But similar to the multiplicative index it is not possible to calculate predicted gain by the summation index. Crosbie et al. (1980), however, found that the same prediction equation used for the multiplicative index provides a reasonably good approximation of the predicted gains for the rank summation index.

Smith et al. (1981) proposed use of heritability (or repeatability) estimates to construct a selection index. The heritability estimate approach was suggested for recurrent selection programs where the progenies (half-sib, full-sib, or inbred progenies) evaluated were a random sample of genotypes from the different cycle populations sequentially derived from the different cycles of recurrent selection. The same approach can be used for selected cultivars under evaluation to determine the better performing cultivars for continued testing. For selected cultivars from diverse sources and different pedigrees, the estimated parameter is designated as an estimate of repeatability. The methods used to calculate the estimates of heritability and repeatability are the same. The estimates are calculated from the expected mean squares for the ANOVA combined across environments. For estimates of heritability, h^2 is the ratio of $\sigma_G^2/(\sigma^2/re + \sigma_{GE}^2/e + \sigma_G^2)$. The form of the index is $I = \sum_{i=1}^{n} h_i^2 \overline{G}_{ij}$, where h_i^2 is the heritability for the ith trait times the genotype mean $(\overline{G}_{ij})$, obtained by averaging across replications and environments. The weights (h_i^2 *s*) are determined from information of each ANOVA combined across environments and will change depending on the genetic variation among genotypes for each trait. An example for use of the heritability index in maize could be one where genotypes have greatest grain yields (Y), lower grain moisture (M) levels at harvest, and lower incidence of root (RL) and stalk (SL) lodging. The form of the index for this example would be $I = h_1^2(\overline{Y}_i) - h_2^2(\overline{M}_i) - h_3^2(\overline{RL}_i) - h_4^2(\overline{SL}_i)$, h_1^2, h_2^2, h_3^2, and h_4^2 are the estimates for Y, M, RL, and SL, respectively. For this example, genotypes having the greatest index values are selected because we desire to identify genotypes with the greatest grain yield and the lowest grain moisture and incidence of root and stalk lodging. It was assumed that with the use of the heritability index that the genetic correlations between traits were either zero or very low.

Economic values of the various traits that maize breeders wish to select are rarely known and few studies have been conducted to determine how economic values should be assigned to traits. Because of these difficulties, the selection indices of Elston (1963) and Mulamba and Mock (1978) were developed because they are weight-free. The heritability index (Smith et al., 1981) has estimates of h^2 that are relative weights, which can be easily determined from the combined ANOVA. Smith et al. (1981) suggested that traits be assigned economic values according to the goals of the breeding programs instead of according to their relative economic values. This removes emphasis from the economic importance of individual traits but requires breeders to make decisions regarding the goals of the breeding program.

Long-term recurrent selection programs that used selection indices as the criteria of selection have been rare. The more common method for making selections in recurrent selection programs would be similar to the rank summation index. The heritability index has been used in the Iowa recurrent selection programs since 1980 and in the NDSU recurrent selection programs since 1999. Generally, studies designed to compare methods of index selection involve only one or two cycles of selection with the same or different index being used for each cycle of selection. The more common method for comparing selection indices has been to use predicted gains and selection differentials. Predicted gains and selection differentials for a given trait from index selection are usually represented as the percentage of the value obtained for single-trait selection. Cunningham (1969) presented a method for comparing the relative efficiencies of selection indices. Selection indices are constructed to obtain maximum gain in the aggregate genotype. Hence, the relative efficiencies of the different methods of selection are of interest. Hazel and Lush (1942) compared index selection with independent culling levels and tandem selection. They found that the original index selection was at least as efficient as independent culling levels and independent culling levels method was at least as efficient as tandem selection (Baker, 1986). The relative efficiency of any index selection method includes factors, such as progeny type, number of traits included in the index, selection intensity, relative magnitude of the estimates of heritability, genotypic and phenotypic variances and covariances among traits, and, where used, the relative economic weights assigned to the traits. Hazel and Lush (1942) reported that as a general rule, expected genetic gain for one trait (single trait selection) from selection based on an index containing n traits is only $1\sqrt{n}$ times as great for that trait alone. This may seem to be a penalty for use of a selection index. If maize breeders selected only for greater grain yield, ignoring other plant traits required for production within a specific region, greater yield can be attained but maturity may become later (maximize growing season) and root and stalk lodging would increase (redirection of photosynthates from stalks to grain). Greater yield per se would be realized, but the cultivar would not be acceptable in modern production systems because of costs to dry grain and problems in harvesting lodged maize. Hence, maize breeding is an art and science of compromise. Selection indices are used to provide weights for the traits considered to aid in selecting for the best combination of traits. Selection index theory can also be used to make selections based on information of relatives (Henderson, 1963). Information from relatives is used to increase the accuracy of selection defined as the correlation of the sample index with the aggregate genotype. Moreno-Gonzalez and Hallauer (1982) used a procedure to combine information on S_2 progenies per se with information on full-sib families from a RRS program. The resulting selection index was superior to full-sib RRS when the heritability of the trait under selection is low, and the advantage of the selection index is increased when the correlations between the S_2 progenies and full-sib families are large relative to their heritability.

Subandi et al. (1973) compared two versions of the multiplicative index, two Smith–Hazel indices, a base index, and selection for grain yield and percentage of erect plants and dropped ears using predicted gains. They concluded that the multiplicative index would be more useful because correlations were low between traits, no specifications of economic weights were required, and it was parameter free. Compton and Lonnquist (1982) used the multiplicative index for four cycles of intrapopulation full-sib recurrent selection. They reported observed gains of 4.7% per cycle of selection and no significant changes for percentages of nonlodged plants and ear retention, but the trends were in the desired directions. Compton and Lonnquist (1982) concluded that the use of the multiplicative index resulted in grain yield gains similar to other studies where selection emphasized only grain yield. West et al. (1980) conducted replicated S_1 and full-sib recurrent selection and used the multiplicative index for making selections and concluded the index was effective in changing the traits simultaneously in the desired directions.

Widstrom (1974) also compared the relative effectiveness of three selection indices in selection for resistance to ear worm [*Heliothis zea* (Boddie)] damage. He included the corn earworm damage, husk tightness, day-to-50% pollen shed, and husk extension traits to compare the Smith–Hazel index, constructed using estimates of genotypic and phenotypic covariances matrices from information obtained from S_1 progeny trials; a modification of Smith–Hazel index where standardized direct and correlated responses were used rather than the genotypic co-variance matrix; and the rank summation index for their effectiveness to increase corn earworm resistance. The experimental evidence supported the conclusion that the two indices derived from realized gains are expected to be as efficient as the Smith–Hazel index when expected and observed results are similar. Widstrom et al. (1982) evaluated response to four cycles of S_1 recurrent selection using the same three indices. There were differences among indices for husk tightness and husk extension, but the populations were 1–2 days earlier and the correlations between traits included in the indices changed over cycles of selection for all indices, which were attributed to breakup of genetic linkages. Widstrom et al. (1982) suggested that new indices should be constructed for future cycles of selection. Except for the parameter-free selection indices, this would require reestimation of genetic variances and covariances. For cyclical selection programs, it seems the use of the simpler selection indices would be more appropriate. If selection is effective, the frequency of alleles will change, which in turn affect the estimates of genetic variances and covariances. The heritability index suggested by Smith et al. (1981) uses components of variance estimated from the evaluation trials for each cycle of selection under specific environmental conditions. If significant allele frequencies are modified during selection, the changes would be reflected in the estimates of heritability, assuming other factors are not affecting progeny performance.

Suwataradon et al. (1975) compared the Smith–Hazel index, base index (Pesek and Baker, 1969), and the desired gain index (Pesek and Baker, 1969). Two sets of arbitrary economic values were used for the Smith–Hazel index and in both instances the response of the index was less than satisfactory. The base index was 95–97% as efficient as the Smith–Hazel index, and they recommended the base index instead of use of Smith–Hazel index when the economic weights are known, heritabilities are relatively high, and the correlations between traits are low. With use of the desired gain index, the goals of the S_1 recurrent selection would be attained after 14 cycles. They preferred use of the base index when relative economic values are difficult to specify, usually the case.

The response to use of selection indices is dependent on obtaining relatively precise estimates of genotypic and phenotypic variances and covariances. In most instances, relative precise estimates of genetic and phenotypic variance and covariances require extensive and expensive studies in order to minimize G $\times$ E interactions. Moll et al. (1975) evaluated the performance of five selection indices and concluded that nonlinear relations between traits will have significant effect on the prediction of correlated selection responses. The responses of traits included in the index were more variable than responses to index selection. Kauffmann and Dudley (1979) evaluated seven selection indices constructed to improve grain yield, percentage protein, and kernel weight simultaneously. They reported good agreement between observed and predicted responses to index selection and concluded that estimates of genetic variances and covariances obtained from 200 half-sib families were sufficiently precise in development of selection indices. Miles et al. (1981) had similar conclusions, but they also found that use of selection indices to simultaneously improve resistance to four maize diseases was no more effective than selection for disease score per se.

Crosbie et al. (1980) compared predicted gains for three cold tolerance traits of maize with use of different selection indices. The Smith–Hazel and base indices ranked the lines similarly, but there were problems with both indices because of substantially different variances for the three cold tolerance traits and these indices were designed to maximize gain in the aggregate genotype. The authors recommended use of indices, such as the multiplicative and rank-summation indices, to improve composite traits, such as cold tolerance, which are composed of component traits (rate of emergence, percentage emergence, seedling vigor, etc.) that do not have easily identified economic values. Crosbie et al. (1980) recommended that indices that are parameter free, easy to use, and do not require specification of economic values should be used.

The relative rankings of testcrosses for the BS29(R)C2 cycle of RRS with BS28 (R)C2 as the tester are illustrated in Table 16 for three weight-free indices. The heritability index proposed by Smith et al. (1981) was used to make selections of S_1 progenies included for intermating to form BS29(R)C3 cycle. The heritability index values of the best 10 selections ranged from 14.2 to 23.6. The entry 208 was clearly the best testcross because it had above average yield and was below average for grain moisture and incidence of root and stalk lodging, which was in the desired direction for the four traits. Entry 019 was ranked second best selection: grain yield was 4.9 q ha^{-1} less than third ranked entry 061 but entry 019 had the lowest incidence of root and stalk lodging of all testcrosses evaluated. The rank-summa-

Table 16 Agronomic data for 20 of 144 BS29(R)C2 S_1 testcrosses evaluated in two replications at three locations for 1 year that were ranked by heritability index values with remnant S_1 seed of the best ten testcrosses used for intermating to form the C_3

Entry	Grain		Lodging		Index[a]		
	Yield (q ha^{-1})	Moisture (%)	Root (%)	Stalk (%)	1	2	3[b]
BS29(R)C2-208	67.5	18.3	1.5	11.4	23.6 (1)	18 (1)	28.7 (5)
019	58.9	20.3	0.5	9.1	18.0 (2)	44	49.9 (10)
061	63.8	19.0	1.8	15.1	17.9 (3)	38 (5)	60.9
148	64.1	19.9	1.0	15.1	17.5 (4)	43 (10)	121.4
166	68.1	18.8	2.3	22.2	15.8 (5)	37 (4)	43.3 (8)
162	61.6	18.8	1.0	17.4	15.0 (6)	35 (3)	33.4 (6)
120	60.5	19.2	2.9	15.9	14.9 (7)	54	39.2 (7)
067	66.5	19.0	0.0	22.2	14.7 (8)	31 (2)	77.8
123	57.6	18.9	3.6	13.9	14.6 (9)	50	8.2 (1)
149	66.3	18.9	0.9	22.8	14.2 (10)	40 (7)	67.5
BS29(R)C2-172	59.6	19.7	0.9	15.8	14.1	48	45.4
235	60.7	18.5	1.9	18.0	14.1	42 (9)	17.5 (2)
244	67.4	19.1	0.9	23.9	14.0	41 (8)	90.4
047	67.2	19.9	0.0	23.3	13.7	43 (10)	155.9
070	65.3	19.3	1.4	22.5	13.4	50	87.4
182	61.1	19.6	0.5	19.3	12.7	52	41.8 (9)
188	64.4	18.4	1.9	23.8	12.5	42 (9)	59.1
133	60.2	18.8	0.0	21.0	11.5	38 (6)	23.4 (3)
024	60.6	19.3	1.0	20.8	11.4	52	41.8 (9)
012	61.0	19.8	0.5	20.8	11.4	48	63.8
$\overline{X}$	56.5	18.8	1.7	29.6			
Standard error	8.9	0.8	2.4^2	9.8^2			

The rankings by the rank-summation and multiplicative indexes are shown for comparison
Index 2 is the rank summation index suggested by Mulumba and Mock (1978)
Index 3 is the multiplicative index suggested by Elston (1963)
[a]Index 1 is the heritability index suggested by Smith et al. (1981) as $I = 0.68$ yield − 0.75 moisture − 0.04 root lodging − 0.72 stalk lodging
[b]Index values were divided by 1,000 for values shown. Means were used for grain yield and nonlodged plants for root and stalk lodging. Grain moisture was one standard error below experiment mean

tion index included seven and the multiplicative index included six of the top ten selections of ten entries identified by the heritability index. Although there were differences in the rankings of the top ten testcrosses among the three selection indices, each index would have included selections among the 20 selections with the greater index values with the heritability index. Each index was effective for identifying the above average testcrosses. Each of the three indexes is relatively easy to use and seem to be equally effective in identifying the better selections. The choice of index depends on the data available and the emphasis given to each trait during selection.

Multitrait selection, multistage testing, and multiprogeny tests are common in maize breeding. In each instances, maize breeders will need to develop selection indices that facilitate selection to identify elite cultivars. Most of the empirical research conducted in maize involving comparisons of selection indices have generally preferred use of indices that are parameter free and do not require specification of relative economic values. Maize breeding rapidly evolves over time and the same parameters and economic values used to develop certain indices may not be appropriate for the next cross or the next year. Successful application of selection indices in maize improvement will depend on the goals of the breeding programs, genetic materials under selection, traits considered in selection, and the definition of the aggregate genotype. Selection indices are used on a regular basis in maize breeding. It becomes the breeders' choice to either choose or develop an index appropriate for their situations. The empirical evidence in the literature suggests that it is not possible to make general statements for what may be considered the best selection index.

14 Summary

The fundamentals of maize breeding were described by Shull (1910) as the development of pure lines by self-pollination, production of crosses between the derived inbred lines, evaluation of crosses to determine the greatest yielding cross, and production of the greatest yielding cross for use by the farmer. The goals of maize breeding were stated by Hartley (1908): "The wise maize breeder will consider all visible characters, that is, those variations that are apparent to the eye, as secondary to the inherent ability of the individual to produce heavily and to transmit its high yielding character to its offspring. ... The most variable characters can be determined only by actual field tests, and it is imperative for purpose of just comparison that these performance testers be ascertained under like normal conditions. ... We must resist the tendency to breed toward artificial standards and must breed maize for the purpose for which maize is grown, namely, profit. ... It is the accumulation and perpetuation of desirable variations, such as high-yielding power, early maturity, proper form, etc. coupled with better cultural methods that has made possible the growing of millions of bushels of maize where 25 years ago it was considered impossible to grow maize." The fundamental methods of maize breeding (Shull,

1910) and the goals of breeding (Hartley, 1908) have essentially remained the same for nearly 100 years. But the advances made in the methods of maize breeding to achieve the goals of maize breeding have exceeded those of Shull (1910) and Hartley (1908). Hartley (1908) emphasized that methods should developed to achieve the objective of 62.5 q ha^{-1}; average US maize yield was 16.8 q ha^{-1} in 1908 and average US maize yield of 62.5 q ha^{-1} was attained in 1978, 70 years later. Continued genetic improvements of hybrids since 1978 have increased to where farmers presently often produce yields that exceed 125.0 q ha^{-1}. Only with applied maize breeding as a leading force of biotechnology and molecular genetics will maize grain yields exceed 187.5 q ha^{-1} within the next 20–30 years as predicted with the use of molecular markers.

Similar to other scientific disciplines, rapid advances in technology have impacted either directly or indirectly the genetic improvements realized from maize breeding. Technological advances in mechanization for conducting research, development of statistical methods and computer hardware and software for better analysis of experimental data, and improvement of husbandry practices in maize production, for example, have had significant impacts on the advancements made for increased grain yields (Hallauer, 2006). Since the acceptance of double-cross hybrids by the producers during the 1930s and 1940s, research has been conducted for the potential of mutation breeding, the study of the inheritance of quantitative traits, the use of recessive mutants for specific traits, the study of the physiological functions within maize plants and developing models of ideotypes for maximum yield, function and importance of transposable elements within the maize genome, and the rapid developments of molecular genetics to further our knowledge of the maize genome, recently sequenced. Some have had greater impacts than others. Gardner (1961) compared response to mass selection within the same population that was either irradiated or nonirradiated; response to selection was not enhanced by irradiation. Mutation breeding has not been an important component of maize breeding because persistent genetic variation has prevailed to provide continued genetic improvements. The predominant maize breeding method continues to include crossing elite by elite to generate genetic variability and selfing to derive inbred lines. Transposable elements seem pervasive in the maize genome and may have some impact on generating genetic variation, but the importance of transposable elements on maize improvement are difficult to quantify (Peterson, 1986, 1987). Breeding methods for the successful development of double-cross hybrids were determined empirically. The consistent expression of heterosis in maize crosses was the basis for the interest in hybrids, but the genetic basis of heterosis, however, has proven to be elusive (Gowen, 1952; Coors and Pandey, 1999). Because of the interest in determining the genetic basis of heterosis, extensive quantitative genetic studies were conducted to determine the types of genetic effects important in the expression of quantitative traits, particularly for grain yield. Estimates of additive genetic and dominance variances for different types of maize populations suggested that additive genetic effects were of greater importance, and that selection should be effective (Hallauer and Miranda, 1988). Studies to quantify the relative importance of epistatic effects have not been successful.

Significant estimates of epistatic effects from comparisons of means for different generations and types of hybrids were generally found in all instances for different plant and ear traits. Although the estimates of additive genetic effects generally seemed of greater importance than the dominance and epistatic effects, within populations, the expression of heterosis depends on the presence and relative magnitude of nonadditive genetic effects. The specific combinations of dominance and epistatic effects are most certainly different for each hybrid. Similar to specific combining ability estimates, the heterotic effects will be unique for each hybrid. To enhance the expression of heterosis, heterotic groups have been identified, and inbred lines are developed within heterotic groups with the hybrids produced and tested that involve lines for each of the heterotic groups. Quantitative genetic studies do not develop inbred lines directly, but the theoretical and empirical information derived from quantitative genetic studies have provided guidelines in developing selection and breeding strategies, including selection methods, testing methods, inbred line development, estimation of heritability, and genetic effects important in heterosis (Hallauer, 2006). Most of the economically important traits considered in maize breeding are inherited quantitatively. Their importance is recognized by molecular geneticists with emphasis given to identifying QTL and use of molecular markers to assist in the improvement of quantitatively inherited traits. The use of molecular markers transfers emphasis of selection based on phenotype to greater emphasis at the DNA level. Lamkey and Lee (1993), Eathington et al. (1997), Johnson (2004), Bernardo and Charcosset (2006), and others have discussed how molecular markers can complement other breeding methods.

Numerous mutants of maize were identified in the early genetic studies (Neuffer et al., 1968). Mutants were important in genetic studies to determine the expression and functions of specific loci for different plant and ear traits. Except for specific uses of maize, most recessive mutants have not had a major impact on breeding strategies for improvement of quality and quantity of maize. During the 1960s and 1970s, suggestions for use of dwarf (dw_i) genes to reduce plant size, terminal ear (*te*) and tassel seed (ts_i) to increase number of kernels, brown midrib (bm_i) mutants to improve silage quality, leafy gene (*Lfy*) to increase light interception, liguleless mutanats (lg_i) to change leaf orientation, waxy (*wx*) allele to modify starch, and opaque-2 (o_2) and floury (fl_2) mutants to improve kernel quality are some examples of mutants that have been studied. Initially, there was considerable interest in their potential, but interest waned when the recessive mutants were incorporated into elite inbred lines. In nearly all instances, a 5–15% grain yield loss was a common experience. Maize starch is generally considered low quality if an important component of the human diet but has recently increased its interest due to the corn–ethanol relationship. Mertz et al. (1964) reported that the o_2 gene improved grain quality. Interest in changing the quality of maize kernel was to insert the o_2 allele in widely used inbred lines. The results were generally not satisfactory because of the correlated effects of grain yield loss, higher moisture retention in the grain, greater harvest losses due to soft kernels, and greater susceptibility to common ear pests. Interest decreased and most programs were terminated. It was not until breeding methods included selection for modifiers to improve kernel texture, accompanied by laboratory analyses at all stages of breeding, that acceptable inbred lines and hybrids were developed (Vasal, 2001). It required ~30 years of breeding and selection to attain acceptable agronomic and quality standards. In areas where maize is consumed directly, quality protein maize (QPM) has an important place in maize production. Coors and Lauer (2001) were not as successful in developing superior silage quality with use of brown midrib mutant. Similar efforts were not given to most of the other mutants. Sweet corn and popcorn are two types of maize that are successfully used as special niche crops for human consumption. Sweet corn breeders have relied on mutant alleles (su_2 and sh_2, etc.) to enhance the taste of sweet corn. The recessive waxy (*wx*) allele was used effectively to modify the starch to produce only amylopectin, which has important properties for uses in foods and textiles.

Maize physiologists have studied maize for methods of increasing rates of photosynthesis and deposition of sugars in the kernels and uptake of applied nutrients to enhance grain production. Because maize has separate, distinct male and female inflorescences, interest also was given to the relative importance of source – sink relations, biomass, harvest index, the fate of CO_2 from photosynthesis and respiration, leaf area indices, and the hormones in plant development (Westgate et al., 2004). Mock and Pearce (1975) proposed an ideotype as the prototype maize plant to maximize yield; their ideotype included traits, such as smaller tassel size, upright leaf orientation above top ear and flatter orientation below top ear, prolificacy, darker green plant color, simultaneous silking and pollen shedding, greater efficiency of CO_2 production and use, and more kernels per ear. Information from the physiology studies has indirectly impacted maize breeding. Maize breeders recognized the importance of the different plant and ear traits physiologically. Greatest emphasis is usually for greater grain yield. The greater yielding cultivars incorporate many of the suggestions made in the ideotype because newer cultivars have smaller tassels, more upright leaf orientation, greater tendency for prolificacy, etc.

The latest technology to impact maize breeding is molecular genetics. Molecular genetics has already contributed to maize improvement with use of transgenes to reduce losses due to pests and weeds. Molecular genetics has been used to identify ownership of genotypes, to monitor seed purity, to determine the essential derivation of lines from the original lines, to fingerprint inbred line and hybrid genotypes, to monitor changes with selection, to precisely follow gene transfer in backcrossing programs, and to classify inbred lines in appropriate heterotic groups. One can speculate that the greatest use of molecular genetics will be the development of adequate set of molecular markers and/or ideally genes to effectively develop and use marker and/or gene assisted selection schemes for the genetic improvement of quantitative traits, such as grain yield. However, the use of molecular markers seems would be more important on simple inherited traits that are still difficult to accurately phenotype. The potential of molecular genetics in maize breeding can only be visualized. As the technology develops and techniques become more cost effective, molecular genetics should have even more practical impact in maize breeding programs. The information derived from molecular genetics increases daily with the complete sequence of the maize genome just finished. The major task will be to incorporate the molecular biology information to complement the selection and testing methods at the phenotypic level. Kaeppler (2004) has summarized the potential of molecular genetics in maize breeding.

Information in maize genetics has expanded exponentially during the past 50 years. The major factor for enhanced grain yields remains the same – germplasm. The most sophisticated breeding methods that encompass all of the genetic information currently available will have limited success if poor choices of source germplasm are made in initiating inbred line and hybrid development programs. Because of past selection an elite pool of germplasm has been developed. The current methods of inbreeding, selection, and testing of progenies derived from elite line crosses have been successful (Russell, 1991; Duvick, 1992, 2004; Fig. 1). Bauman (1981) reported that genetically broad-based populations (e.g., open-pollinated, composite, and synthetic cultivars) received limited attention as sources of breeding germplasm, and they are even less at the present time. The probabilities of developing a new, unique inbred line are low (Lindstrom, 1939; Hallauer and Miranda, 1988; Hadi, 2004). Development of improved inbred lines via recycling of elite inbred lines is the normal mode of present-day maize breeding. Although Hadi (2004) emphasized that the development of useful inbred lines from genetically broad-based population is extremely limited, I205, C103, Oh43, B14, B37, B73, and B84 have been very useful foundation materials in recycling of newer inbreds. Many inbred lines trace their origin to important first-cycle lines that have provided useful gene complexes to the recycled lines (Mikel and Dudley, 2006). It seems, therefore, that although the odds are very limited, that germplasm enhancement with use of recurrent selection procedures should be continued especially for creating useful genetic diversity and preventing genetic vulnerability. The GEM program, for example, may identify lines that can contribute useful genes to elite inbred lines included in line recycling and the program has already expanded to the central and northern US Corn Belt, eastern US, and Europe. The main concern is that the introduced DNA will not disrupt the finely tuned linkage combinations that were gradually accumulated during recycling. Despite the obvious limitations of recycling within adapted populations or populations with introduced exotic germplasm, it seems necessary that prebreeding and recurrent selection of advanced populations be continued to provide backup germplasm for possible future needs. The odds are not good for developing a truly superior inbred line, but it takes only one (e.g., B73 developed from an improved version of a genetically broad-based synthetic variety) to make a significant impact. Good choices of germplasm are a constant component of successful maize breeding programs. Different breeding methods, different testing procedures, advances in technology, and greater genetic information can vary among individuals and programs, but the ultimate success depends on the germplasm used.

References

Acosta, A.E. and P.L. Crane. 1972. Further selection for lower ear height in maize. Crop Sci. 12:165–167.

Adams, M.W. and D.B. Shank. 1959. The relation of heterozygosity to homeostasis in maize hybrids. Genetics 44:777–786.

Albrecht, B. and J.W. Dudley. 1987. Evaluation of four maize populations containing different proportions of exotic germplasm. Crop Sci. 27:480–486.

Anonymous. 1972. Genetic vulnerability of major crops. NRC-NAS, Washington, DC.

Bailey, T.B., Jr. 1977. Selection limits in self-fertilizing populations following the cross of homozygous lines. pp. 399–412. *In* E. Pollak, O. Kempthorne, and T.B. Bailey, (eds.) Proceedings of the International Conference on Quantitative Genetics. Iowa State University Press, Ames, IA.

Bailey, T.B., Jr. and R.E. Comstock. 1976. Linkage and the synthesis of better genotypes in self-fertilizing species. Crop Sci. 16:363–370.

Baker, R.J. 1986. Selection Indices in Plant Breeding. CRC Press, Boca Raton, FL.

Baker, R.J. 1974. Selection indices without economic weights for animal breeding. Can. J. Anal. Sci. 54:1–8.

Barata, C. and M.J. Carena. 2006. Classification of North Dakota maize inbred lines into heterotic groups based on molecular and testcross data. Euphytica 151:339–349.

Bauman, L.F. 1959. Evidence of non-allelic gene interaction in determining yield, ear height, and kernel row number in corn. Agron. J. 51:531–534.

Bauman, L.F. 1981. Review of methods used by breeders to develop superior corn inbreds. Proc. Annu. Corn Sorghum Ind. Res. Conf. 36:199–208.

Beal, W.J. 1880. Indian corn. pp. 279–289. Rep. Michigan Board Agric.

Bernardo, R. 1990. Identifying populations useful for improving parents of a single cross based on net transfer of alleles. Theor. Appl. Genet. 80:349–352.

Bernardo, R. 1991. Correlation between testcross performance of lines at early and late selfing generations. Theor. Appl. Genet. 82:17–21.

Bernardo, R. 1992. Retention of genetically superior lines during early-generation of testcrossing in maize. Crop Sci. 32:933–937.

Bernardo, R. 2001. What if we knew all the genes for a quantitative trait in hybrid crops? Crop Sci. 41:1–4.

Bernardo, R. and A. Charcosset. 2006. Usefulness of gene information in marker-assisted recurrent selection: A simulation appraisal. Crop Sci. 46:614–621.

Betra'n, F.J. and A.R. Hallauer. 1996. Hybrid improvement after reciprocal recurrent selection in BSSS and BSCB1 maize populations. Maydica 41:25–33.

Brown, W.L. 1953. Sources of germplasm for hybrid corn. Proc. Annu. Corn Ind. Res. Conf. 8:11–16.

Brown, W.L. 1975. A broader germplasm base in corn and sorghum. Proc. Annu. Corn Sorghum Ind. Res. Conf. 30:81–89.

Burton, G.W. 1951. Quantitative inheritance in pearl millet (*Pennisetum glaucum*). Agron. J. 43:409–417.

Campbell, M.R., L.M. Pollak, and P.J. White. 1995a. Genetic variation for starch thermal and functional properties among nonmutant maize inbreds. Cereal Chem. 77:281–286.

Campbell, M.R., P.J. White, and L.M. Pollak. 1995b. Properties of sugary-2 maize starch: Influence of exotic background. Cereal Chem. 72:389–392.

Carena, M.J. 2005a. Maize commercial hybrids compared to improved population hybrids for grain yield and agronomic performance. Euphytica 141:201–208.

Carena, M.J. 2005b. Registration of NDSAB(MER-FS)C13 maize Germplasm. Crop Sci. 45:1670–1671.

Carena, M.J. 2008. Increasing the genetic diversity of northern U.S. maize hybrids: Integrating pre-breeding with cultivar development. In Conventional and molecular breeding of field and vegetable crops. Nov. 24-28. Novi Sad, Serbia.
Carena, M.J., C. Eno, and D.W. Wanner. 2008. Registration of NDBS11(FR-M)C3, NDBS1011, and NDBSK(HI-M)C3 maize germplasms. J. Plant Reg. 2:132–136.
Carena, M.J. and A.R. Hallauer. 2001a. Response to inbred progeny selection in Leaming and Midland Yellow Dent maize populations. Maydica 46:1–10.
Carena, M.J. and A.R. Hallauer. 2001b. Expression of heterosis in Leaming and Midland Corn Belt dent populations. J. Iowa Acad. Sci. 108:73–78.
Carena, M.J. and Z.W. Wicks III. 2006. Maize early maturing hybrids: An exploitation of U.S. temperate public genetic diversity in reserve. Maydica 51:201–208.
Carena, M.J., I. Santiago, and A. Ordás. 1998. Direct and correlated response to recurrent selection for prolificacy in maize at two planting densities. Maydica 43:95–102.
Castellanos, J.S., A.R. Hallauer, and H.S. Cordova. 1998. Relative performance of testers to identify elite lines of corn (*Zea mays* L.). Maydica 43:217–226.
Caton, H.G. 1999. Conversion of elite subtropical maize germplasm to temperate environments. Ph.D. Diss. Iowa State University, Ames, IA (Diss. Abstr. DAI-B 60/8, p. 3629, February 2000).
Chase, S.S. 1949. Monoploid frequencies in a commercial double cross hybrid maize, and its component single cross hybrids and inbred lines. Genetics 34:328–332.
Chase, S.S. 1952. Production of homozygous diploids of maize from monoploids. Agron. J. 44:263–267.
Christensen, D.W. 1984. Utility of zein-isoelectric focusing for predicting heterosis in maize. Ph. D. Diss. University Minnesota, St. Paul, MN (Diss. Abstr. DAI-B 45/05, p. 1374, November 1984).
CIMMYT. 1984. CIMMYT research highlights 1984. CIMMYT, El Batan, Mexico.
Cockerham, C.C. 1954. An extension of the concept of partitioning hereditary variance for analysis of covariance among relatives when epistasis is present. Genetics 39:859–882.
Cockerham, C.C. 1956a. Analysis of quantitative gene action. Brookhaven Symp. Biol. 9:53–68.
Cockerham, C.C. 1956b. Effects of linkage on the covariances between relatives. Genetics 41:138–141.
Cockerham, C.C. 1961. Implications of genetic variances in a hybrid breeding program. Crop Sci. 1:47–52.
Cockerham, C.C. 1963. Estimation of genetic variances. pp. 53–94. *In* W.D. Hanson and H.F. Robinson (eds.) Statistical Genetics and Plant Breeding. NAS-NRC Publication 982.
Collins, G.N. 1909. The importance of broad breeding in corn. USDA Bureau Plant Ind. Bull. 141. Part 4:33–42.
Compton, W.A. and R.E. Comstock. 1976. More on modified ear-to-row selection. Crop Sci. 16:122.
Compton, W.A. and J.H. Lonnquist. 1982. A multiplicative selection index applied to four cycles of full-sib recurrent selection in maize. Crop Sci. 22:981–983.
Comstock, R.E. and H.F. Robinson. 1952. Estimation of average dominance of genes. pp. 496–516. *In* J.W. Gowen (ed.) Heterosis. Iowa State University Press, Ames, IA.
Comstock, R.E., H.F. Robinson, and P.H. Harvey. 1949. A breeding procedure designed to make maximum use of both general and specific combining ability. Agron. J. 41:360–367.
Cook, K.A. 1998. Linkage disequilibrium in maize F_2 populations. Ph.D. Diss. Iowa State University, Ames, IA.
Coors, J.G. and J.G. Lauer. 2001. Silage corn. pp. 347–392. *In* A.R. Hallauer (ed.) Special Corns, 2nd ed. CRC Press, Boca Raton, FL.
Coors, J.G. and S. Pandey (eds.). 1999. Genetics and Exploitation of Heterosis in Crops. ASA, CSSA, SSSA, Madison, WI.
Covarrubias-Prieto, J., A.R. Hallauer, and K.R. Lamkey. 1989. Intermating F_2 populations of maize. Genetika 21:111–126.
Crabb, A.R. 1947. The Hybrid Corn Makers. Rutgers University Press, New Brunswick, NJ.

Cress, C.E. 1967. Reciprocal recurrent selection and modifications in simulated populations. Crop Sci. 7:561–567.

Crosbie, T.M., J.J. Mock, and O.S. Smith. 1980. Comparison of gains predicted by several selection methods for cold tolerance traits of two maize populations. Crop Sci. 20:649–655.

Cross, H.Z., J.R. Chyle, Jr., and J.J. Hammond. 1987. Divergent selection for ear moisture in early maize. Crop Sci. 27:914–918.

Crossa, J. 1989. Theoretical considerations for the introgression of exotic germplasm into adapted maize populations. Maydica 34:53–62.

Crossa, J. and C.O. Gardner. 1987. Introgression of an exotic germplasm for improving an adapted maize population. Crop Sci. 27:187–190.

Crossa, J., C.O. Gardner, and R.F. Mumm. 1987. Heterosis among populations of maize (*Zea mays* L.) with different levels of exotic germplasm. Theor. Appl. Genet. 73:445–450.

Crossa, J., S. Taba, and E.J. Wellhausen. 1990. Heterotic patterns among Mexican races of maize. Crop Sci. 30:1182–1190.

Cunningham, E.P. 1969. The relative efficiencies of selection indexes. Acta. Agric. Scandinavica 19:45–48.

Darrah, L.L. 1986. Evaluation of population improvement in the Kenya maize breeding methods study. pp. 160–175. *In* B. Gelaw (ed.) To Feed Ourselves. Proc. First Eastern, Central, and Southern Africa Regional Maize Workshop, Lusaka, Zambia. March 10–17, 1985. CIMMYT, Mexico, D.F.

Darrah, L.L. and L.H. Penny 1975. Inbred line extraction from improved breeding populations. E. Afr. Agric. For. J. 41:1–8.

Darrah, L.L., S.A. Eberhart, and L.H. Penny. 1978. Six years of maize selection in Kitale Synthetic II, Ecuador 573, and Kitale Composite A using methods of the comprehensive breeding systems. Euphytica 27:191–204.

Dekker, J.C.M. and P. Settar. 2004. Long-term selection with known quantitative trait loci. pp. 311–335. *In* J. Janick (ed.) Long-term Selection: Maize. Part 1. Plant Breed. Rev. Vol. 24. John Wiley & Sons, Inc., Hoboken, NJ.

DeKruif, P. 1928. Maize. pp. 165–259. *In* Hunger Fighters. Harcourt, Brace, and World, Inc., New York. NY.

Dhillon, B.S. and A.S. Khehra. 1989. Modified S_1 recurrent selection in maize improvement. Crop Sci. 29:226–228.

Douglas, A.G., J.W. Collier, M.F. El-Ebrasly, and J.S. Rogers. 1961. An evaluation of three cycles of reciprocal recurrent selection in a corn improvement program. Crop Sci. 1:157–161.

Dowswell, C.R., R.L. Paliwal, and R.P. Cantrell. 1996. Maize in the Third World. Westview Press, Boulder, CO.

Dudley, J.W. 1982. Theory for transfer of genes. Crop Sci. 22:631–637.

Dudley, J.W. 1984a. Theory for the identification and use of exotic germplasm maize breeding programs. Maydica 29:391–407.

Dudley, J.W. 1984b. A method of identifying lines for use in improving parents of a single cross. Crop Sci. 24:355–357.

Dudley, J.W. 1987. Modification of methods for identifying inbred lines useful for improving parents of elite single crosses. Crop Sci. 27:940–943.

Dudley, J.W. 1997. Quantitative genetics and plant breeding. Adv. Agron. 59:1–23.

Dudley, J.W. and R.J. Lambert. 2004. 100 generations of selection for oil and protein in corn. pp. 79–100. *In* J. Janick (ed.) Plant Breed. Rev. 24, Part 1. John Wiley & Sons, Inc., Hoboken, NJ.

Dudley, J.W. and R.H. Moll. 1969. Interpretations and uses of estimates of heritability and genetic variances in plant breeding. Crop Sci. 9:257–262.

Duvick, D.N. 1992. Genetic contributions to advances in yield of U.S. maize. Maydica 37:69–79.

Duvick, D.N. 1999. Heterosis: Feeding people and protecting natural resources. pp. 19–29. *In* J.G. Coors and S. Pandey (eds.) Genetics and Exploitation of Heterosis in Crops. ASA, CSSA, SSSA, Madison, WI.

Duvick, D.N. 2004. Long-term selection in a commercial hybrid maize breeding program. pp. 109–151. *In* J. Janick (eds.) Plant Breed. Rev. 24, Part 2. John Wiley & Sons, Inc., Hoboken, NJ.
Duvick, D.N., J.S.C. Smith, and M. Cooper. 2004. Changes in performance, parentage, and genetic diversity of successful corn hybrids, 1930–2000. pp. 65–97. *In* C.W. Smith, J. Betra'n, and E.C.A. Runge (eds.) Corn: Origin, History, and Production. John Wiley & Sons, Hoboden, NJ.
Eagles, H.A. and A.K. Hardacre. 1990. Genetic changes from introgression of highland Mexican germplasm into a Corn Belt dent population of maize. Theor. Appl. Genet. 79:543–549.
East, E.M. 1908. Inbreeding in corn. pp. 419–428. *In* Connecticut Agricultural Experiments Station Report 1907.
East, E.M. and H.K. Hayes. 1911. Inheritance in corn. Conn. Agr. Exp. Stn. Bull. 167:1–41.
Eathington, S.R., J.W. Dudley, and G.F. Rufener II. 1997. Usefulness of marker-QTL associations in early generation selection. Crop Sci. 37:1686–1693.
Eberhart, S.A. 1964. Least squares method for comparing progress among recurrent selection methods. Crop Sci. 4:230–231.
Eberhart, S.A. 1970. Factors affecting efficiencies of breeding methods. African Soils 15:669–680.
Eberhart, S.A. and W.A. Russell. 1966. Stability parameters for comparing varieties. Crop Sci. 6:36–40.
Eberhart, S.A., M.N. Harrison, and F. Ogada. 1967. A comprehensive breeding system. Züchter 37:169–174.
Echandi, C.R. and A.R. Hallauer. 1996. Evaluation of U.S. Corn Belt and adapted maize cultivars and their diallel crosses. Maydica 41:317–324.
Eckhardt, R.C. and A.A. Bryan. 1940. Effect of method of combining the four inbred lines of a double-cross of maize upon the yield and variability of the resulting double crosses. J. Am. Soc. Agron. 32:347–353.
Elston, R.C. 1963. A weight free index for the purpose of ranking or selection with respect to several traits at a time. Biometrics 19:85–97.
Eyhe'rabide, G.H. and A.R. Hallauer. 1991. Reciprocal full-sib selection in maize. I. Direct and indirect responses. Crop Sci. 31:952–959.
Falconer, D.S. 1960. Introduction to Quantitative Genetics. Ronald Press, New York. NY.
Findlay, K.W. and G.N. Wilkinson. 1963. The analysis of adaptation in a plant-breeding programme. Australian J. Agric. Res. 14:742–754.
Fountain, M.O. and A.R. Hallauer. 1996. Genetic variation within maize breeding populations. Crop Sci. 36:26–32.
Freeman, G.H. and J.M. Perkins. 1971. Environmental and genotype-environmental components of variability. VIII. Relation between genotypes grown in different environments and measures of these environments. Heredity 27:15–23.
Galinat, W.C. 1988. The origin of corn. pp. 1–31. *In* G.F. Sprague and J.W. Dudley (eds.) Corn and Corn Improvement. 3rd ed. ASA, CSSA, SSSA, Madison, WI.
Gama, E.E.G. and A.R. Hallauer. 1977. Relation between inbred and hybrid traits in maize. Crop Sci. 17:703–706.
Gamble, E.E. 1962. Gene effects in corn (*Zea mays* L.). I. Selection and relative importance of gene effects for yield. Can. J. Plant Sci. 42:339–348.
Gardner, C.O. 1961. An evaluation of effects of mass selection and seed irradiation with thermal neutrons on yields of corn. Crop Sci. 1:241–245.
Gardner, C.O. 1963. Estimates of genetic parameters in cross-fertilizing plants and their implications in plant breeding. pp. 225–248. *In* W.D. Hanson and H.F. Robinson (eds.) Statistical Genetics and Plant Breeding. NAS-NRC Publication 982. Washington, DC.
Gardner, C.O. 1976. Quantitative genetic studies and population improvement in maize and sorghum. pp. 475–489. *In* E. Pollak, O. Kempthorne, and T.B. Bailey (eds.) Proceedings of the Conference on Quantitative Genetics. Iowa State University Press, Ames, IA.
Genter, C.F. 1976. Mass selection in a composite of intercrosses of Mexican races of maize. Crop Sci. 16:556–558.

Gepts, P. 2004. Crop domestication as a long-term selection experiment. pp. 1–44. *In* J. Janick (ed.) Long-term Selection: Crops, Animals, and Bacteria. Plant Breed. Rev. Part 2, Vol. 24. John Wiley & Sons, Inc., Hoboken, NJ.
Gerdes, J.T., C.F. Behr, J.G. Coors, and W.F. Tracy. 1994. Compilation of North American Maize Breeding Germplasm. CSSA, Madison, WI.
Gerloff, J.E. and O.S. Smith. 1988. Choice of method for identifying germplasm with superior alleles. Theor. Appl. Genet. 76:209–216.
Gerrish, E.E. 1983. Indications from a diallel study for interracial maize hybridization in the Corn Belt. Crop Sci. 23:1082–1084.
Gethi, J.G., J.A. Labate, K.R. Lamkey, M.E. Smith, and S. Kresovich. 2002. SSR variation in important U.S. inbred lines. Crop Sci. 42:951–957.
Getschman, R.J. and A.R. Hallauer. 1991. Genetic variation among and within S_1 progenies of maize. J. Iowa Acad. Sci. 98:127–133.
Gevers, H.O. 1975. Three cycles of reciprocal recurrent selection in maize under two systems of plant selection. Agroplante 7:107–108.
Goldman, I.L. 1999. Inbreeding and outbreeding in the development of a modern heterosis concept. pp. 7–18. *In* J.G. Coors and S. Pandey (eds.) Genetics and Explortation of Heterosis in Crops. ASA, CSSA, and SSSA, Madison, WI.
Good, R.L. and A.R. Hallauer. 1977. Inbreeding depression in maize by selfing and full-sibbing. Crop Sci. 17:735–740.
Goodman, M.M. 1965. Estimates of genetic variance in adapted and exotic populations of maize. Crop Sci. 5:87–90.
Goodman, M.M. 1985. Exotic maize germplasm: Status, prospects, and remedies. Iowa State J. Res. 59:497–527.
Goodman, M.M. 1999a. Developing temperate inbreds from tropical germplasm: Rationale, results, and conclusions. Illinois Corn Breeders' School 35:1–19.
Goodman, M.M. 1999b. Broadening the genetic diversity in maize breeding by use of exotic germplasm. pp. 139–148. *In* J.G. Coors and S. Pandey (eds.) Genetics and Exploitation of Heterosis in Crops. ASA, CSSA, and SSSA, Madison, WI.
Goodman, M.M. and W.L. Brown. 1988. Races of corn. pp. 33–79. *In* G.F. Sprague and J.W. Dudley (eds.) Corn and Corn Improvement. 3rd ed. ASA, CSSA, SSSA, Madison, WI.
Goodsell, S. 1961. Male sterility in corn by androgenesis. Crop Sci. 1:227–228.
Gowen, J.W. (ed.). 1952. Heterosis. Iowa State University Press, Ames, IA.
Griffing, B. and E.W. Lindstrom. 1954. A study of combining abilities of corn inbreds having varying proportions of Corn Belt and non-Corn Belt germplasm. Agron. J. 46:845–852.
Gutierrez-Gaitan, M.A., H. Cortez-Mendoza, E.N. Wathika, C.O. Gardner, M. Oyervides-Garacia, A.R. Hallauer, and L.L. Darrah. 1986. Testcross evaluations of Mexican maize populations. Crop Sci. 26:99–104.
Hadi, G. 2004. Improvement of commercial maize lines through the integration of genes and gene combinations from elite lines. Acta Agron. Hungarica 52:69–74.
Hainzelin, E. 1998. Exotic introgression incidence on two elite tropical maize populations. Maydica 43:19–26.
Hallauer, A.R. 1985. Compendium of recurrent selection methods and their application. Critical Rev. Plant Sci. 3(1):1–33.
Hallauer, A.R. 1987. Maize. pp. 249–294. *In* W.R. Fehr (ed.) Principles of Cultivar Development. Vol. 2. Macmillan Publishing Co., New York.
Hallauer, A.R. 1990. Methods used in developing maize inbreds. Maydica 55:1–16.
Hallauer, A.R. 1992. Recurrent selection in maize. Plant Breed. Rev. 9:115–179.
Hallauer, A.R. 1994. Corn genetics and breeding. Encycl. Agric. Sci. 1:455–467. Academic Press, Inc., New York. NY.
Hallauer, A.R. 1999a. Temperate maize and heterosis. pp. 353–36. *In* J.G. Coors and S. Pandey (eds.) The Genetics and Exploitation of Heterosis in Crops. ASA, CSSA, SSSA, Madison, WI.

Hallauer, A.R. 1999b. Conversion of tropical germplasm for temperate area use. Illinois Corn Breeders' School 35:20–36.
Hallauer, A.R. (ed.). 2001. Specialty Corns. 2nd ed. CRC Press LLC, Boca Raton, FL.
Hallauer, A.R. 2002. Integration of germplasm improvement with corn breeding. Illinois Corn Breeders' School 38:11–41.
Hallauer, A.R. 2003. Conversion of tropical germplasm for temperate area. Proc. Annual Corn Sorghum Ind. Res. Conf. 58:187–200.
Hallauer, A.R. 2004. Specialty corns. pp. 897–933. *In* C.W. Smith, J. Betra'n, and E.C.A. Runge (eds.) Corn: Origin, History, Technology, and Production. John Wiley & Sons, Inc., Hoboken, NJ.
Hallauer, A.R. 2006. Defining and achieving plant-breeding goals. pp. 73–89. *In* K.R. Lamkey and M. Lee (eds.) Plant Breeding: The Arnel R. Hallauer International Symposium. Blackwell Publishing, Oxford, UK.
Hallauer, A.R. 2006. History, contribution and future of quantitative genetics in plant breeding: Lessons from maize. *In* R. Bernardo (ed.) International Plant Breeding Symposium. August 20–26, 2006, Mexico City.
Hallauer, A.R. and S.A. Eberhart. 1970. Reciprocal full-sib selection. Crop Sci. 10:315–316.
Hallauer, A.R. and E. Lopez-Perez. 1979. Comparisons among testers for evaluating lines of corn. Proc. Annu. Corn Sorghum Ind. Res. Conf. 34:57–75.
Hallauer, A.R. and J.B. Miranda, 1988. Quantitative Genetics in Maize Breeding. 2nd ed. Iowa State University Press. Ames, IA.
Hallauer, A.R. and J.H. Sears. 1969. Mass selection for yield in two varieties of maize. Crop Sci. 9:47–50.
Hallauer, A.R. and J.H. Sears. 1973. Changes in quantitative traits associated with inbreeding in a synthetic variety of maize. Crop Sci. 13:327–330.
Hallauer, A.R., W.A. Russell, and O.S. Smith. 1983. Quantitative analysis of Iowa Stiff Stalk Synthtic. pp. 105–118. *In* J.P. Gustafson (ed.) 15th Stadler Genetics Symposium. Columbia, MO.
Hallauer, A.R., W.A. Russell, and K.R. Lamkey. 1988. Corn breeding. pp. 463–564. *In* G.F. Sprague and J.W. Dudley (eds.) Corn and Corn Improvement. 3rd ed. CSSA, Madison, WI.
Hallauer, A.R., A.J. Ross, and M. Lee. 2004. Long-term divergent selection for ear length in maize. pp. 153–168. *In* J. Janick (ed.) Plant Breed. Rev. Vol. 24, Part 2. John Wiley & Sons, Inc., Hoboken, NJ.
Hameed, A., L.M. Pollak, and P.N. Hinz. 1994. Evaluation of Cateto maize accessions for grain yield and other agronomic traits in temperate and tropical environments. Crop Sci. 34:270–275.
Hammond, J.J. and M.J. Carena. 2008. A breeding plan with molecular markers. *In* Agronomy Abstracts [CD-ROM computer file]. ASA, Madison, WI.
Han, G. and A.R. Hallauer. 1989. Estimates of genetic variability in F_2 maize populations. J. Iowa Acad. Sci. 96:277–313.
Hanson, W.D. 1959. The breakup of initial linkage blocks under selected mating systems. Genetics 44:857–868.
Hanson, W.D. 1963. Heritability. pp. 125–140. *In* W.D. Hanson and H.F. Robinson (eds.) Statistical Genetics and Plant Breeding. NAS-NCR Publication 982.
Harrison, M.N. 1967. Selection of source material and methods of formation of composites for maize breeding. Paper presented at Eastern African Cereals Research Conference.
Hartley, C.P. 1908. Utilization of Corn Variations. pp. 13–27. Ohio Crop Imp. Association Annual Report.
Hawbaker, M.S., W.H. Hill, and M.M. Goodman. 1997. Application of recurrent selection for low moisture content in tropical maize (*Zea mays* L.). I. Testcross yield trials. Crop Sci. 37:1650–1655.
Hayes, H.K. 1956. I Saw Hybrid Corn Develop. Proc. Annu. Corn Sorghum Ind. Res. Conf. 11:48–55.

Hayman, B.I. 1958. The separation of epistatic from additive and dominance in generation means. Heredity 12:371–390.

Hayman, B.I. 1960. The separation of espistatic from additive and dominance variation in generation means. II. Genetica 31:133–146.

Hazel, L.N. and J.L. Lush. 1942. The efficiency of three methods of selection. J. Hered. 33:393–399.

Henderson, C.R. 1963. Selection index and expected genetic advances. pp. 141–163. *In* W.D. Hanson and H.F. Robinson (eds.) Statistical Genetics and Plant Breeding. NAS-NRC Publication 982.

Hofbeck, M.D., S.J. Openshaw, J.L. Geadelmann, R.H. Peterson, and D.D. Stuthman. 1995. Backcrossing and intermating in an exotic adapted cross of maize. Crop Sci. 35:1359–1364.

Holland, J.B. and M.M. Goodman. 1995. Combining ability of tropical maize accessions with U.S. germplasm. Crop Sci. 35:767–773.

Holley, R.N. and M.M. Goodman. 1988a. Stalk quality and stalk rot resistance of tropical hybrid maize derivatives. Plant Dis. 72:321–324.

Holley, R.N. and M.M. Goodman. 1988b. Yield potential of tropical maize derivatives. Crop Sci. 28:213–218.

Holley, R.N., P.B. Hamilton, and M.M. Goodman. 1989. Evaluation of tropical germplasm for resistance to kernel colonization by *Fusarium moniliforme*. Plant Dis. 73:578–580.

Hopkins, C.G. 1899. Improvement in the chemical composition of the corn kernel. Bull. Ill. Agric. Exp. Stn. 55:205–240.

Horner, E.S., E. Magloire, and J.A. Morera. 1989. Comparison of selection for S_2 progeny vs. testcross performance for population improvement in maize. Crop Sci. 29:868–874.

Hudson, J.C. 1994. Making the Corn Belt. Indiana University Press, Bloomington & Indianapolis, IN.

Hull, F.H. 1945. Recurrent selection and specific combining ability in corn. J. Am. Soc. Agron. 37:134–145.

Ito, G.M. and J. Brewbaker. 1981. Genetic advance through mass selection for tenderness in corn. J. Am. Soc. Hort. Sci. 106:496–499.

Janick, J. (ed.). 2004. Long-term selection: Maize. Plant Breed. Rev. Vol. 24, Part 1. John Wiley & Sons, Inc., Hoboken, NJ.

Jenkins, M.T. 1928. Correlation studies with inbred and crossbred strains of maize. J. Agric. Res. 39:677–721.

Jenkins, M.T. 1934. Methods of estimating the performance of double crosses in corn. J. Am. Soc. Agron. 26:199–204.

Jenkins, M.T. 1935. The effect of inbreeding and of selection within inbred lines of maize upon hybrids made after successive generations of selfing. Iowa State Coll. J. Sci. 9:429–450.

Jenkins, M.T. 1936. Corn improvement. pp. 455–522. *In* Yearbook of Agriculture, 1936. U.S. Gov. Print. Office, Washington, DC.

Jenkins, M.T. 1940. The segregation of genes affecting yield of grain in maize. J. Am. Soc. Agron. 32:55–63.

Jenkins, M.T. and A.M. Brunson. 1932. Methods of testing inbred lines of maize in crossbred combinations. J. Am. Soc. Agron. 24:523–530.

Jenkins, M.T., A.L. Robert, and W.F. Findley, Jr. 1954. Recurrent selection as a method for concentrating genes for resistance to *Helminthosporium turcicum* leaf blight in corn. Agron. J. 46:89–94.

Jensen, S.D., W.E. Kuhn, and R.L. McConnell. 1983. Combining ability in elite U.S. maize germplasm. Proc. Annu. Corn Sorghum Ind. Res. Conf. 38:87–96.

Johnson, G.R. 1982. Two-locus theory in recurrent selection for general combining ability in maize. Theor. Appl. Genet. 61:279–283.

Johnson, G.R. 2004. Marker-assisted selection. pp. 293–309. *In* J. Janick (ed.) Plant Breed. Rev. 24, Part 1. John Wiley & Sons, Inc., Hoboken, NJ.

Johnson, E.C., K.S. Fisher, G.O. Edmeades, and A.F.E. Palmer. 1986. Recurrent selection for reduced plant height in lowland tropical maize. Crop Sci. 26:253–260.

Jones, D.F. 1918. The effects of inbreeding and cross breeding upon development. Connecticut Agric. Exp. Stn. Bull. 207:5–100.

Jones, L.P., W.A. Compton, and C.O. Gardner. 1971. Comparison of full-sib and half-sib recurrent selection. Theor. Appl. Genet. 41:36–39.

Jumbo, M.B., and M.J. Carena. 2007. Combining ability, maternal, and reciprocal effects of elite early-maturing maize population hybrids [online]. Euphytica. Available at http://springerlink.com/content/x4455x4657455803.

Kaeppler, S. 2004. Biotechnology: New horizons. pp. 399–425. *In* C.W. Smith, J. Betra'n, and E.C.A. Runge (eds.) Corn: Origin, History, and Production. John Wiley & Sons, Hoboken, NJ.

Kauffmann, K.D. and J.W. Dudley. 1979. Selection indices for corn grain yield, percent protein and kernel weight. Crop Sci. 19:583–588.

Kauffmann, K.D., C.W. Crum, and M.F. Lindsey. 1982. Exotic germplasm in a corn breeding program. Illinois Corn Breeders' School 18:6–39.

Keeratinijakal, V. and K.R. Lamkey. 1993. Responses to reciprocal recurrent selection in BSSS and BSCB1 maize populations. Crop Sci. 33:73–77.

Kermicle, J.L. 1969. Androgenesis conditioned by a mutation in maize. Science 166:1422–1424.

Kim, S.K., J.L. Brewbaker, and A.R. Hallauer. 1988. Insect and disease resistance from tropical maize for use in temperate zone hybrids. Proc. Annu. Corn Sorghum Ind. Res. Conf. 43:194–226.

Knapp, S.J. 1998. Marker-assisted selection as a strategy for increasing the probability of selecting superior genotypes. Crop Sci. 38:1164–1174.

Kraja, A., J.W. Dudley, and D.G. White. 2000. Identification of tropical and temperate maize populations having favorable alleles for disease resistance. Crop Sci. 40:948–954.

Kramer, H.H. and A.J. Ullstrup. 1959. Preliminary evaluations of exotic maize germplasm. Agron. J. 51:687–689.

Lambert, R.J. 1984. Reciprocal recurrent selection of maize in a high-yield environment. Maydica 29:419–430.

Lamkey, K.R. and M. Lee. 1993. Quantitative genetics, molecular markers, and plant improvement. pp. 104–115. *In* B.C. Imire and J.B. Hacker (eds.) Focused Plant Improvement; Towards Responsible and Sustainable Agriculture. Proceedings of the 10th Australian Plant Breeding Conference. April 18–23, 1993. Gold Coast, Canberra.

Larkin, P.J. and W.R. Scowcroft. 1981. Somaclonal variation – A novel source of variability from cell cultures for plant improvement. Theor. Appl. Genet. 60:197–214.

Lee, M. 1995. DNA markers and plant breeding programs. Adv. Agron. 55:265–344.

Lee, M., J.L. Geadelmann, and R.L. Phillips. 1988. Agronomic evaluation of inbred lines derived from tissue cultures of maize. Theor. Appl. Genet. 75:841–849.

Lee, E.A., A. Singh, M.J. Ash, and B. Good. 2006. Use of sister lines and the performance of modified single-cross maize hybrids. Crop Sci. 46:312–320.

Lile, S.M. and A.R. Hallauer. 1994. Relation between S_2 and later generation testcrosses of two corn populations. J. Iowa Acad. Sci. 101:19–23.

Lindstrom, E.W. 1939. Analysis of modern maize breeding principles and methods. Proc. 7th Intl. Genet. Congress 7:191–196.

Lonnquist, J.H. 1964. A modification of the ear-to-row procedure for the improvement of maize populations. Crop Sci. 4:227–228.

Lonnquist, J.H. 1967. Mass selection for prolificacy in maize. Züchter 37:185–188.

Lonnquist, J.H. 1974. Consideration and experiences with recombinations of exotic and Corn Belt maize germplasm. Proc. Annu. Corn Sorghum Ind. Res. Conf. 29:102–117.

Lush, J.L. 1945. Animal Breeding Plans. Iowa State University Press, Ames, IA.

Mahmud, I. and H.H. Kramer. 1951. Segregation for yield, height, and maturity following a soybean cross. Agron. J. 43:605–609.

Majaya, M.J. and R.J. Lambert. 1992. Improving performance of adapted corn hybrid using exotic inbred. Indonesian J. Crop Sci. 7:55–62.

Mareck, J.H. and C.O. Gardner. 1979. Responses to mass selection in maize (Zea Mays L.) cultivars Hays Golden and Nebraska-5010. Crop Sci. 19:779–783.

Mangelsdorf, P.C. 1974. Corn: Its Origin, Evolution, and Improvement. Harvard University Press, Cambridge, MA.

Marks, R.E. 1993. Cortés: The Great Adventurer and the Fate of Aztec Mexico. A.A. Knopf. New York. NY.

Marquez-Sanchez, F. 1982. Modifications to cyclic hybridization in maize with single-eared plants. Crop Sci. 22:314–319.

Marquez-Sanchez, F. and N. Gomez-Montrel. 1988. Seed increase of half-sib families in the modified ear-to-row method of selection in maize with additional genetic gain. Maydica 33:223–226.

Marquez-Sanchez, F. and A.R. Hallauer. 1970a. Influence of sample size on the estimation of genetic variances in a synthetic variety of maize. I. Grain yield. Crop Sci. 10:357–361.

Marquez-Sanchez, F. and A.R. Hallauer. 1970b. Influence of sample size on the estimation of genetic variances in a synthetic variety of maize. II. Plant and ear characters. Iowa State J. Sci. 44:423–436.

Mather, K. 1949. Biometrical Genetics. Methuen, London.

Melani, M.D. and M.J. Carena. 2005. Alternative maize heterotic patterns for the Northern Corn Belt. Crop Sci. 45:2186–2194.

Melchinger, A.E., M.M. Messmer, M. Lee, W.L. Woodman, and K.R. Lamkey. 1991. Diversity and relationships among U.S. maize inbreds revealed by restriction fragment length polymorphisms. Crop Sci. 31:669–678.

Menz, M.A. and A.R. Hallauer. 1997. Reciprocal recurrent selection of two tropical corn populations adapted to Iowa. Maydica 42:239–246.

Menz, M.A., A.R. Hallauer, and W.A. Russell. 1999. Comparative response to selection of two reciprocal recurrent selection procedures in BS21 and BS22 maize populations. Crop Sci. 39:89–97.

Mertz, E.T., L.S. Bates, and O.E. Nelson. 1964. Mutant gene that changes protein composition and increases lysine content of maize endosperm. Science 145:279–280.

Michelini, L.A. and A.R. Hallauer. 1993. Evaluation of exotic and adapted maize (*Zea mays* L.) germplasm crosses. Maydica 38:275–282.

Mikel, M.A. 2006. Availability and analysis of proprietary dent corn inbred lines with expired U.S. Plant Variety Protection. Crop Sci. 46:2555–2560.

Mikel, M.A. and J.W. Dudley. 2006. Evolution of North American dent corn from public to proprietary germplasm. Crop Sci. 46:1193–1206.

Miles, J.W., J.W. Dudley, D.G. White, and R.J. Lambert. 1981. Response to selection for resistance to four diseases in two corn populations. Crop Sci. 21:980–983.

Mishra, T.N. 1977. Classification of maize populations into heterotic groups. M. Ag. University Minnesota., St. Paul, MN.

Mock, J.J. and R.B. Pearce. 1975. An ideotype of maize. Euphytica 24:613–623.

Moll, R.H. and W.D. Hanson. 1984. Comparisons of effects of intrapopulation vs. interpopulation selection in maize. Crop Sci. 24:1047–1052.

Moll, R.H. and H.F. Robinson. 1966. Observed and expected response in four selection experiments in maize. Crop Sci. 319–324.

Moll, R.H. and C.W. Stuber. 1971. Comparisons of response to alternative selection procedures initiated with two populations of maize. Crop Sci. 11:706–711.

Moll, R.H. and C.W. Stuber. 1974. Quantitative genetics. Empirical results relative to plant breeding. Adv. Agron. 26:277–313.

Moll, R.H., W.S. Salhuana, and H.F. Robinson. 1962. Heterosis and genetic diversity in variety crosses of maize. Crop Sci. 2:197–198.

Moll, R.H., J.H. Lonnquist, J.V. Fortuno, and E.J. Johnson. 1965. The relationship of heterosis and genetic divergence in maize. Genetics 42:139–144.

Moll, R.H., C.W. Stuber, and W.D. Hanson. 1975. Correlated responses and responses to index selection involving yield and ear height of maize. Crop Sci. 15:243–248.

Moll, R.H., A. Bari, and C.W. Stuber. 1977. Frequency distribution of maize yield before and after reciprocal selection. Crop Sci. 17:794–796.

Mongoma, C. and L.M. Pollak. 1988. Heterotic patterns among 10 Corn Belt and exotic maize populations. Crop Sci. 28:500–504.

Moreno-Gonzalez, J. and J.W. Dudley. 1981. Epistasis in related and unrelated maize hybrids determined by three methods. Crop Sci. 21:644–651.

Moreno-Gonzalez, J. and A.R. Hallauer. 1982. Combined S_2 and crossbred family selection in full-sib reciprocal recurrent selection. Theor. Appl. Genet. 61:353–358.

Mulamba, N.N. and J.J. Mock. 1978. Improvement of yield potential of the Eto Blanco maize (*Zea mays* L.) population by breeding for plant traits. Egypt. J. Genet. Cytol. 7:40–51.

Narro, L.N. 1990. Direct and correlated response for earliness in a tropical maize population. Ph.D. Diss. Iowa State University, Ames, IA (Diss. Abstr. DAI-B 51/07, p. 3196, January 1991).

Nelson, H.G. 1972. The use of exotic germplasm in practical corn breeding programs. Proc. Annu. Corn Sorghum Ind. Res. Conf. 27:115–118.

Neuffer, M.G., L. Jones, and M.S. Zuber. 1968. Mutants of Maize. CSSA, Madison, WI.

Obaidi, M., B.B. Johnson, L.D. Van Vleck, S.D. Kachman, and O.S. Smith. 1998. Family per se response in selfing and selection in maize based on testcross performance: A simulation study. Crop Sci. 38:367–371.

Obilana, A.T. and A.R. Hallauer. 1974. Estimation of variability of quantitative traits in BSSS by using unselected maize inbred lines. Crop Sci. 14:99–103.

Odhiambo, M.O. and W.A. Compton. 1987. Twenty cycles of divergent mass selection for seed size in corn. Crop Sci. 27:113–116.

Oyervides-Garcia, M., A.R. Hallauer, and H. Cortez-Mendoza. 1985. Evaluation of improved maize populations in Mexico and U.S. Corn Belt. Crop Sci. 25:115–120.

Pandey, S. and C.O. Gardner. 1992. Recurrent selection for population, variety, and hybrid improvement in tropical maize. Adv. Agron. 48:1–87.

Paterniani, E. and R. Vencovsky. 1977. Reciprocal recurrent selection (*Zea mays* L.) based on testcrosses of half-sib families. Maydica 22:141–152.

Paterniani, E. and R. Vencovsky. 1978. Reciprocal recurrent selection based on half-sib progenies and prolific plants in maize (*Zea mays* L.). Maydica 23:209–219.

Peiris, B.L. and A.R. Hallauer. 2005. Comparison of half-sib and full-sib reciprocal recurrent selection and their modifications in simulated populations. Maydica 50:1–13.

Pesek, J. and R.J. Baker. 1969. Desired improvement in relation to selection indices. Can. J. Plant Sci. 49:803–804.

Peterson, P.A. 1986. Mobile elements in maize: A force in evolutionary and plant breeding processes. pp. 47–78. *In* J.P. Gustafson, G.L. Stebbens, and F.J. Ayala (eds.) Genetics, Development, and Evolution: 17th Stadler Genet. Sym. Plenum Press, New York. NY.

Peterson, P.A. 1987. Mobile elements in plants. CRC Crit. Rev. Plant Sci. 6(2):105–208.

Plaisted, R.L. and L.C. Peterson. 1959. A technique for evaluating the ability of selections to yield consistently in different locations or seasons. Potato J. 36:381–385.

Pollak, L.M. 2003. The history and success of the public-private project on germplasm enhancement of maize (GEM). Adv. Agron. 78:45–87.

Pollak, L.M. and W. Salhuana. 1999. Lines for improved yield and value-added traits – Results from GEM. Proc. Annu. Corn Sorghum Ind. Res. Conf. 53:143–158.

Rawlings, J.O. 1970. Present status of research on long and short term recurrent selection in finite populations – Choice of population size. pp. 1–15. *In* Proc. 2nd Meet. Work Group Quant. Genet., Sec. 22. IUFRO, Raleigh, NC.

Rawlings, J.C. and D.L. Thompson. 1962. Performance level as criterion for the choice of maize testers. Crop Sci. 2:217–220.

Richey, F.D. 1922. The experimental basis for the present status of corn breeding. J. Am. Soc. Agron. 14:1–17.

Richey, F.D. 1944. The shattered dreams of a corn breeder. Agron. J. 44:267–268.
Rinke, E.H. and J.C. Sentz. 1962. Moving Corn Belt Dent germplasm northward. Proc. Hybrid Corn Ind. Res. Conf. 16:53–56.
Richey, F.D. 1950. Corn breeding. Adv. Genet. 3:159–192.
Rodriguez, O.A. and A.R. Hallauer. 1991. Variation among full-sib families of corn in different generations of inbreeding. Crop Sci. 31:43–47.
Russell, W.A. 1971. Types of gene action at three gene loci in sublines of a maize inbred line. Can. J. Genet. Cytol. 13:322–334.
Russell, W.A. 1991. Genetic improvements in maize yields. Adv. Agron. 46:245–298.
Russell, W.A. and S.A. Eberhart. 1970. Effects of three gene loci in the inheritance of quantitative characters in maize. Crop Sci. 10:165–169.
Russell, W.A. and S.A. Eberhart. 1975. Hybrid performance of selected maize lines from reciprocal recurrent and testcross selection programs. Crop Sci. 15:1–4.
Russell, W.A. and A.R. Hallauer. 1980. Corn. pp. 299–312. *In* W.R. Fehr and H.H. Hadley (eds.) Hybridization of Crop Plants. ASA and CSSA, Madison, WI.
Salhuana, W., R. Sevilla, and S.A. Eberhart. 1997. Latin American Maize Project (LAMP) Final Report. Pioneer Hi-Bred International Spec. Publ.
Schmidt, D.H. and A.R. Hallauer. 1995. Inheritance and number of genes affecting quantitative traits within F_2 maize populations. J. Iowa Acad. Sci. 102:32–40.
Schnell, F.W. 1975. Type of variety and average performance in hybrid maize. Z. Pflanzenzüchtg 74:177–188.
Schnicker, B.J. and K.R. Lamkey. 1993. Interpopulation genetic variance after reciprocal recurrent selection in BSSS and BSCB1 maize populations. Crop Sci. 33:90–95.
Shamel, A.D. 1905. The effect of inbreeding in plants. pp. 377–392. U.S. Dep. Agric. Yearbook.
Shank, D.B. and M.W. Adams. 1960. Environmental variability within inbred lines and single crosses of maize. J. Genet. 57:119–126.
Shull, G.H. 1908. The composition of a field of maize. Am. Breeders' Assoc. Rep. 4:296–301.
Shull, G.H. 1909. A pure-line method of corn breeding. Am. Breeders' Assoc. Rep. 5:51–59.
Shull, G.H. 1910. Hybridization methods in corn breeding. Am. Breeders' Mag. 1:98–107.
Silva, J.C. and A.R. Hallauer. 1975. Estimation of epistatic variance in Iowa Stiff Stalk Synthetic maize. J. Heredity 66:290–296.
Smith, H.F. 1936. A discrimnant function for plant selection. Annu. Eugen. 7:240–250.
Smith, O.S. 1979. A model for evaluating progress from recurrent selection. Crop Sci. 19:223–226.
Smith, O.S. 1983. Evaluation of recurrent selection in BSSS, BSCB1, and BS13 maize populations. Crop Sci. 13:35–40.
Smith, O.S. 1986. Covariance between line per se and testcross performance. Crop Sci. 26:540–543.
Smith, J.S.C. 1988. Diversity of United States hybrid maize germplasm; isozymic and chromatographic evidence. Crop Sci. 28:63–69.
Smith, J.S.C. and O.S. Smith. 1991. Restriction fragment length polymorphisms can differentiate among U.S. maize hybrids. Crop Sci. 31:893–899.
Smith, D.R. and D.G. White. 1988. Diseases of corn. pp. 687–766. *In* G.F. Sprague and J.W. Dudley (eds.) Corn and Corn Improvement. 3rd ed. ASA, CSSA, and SSSA, Madison, WI.
Smith, O.S., A.R. Hallauer, and W.A. Russell. 1981. Use of index selection in recurrent selection programs in maize. Euphytica 30:611–618.
Smith, J.S.C., M.M. Goodman, and C.W. Stuber. 1985. Genetic variability within U.S. maize germplasm. II. Widely-used inbred lines 1970 to 1979. Crop Sci. 25:681–685.
Snape, J.W. 1989. Doubled haploid breeding: Theoretical basis and practical applications. pp. 19–30. *In* A. Mujeeb-Kazi and L.A. Sitch (eds.) Rev. Adv. in Plant Biotech., 1985–1988. 2nd Int'l. Sym. Genet. Manipulation in Crops. CIMMYT Mexico and IRRS, Philippines.
Spitko, T., L. Sagi, J. Printer, L.C. Marton, and B. Barnabas. 2006. Haploid regeneration aptitude of maize (*Zea mays* L.) lines of various origin and of their hybrids. Maydica 34:537–542.

Sprague, G.F. 1946a. Early testing of inbred lines. J. Am. Soc. Agron. 38:108–117.
Sprague, G.F. 1946b. The experimental basis for hybrid maize. Biol. Rev. 21:101–120.
Sprague, G.F. 1955. Corn breeding. pp. 221–292. *In* G.F. Sprague (ed.) Corn and Corn Improvement. Academic Press, New York. NY.
Sprague, G.F. and W.T. Federer. 1951. A comparison of variance components in corn yield trials. II. Error, year × variety, location × variety, and variety components. Agron. J. 43:535–541.
Sprague, G.F. and L.A. Tatum. 1942. General vs. specific combining ability in single crosses of corn. J. Am. Soc. Agron. 34:923–932.
Stringfield, G.H. 1947. Grouping of inbred lines for breeding purposes. p. 25. *In* Rep. 9th Corn Impr. Conf. North Central Region. Cincinnati, OH. November 16, 1947.
Stringfield, G.H. 1974. Developing heterozygous parent stocks for hybrids. DeKalb AgResearch, DeKalb, IL.
Stuber, C.W. 1978. Exotic sources for broadening genetic diversity in corn breeding programs. Proc. Annu. Corn Sorghum Ind. Res. Conf. 33:35–47.
Sturtevant, E.L. 1899. Varieties of corn. USDA Off. Exp. Stn. Bull. 57:1–108.
Subandi, W.A. Compton, and L.T. Empig. 1973. Comparison of the efficiencies of selection indices for three traits in two variety crosses of corn. Crop Sci. 13:184–186.
Suwataradon, K., S.A. Eberhart, J.J. Mock, J.C. Owens, and W.D. Guthrie. 1975. Index selection for several agronomic traits in the BSSS2 maize population. Crop Sci. 15:827–833.
Tallury, S.P. and M.M. Goodman. 1999. Experimental evaluation of the potential of tropical germplasm for temperate maize improvement. Theor. Appl. Genet. 98:54–61.
Tanksley, S.D. 1993. Mapping polygenes. Annu. Rev. Genet. 27:205–233. Palo Alto, CA.
Tanner, A.H. and O.S. Smith. 1987. Comparison of half-sib and S_1 recurrent selection in Krug Yellow Dent maize populations. Crop Sci. 27:509–513.
Tatum, L.A. 1971. The southern corn leaf blight epidemic. Science 171:1113–1116.
Thompson, D.L. 1954. Combining ability of homozygous diploids of corn relative to lines derived by inbreeding. Agron. J. 46:133–136.
Thompson, D.L. 1968. Silage yield of exotic corn. Agron. J. 60:579–581.
Torregroza, M. 1973. Response of a highland maize synthetic to eleven cycles of divergent mass selection for ears per plant. p. 16. Agron. Abstr., ASA, Madison, WI.
Troyer, A.F. 1999. Background of U.S. hybrid corn. Crop Sci. 39:601–626.
Troyer, A.F. 2001. Temperate corn – Background, behavior, and breeding. pp. 393–466. *In* A.R. Hallauer (ed.) Specialty Corns. 2nd ed. CRC Press, Boca Raton, FL.
Troyer, A.F. 2004a. Background of U.S. hybrid corn. II: Breeding, climate and food. Crop Sci. 44:370–380.
Troyer, A.F. 2004b. Persistent and popular germplasm in seventy centuries of corn evolution. pp. 133–231. *In* C.W. Smith, J. Betra'n, and E.C.A. Runge (eds.) Corn: Origin, History, and Production. John Wiley & Sons, Hoboken, NJ.
Troyer, A.F. 2006. Adaptedness and heterosis in corn and mule hybrids. Crop Sci. 46:528–543.
Troyer, A.F. and W.L. Brown. 1972. Selection for early flowering in corn. Crop Sci. 12:301–304.
Troyer, A.F. and W.F. Brown. 1976. Selection for early flowering in corn: Seven late synthetics. Crop Sci. 16:767–772.
Uhr, D.V. and M.M. Goodman. 1995. Temperate maize inbreds from tropical germplasm. I. Testcross yield trials. Crop Sci. 35:779–784.
Vasal, S.K., A. Ortega, and S. Pandey. 1982. CIMMYT's maize germplasm management, improvement, and utilization program. CIMMYT, El Batán, Mexico.
Vasal, S.K., H. Cordova, S. Pandey, and G. Srinivasan. 1999. Tropical maize and heterosis. pp. 363–373. *In* J.G. Coors and S. Pandey (eds.) Genetics and Exploitation of Heterosis in Crops. ASA, CSSA, and SSSA, Madison, WI.
Vasal, S.K. 2001. High quality protein corn. pp. 85–129. In A.R. Hallauer (Ed.) Specialty corns. 2nd Ed., CRC Press LLC, Boca raton, FL.
Wallace, W.A. and W.L. Brown. 1988. Corn and Its Early Fathers. Iowa State University Press (Rev. ed.), Ames, IA.

Warner, J.N. 1952. A method for estimating heritability. Agron. J. 44:427–430.
Weatherspoon, J.H. 1970. Comparative yields of single, three-way and double crosses of maize. Crop Sci. 10:157–159.
Weber, C.R. and B.R. Morthy. 1952. Heritable on nonheritable relationships and variability of oil content and agronomic characters in the F_2 generation of soybean crosses. Agron. J. 44:204–209.
Wellhausen, E.J. 1956. Improving American corn with exotic germplasm. Proc. Annu. Corn Sorghum Ind. Res. Conf. 11:85–96.
Wellhausen, E.J. 1965. Exotic germplasm for improvement of Corn Belt maize. Proc. Annu. Corn Sorghum Ind. Res. Conf. 20:31–45.
West, D.R., W.A. Compton, and M.A. Thomas. 1980. A comparison of replicated S_1 per se vs. reciprocal full-sib selection in corn. I. Indirect response to population densities. Crop Sci. 20:35–42.
Westgate, M.E., M.E. Otegui, and F.H. Andrade. 2004. Physiology of the corn plant. pp. 235–303. *In* C.W. Smith, J. Betra'n, and E.C.A. Runge (eds.) Corn: Origin, History, and Produciton. John Wiley & Sons, Hoboken, NJ.
Weyhrich, R.A., K.A. Lamkey, and A.R. Hallauer. 1998. Responses to seven methods of recurrent selection in the BS11 maize population. Crop Sci. 38:308–321.
Whitehead, F.C. 2002. Backcross introgression and two-stage testing for conversion of improved tropical germplasm to temperate environments. Ph.D. Diss. Iowa State University, Ames, IA (Diss. Abstr. DAI-B 63/04, p. 1617, Oct. 2002).
Whitehead, F.C., H.G. Caton, A.R. Hallauer, S. Vasal, and H. Cordova. 2006. Incorporation of elite subtropical and tropical maize germplasm into elite temperate germplasm. Maydica 51:43–56.
Widstrom, N.W. 1974. Selection indices for resistance to corn earworm based on realized gains in corn. Crop Sci. 14:673–675.
Widstrom, N.W., B.R. Wiseman, and W.W. McMillan. 1982. Response to index selection in maize resistant to ear damage by corn earworm. Crop Sci. 22:843–846.
Wilkes, G. 2004. Corn, strange, and marvelous: But is a definitive origin known? pp. 3–63. *In* C.W. Smith, J. Betra'n, and E.C.A. Runge (eds.) Corn: Origin, History, Technology, and Production. Wiley & Sons, Inc., Hoboken, NJ.
Williams, W.P. and F.M. Davis. 1983. Recurrent selection for resistance in corn to tunneling by the second-brood southwestern corn borer. Crop Sci. 23:169–171.
Williams, T.R. and A.R. Hallauer. 2000. Genetic diversity among maize hybrids. Maydica 45:163–171.
Wolf, D.P. and A.R. Hallauer. 1997. Triple testcross analysis to detect epistasis in maize. Crop Sci. 37:763–770.
Wolf, D.P., L.A. Peternelli, and A.R. Hallauer. 2000. Estimates of genetic variance in an F_2 population. J. Hered. 91:384–391.
Wricke, W. 1962. Uber eine method zurerlassung der kologischen kologischen streubreite in felduersuchen. Z. Pflazurcht 47:92–96.
Zuber, M.S., M.L. Fairchild, A.J. Keaster, V.L. Fergason, G.F. Krause, E. Hildebrant, and P.J. Loesch, Jr. 1971. Evaluation of 10 generations of mass selection for corn earworm resistance. Crop Sci. 11:16–18.

Rice Breeding

Elcio P. Guimarães

Abstract This chapter deals with breeding aspects of one of the most important crops for food security in the world. Initially it shows how diverse rice is with 22 species, different levels of ploidy and six diversity groups. The choice of parents for crossing, when having such wide genetic diversity available, requires careful characterization and evaluation of the germplasm as well as good knowledge and breeding skills to make the right decisions. Rice breeders have been very successful in improving the crop. Some milestones are: the contribution to the green revolution with the semi-dwarf varieties, the new rice plant type, hybrid rice, and the NERICA rice. Even though there was a series of breakthroughs the main breeding goals in most national programs remain similar since a long time ago: increasing grain yield potential, resistance to blast disease, grain quality, and drought tolerance. The main breeding method used to improve rice is the pedigree, but development of hybrids and population improvement were added to the breeder's portfolio. Breeders have been taking advantage of biotechnology tools to enhance their breeding capacity; however, many national programs are still struggling on how to integrate them into the breeding programs and how to balance the allocation of resources between conventional and modern tools. The chapter closes with information on the rice breeding capacity around the world, showing that rice breeders are widely distributed across all regions and the existing capacity, using the above mentioned information, will still be able to cope with the challenge of making genetic progress for one of the most important food security crops.

1 Introduction

Rice is the world's most important food crop with a total production around 600 million ton occupying 11% of the world's total arable land; it supplies 2,808 calories/person/day, which represents 21% of the total calorie supply. It is source

E.P. Guimarães
Food and Agriculture Organization of the United Nations (FAO), Viale delle Termi di Caracalla, Crop and Grassland Service (AGPC) – Room C-778, 00153 Rome, Italy,
e-mail: elcio.guimaraes@fao.org

M.J. Carena (ed.), *Cereals,The Banks and the Italian Economy*
DOI: 10.1007/978-0-387-72297-9,

of income for more than 100 million householders around the world (IRRI, 2002). It is one of the crops responsible for the so-called green revolution that happened in the 1960s and 1970s. In addition of having strong breeding programs in all different regions around the world, this crop has three Consultative Groups on International Agricultural Research (CGIAR) centers with the mandate to work with rice: the International Rice Research Institute (IRRI), with global mandate; the West Africa Rice Development Association (WARDA), with mandate to work in West Africa; and the International Centre for Tropical Agriculture (CIAT), with the regional mandate for Latin America.

International centers made a tremendous effort to educate and train rice breeders at the time of the green revolution. Today, 25–35 years later most of the rice breeders working in national programs around the world represent that period. The international germplasm evaluation nurseries [International Rice Testing Program (IRTP) and International Network for the Genetic Evaluation of Rice (INGER)] were excellent tools to provide new breeders with improved breeding lines as well as additional opportunities for training, including hands-on exercises on breeding techniques.

This chapter aims at proving general information on the following matters: the sources of genetic diversity available to breeders; criteria to be considered when selecting parental material to generate genetic variability for variety development; the most relevant breeding achievements; rice breeding methods used around the world, how biotechnology has been integrated into breeding programs, genetic seed production strategy; and elements related to the world's capacity to carry out rice breeding programs.

2 Genetic Diversity

The success of the breeding strategies relies heavily on the genetic diversity of the crop. Rice gene banks around the world exhibit a very large amount of genetic diversity present in farmers' cultivars, landraces, as well as in the genetic make up of the 22 *Oryza* species. At the IRRI, in Manila, Philippines, there are more than 108,000 accessions conserved (Jackson and Lettington, 2003); in addition, there are hundreds of rice accessions held in trust in other CGIAR centers; WARDA; CIAT; and International Institute for Tropical Agriculture (IITA). Almost as many accessions are preserved in genebanks in other Asian countries such as China, India, Indonesia, Philippines, and Thailand (Jackson et al., 1997). Furthermore, considering that the International Rice Genome Sequencing Project has identified more than 80,000 genes in the rice genome and that each gene has an unknown number of alleles, the conclusion is that breeders will continue to have useful genetic diversity to draw on for many generations to come as long as there is a good choice germplasm.

Rice belongs to the genus *Oryza* and the various attempts to classify it no agreement was obtained regarding the number and the names of the species belonging to this genus. In 1994, Vaughan (1994) published a handbook indicating that the genus has 22 species. However, only *O. sativa* and *O. glaberrima* are cultivated. The number of chromosomes of the cultivated rice and its related species varies from 24 to 48, with the "*n*" number equal to 12. According to Morishima (1984), based on the chromosome paring in the meiosis, rice has the following genomes: AA, BB, CC, EE, and FF for the diploid species and BBCC and CCDD for the tetraploid species.

The two cultivated species, which are diploid ($2n = 24$), were domesticated under different environmental conditions. *O. sativa* was domesticated in South and Southeast Asia and has the species *O. rufipogon* and *O. nivara* as its direct ancestors. *O. glaberrima* comes from tropical West Africa and has *O. barthii* as progenitor. The former is cultivated throughout all the rice growing environments around the world. However, cultivation of the African species is confined to its region of origin.

Morishima and Oka (1981) divided the cultivated species in two groups called *indica* and *japonica* based on principal component analyses of 11 variety characteristics. Their study indicated that there was no specific characteristic that clearly distinguished the two groups; however, the existence of the two groups can be proved by combining two or more characteristics. Khush et al. (1984), using trisomics, observed a complete correspondence between *japonica* and *indica* linkage blocks; meaning, there was not a single case where genes located in one *japonica* chromosome were found in a different *indica* chromosome. Harushima et al. (2002) believe that the domestication process caused the differences between the two groups including their reproductive barriers. The evolutionary process gave these two groups their distinct characteristics such as tolerance to low temperatures and drought stress, responsiveness to fertilizers, ability to compete with neighboring plants, and photosynthetic capacity, among other things.

Based on geographical distribution, Morinaga (1954) described three morphological groups called *japonica*, *javanica*, and *indica*. Oka (1958) indicated that *japonica* and *javanica* groups can be considered as tropical and temperate *japonicas*, respectively. The former has tall varieties with heavy panicles (Glaszmann and Arraudeau, 1986). The ideal plant type designed by Khush (1994), which can help boost the rice yields by up to 30%, capitalizes on the genetic variability of this group.

A breakthrough related to rice genetic diversity groups was made by Glaszmann (1987) who analyzed 1,688 Asian traditional varieties using isozymes and identified six genetically distinct groups, which were called groups I–VI. Group I encompasses the varieties of the tropical regions classified by Oka (1958) as the "*indicas*". At the other extreme is group VI where there are the varieties adapted to temperate climates called "*japonicas*" by Oka (1958). The latter group includes most of the upland rice varieties.

As rice is cultivated all over the world, its diversity is also due to the wide range of ecosystems the crop is adapted to. According to IRRI (2002) one way of categorizing it is to distribute the ecosystems in four broad categories: irrigated, lowland, upland, and flood-prone. A combination of these ecosystems with different agro-ecological zones gives a very complex matrix in which rice genetic diversity has become available adapting naturally to meet farmers' demands.

Even though rice is a rich crop in terms of its genetic diversity, there are several reports in the literature indicating that the varieties released by the breeding programs in different parts of the world have a narrow genetic diversity basis. Cuevas-Pérez et al. (1992) and Montalban et al. (1998) presented results on irrigated and upland rice in Latin America, respectively, indicating that commercial varieties released for both systems had a narrow genetic base. Guimarães (2002) dissected the Brazilian rice varieties and arrived at the same conclusion. Mishra (2002), considering the breeding approaches used in India and the varieties released in the last 30 years, concluded: "the genetic base is narrowing and this is a matter of concern". Evidence was added by Rai (2003) when analyzing the 29 varieties released in the Indian Kerala State. He also pointed out that in Nigeria there is genetic uniformity within the upland rice varieties. Dilday (1990) showed similar results when analyzing the genetic diversity of the rice varieties released in the USA as well as Kaneda (1985) in Japan. However, as indicated before, these results do not suggest that this is true when looking at the rice species. For example, Sun et al. (2001) analyzed *O. rufipogon* and *O. sativa* using molecular markers and concluded that they still have a wide genetic variability with regards of number of alleles, number of genotypes, heterozygosity, and diversity among genes. In the cultivated species, they found only 4 exclusive alleles but in the wild species there were 78 of these alleles. These results indicate the wide genetic variability still present in the species, mainly in the wild relatives.

Second (1982) found large differences in allelic frequency between the *indica* and *japonica* species. Oka (1964) concluded that the genetic diversity is maintained within groups independent of whether there are crosses and recombination in the segregating generations. Junjian et al. (2002) used simple sequence repeat (SSR) markers and studied the genetic diversity between *indica* and *japonica*. They found similar average numbers of alleles: 4.4 and 4.3, respectively. However, the average genetic distance was greater for the *indicas*, which suggested a higher level of genetic variation for this group in relation to the *japonica*. The two wild species included in the study (*O. rufipogon* and *O. nivara*) fall outside of the range of cultivated species, suggesting the presence of unique alleles still to be used by breeders to exploit the between species variability. This knowledge has been exploited to increase the yield potential of commercial varieties (Xiao, Li, Grandillo, Ahn, Yuan, Tanksley, and McCouch 1998; Moncada, Martínez, Borrero, Gauch, Guimarães, Tohme, and McCouch 2001; Brondani, Rangel, Brondani, and Ferreira 2002).

The importance of having genetic diversity available is the possibility of making it useful to develop products that will have an impact at farmers' field level. Rice breeders have been exploiting this potential in many different and creative ways. There are a few examples where exploitation of wild relatives has produced significant results. The first opportunity to take advantage of the wild relatives' unique characteristics was through exploiting the existence of genes for disease and insect resistance. Khush (1977) used *O. nivara* as a source of resistance to grass stunt virus resistance and introduced it into IR28, IR29, and IR30 cultivars; also, from *O. rufipogon* the resistance to the viral disease called "tungro" was obtained.

Rice bacterial blight resistance was obtained from *O. longistaminata* and introduced in the commercial variety IR24 (Khush et al., 1990).

The first hybrid rice was developed based on a genetic cytoplasm male sterility system identified in the *O. sativa* L. *f. spontanea* in China (Shih-Cheng and Loung Ping, 1980). Khush (1994) combined different genetic groups (tropical *japonicas*, temperate *indicas*, and *japonicas*) in order to create a new rice type expected to increase the grain yield of commercial varieties by ~30%.

Rice is rich in genetic diversity and breeders have a wide choice when looking for parental materials.

3 Choice of Germplasm

One of the most difficult tasks in carrying out a successful breeding program is the choice of germplasm. To be able to develop a variety with a set of desirable characteristics rice breeders need to be sure that the source germplasm has desirable genetic variability. After the parents are chosen and the crosses are made there are almost no chances of new alleles appearing in the segregating populations.

To make the right choice of parental material to be used in a breeding program, breeders must clearly know the type of product to be developed; the characteristics of the species to be bred; the combining ability of the parents in case of hybrid cultivars; the environmental conditions of the target area; the social and economic aspects of the farmers and markets; and the different breeding approaches available to achieve the proposed goals. Today, an additional element to be considered is the legal aspect in relation to the materials to be used as parents.

In general, rice breeding programs have two major different end products. The first and the most common one is a pure line, which will be evaluated and released as a commercial variety. The second one is an inbred line that will be the parent of a commercial hybrid. An intermediate product may be a population with certain desirable characteristics that could be used for further improvement, for cultivars per se, or for line extraction.

If the target of breeders is to develop pure lines it is important to know the ability of the progenitor to transfer its characteristics to the segregating populations. An interesting example is the rice variety "Ceysvoni" from Surinam; it has a high level of resistance to blast and a good plant type as well as excellent grain type to Brazilian standards. However, when used in crosses with other upland and irrigated varieties, the different combinations do not produce high quality breeding lines. In general, all segregating populations are discarded before reaching homozygosis. It seems that some materials when used in crosses produce undesirable changes in the genetic composition of the resulting population. Another example is the cultivar BG90–2, a high-yielding irrigated rice variety. Every time this genotype was used in crosses, the resulting segregating populations did not seem to retain its yield potential and they were discarded before producing homozygous lines. Brondani et al. (2002) proposed combining this cultivar with wild species to identify yield-related quantitative trait loci (QTLs). More information on combining ability of the progenitors is desirable to achieve the proposed goal.

In general, rice breeders, as well as breeders for other crops, tend to recycle and cross high-performing parents (e.g., elite × elite) among themselves and conduct maximum inbreeding when aiming at developing new commercial varieties. This strategy is based on the concept that self-pollinated crops have a large part of their genetic variance as additive variance. In addition, high-performing parents with reduced genetic variance present a higher probability of generating superior genotypes.

Another important aspect to consider is the genealogy of the parental material. In general, breeders avoid crossing parents with similar genetic make-up, because their combination will not produce a broad genetic variability limiting the possibilities of obtaining desirable gene combinations (desired genotype). If the final product is a hybrid, knowledge about the genealogy of the parents is crucial, since heterosis relies on genetic differences. Hybrid rice is produced based on cytoplasmic-genetic male sterility. It requires three types of breeding lines: a cytoplasmic-genetic male sterile line (A line), a maintainer line (B line), and a restore line (R line). Therefore, if the objective is to produce hybrid rice, knowledge of the general and the specific combining ability of lines becomes essential. Heterosis is based on genetic differences (among other factors), thus information on the genetic distance between A and R lines is fundamental to produce high-performing hybrids. Therefore, knowledge about the genealogy of the parental material is crucial.

To facilitate the development of economically high-yielding hybrids with all of the desirable agronomic traits it is also important to consider other characteristics when selecting the parental material such as aspects related to difference in grain type and shape, plant height, resistance to biotic and abiotic stresses. Hybrid rice seed production depends on a series of factors, including the coincidence in the flowering period of the male and female lines. It is important to have parents that complement each other well, with good specific and general combining abilities. This terminology was introduced by Sprague and Tatum (1941) to differentiate between the mean performance of a parent in crosses (general combining ability) and the deviations of individual combinations from the mean (specific combining ability). Knowledge of this behavior is more important when the non-additive effects are predominant.

To create a rice population, it is important to choose parental materials with high levels of genetic differences for the traits under selection. However, it is desirable to have also a low genetic divergence for the traits the breeder wants to keep in the population. An example of the use of such information in deciding between rice parents with tolerance to iron toxicity was presented by Khatiwada et al. (1996). "Azucena", IRAT 104, and "Moroberekan" were the cultivars with the best general combining ability for iron toxicity tolerance and were the recommended ones for elite × elite crosses.

The choice of parental material depends on the breeders' objectives, the desired type of product, the existing genetic diversity, and the information available, as well as the combining ability of the parents.

4 Major Breeding Achievements

4.1 The Rice Green Revolution

In the 1960s, scientists quickly realized that most tall traditional rice varieties lodged easily when nitrogen fertilization was applied, which was the major limitation to grain yield (Khush et al., 2001). The semi-dwarf (sd_1) IR8 was the first high-yielding rice variety developed from a combination between the Indonesian variety "Peta" and "Dee Geo Woo Gen" from Taiwan. The key factor responsible for the increase in yield potential was the improvement of the harvest index. However, even though IR8 had a major drawback regarding its poor grain quality, it still became the symbol of the green revolution in rice. Within a few years, many countries around the world were replacing their traditional cultivars with the modern high-yielding varieties.

The icon of the rice green revolution, when compared to traditional varieties, exhibits certain distinct characteristics; it has shorter stature, a shorter growth cycle, higher tillering ability, higher photosynthetic capacity, responsiveness to fertilizers (mainly nitrogen), and consequently much higher yield potential to high-input environments.

In the following decades IRRI developed IR36, which became the most widely planted variety in the 1980s and IR64 was the most used in the 1990s (Peng and Khush, 2003). In addition to these varieties, IRRI released a large series of IR-coded varieties. However, while these newer materials were characterized by their resistance to disease and insects, they did not contribute significantly to genetic gains for grain yield. Scientists then believed that a new breakthrough in yield potential had to come through a new plant type.

4.2 The New Plant Type

Donald (1968) was one the pioneers in the discussion of breeding for ideotype plants. Yang et al. (1996) suggested that in order to develop super high-yielding rice varieties it was essential to increase the biological yield. Searching for a second green revolution IRRI had been working on a new rice ideotype or new plant type (NPT) with a harvest index of 0.6 (60% grain: 40% straw weight) and with an increased ability for photosynthesis to increase total biological yield. Peng et al. (2005) considered the following components on this NPT: low tillering capacity, few unproductive tillers, from 200 to 250 grains per panicle, from 90 to 100 cm of plant height, thick and strong stems, vigorous root system, and from 100 to 130 days of growth cycle. These traits would allow the rice plant to transform more energy into grain production, increasing the yield potential by about 20% but with more input and cost.

Even before IRRI, Japan was the first country to pursue research on the NPT idea. In 1981, Japan launched a project aiming at combining varieties from *indica* and *japonica* groups to develop a super high-yielding rice cultivar (Wenfu et al., 2001). Dingkuhn et al. (1991) carried out physiological studies to understand the yield potential limitations of the *indica* varieties. They observed that under direct-seeded systems rice plants produced an excessive leaf area, which caused mutual shading and reduction in the canopy photosynthesis and sink size. In addition, they developed a large number of unproductive tillers.

The development of this NPT was based on tropical *japonica* germplasm derived from Indonesia, being the source of low tillering, large panicles, thick stems, vigorous root system, and short stature. According to Peng et al. (2005) the process of developing the NPT was more complicated than originally thought. The first generation of breeding lines with the above mentioned traits did not perform as expected. New crosses were made combining the tropical *japonicas* with elite *indica* breeding lines. The expectation was that lines coming out of these crosses would increase the yield potential of irrigated lowland rice by about 10%.

The development of super high-yielding rice varieties following the concepts proposed by Khush and Peng (1996) have encountered various technical difficulties. However, the basic principles remain the same. Horie (2001), Sheehy et al. (2001), and Murchie et al. (2001) looked at several of these physiological limitations and made technical arguments aimed at addressing them. Even though the results of these strategies are not yet producing an impact at the farmers' field level, it is important to highlight how rice breeders have been combining knowledge in creative ways on genetic diversity, plant physiology, and rice breeding methods to address these challenges.

4.3 Hybrid Rice

The hybrid rice technology concept dates back to 1964 in China. However, only in 1970, when a wild abortive pollen plant was identified in Southern China, did the idea begin to materialize. In 1980, Shih-Cheng and Loung Ping (1980) published one of the first articles indicating the potential of hybrid rice. The proposed strategy then relied on the male sterility produced by the abortive pollen system identified in the wild species *O. sativa* L. *f. spontanea*. Hybrid rice would then be produced through a so-called three-line system, where one line would have the genetic–cytoplasmic male sterility; the second line would be responsible for maintaining the sterility, and a third one would be used as the matching parent for the hybrid with the responsibility of restoring the fertility.

The first set of genetic–cytoplasmic male sterile lines was produced in 1970, while the first hybrid rice was released in 1974, with the hybrids outyielding, on average, the conventional rice varieties by 20%. In 1999, the area planted to hybrids was about 15.5 million ha, representing 50% of the total rice area and 60% of the total Chinese rice production (Guohui and Longping, 2003). Since 1994, hybrids have been released in India, Philippines, Vietnam, Bangladesh, and Indonesia. The yield gains of the released hybrids in relation to the conventional varieties vary from 20% in Philippines to 30.2% in Vietnam (Virmani, 2003). India has released six hybrids since 1989, however, the pace of adoption by farmers has been slower than expected and only 200,000 ha are cultivated (Mishra et al., 2003). Among the above mentioned countries, Vietnam was the first to begin releasing hybrids, initially in 1979. By 2001, it had around 480,000 ha planted to hybrids (Hoan and Nghia, 2003). Indonesia began its hybrid production in 1998 and has released two public and five private hybrids. The projection is for more than 500,000 ha to be planted in the years to come (Suwarno et al., 2003). Bangladesh followed in close collaboration with IRRI, and released two IRRI hybrids. While the area planted is not significant yet, the government has put in place a hybrid rice master plan to boost its adoption (Julfiquar et al., 2003).

To simplify the hybrid rice production system, the concept of environmental genetic male sterility (EGMS) was introduced. The two environmental factors considered were the photoperiod (PGMS) and the temperature (TGMS) sensitivities, which are controlled by recessive nuclear genes. This technology allows, according to Mou et al. (2003), the use of any genotype with good traits as male parent, to obtain *japonica* hybrids (e.g., it is difficult to identify restorers for this group), and to develop inter-group hybrids such as *indica*/*japonica* (e.g., there is no restriction regarding the restorer–maintainer relationship). The first two-line hybrid was released in China. It represented 17.2% of the total hybrid rice area in the country in 2001, some 2.67 million ha (Guohui and Longping, 2003).

4.4 NERICA Rice

Upland and lowland dry land environments are the two most important rice production ecosystems in Africa, where it is staple food for the sub-Saharan population. Certain challenging problems and environmental conditions as well as production practices common to these ecosystems limit rice production, such as weeds, diseases, and insect pressure, soil fertility decline, soil acidity, and drought stress. WARDA began a program to combine the two cultivated rice species *O. sativa* and *O. glaberrima* in 1991. Their genetic dissimilarity needed the use of a different breeding approach. Embryo rescue technique was employed to obtain viable segregating populations (Jones et al., 1997). The newly developed materials were called "new rice for Africa" and were popularized as NERICA varieties. There are not many technical publications about the development of these varieties. Information is gathered in the form of press releases, on the WARDA web page (www.warda.cgiar.org/) and in articles such as that of Jones and Wopereis-Pura (2001). The main features of these new varieties, when compared to the traditional *O. glaberrima*, cultivated by farmers, are their improved ability to compete with weeds, their larger panicles with around 400 grains and a higher yield potential. In addition, shattering is reduced, stems are stronger thus preventing lodging, maturity occurs around 30 days earlier than other conventional cultivars, and they have greater resistance to the most common biotic and abiotic stresses, as well as improved adaptability to the poor African rice growing soils. The success story of the NERICA varieties includes a strong participation of the farmers in the process of evaluation of the breeding lines as well as in the development of the materials (e.g., farmer–breeder inititatives, participatory plant breeding, see Chapter 14). Information on the impact of this technology can be found at WARDA (2003).

5 Current Breeding Goals

An increase in productivity is always one of the main goals of any crop breeding program including rice. However, a long list of goals can be identified for this crop varying in importance from region to region, country to country, and even within a given country. A few examples of current breeding goals are described in the sequence.

Increase grain yield potential is the major goal of almost all rice breeders programs around the world. The major impacts, listed elsewhere in this chapter, are related to the development of new strategies to increase the genetic grain yield potential of the varieties. Nonetheless, there are still regions in the world where rice leaves and straws play a major role in farmers' livelihoods.

Resistance to blast disease has been among one of the most researched rice breeders' goals for decades. This disease is the most widespread pest of rice. It is present in almost all countries and agro-ecological zones where rice is grown. It causes leaf and panicle damages. Improvement strategies have to rely either on gene pyramiding or multiple long-term resistance and/or tolerance because the fungus has a complex set of races and single gene resistance is frequently overcome in a very short time by the pathogen.

Grain quality characteristics vary from region to region and market requirements. Very often varieties are discarded by farmers because they do not meet their required quality standards. One example of the importance of the trait in determining the success of a breeding program is upland rice in Brazil. Until the 1980s, the most wanted grain type in the Brazilian mid-west region was the upland type (medium to long bold grains). However, due to the market pressure made by the industry from the Brazilian south (irrigated rice grain type – long and slender grain type) the upland grain type lost market. Thc upland rice program had to quickly shift its grain type objective and only when upland varieties with long and slender grains were released, only then upland rice became popular again. Specialty grain quality rice types are also an objective of many breeding programs around the world today.

Drought tolerance is another trait highly researched in rice. The increase trend in global water scarcity, the gradual seriousness of water shortage around the world due to climate change, and the high water demand of rice varieties make this a highly important objective of rice breeding programs. In addition, due to the urbanization and pressure from other more important cash crops rice cultivation has been pushed to less favorable areas with larger water availability problems. The complexity of the trait and the difficulties in developing a reliable and simple screening system make the development of tolerant varieties an important challenge. The use of biotechnology tools is making a significant contribution to identify genes and strategies to incorporate them in new varieties. However, the progress is still below the required level to produce significant impact in rice production due to the genetic complexity of the trait.

An important point to make is that breeders have a tremendous challenge to cope with farmers and markets demands. Other challenging goals include resistance to bacterial and sheath blight, and several viruses; resistance to insects such as brown plant hopper, green leafhopper (vector of *tungro* viruses) and gall midge; and tolerance to salinity, iron toxicity, and low temperatures.

6 Breeding Methods and Techniques

6.1 Conventional Rice Breeding Methods

If one makes a global literature review on the breeding methods commonly used to develop rice varieties around the world pedigree selection is always at the top. More than 85% of the released rice varieties published in Crop Science Society of America have been developed through pedigree selection. When there are possibilities to carry out more than one generation per year (e.g., winter nurseries) the method is combined with modified bulk or even single-seed descent to speed up the process of having pure lines for agronomic evaluation.

This chapter will focus on methods that are more unique or that can bring new elements into the attention of the readers.

6.2 Population Improvement Through Recurrent Selection

This section will not try to dissect rice population improvement through recurrent selection but it will highlight the experiences of using such methods in Latin America where the method has been employed for more than 15 years. There are breeding programs with different capacities run by international organizations such as the CIAT, the "Centre de cooperation internationale en recherche agronomique pour le developpement" (Cirad), and several national programs such as the Brazilian Agricultural Research Corporation (Embrapa), and the "Fundación para la Investigación Agricola" (Danac), among others.

A question one could ask is why population improvement strategies including genetically broad-based populations should be considered for a self-pollinated crop such as rice. The answer to this question is simple: several reports indicate that the genetic gains obtained by different breeding programs around the world and particularly in Latin America (Santos et al., 1997; Muralidharan et al., 2002) are stagnating or even decreasing. In addition, other reports indicate that the genetic base of the rice varieties is narrowing (Dilday, 1990; Cuevas-Pérez et al., 1992; Rangel et al., 1996).

Population improvement through recurrent selection is a traditional breeding method that has been used in maize for over 50 years (Hull, 1945; Dudley and Lambert, 2004). However, it has not been a common breeding methodology choice in self-pollinated crops. Fujimaki (1979) suggested its application in rice using male sterility. In soybean, Werner and Wilcox (2004) reported the results of male sterility facilitated population improvement for yield characteristics. Wang et al. (1996) used the *Tai Gu* gene to induce sterility and apply recurrent selection in wheat. These are other successful examples of the utilization of population improvement facilitated by the existence of male sterile genes. However, fewer cases are shown in the literature in which male sterility was induced by the application of chemical products (Picard et al., 2004).

Hand crossing in rice is a laborious task as described by Guimarães (1999). Some of the requirements when using recurrent selection methods are to produce progenies (sometimes crossing when using full or half-sib families) and recombine the selected ones after replicated experiment trials across environments. Therefore, the utilization of population improvement methods in rice only became feasible after the discovery of the male sterile gene obtained by Singh and Ikehashi (1981) through induced mutation of the rice variety "IR36". The recessive male sterile gene was employed in 1984 by Embrapa and Cirad to create populations with broad genetic bases (Taillebois and Guimarães, 1989; Rangel and Neves, 1997). Moreover, the simplification of the crossing method developed by Taillebois and Castro (1986) and described by Sarkarung (1991) made a significant contribution to promoting the use of breeding methods that require a large number of crosses each year.

Population improvement through recurrent selection in rice is a methodology widely used in Latin America; however, it is not as popular elsewhere. In rice, as probably happens in almost all self-pollinated crops, breeders tend to use pedigree selection which is a complement to recurrent selection if well managed. The genetic improvement process is cyclical, aiming at taking advantage of the progress made in the previous years. In general, each year breeders select the best breeding lines to make new elite crosses between them and/or with new germplasm. Guimarães et al. (1996), analyzing the upland rice breeding program at CIAT, which is based on pedigree and modified bulk selection, found that "even though CIAT did not follow the recurrent selection method, a modified approach, similar to the proposed methodology, was used" during the period 1984–1993. The aim of having such cycles is to capitalize on the genetic gains made in previous years; however, through pedigree selection this is done in a non-systematic way.

The main feature of recurrent selection is to increase the frequencies of the favorable alleles, as was pointed out by Hull (1945) when describing the process of recurrent selection. Thus, by applying the recurrent selection method in rice, breeders are following the same principle but in a systematic and long-term way. Therefore, recurrent selection allows defined and shorter breeding cycles, the possibility of a more precise follow-up of genetic gains, and opportunities to develop breeding lines with a wide genetic make-up.

Population genetic improvement through recurrent selection in Latin America is, unlike maize, recent and dates back to 1996. This is due in part to the support of the Organization of American States (OSA), Embrapa, Cirad, and CIAT who offered the first training course to introduce the subject to breeders in the region. As the next step, broad genetic base populations were made available to them by Embrapa and CIAT. In addition, a close follow-up was provided by more experienced breeders from both institutions. Moreover, Chatel and Guimarães (1997) prepared a handbook providing guidance to rice population development and improvement. Guimarães (1997) was the first to report the progress made by the different breeding programs in the region. Similar reports were produced in 2000 (Guimarães, 2000) and 2004 (Guimarães, 2005). During the last 15 years these programs have made significant progress. There are currently more than 50 genetically broad base rice populations developed in the region (GRUMEGA Grupo de Mejoramiento Genético Avanzado en Arroz, 2006a). In Argentina, breeding programs developed *indica* populations PARG-1 and PARG-2 (Marassi et al., 2000) and PARG-3 (Marassi et al., 2004) aiming at improving cold tolerance. For the same trait, breeding programs in Chile developed *japonica* populations PQUI-1 and PQUI-2 (Hernaiz-L et al., 2004). Graterol (2000) described how PFD-1 and PFD-2 populations were produced in Venezuela to adapt to two different environmental conditions (winter and summer growing seasons). In Cuba, the national program developed the populations IACuba-1 and IACuba-2 searching for a genetically diverse population adapted to local challenges (Pérez-Polanco et al., 2000). The cultivar CG-91 with resistance to rice blast for upland conditions was developed by Guimarães and Correa-Victoria (2000). Similar research has been conducted by Courtois et al. (1997) for upland rice.

Breeders in Latin America highlighted the following advantages of using population genetic improvement through recurrent selection: (a) the possibility of creating and managing their own segregating populations without incurring any additional expenses necessary to evaluate potential parents every year; having a structured crossing program; keeping detailed information on lines and parents; (b) the possibility of having improved and diverse breeding lines available at the end of every recurrent cycle as well as continuing to make progress in increasing the frequency of favorable genes in the population; (c) the national programs can have more than one population improvement program with minimal additional resources, avoiding the duplication of similar activities in a given year (instead of having to evaluate hundreds of line of two populations in one year, the program can be organized in such a way that in a given year one population is in the recombination phase and the other in line evaluation phase); and (d) as in (b) varietal development process can be integrated with the population improvement program and become a unique and more powerful project.

An extra advantage that is worth highlighting is the possibility of integrating different rice breeding programs within countries. In Brazil the evaluation phase of the different populations being managed for the irrigated and the upland ecosystems are shared among state organizations, Embrapa units, and universities. Each partner carries out the evaluation of progenies at fewer locations than in evaluations carried out by single breeding programs. All locations are pooled and a combined analysis is performed. Selection of the best families for the recombination phase is done through discussion of the results in a joint meeting. The best lines at each location are kept by the local breeding program for further selection and line development. A similar strategy was adopted in Venezuela through the "Fundación Danac", who developed populations PFD-1 and PFD-2 and evaluates families through the national program, universities, and the private sector.

Many rice programs in Latin America did not have a fully operational breeding project. In general, they were dependent on ad hoc introductions of pure lines from other stronger national programs or international organizations. Today, these programs are releasing improved varieties obtained from segregating lines used for recombination and development of new recurrent cycles. A good example of this progress can be seen in Bolivia, which recently released the variety "Esperanza" (GRUMEGA, 2006b). Chile has collaborated closely with CIAT for a long time and has always had a strong breeding program. In 1990 the country decided to add to its portfolio of breeding methods recurrent selection for genetic improvement (Alvarado-A, 1997), and in 2007 it has released R-Quila 23 (GRUMEGA, 2006c). In the region, countries are carrying out their population improvement programs using different recurrent selection strategies for genetic improvement. Most countries utilize the S_1–S_2 recurrent selection procedure, evaluating and selecting S_0 plants, advancing to the $S_{0:1}$ generation outside the normal growing season, and evaluating and selecting $S_{0:2}$ families for recombination. In temperate countries, where two growing seasons per year are not feasible unless using winter nurseries, scientists evaluate and select $S_{0:1}$ and $S_{1:2}$ families. Very seldom do scientists in these countries use a different selection scheme, therefore, non-additive effects are not considered in genetic improvement. In almost all cases there have been reported genetic gains for the target traits when comparing more advanced generations with the original populations or less advanced recurrent cycles. Brazil has a very strong rice breeding program and has been one of the promoters of this methodology. The first variety derived from a genetically broad-based population under recurrent selection was released in 2002 (GRUMEGA, 2006d). The breeding programs are currently managing five populations for irrigated conditions (Rangel et al., 2000) and eight for upland ecosystems (Castro et al., 2000). Evaluation studies were carried out in different populations in order to assess the efficacy of the method in rice. Rangel et al. (2005) reported 6.65% genetic gain after evaluating two cycles of recurrent selection in the irrigated rice population CNA-IRAT 4. Badan et al (2005) reported 6.2% gains after selecting for rice blast resistance when comparing cycles 1 and 2 of the upland rice population CNA-7. Moreover, evaluation of three cycles of recurrent selection for grain yield and neck blat in CG-3 upland rice population showed 3.6% and 3.4% genetic gain per year, respectively (Morais et al., 2008).

6.3 *Hybrid Rice*

To cope with future demand for rice production, yield per unit area has to be increased rapidly in the major rice producing countries. As stated before, hybrid rice was first released in China in 1974 with the promise of increasing yield potential beyond the level of the traditional varieties through the exploitation of heterosis. The initial breeding strategy to produce hybrids relied on three breeding lines known as A line (the male sterile line), B line (responsible to maintain the genetic male sterility of the A line), and R line (used to restore the fertility of the A line and to produce the hybrid seed). The technique evolved to a two-line process using environmental genetic male sterility (EGMS) counting on photoperiod (PGMS) and temperature (TGMS) sensitivity to induce sterility. The ideal system for these and other cross-pollinated crops would be the one-line method utilizing the apomixis system that allows preserving the right cultivar.

According to Virmani et al. (1997) the development of a hybrid breeding program has to go through the following stages of identification and development of the A, B, and R breeding lines: identification and evaluation of male-sterile lines and their restorers; testcross phase to select heterotic combinations and to initiate conversion of maintainer lines into male-sterile lines; backcrosses to transfer the cytoplasmic male-sterility to elite maintainer lines; trials to study the combining ability (general and specific) of the parental lines; and foundation seed production of all three lines. Production of breeding lines for the three- or two-line methods is still a difficult task for most of the breeding programs outside of China. One of the bottle necks for spreading the technology worldwide is the seed production process. In general, the production system relies on planting a few rows of the male line (R line) and rows of the female (A line) in such way that the maximum hybrid seed production per unit area occurs in the A line. Few ratios of female to male line seed production have been used (e.g., the 6:2, 8:2, and/or the 10:2 ratios). Mao (2001) reported that the average hybrid seed production in rice lines is between 2.5 and 3.0 t ha^{-1}, however, outside of China it is much lower (1.0–1.5 t ha^{-1}, according to Virmani et al., 2001). In Argentina, RiceTec is producing hybrid rice for southern South America. The company is using three- and two-line systems. Their seed production varies tremendously from year to year with an average around 1.2 t ha^{-1}.

In addition to the seed production of the female parent, the success of a hybrid is dependent on the level of heterosis it can express after crossing their parental varieties. The combination between different varieties is the first step to obtain heterosis, but its expression improves as combinations between varieties belonging to different groups (*indica* and *japonica*) are explored. An alternative to develop hybrids with higher potential might be the use of yield enhancing genes from other species (Yuan, 2003). Molecular markers are trying to identify restorer genes in *japonica* background (Tan et al., 1998) and thermo-sensitive genetic male sterility genes (Yamaguchi et al., 1997; Latha et al., 2004). In addition, marker-assisted selection (MAS) has been reported to assist in the development of hybrids with disease and insect resistance in China (He et al., 2004).

6.4 Mutation Breeding

The use of different sources derived from induced mutations was a popular choice to generate genetic diversity for specific traits in rice in the 1980s. Today the technique became part of the tools kit breeders have to enhance specific rice characteristics in well-adapted varieties. The intention of this section is not to discuss all aspects related to the use of mutation breeding, but to highlight a few successes and flag the importance of mutations to rice improvement.

According to Wang (1992) during the period 1966–1990, there were 78 varieties released in China originated from mutation breeding. More recently, from 1991 to 2004, there was a similar number (77) of new releases coming from application of mutation (Chen et al., 2006). The most popular mutagen is still the gamma rays and the mutated characteristics are the ones responsible for the expression of agronomic (e.g., resistance to pests) and grain quality phenotypes.

In Indonesia the first mutant variety (Atomita 1) was released in 1982 and up to today there are 14 officially released varieties, 13 of them were improved for biotic stresses such as resistance to brown plant hopper; in all cases the mutagen agent was the gamma rays (Ismachin and Sobrizal, 2006).

Vietnam is one of the most important rice producing country in the world. Reports from Tran et al. (2006) indicated that during the period 1990 and 2002 the Agricultural Genetic Institute developed and released 10 varieties, most of them have better grain quality, in addition to other agronomic traits; once more the gamma rays were the most common mutagen agent used.

Maluszynski et al. (1998) summarized the number officially released mutant varieties and came up with "cereals" as the group with the largest numbers followed by legumes and industrial crops. Among cereals rice presented the highest number with barley in second. In rice the main improved traits were early maturity, plant height, and disease resistance. It is worth mentioning that the famous gene sd_1 (see section on major breeding achievements) is a mutant. However, the most commonly mutated trait over all crops was "semi-dwarfness". Table 1 summarizes the number of varieties released around the world, which were developed by the use of mutagens. The Food and Agriculture Organization of the United Nation (FAO)/ International Atomic Energy Agency (IAEA) Mutant Varieties Database indicates that there were 2,541 releases up to March 2007. The largest numbers are from cereals (1,212), followed by legumes and industrial crops. Among cereals rice presented the highest number (525) with barley in second (303) and wheat in third (200).

Table 1 Officially released mutant varieties of rice in the FAO/IAEA Mutant Varieties Database, March 2007

Country	Varieties released (#)	Country	Varieties released (#)
Bangladesh	5	Japan	70
Brazil	28	Korea	7
Burkina Faso	3	Myanmar	5
China	222	Nigeria	3
Costa Rica	2	Pakistan	10
Ivory Coast	26	Philippines	8
France	5	Portugal	1
Guyana	26	Romania	1
Hungary	3	Senegal	2
India	40	Sri Lanka	1
Indonesia	6	Thailand	4
Iraq	3	USA	23
Italy	2	Vietnam	28

7 Integration of New Biotechnologies in Breeding Programs

The first and most important aspect to successfully take advantage of the variety of biotechnology tools available to rice breeders is to have a well-structured, efficient, and effective breeding program. This statement may seem obvious for many readers but it does not reflect the reality of a large portion of the rice breeding programs in developing countries around the world.

FAO has started a worldwide plant breeding and associated biotechnology assessment in 2002. This work has been concluded in a sample of more than 50 developing countries in all the different rice growing regions. Among other things, the results indicate that almost every country has made investments in the area of biotechnology recently. However, only a very limited number of them have reinforced their breeding activities and worse still, the great majority do not even have well-structured and fully operational breeding programs that can incorporate biotechnology tools. To add to this, very seldom have they ensured linkages between biotechnology efforts and breeding priorities or strategies.

Anther culture is a simple biotechnology tool that has been around for quite a long time. The technique allows the development of double haploid lines or true breeding lines, which shortens the breeding cycle and helps produce new rice varieties. One of the main uses of double haploid lines is for the development of mapping populations for molecular analysis and mapping of DNA markers (Lu et al., 1996).

As mentioned previously in this chapter, rice has a series of species that can and have been used to address specific breeding problems such as resistance to pests and tolerance to abiotic stresses. However, one of the main limitations on the use of wild relatives in breeding programs is the lack of crossability between species due to chromosomal and genetic differences. One alternative to overcome these sexual barriers is to use embryo rescue and protoplast fusion, which are simple biotechnology techniques that have been used successfully in rice. Fertile *O. sativa* and *O. glaberrima* progenies were obtained through backcrossing and double haploid production by Jones et al. (1997). The NERICA varieties, mentioned elsewhere in this chapter, provide a good example of how these techniques were used to help address some specific breeding objectives.

Plant breeders want to use of molecular markers. Several different types of markers are being used in rice, among which one may find the following: restriction fragment length polymorphisms (RFLPs); randomly amplified polymorphic DNA markers (RAPDs); amplified fragment length polymorphisms (AFLPs); diversity array technology (DArT); simple sequence length polymorphisms (SSLPs) also known as SSRs or microsatellites; transposable elements (TEs); and/or single nucleotide polymorphisms (SNPs). If genes of interest are identified and linked to some of these markers they can be used to aid selection in a process known as MAS. Knowledge of gene and marker location, linkage strength, and stability is essential. Therefore, basic information is required and molecular linkage maps play a major role. Rice maps have been developed to that end; the first RFLP map was published in 1988 and was constructed at Cornell University by McCouch et al. (1988). Breeders are interested in transferring genes of interest from one parent to the other. This process can be facilitated by tagging such genes, which means identifying a tight linkage between the targeted gene and a molecular marker. By selecting the marker the breeder is indirectly selecting the trait of interest using MAS with the limitations of indirect selection. In the literature, there are examples of application of MAS in rice to aid backcrossing programs; in fact, theoretical studies have indicated that MAS can help reduce from 6 to 3 the number of backcrosses necessary to transfer a targeted gene (Frisch et al., 1999). In hybrid rice, Chen et al. (2000) transferred a resistance gene for bacterial leaf blight into a widely used parent. Huang et al. (1997b) successfully pyramided four bacterial blight resistance genes through MAS into a rice variety. Nevertheless, the application of this tool in conventional breeding programs has been limited. The rice genome is one of the most studied by scientists around the world. Arumuganathan and Earle (1991) described it as having 430 Mb. Chen et al. (2002) described it as 400 Mb once re-evaluated. Goff et al. (2002) sequenced the *japonicas* genome and Yu et al. (2002) did the same for the *indicas*. Having the rice genome sequenced brings a new and more important challenge that is to use this information to identify the biological functions of these genes and their interactions with other genes and environments. Therefore, the matching between genotyping and phenotyping plays an important role and the existence of breeding programs with excellent screening techniques and capable breeders are essential to capture the best advances of modern biotechnology and discard the rest. The introduction of an alien gene into rice by production genetically engineered rice allows breeders to target problems that without this technology it was not feasible. The golden rice is the most well known case of application of genetic engineering in rice in the 1990s. This specific project genetically engineered the provitamin A pathway into rice. Most cases, however, were related to the production of transgenic rice for resistance to diseases, insects, and abiotic stresses. Khush and Brar (2003) presented a table with 19 examples of transgenic rice.

In closing this section it is not redundant to reiterate the need of integrating biotechnology tools and certain successful techniques into the existing breeding programs. Decision-makers responsible for allocation of resources for research should not have two choices – biotechnology or plant breeding – but only one integrative way forward which is to ensure the integration of these activities towards producing improved varieties to solve farmers' demands.

8 Foundation Seed Production

Seed production is one of the key steps for the success of a variety. Thus, it has to be considered as an integral part of the breeding programs. The objective of this section is to indicate the linkages between foundation seed production and the phases involved in a breeding program. As an example of how this can be done I will use the past experience of the Embrapa Rice and Beans Centres in Brazil.

The segregating populations produced by the hundreds of crosses made every year are taken to the field and advanced through pedigree selection or modified bulk selection. As soon as the breeders identify potential breeding lines in the F_4, F_5, or F_6 generations they are included in the observational trials, which are planted across locations throughout the country. The best 50–100 lines are promoted to preliminary yielding tests planted across several locations also throughout the country. This is the stage when the breeders start considering lines for the foundation seed production. Headquarters seed specialists, together with breeders, select around 100 panicles to initiate the seed production process using the panicle-row process. As the breeding lines move from preliminary yield trials to advanced and regional yield trials the seed multiplication process advances from 2 to 3 kg of foundation seed to the required amount of high quality seeds necessary to attend the seed producers. This strategy requires high resource mobilization since it starts based on 20–30 breeding lines with potential to be released as varieties and ends on one or two released varieties. However, it speeds up the process of varietal release allowing to arriving at the moment of release with a large amount of high quality seeds. In addition it links breeders and seed specialists in the early stages of the seed production process ensuring the high quality of the final product.

9 Rice Breeding Capacity Around the World

FAO, in collaboration with CGIAR centers and other stakeholders has been assessing the national plant breeding and related biotechnology capacity, as proposed in the Global Plant of Action (FAO, 1996) of the International Treaty on Plant Genetic

Resources for Food and Agriculture (FAO, 2002). The mechanism to gather information on countries' capacity is a survey focusing on several breeding and biotechnology issues. For this chapter the analysis will consider the number of full-time equivalent[1] (FTE) plant breeders[2] available in all private and public institutions in each surveyed country and the resource allocation per crop (rice in this case). The organizations were asked to provide the total number FTE breeders and the percentage of the total resources that was allocated to rice breeding activities. The numbers in Table 2 were obtained by multiplying the total number of FTEs by the percentage of resource allocations to rice. Preliminary survey results for a sample of countries in Central Asia were published by Guimarães et al. (2006a) and in Africa by Guimarães et al. (2006b). These results covered all crops; however, in this chapter the focus will be on rice only (Table 2). As one might expect, because of the importance of the rice crop for the Asian population, the largest number of FTE breeders was observed in Asia, even though that was the region with the least number of countries surveyed (only four countries were sampled). The results indicated that there were 84.9 rice breeders, representing 21.6% of the total FTE breeders in the region, the highest percentage of all the regions. The next highest result was found for Latin America, which had 46 FTE rice breeders, some 17.1% of the total number of FTE breeders in the region. As rice is the staple food for the majority of the countries in these two regions, these results reflect the importance that the national programs give to the development of improved varieties.

The total number of rice breeders in Brazil represents 50% of the number of rice breeders present in all seven countries sampled in Latin America. Embrapa has two-thirds of the total number of FTE rice breeders in the country. The state organizations follow with much lower numbers while the private sector has only two breeders working in the country. Considering the whole country's breeding capacity, rice represents only 4.4% (Table 2).

In Africa, for many countries, mainly in West Africa, rice is the staple food and one of the most important sources of calories. The results in Table 2 reflect its importance by the total number of FTE rice breeders working in Africa. An important part of the 28.4 FTEs rice breeders are in West Africa. The West African countries all have, with the exception of Niger and Senegal, more than two breeders working in national rice breeding programs. However, looking at the total number of FTE breeders in Africa rice represents a very small fraction (3.6%). The potential that rice has in West African countries is due to its increasing popularity in consumption patterns although the gap between supply and demand is still significant. Nevertheless, some countries have invested in rice breeding to contribute to the growth of local rice production (Oladele and Sakagami, 2004).

[1]A Full Time Equivalent (FTE) is the work done by a person who has any responsibility linked to plant breeding (genetic enhancement, line development, line evaluation, or genetic studies) during one year (365 days).

[2]The survey considered as plant breeders all scientific personnel with a plant breeding degree and also the ones directly involved in plant breeding activities.

Table 2 Distributions of the number of Full Time rice breeders' equivalent of selected organizations obtained through a survey carried out in Brazil in 2005

Organization	Rice breeders	Breeders for other crops	Breeders (#)	Rice breeders (%)
Brazil	20.5	446.5	467.0	4.4
Embrapa[a]	13.0	201.0	214.0	6.1
State Institutions[b]	4.0	123.0	127.0	3.1
Universities[b]	1.5	89.5	91.0	1.6
Private companies[b]	2.0	33.0	35.0	5.7
Latin America[c]	46	222.6	268.6	17.1
Africa[d]	28.4	770.6	799.0	3.6
Near East and North Africa[e]	6.1	595.5	601.6	1.0
Central Asia and Caucasus[f]	16.9	1,240.0	1,257.0	1.3
Asia[g]	84.9	307.5	392.4	21.6
East Europe[h]	7.7	1,022.0	1,030.0	0.7

[a]Embrapa is the largest public research organization in the country with 37 research centers (www.embrapa.br)

[b]The sample included 8 state institutions, 20 universities, and 7 private companies distributed through out the whole country

[c]The eight countries sampled were Argentina, Bolivia, Costa Rica, Dominican Republic, Ecuador, Nicaragua, Uruguay, and Venezuela. All data refer to 2004 except for Venezuela that has data for 2001

[d]The 15 countries sampled were Angola, Cameroon, Ethiopia, Ghana, Kenya, Malawi, Mali, Mozambique, Niger, Nigeria, Senegal, Sierra Leone, Uganda, Zambia, and Zimbabwe. All countries refer to 2001, with the exception of Ethiopia and Sierra Leone (2004) and Angola and Cameroon (2003)

[e]The seven countries sampled were Algeria, Sudan, Jordan, Lebanon, Oman, Tunisia, and Turkey. All countries have data for 2004 except for Algeria and Sudan (2001)

[f]The seven countries sampled were Kazakhstan, Kyrgyzstan, Tajikistan, Uzbekistan, Armenia, Georgia, and Azerbaijan. All data are for 2004

[g]The four countries sampled were Bangladesh, The Philippines, Thailand, and Sri Lanka. All data are for 2004

[h]The five countries sampled were Albania, Bulgaria, Macedonia, Moldova, and Slovak Republic. All data are for 2004

Rice is not an important crop in Central Asia and Caucasus which is why only 1.3% of the total FTE breeders in the region are working with the crop. Kazakhstan, the largest producer in the region, cultivated only 83,000 ha in 2006 (FAOSTAT, 2006) and had 12.1 FTE rice breeders in 2004. The smallest numbers were found for Eastern Europe, and Near East and North Africa regions, yet they have a sizeable total number of FTE breeders. Rice production does not have the same high priority in these regions as it does in Asia or Latin America. Therefore, the resources allocations for rice breeding activities can be expected to be limited compared with other crops such as wheat and maize.

In conclusion, different regions allocate their breeding resources according to their crop priorities. Moreover, rice breeders are widely distributed across all regions. What is more important, the aforementioned genetic diversity and the wise choice of germplasm help bring about food security in countries where rice is a staple. This, however, depends on applying creativity to different breeding approaches and the wise use of biotechnology tools.

References

Alvarado-A, J.R. (1997) Mejoramiento de riego en Chile y utilización de las selección recurrente. In: Guimarães, E.P. (Ed.) *Selección recurrente en arroz*. Centro Internacional de Agricultura Tropical, Cali, Colombia, pp. 117–123.

Arumuganathan, K. and Earle, E.D. (1991) Nuclear DNA content of some important species. Plant Mol. Biol. Rep. 9, 208–218.

Badan, A.C de., Guimarães, E.P. and Ramis, C. (2005) Genetic gain for resistance to blast in a rice population. In: Guimarães, E.P. (Ed.) *Population improvement, A way of exploiting rice genetic resources in Latin America*. Food and Agriculture Organization of the United Nations (FAO), Rome, Italy, pp. 299–329.

Brondani, C., Rangel, P.H.N., Brondani, R.P.V. and Ferreira, M.E. (2002) QTL, mapping and introgression of yield-related traits from *Oryza glumaepatula* to cultivated rice (*Oryza sativa*) using microsatellite markers. Theor. Appl. Genet. 104, 1192–1203.

Castro, E. Da M. de., Morais, O.P. de., Sant'Ana, E.P., Breseghello, F. and Neto, F.P. de M. (2000) Mejoramiento poblacional de arroz de tierras altas en Brasil. In: Guimarães, E.P. (Ed.) *Avances en el mejoramiento poblacional en arroz*. Embrapa Arroz e Feijao, Santo Antonio de Goias, Brasil, pp. 221–240.

Chatel, M. and Guimarães, E.P. (1997) *Recurrent selection in rice, using a male-sterile gene*. Centro Internacional de Agricultura Tropical and Centre de coopération internationale en recherche agronomique pour le développement, Cali, Colombia. pp. 1–70.

Chen, S., Lin, X.H., Xu, C.G. and Zhang, Q.F. (2000) Improvement of bacterial blight resistance of 'Minghui 63', an elite line of hybrid rice, by molecular marker-assisted selection. Crop Sci. 40, 239–244.

Chen, M., Presting, G., Barbazuk, W.B., Goicoechea, J.L., Blackmon, B., Frang, G., Kim, H., Frisch, D., Yu, Y., Sun, S., Higingbottom, S., Phimphilai, J., Phimphilai, D., Thurmond, S., Gaudette, B., Li, P., Lui, J., Hatfield, J., Main, D., Farrar, K., Henderson, C., Barnett, L., Costa, R., Williams, B., Walser, S., Atkins, M., Hall, C., Budiman, M.A., Tomkins, J.P., Luo, M., Bancroft, I., Salse, J., Regad, F., Mohapatra, T., Singh, N.K., Tyagi, A.K., Soderlund, C., Dean, R.A. and Wing, R.A. (2002) An integrated physical and genetic map of rice genome. Plant Cell 14, 537–545.

Chen, X., Liu, X., Wu, D. and Shu, Q.Y. (2006) Recent progress of rice mutation breeding and germplasm enhancement in China. Plant Mutat. Rep. 1(1), 4–6.

Courtois, B., Nelson, R. and Roumen, E. (1997) Creacion de un acervo genetico para mejorar la resistancia parcial a Piricularia en el arroz de secanano, mediante la seleccion recurrente. In: Guimarães, E.P. (Ed.) *Selección recurrente en arroz*. Centro Internacional de Agricultura Tropical, Cali, Colombia, pp. 189–202.

Cuevas-Pérez, F.E., Guimarães, E.P., Berrio, L.E. and Gonzalez, D.I. (1992) Genetic base of irrigated rice in Latin America and the Caribbean, 1971 to 1989. Crop Sci. 32, 1054–1059.

Dilday, R.H. (1990) Contribution of ancestral lines in the development of new cultivars of rice. Crop Sci. 32, 1054–1058.

Dingkuhn, M., Penning De Vries, F.W.T., De Datta, S.K. and van Laar, H.H. (1991) Concepts for a new plant type for direct seeded flooded tropical rice. In: Direct seeded flooded rice in the tropics. International Rice Research Institute, Manila, Philippines, pp. 17–38.

Donald, C.M. (1968) The breeding of crop ideotypes. Euphytica 17(3), 385–403.

Dudley, J.W. and Lambert, R.J. (2004) 100 generations of selection for oil and protein in corn. Plant Breed. Rev. 24(1), 79–110.
FAO (1996) Global Plan of Action of the Conservation and Sustainable Utilization of the Plant Genetic Resources for Food and Agriculture. Rome, Italy.
FAO (2002) International Treaty on Plant Genetic Resources for Food and Agriculture. Rome, Italy.
FAOSTAT (2006) FAO statistical database on food and agriculture. Available at http://faostat.external.fao.org/default. Consulted on 05 December 2006.
Frisch, M., Bohn, M. and Melchinger, A.E. (1999) Comparation of selection strategies for marker-assisted backcrossing of a gene. Crop Sci. 39, 1295–1301.
Fujimaki, H. (1979) Recurrent selection by using male sterility for rice improvement. Jpn. Agric. Res. Qua. 13, 153–156.
Glaszmann, J.C. (1987) Isozymes and classification of Asian rice varieties. Theor. Appl. Genet. 74, 21–30.
Glaszmann, J.C. and Arraudeau, M. (1986) Rice plant type variation, "japonica" – "javanica" relationship. Rice Genet. Newslett. 3, 41–43.
Goff, S.A., Ricke, D., Lan, T.H., Presting, G., Wang, R., Dunn, M., Glazebrook, J., Sessions, A., Oeller, P., Varma, H., Hadley, D., Hutchison, D., Martin, C., Katagiri, F., Lange, B.M., Moughamer, T., Xia, Y., Budworth, P., Zhong, J., Miguel, T., Paszkowski, U., Zhang, S., Colbert, M., Sun, W.L., Chen, L., Cooper, B., Park, S., Wood, T.C., Mao, L., Quail, P., Wing, R., Dean, R., Yu, Y., Zharkikh, A., Shen, R., Sahasrabudhe, S., Thomas, A., Cannings, R., Gutin, A., Pruss, D., Reid, J., Tavtigian, S., Mitchell, J., Eldredge, G., Scholl, T., Miller, R.M., Bhatnagar, S., Adey, N., Rubano, T., Tusneem, N., Robinson, R., Feldhaus, J., Macalma, T., Oliphant, A. and Briggs, S. (2002) A draft sequence of the rice genome (*Oryza sativa* L. ssp. *japonica*). Science 296, 92–100.
Graterol, M.E.J. (2000) Caracterización de poblaciones e introducción de variabilidad para iniciar un programa de mejoramiento poblacional de arroz en Venezuela. In: Guimarães, E.P. (Ed.) *Avances en el mejoramiento poblacional en arroz*. Embrapa Arroz e Feijao, Santo Antonio de Goias, Brasil, pp. 87–103.
GRUMEGA [Grupo de Mejoramiento Genético Avanzado en Arroz] (2006a) Catalogue Registration of Gene Pools and Populations, for Rice Improvement by Recurrent Selection. Available at http://iserver.ciat.cgiar.org/grumega/tiki-download_wiki_attachment.php?attId = 56. Consulted on 6 December 2006.
GRUMEGA [Grupo de Mejoramiento Genético Avanzado en Arroz] (2006b) Bolivia, "Esperanza" (2006) Nueva variedad de arroz para condiciones de secano. Available at http://iserver.ciat.cgiar.org/grumega/tiki-download_wiki_attachment.php?attId = 18. Consulted on 6 December 2006.
GRUMEGA [Grupo de Mejoramiento Genético Avanzado en Arroz] (2006c) Chile, "R Quila 23" (2006) Available at http://iserver.ciat.cgiar.org/grumega/tiki-download_wiki_attachment.php?attId = 16. Consulted on 6 December 2006.
GRUMEGA [Grupo de Mejoramiento Genético Avanzado en Arroz] (2006d) Cultivar de arroz irrigado da EPAGRI, SCS 113-Tio Taka. Available at http://iserver.ciat.cgiar.org/grumega/tiki-download_wiki_attachment.php?attId = 58. Consulted on 6 December 2006.
Guimarães, E.P. (1997) *Selección recurrente en arroz.* Centro Internacional de Agricultura Tropical, Cali, Colombia. pp. 1–240.
Guimarães, E.P. (1999) Hibridacao em arroz. In: Borem, A. (Ed.) *Hibridação artificial de plantas.* Universidade Federal de Vicosa (UFV), Vicosa, Brasil, pp. 101–120.
Guimarães, E.P. (2000) *Avances en el mejoramiento poblacional en arroz.* Embrapa Arroz e Feijao, Santo Antonio de Goias, Brasil. pp. 1–311.
Guimarães, E.P. (2002) Genetic diversity of rice production in Brazil. In: Nguyen, V.N. (Ed.) *Genetic diversity in rice production, Case studies from Brazil, India and Nigeria.* Food and Agriculture Organization of the United Nations (FAO), Rome, Italy, pp. 11–35.

Guimarães, E.P. (2005) *Population improvement, A way of exploiting rice genetic resources in Latin America*. Food and Agriculture Organization of the United Nations (FAO), Rome, Italy. pp. 1–350.

Guimarães, E.P. and Correa-Victoria, F. (2000) Utilizacion de la seleccion recurrente para desarrollar resistencia a *Pyricularria grisea* Sacc. en arroz. In: Guimarães, E.P. (Ed.) *Avances en el mejoramiento poblacional en arroz*. Embrapa Arroz e Feijao, Santo Antonio de Goias, Brasil, pp. 165–175.

Guimarães, E.P., Borrero, J. and Ospina-Rey, Y. (1996) Genetic diversity of upland rice germplasm distributed in Latin America. Pesq. Agropec. Bras. 31(3), 187–194.

Guimarães, E.P., Bedoshvili, D., Morgounov, A., Baboev, S., Iskakov, A., Muninjanov, H., Kueneman, E. and Paganini, M. (2006a) Plant breeding and related biotechnology competence in Central Asia and recommendations to strengthen regional capacity. Agromeridian Theor. Appl. Agric. Res. J. 2, 137–143.

Guimarães, E.P., Kueneman, E. and Carena, M.J. (2006b) Assessment of national plant breeding and biotechnology capacity in Africa and recommendations for future capacity building. Hort. Sci. 41, 50–52.

Guohui, M. and Longping, Y. (2003) Hybrid rice achievements and development in China. In: Virmani, S.S., Mao, C.X. and Hardy, B. (Eds.) *Hybrid rice for food security, poverty alleviation, and environmental protection*. International Rice Research Institute, Manila, Philippines, pp. 247–256.

Harushima, Y., Nakagahra, M., Yano, M., Sasaki, T. and Murata, N. (2002) Diverse variation of reproductive barriers in three intraspecific rice crosses. Genetics 160, 313–322.

He, Y., Li, X., Zhang, J., Jiang, G., Liu, S., Chen, S., Tu, J., Xu, C. and Zhang, Q. (2004) Gene pyramiding to improve hybrid rice by molecular marker techniques. International Crop Science Congress Brisbane, Australia, 26 Sep–1 Oct. http://www.cropscience.org.au/icsc2004/poster/3/4/4/1041_heyuqing.htm. Consulted on 06 April 2007.

Hernaiz-L, S.I., Alvarado, J.R., Chatel, M., Castillo, D. and Ospina, Y. (2004) Improving irrigated rice populations for temperate climate in Chile. In: Guimarães, E.P. (Ed.) *Population improvement, A way of exploiting rice genetic resources in Latin America*. Food and Agriculture Organization of the United Nations (FAO), Rome, Italy, pp. 129–143.

Hoan, N.T. and Nghia, N.H. (2003) Hybrid rice development and use in Vietnam. In: Virmani, S. S., Mao, C.X. and Hardy, B. (Eds.) *Hybrid rice for food security, poverty alleviation, and environmental protection*. International Rice Research Institute, Manila, Philippines, pp. 357–371.

Horie, T. (2001) Increasing yield potential in irrigated rice: breaking the yield barrier. In: Peng, S. and Hardy, B. (Eds.) *Rice research for food security and poverty alleviation*. International Rice Research Institute, Manila, Philippines, pp. 3–25.

Huang, N., Angeles, E.R., Domingo, J., Magpantay, G., Singh, S., Zhang, G., Kumaravadivel, N., Bennet, J. and Khush, G.S. (1997a) Pyramiding of bacterial blight resistance genes in rice: marker assisted selection using RFLP and PCR. Theor. Appl. Genet. 95, 313–320.

Huang, N., Parco, A., Mew, T., Magpantay, G., McCouch, S., Guiderdoni, E., Xu, J., Subudhi, P., Angeles, E.R. and Khush, G.S. (1997b) RFLP mapping of isozymes, RAPD and QTLs for grain shape, brown planthopper resistance in a doubled haploid rice population. Molecular Breeding 3(2), 105–113.

Hull, F.H. (1945) Recurrent selection for specific combining ability in corn. J. Am. Soc. Agron. 37 (2), 134–145.

Ismachin, M. and Sobrizal (2006) A significant contribution of mutation techniques to rice breeding in Indonesia. Plant Mutat. Rep. 1(1), 18–22.

IRRI (2002) *Rice almanac, source book for the most important economic activity on earth. Third edition.* Maclean, J.L., Dawe, D.C., Hardy, B., and Hettel, G.P. (Eds.) International Rice Research Institute, Manila, Philippines. pp. 1–253.

Jackson, M.T., Loresto, G.C., Appa Rao, S., Jones, M., Guimarães, E.P. and Ng, N.Q. (1997) Rice. In: Fuccillo, D., Sears, L. and Stapleton, P. (Eds.) *Biodiversity in trust, conservation and use of plant genetic resources in CGIAR Centres*. Cambridge University Press, Cambridge, UK, pp. 273–291.

Jackson, M.T. and Lettington, R.J.L. (2003) Conservation and use of rice germplasm, an evolving paradigm under the International Treaty on Plant Genetic Resources for Food and Agriculture. In: *Sustainable rice production for food security. Proceedings of the 20th session of the International Rice Commission. Bangkok, Thailand, 23–26 July 2002*, pp. 75–88.

Jones, M., Dingkuhn, M., Aluko, G.K. and Semon, M. (1997) Interspecific *Oryza sativa* L. x *O. glaberrima* Steud. progenies in upland rice improvement. Euphytica 94(2), 237–246.

Jones, M. and Wopereis-Pura, M. (2001) *History of Nerica and PVS*. Internal report, West Africa Rice Development Association (WARDA); Bouake, Côte d'Ivoire.

Julfiquar, A.W., Jamil Hasan, M., Azad, A.K., Anwar Hossain, M. and Virmani, S.S. (2003) Hybrid rice research and development in Bangladesh. In: Virmani, S.S., Mao, C.X. and Hardy, B. (Eds.) *Hybrid rice for food security, poverty alleviation, and environmental protection*. International Rice Research Institute, Manila, Philippines, pp. 235–245.

Junjian, N., Colowit, P.M. and Mackill, D.J. (2002) Evaluation of genetic diversity in rice subspecies using microsatellite markers. Crop Sci. 42(2), 601–607.

Kaneda, C. (1985) Development of very high-yielding rice varieties. Farming Japan 19, 25–29.

Khatiwada, S.P., Senadhira, D., Carpena, A.L., Zeigler, R.S. and Fernandez, P.G. (1996) Variability and genetics of tolerance for aluminum toxicity in rice (*Oryza sativa* L.) Theor. Appl. Genet. 93(5–6), 738–744.

Khush, G.S. (1977) Disease and insect resistance in rice. Adv. Agron. 29, 265–341.

Khush, G.S. (1994) Increasing the genetic yield potential of rice, prospects and approaches. Int. Rice Comm. Newsl. 43, 1–8.

Khush, G.S. and Brar, D.S. (2003) Biotechnology for rice breeding: progress and impact. In: *Sustainable rice production for food security. Proceedings of the 20th session of the International Rice Commission, Bangkok, Thailand, 23–26 July 2002*, pp. 41–58.

Khush, G.S. and Peng, S. (1996) Breaking the yield frontier of rice. In: Reynolds, M.P. (Eds.) *Increasing yield potential in wheat: Breaking the barriers. Proceedings of the workshop held on 26–28 March 1996, Ciudad Obregon, Sonora, Mexico*. International Maize and Wheat Improvement Center, El Batan, Mexico. pp. 36–51.

Khush, G.S., Singh, R.J., Sur, S.C. and Librojo, A.L. (1984) Primary trisomics of rice, origin, morphology, cytology and use in linkage mapping. Genetics 107, 141–163.

Khush, G.S., Bacalangco, E. and Ogawa, T. (1990) A new gene for resistance to bacterial blight from *O. longistaminata*. Rice Genet. Newsl. 7, 121–122.

Khush, G.S., Coffman, W.R. and Beachell, H.M. (2001) The history of rice breeding: IRRI's contribution. In: Rockwood, W.G. (Ed.) *Rice research and production in the 21st century: Symposium honouring Robert F. Chandler Jr*. International Rice Research Institute, Manila, Philippines, pp. 117–135.

Latha, R., Thiyaqarajan, K. and Senthilvel, S. (2004) Genetics, fertility behaviour and molecular analysis of a new TGMS line, TS6, in rice. Plant Breed. 123(3), 235–240.

Lu, C., Shen, L., Tan, Z., Xu, Y., He, P., Chen, Y., and Zhu, L. (1996) Comparative mapping of QTLs for agronomic traits of rice across environments using a doubled haploid population. Theor. Appl. Genet. 93(8), 1211–1217.

Mao, C.X. (2001) Improving seed production to speed up the global commercialization of hybrid rice. In: Peng, S. and Hardy, B. (Eds.) *Rice research for food security and poverty alleviation*. International Rice Research Institute, Manila, Philippines, pp. 221–229.

Marassi, J.E., Marassi, M.A., Chatel, M. and Borrero, J. (2000) Desarrollo de poblaciones de arroz en Argentina. In: Guimarães, E.P. (Ed.) *Avances en el mejoramiento poblacional en arroz*. Embrapa Arroz e Feijao, Santo Antonio de Goias, Brasil, pp. 173–186.

Marassi, M.A., Marassi, J.E., Chatel, M. and Ospina, Y. (2004) Exploiting the genetic resources of rice in Argentina through population improvement. In: Guimarães, E.P. (Ed.) *Population improvement, A way of exploiting rice genetic resources in Latin America*. Food and Agriculture Organization of the United Nations (FAO), Rome, Italy, pp. 113–127.

Maluszynski, M., Ahloowalia, B., Ashri, A., Nichterlein, K. and Zanten, L.V. (1998) Induced mutation in rice breeding and germplasm enhancement. In: *Assessment and orientation towards the 21st century. Proceedings of the 19th session of the International Rice Commission, Cairo, Egypt, 7–9 September 1998*, pp. 194–204.

McCouch, S.R., Kochert, G., Yu, Z.H., Wang, Z.Y., Khush, G.S., Coffaman, D.R. and Tanksley, S. D. (1988) Molecular mapping of rice chromosomes. Theor. Appl. Genet. 76, 815–829.

Mishra, B. (2002) Varietal improvement for rice production in India. In: Nguyen, V.N. (Ed.) *Genetic diversity in rice production, Case studies from Brazil, India and Nigeria*. Food and Agriculture Organization of the United Nations (FAO), Rome, Italy, pp. 37–91.

Mishra, B., Viraktamath, B.C., Ilyas Ahmed, M., Ramesha, M.S. and Vijayakumar, C.H.M. (2003) Hybrid rice development and use in India. In: Virmani, S.S., Mao, C.X. and Hardy, B. (Eds.) *Hybrid rice for food security, poverty alleviation, and environmental protection*. International Rice Research Institute, Manila, Philippines, pp. 265–286.

Moncada, P., Martínez, C.P., Borrero, J., Gauch, H. Jr., Guimarães, E.P., Tohme, J. and McCouch, S.R. (2001) Quantitative trait loci for yield and yield components in an *Oryza sativa* x *Oryza rufipogon* BC_2F_2 population evaluated in an upland environment. Theor. Appl. Genet. 102, 41–52.

Montalban, R., Destro, D., Silva, E.F. and Montano, J.C. (1998) Genetic base of Brazilian upland rice cultivars. J. Genet. Breed. 52, 203–209.

Morinaga, T. (1954) Classification of rice varieties on the basis of affinity. In: *International Rice Commission. Working Party on Rice Breeding Rep. 5th Meeting*.

Morishima, H. (1984) Species relationships and the search for ancestros. In: Tsunoda, S. and Takahashi, N. (Eds.) *Biology of rice*. Japan Sci. Soc. Press. Tokio, Japan, pp. 3–30.

Morishima, H. and Oka, H.I. (1981) Phylogenetic differentiation of cultivated rice. XXVII Numerical evaluation of the Indica – Japonica differentiation. JPN. J. Breed. 31, 402–413.

Mou, T.M., Xing-Gui, L., Hoan, N.T. and Virmani, S.S. (2003) Two-line hybrid rice breeding in and outside China. In: Virmani, S.S., Mao, C.X. and Hardy, B. (Eds.) *Hybrid rice for food security, poverty alleviation, and environmental protection*. International Rice Research Institute, Manila, Philippines, pp. 31–52.

Muralidharan, K., Prasad, G.S.V. and Rao, C.S. (2002) Yield performance of rice genotypes in international multi-environment trials during 1976–1997. Curr. Sci. 83(5), 610–619.

Murchie, E.H., Hubbart, S., Peng, S. and Horton, P. (2001) Raising rice yield potential by manipulating photosynthesis: improvement of radiation conversion efficiency. In: Peng, S. and Hardy, B. (Eds.) *Rice research for food security and poverty alleviation*. International Rice Research Institute, Manila, Philippines, pp. 119–128.

Oka, H.I. (1958) Intervarietal variation and classification of cultivated rice. Indian J. Genet. Plant Breed. 18, 79–89.

Oka, H.I. (1964) Pattern of interspecific relationships and evolutionary dynamics in *Oryza*. In: *Rice genetics and cytogenetics*. Elsevier, Amsterdam, Netherlands. pp. 71–90.

Oladele, O.I. and Sakagami, J.I. (2004) Impact of technology innovation on rice yield gap in Asia and West Africa – Technology transfer issues. In: *Rural Poverty Reduction though Research for Development and Transformation, Proceeding of the Deutscher Tropentag, October 5–7, 2004*. Humboldt-Universität, Berlin, Germany. p. 184.

Peng, S. and Khush, G.S. (2003) Four decades of breeding for varietal improvement of irrigated lowland rice in the International Rice Research Institute. Plant Prod. Sci. 6(3), 157–164.

Peng, S., Cassman, K.G., Virmani, S.S., Sheehy, J. and Khush, G.S. (2005) Yield potential of Tropical rice since the release of IR8 and the challenge of increasing rice yield potential. Crop Sci. 39, 1552–1559.

Pérez-Polanco, R., Chatel, M. and Guimarães, E.P. (2000) Mejoramiento poblacional del arroz en Cuba, Situacion actual y perspectives. In: Guimarães, E.P. (Ed.) *Avances en el mejoramiento poblacional en arroz*. Embrapa Arroz e Feijao, Santo Antonio de Goias, Brasil, pp. 131–144.

Picard, E., Hours, C., Grégoire, S., Phan, T.H., and Meunier, J.P. (2004) Significant improvement of androgenetic haploid and double haploid induction from wheat plants treated with a chemical hybridization agent. Theor. Appl. Genet. 74(3), 289–297.

Rai, M. (2003) Genetic diversity in rice production, past contribution and the potential of utilization for sustainable rice production. In: *Sustainable rice production for food security. Proceedings of the 20th session of the International Rice Commission, Bangkok, Thailand, 23–26 July 2002*, pp. 89–115.

Rangel, P.H.N. and Neves, P.C.F. (1997) Selección recurrente aplicada al arroz de riego en Brasil. In: Guimarães, E.P. (Ed.) *Selección recurrente en arroz*. Centro Internacional de Agricultura Tropical, Cali, Colombia, pp. 79–97.

Rangel, P.H.N., Guimarães, E.P. and Neves, P.C.F. (1996) Base genetica das cultivares de arroz (*Oryza sativa* L.) irrigado do Brasil. Pesq. Agropec. Bras. 31(5), 349–357.

Rangel, P.H.N., Zimmermann, F.J.P. and Fagundes, P.R.R. (2000) Mejoramiento poblacional del arroz de riego en Brasil. In: Guimarães, E.P. (Ed.) *Avances en el mejoramiento poblacional en arroz*. Embrapa Arroz e Feijao, Santo Antonio de Goias, Brasil, pp. 65–85.

Rangel, P.H.N., Cordeiro, A.C.C., Lopes, S.I.G., Morais, O.P. de, Brondani, C., Brondani, R.P.V., Yokoyama, S., Schiocchet, M., Bacha, R. and Ishy, T. (2005) Advances in population improvement of irrigated rice in Brazil. In: Guimarães, E.P. (Ed.) *Population improvement, A way of exploiting rice genetic resources in Latin America*. Food and Agriculture Organization of the United Nations (FAO), Rome, Italy, pp. 145–186.

Santos, P.G., Soares, P.C., Soares, A.A., Morais, O.P. and Cornelio, V.M.O. (1997) Estimativa do progresso genetico do programa de mejoramento de arroz irrigado desenvolvido em Minas Gerais no periodo 1974 a 1996. In: *Anais da XXII Reuniao da Cultura do Arroz Irrigado, 1997*. Balneario Camboriu, Santa Catarina, Brasil. pp. 27–20.

Sarkarung, S. (1991) A simplified crossing method for rice breeding. Fondo Latinoamericano de Arroz de Riego (FLAR) and CIAT, Cali, Colombia. pp. 1–32.

Second, G. (1982) Origin of the genetic diversity of cultivated rice (*Oryza* spp): study of the polymorphism scored at 40 isozyme loci. Jpn. J. Genet. 57, 25–57.

Sheehy, J.E., Mitchell, P.L., Dionora, J. and Ferrer, A. (2001) What governs ceiling yield and its rate of attainment? In: Peng, S. and Hardy, B. (Eds.) *Rice research for food security and poverty alleviation*. International Rice Research Institute, Manila, Philippines, pp. 69–77.

Shih-Cheng, L. and Loung Ping, Y. (1980) Hybrid rice breeding in China. In: *Innovative approaches to rice breeding: Selected papers form the 1979 International Rice Research Conference*. International Rice Research Institute, Los Baños, Philippines, pp. 35–51.

Singh, R.J. and Ikehashi, H.I. (1981) Monogenic male sterility in rice, induction, identification and inheritance. Crop Sci. 21(2), 286–289.

Sprague, G.F. and Tatum, L.A. (1941) General vs. specific combining ability in single crosses of corn. J. Am. Soc. Agron. 34, 923–932.

Sun, C.Q., Wang, X.K., Li, Z.C., Yoshimura, A. and Iwata, N. (2001) Comparison of the genetic diversity of common wild rice (*Oryza rufipogon* Griff.) and cultivated rice (*O. sativa* L.) using RFLP markers. Theor. Appl. Genet. 102, 57–162.

Suwarno, Nuswantoro, M.W., Munarso, Y.P. and Direja, M. (2003) Hybrid rice development and use in Indonesia. In: Virmani, S.S., Mao, C.X. and Hardy, B. (Eds.) *Hybrid rice for food security, poverty alleviation, and environmental protection*. International Rice Research Institute, Manila, Philippines, pp. 287–296.

Taillebois, J. and Guimarães, E.P. (1989) CAN-IRAT5 upland rice population. Int. Rice Res. Newsl. 14(3), 8.

Taillebois, J. and Castro, E. da M. de (1986) A new crossing technique. Int. Rice Res. Newsl. 11 (6), 6.

Tan, X.L., Vanavichit, A., Amornsilpa, S. and Trangoorung, S. (1998) Genetic analysis of rice CMS-WA fertility restotation based on QTL mapping. Theor. Appl. Genet. 97(5/6), 994–999.

Tran, D.Q., Dao, T.T.B., Nguyen, H.D., Lam, Q.D., Bui, H.T., Nguyen, V.B., Nguyen, V.X., Le, V.N., Do, H.A. and Phan, P. (2006) Rice mutation breeding in Institute of Agricultural Genetics, Viet Nam. Plant Mutati. Rep. 1(1), 47–49.

Vaughan, D.A. (1994) The wild relatives of rice, A genetic resources handbook. International Rice Research Institute, Manila, Philippines. pp. 1–137.

Virmani, S.S. (2003) Advances in hybrid rice research and development in the tropics. In: Virmani, S.S., Mao, C.X. and Hardy, B. (Eds.) *Hybrid rice for food security, poverty alleviation, and environmental protection*. International Rice Research Institute, Manila, Philippines, pp. 7–20.

Virmani, S.S., Viraktamath, B.C., Casal, C.L., Toledo, R.S., Lopez, M.T. and Manalo, J.O. (1997) Hybrid rice breeding manual. International Rice Research Institute, Manila, Philippines. pp. 1–151.

Virmani, S.S., Mao, C.X., Toledo, R.S., Hossain, M. and Janaiah, A. (2001) Hybrid rice seed production technology for improving seed industries and rural employment opportunities in Asia. Paper presented at the International Workshop on Seed and Seeding Sciences and Technology. Taiwan, China, 10–17 June 2001.

Wang, L.Q. (1992) Advances in plant mutation breeding in China: a full analysis. Bull. Nucl. Agric. Sci. 13, 282–295.

Wang, X.W., Lai, J.R., Fan, L. and Zhang, R.B. (1996) Effects of recurrent selection on populations of various generations in wheat by using the Tai Gu single dominant male-sterile gene. J. Agric. Sci. 126(4), 397–402.

WARDA [West Africa Rice Development Association] (2003) Assessing the impact of NERICA rice varieties: not just surveys and simple mathematics. Available at http://www.warda.cgiar.org/publications/AR2002–03/assessing%20the%20impact.pdf. Consulted on 7 December 2006.

Wenfu, C., Zhenjin, X., Longbu, Z. and Shouren, Y. (2001) Development of the new rice plant type and advances in research on breeding for super high yield. In: Peng, S. and Hardy, B. (Eds.) *Rice research for food security and poverty alleviation*. International Rice Research Institute, Manila, Philippines, pp. 43–50.

Werner, B.K. and Wilcox, J.R. (2004) Recurrent selection for yield in Glycine max using genetic male-sterility. Euphytica 50(1), 19–26.

Xiao, J., Li, J., Grandillo, S., Ahn, S.N., Yuan, L., Tanksley, S.D. and McCouch, S.R. (1998) Identification of trait-improving quantitative trait loci alleles from a wild rice relative, *Oryza rufipogon*. Genetics 150, 899–909.

Yang, S.R., Zhang, L.B. and Chen, W.F. (1996) Theories and methods of rice breeding for maximum yield. Acta Agron. Sin. 22(3), 295–304.

Yamaguchi, Y., Ikeda, R., Hirasawa, H., Minami, M. and Ujikara, A. (1997) Linkage analysis of termosensitive genetic male sterility gene, *tms*$_2$, in rice (*Oryza sativa* L.). Breed. Sci. 47, 371–373.

Yu, J., Hu, S., Wang, J., Wong, K.S., Li, S., Liu, B., Deng, Y., Dai, L., Zhou, Y., Zhang, X., Cao, M., Liu, J., Sun, J., Tang, J., Chen, Y., Huang, X., Lin, W., Ye, C., Tong, W., Cong, L., Geng, J., Han, Y., Li, L., Li, W., Hu, G., Huang, X., Li, W., Li, J., Liu, Z., Li, L., Liu, J., Qi, Q., Liu, J., Li, L., Li, T., Wang, X., Lu, H., Wu, T., Zhu, M., Ni, P., Han, H., Dong, W., Ren, X., Feng, X., Cui, P., Li, X., Wang, H., Xu, X., Zhai, W., Xu, Z., Zhang, J., He, S., Zhang, J., Xu, J., Zhang, K., Zheng, X., Dong, J., Zeng, W., Tao, L., Ye, J., Tan, J., Ren, X., Chen, X., He, J., Liu, D., Tian, W., Tian, C., Xia, H., Bao, Q., Li, G., Gao, H., Cao, T., Wang, J., Zhao, W., Li, P., Chen, W., Wang, X., Zhang, Y., Hu, J., Wang, J., Liu, S., Yang, J., Zhang, G., Xiong, Y., Li, Z., Mao, L., Zhou, C., Zhu, Z., Chen, R., Hao, B., Zheng, W., Chen, S., Guo, W., Li, G., Liu, S., Tao, M., Wang, J., Zhu, L., Yuan, L., and Yang, H. (2002) A draft sequence of the rice genome (*Oryza sativa* L. ssp. *indica*). Science 296, 79–92.

Yuan, L.P. (2003) The second generation of hybrid rice in China. In: *Sustainable rice production for food security. Proceedings of the 20th session of the International Rice Commission. Bangkok, Thailand, 23–26 July 2002*, pp. 117–120.

Spring Wheat Breeding

M. Mergoum, P.K. Singh, J.A. Anderson, R.J. Peña, R.P. Singh, S.S. Xu, and J.K. Ransom

Abstract Wheat (various species of the genus *Triticum*) is a grass originating from the Levant area of the Middle East. However, only hexaploid common wheat (*Triticum eastivum*), and tetraploid durum wheat (*Triticum turgidum* ssp. *durum*) are presently cultivated worldwide. Not only is wheat an important crop today, it may well have influenced human history. Wheat was a key factor enabling the emergence of civilization because it was one of the first crops that could be easily cultivated on a large scale, and had the additional advantage of yielding a harvest that provides long-term storage of food. Today, there are different classes and uses of wheat. Although, it is mainly used as a staple food to make flour for leavened, flat and steamed breads, wheat can also be used as livestock feed, for fermentation to make beer and other alcoholic liquids, and recently, as a source of bio-energy. Global wheat production must increase at about 2% annually to meet future demands. The potential of increasing the global arable land is limited; hence, future increases in wheat production must be achieved by enhancing the wheat productivity to the land already in use. The objectives of most breeding programs include: high and stable yields, superior end-use quality, desirable agronomic characteristics, biotic (mainly, pests) resistance, and abiotic (environmental stresses) tolerance. While it is virtually impossible to combine all these characteristics into a single "perfect" variety, continuous breeding efforts toward achieving these objectives will ensure that new varieties possess as many desirable and economic traits as possible. Details of the different breeding approaches to enhance modern wheat breeding are discussed in this chapter.

M. Mergoum(✉)
Department of Plant Sciences, Loftsgard Hall, P.O. Box 6050, North Dakota State University Dept. 7670, Fargo, ND, 58108-6050, USA

M.J. Carena (ed.), *Cereals*,
DOI: 10.1007/978-0-387-72297-9,

1 Introduction

Wheat (*Triticum aestivum* L.) is a hexaploid ($2n = 6X = 42$ = AABBDD genomes), annual, and self-pollinated cereal which is grown worldwide. Similar to many crops of the Old World, wheat evolved in the Fertile Crescent of the Middle East. The origin of modern day wheat must have taken place by a spontaneous hybridization between tetraploid cultivated wheat of the *T. turgidum* complex and the wild species *Aegilops tauschii* within the main centre of diversity in the Fertile Crescent comprising Lebanon, Syria, Jordan, and Iraq. Wheat was one of the first domesticated food crops and has become a basic staple food of the present day human population.

Today, wheat is grown on more land area than any other commercial crop and continues to be the most important food grain source for humans. Although, wheat is most successful between the latitudes of 30°N and 60°N and 27°S and 40°S, it has been successfully grown beyond these limits. The optimum growing temperature is about 25°C, with minimum and maximum growth temperatures of 3–4 and 34–36 °C, respectively. Wheat is adapted to a broad range of moisture conditions although about three-fourths of the land area where wheat is grown receives an average of between 375 and 875 mm of annual precipitation, it can be grown in most locations where precipitation ranges from 250 to 1,750 mm. Wheat is being harvested somewhere in the world in any given month, but the largest volumes are harvested in the temperate zones between April and September in the Northern Hemisphere and between October and January in the Southern Hemisphere.

Worldwide wheat production for years 2005–2006, 2006–2007, and 2007–2008 (estimated) was 622.0, 594.0, and 610.2 million metric tons, respectively (Source: N. D. Wheat Commission, http://www.ndwheat.com/uploads/resources/546/ freelance-graphics—cworld.pdf). For the same years the wheat utilization figures were 624.4, 621.0, and 620.1 million metric tons, respectively. Of this quantity, spring wheat is the largest component followed by winter and durum wheat. China is the single largest wheat producing country. The leading production centers/countries for wheat are summarized in Table 1.

Table 1 Leading countries/groups for wheat production in the world (million metric tons)

S. No.	Country	2005–2006	2006–2007	2007–2008 (estimate)
1	European Union	122.7	124.8	127.3
2	China	97.5	103.5	100.0
3	Former Soviet Union	92.2	85.9	84.6
4	India	69.0	69.0	73.7
5	USA	57.3	49.3	59.0
6	Canada	26.8	27.3	24.5
7	Australia	24.5	10.5	22.1
8	Argentina	13.8	14.2	14.0
	Worldwide	622.0	594.0	610.2

1.1 Wheat Uses

Worldwide, the main uses of wheat are for human food and animal feed. However, wheat can be fractionated into grain components (starch, gluten, and oil), non-grain components (straw), or minor components associated with flour milling by-products, particularly wheat bran. Three-phase centrifugal separation of starch, gluten, pentosans (hemicellulose), and fiber from wet-milled wheat flour results in wheat components that have diverse uses in food and non-food products (Bergthaller, 1997).

Wheat starch can be chemically modified or hydrolyzed to yield many functional products that can be used in the production of paper, adhesives, plastic films, sweeteners, thickeners, cosmetic powders and creams, packaging materials, and foams (Maningat and Seib, 1997). Yeast fermentation of starch to produce bio-fuel and industrial alcohol is gaining interest worldwide (Maningat and Seib, 1997), particularly now that fossil fuel prices are high and unpredictable. Vital gluten, a co-product of wheat starch extraction, has been used for decades in the food industry as a foaming agent, surfactant, labeling or packaging adhesive, and coating or a film-forming membrane for food and non-food materials such as paper and cardboard (Bergthaller, 1997; Popineau et al., 2002).

Wheat straw is rich in fibrous materials and is used in addition to traditional uses, for making textiles, filters, sorbents, structural composites, molded products, and packaging materials. Wheat straw may also be used as a relatively "clean" energy source considering that when burned, its gas emissions are low (Culshaw, 1997). Wheat germ contains around 11% oil, which is a good source of vitamin E. Wheat germ oil is used in foods, insect repellents, pharmaceuticals, and cosmetics (Kahlon, 1989). Both wheat germ oil and bran (flour milling by-products) are rich in polyunsaturated fatty acids and bioactive compounds, such as a mixture of long-chain aliphatic primary alcohols known as policosanol that are effective for the prevention and treatment of cardiovascular diseases and have been associated with increased physical endurance and fitness (Varady et al., 2003). However, wheat is still mainly used for human consumption in various forms of products.

1.2 Breads

Consumed globally, bread is common in human diets in the Western Hemisphere and a staple food in North Africa, the Middle East, and West-Central Asia (Prior, 1997). Bread consumption is low in rural areas of Southeast Asia but tends to increase among the urban population with improved income (McKee, 2006). All wheat-based bread types (leavened, flat, and steamed) are made with flour doughs showing different viscoelastic properties. Leavened breads are made from wheat (and/or wheat-rye blends in Central and Eastern Europe) viscoelastic flour dough, leavened by yeast and/or other fermenting agents. Flat breads are traditionally consumed in Northern Europe, North Africa, the Middle East, South Asia, and North and Central America (Faridi, 1988) and are prepared with fermented or unfermented doughs made from unmixed wheat flour, or composite flours including wheat and other cereals. Steamed bread, very popular in China and Southeast Asia (He et al., 2003), is spherical and have a spongy crumb and no crust.

1.3 Flour Noodles

Wheat flour noodles are a staple food in Northern China and are widely consumed all over the Far East (Liu et al., 2003). Flour noodles are made from sheeted stiff flour dough that is cut into noodle strands and sold fresh or dried. Mechanized noodle production predominates in Japan, while semi-mechanized and handmade production predominates in China.

1.4 Breakfast Cereals and Cereal Bars

Wheat is used to manufacture ready-to-eat breakfast cereals and cereal bars, convenience foods of increasing popularity as regular and calorie-low sources of nutrients and dietary fiber to reduce obesity-related health problems (Palazzolo, 2003).

1.5 Cookies and Cakes

Produced worldwide, soft wheat-based foods come in wide variety of shapes, textures, sizes, and flavors. Cookies are usually made with inelastic stiff dough, but some cookies, cakes, pancakes, and waffles require thick or thin viscous batters.

1.6 Blending

Hard Red Spring (HRS) wheat produced in USA and Canada is considered the premium standard in the world market for baking bread and other products because of its high protein and superior gluten functionality. HRS wheat from these two countries is extensively used for blending with and improving the functional quality of other cheaper but inferior wheat classes/other cereal flours in different parts of the world. HRS wheat's use in flour blends help flour mills meet their customer's demands for a specific level of protein content, water absorption, and mixing stabilities. The blended products also have improved moistness, softness, and increased shelf-life, making it more desired to the local consumer's demand.

2 Genetics

Spring wheats botanically belong to common wheat (*Triticum aestivum* L., $2n = 6X = 42$). Spring growth habit of common wheat is genetically controlled by the dominant vernalization gene *Vrn1* (Yan et al., 2003). Common wheat is an allohexaploid with the A, B, and D genomes. The 21 pairs of chromosomes are grouped into 7 homoeologous groups and each chromosome has 2 homoeologues in the other 2 genomes (Sears, 1954, 1966). Thus, many important characters of common wheat are genetically controlled by orthologous gene sets located on colinear regions of three homoeologous chromosomes such as the vernalization gene *Vrn1* on chromosome 5A, 5B, and 5D (Yan et al., 2003) and storage protein genes *Glu* and *Gli* on 1A, 1B, and 1D. Hexaploid common wheat behaves as a diploid organism at meiosis due to the *Ph1* (pairing homoeologous) gene on the long arm of chromosome 5B, which prohibits pairing between homoeologous chromosomes (Riley and Chapman, 1958). Since the absence or mutation of the *Ph1* gene induces homoeologous pairing, manipulation of the *Ph1* gene has been a major approach for transferring desirable genes from related wheat species to common wheat. A large number of translocation lines containing alien genes for resistance to various abiotic and biotic stresses have been developed using the *ph1b* mutant.

The allopolyploid nature allows common wheat to tolerate structural and numerical changes of chromosomes. Various types of aneuploid and genetic stocks have been established in spring wheat "Chinese Spring" (CS) and other cultivars. The CS aneuploids and genetic stocks that are currently available include sets of monosomics ($2n - 1$), trisomics ($2n + 1$), tetrasomics ($2n + 2$), nullisomics ($2n - 2$), nullisomic-tetrasomics ($2n - 2 + 2$), ditelosomics, ditelo-monotelosomics, double-ditelosomics, monoisosomics, and segmental deletion lines (Sears, 1954, 1966; Gill et al., 2004). These aneuploids have been widely used to locate individual genes or molecular marker loci to specific chromosomes and chromosome intervals.

Common wheat possesses a very complex and huge genome with a size of $\sim 16 \times 10^9$ bp, which consists of about 90% repeated sequences (Li et al., 2004). Such a large genome with excess repetitive sequences seriously hampers genomic analysis and whole-genome sequencing for gene discovery at the present time (Janda et al., 2004). However, enormous progress in exploring the wheat genome has been made through molecular and physical mapping and functional and comparative genomic studies. Thus far, about 5,000 molecular markers have been mapped onto more than 20 genetic maps in wheat and its relative species (http://wheat.pw.usda.gov/ggpages/mapsframe.html), from which a composite linkage map has been compiled (http://wheat.pw.usda.gov/ggpages/wgc) (Gill et al., 2004). A high-density microsatellite consensus linkage map consisting of 1,235 microsatellite marker loci has been established (Somers et al., 2004). More than 855,000 Expressed Sequence Tags (ESTs) have been generated from common wheat (http://www.ncbi.nlm.nih.gov/dbEST) and about 7,000 unigenes have been mapped to 159 bins across the 21 chromosomes (http://wheat.pw.usda.gov/nsf; Gill et al., 2004). Two bacterial artificial chromosome (BAC) libraries consisting of

120,000 and 2,000,000 clones have been constructed from the genomes of common wheat CS and "Renan", respectively (Gill et al., 2004). A sub-genomic BAC library for chromosomes 1D, 4D, and 6D and two chromosome-specific BAC libraries for chromosomes 3B and 1BS have been constructed using chromosome sorting (Janda et al., 2004; Šafář et al., 2004).

The abundance of genetic and genomic resources in common wheat provides great facilitation in characterizing the wheat genome and individual genes. The well-characterized aneuploid stocks, molecular markers and maps, deletion-mapped ESTs, and BAC libraries have considerably promoted molecular mapping and tagging of genes from common wheat. A large number of important genes have been precisely mapped and several genes including the leaf rust resistance genes *Lr21* and *Lr10*, powdery mildew resistance genes *Pm3*, *Vrn1*, and domestication gene *Q* have been isolated by map-based cloning (Gill et al., 2004). Rapid advances in wheat genomics would further deepen our understanding of the wheat genome and genetics and strengthen improvement of cultivated wheat through innovative paradigms of molecular breeding.

3 Wheat Gene Pools

Wheat has one of the largest gene pool among the cereal crops and is notable for its diversity. Modern day wheat (*Triticum aestivum* L.) is a hexaploid composed of AABBDD genome. The genome donors are *T. urartu* (A), the *Sitopsis* section of *Aegilops* (B), and *Aegilops tauschii* (D) (Dvořák, 1998). Wheat belongs to tribe "Triticeae" of the family "Gramineae". There are three major gene pools of wheat (Mujeeb-Kazi and Rajaram, 2002). Primary gene pool members cross readily among one another and consist of all *Triticum* species. The primary gene pool species include the common cultivated and landraces of hexaploid wheat, cultivated and landraces of tetraploid wheat, wild *T. dicoccoides* and diploid donors of the A and D genomes to durum and bread wheats. Genetic transfers from these two genomes occur as a consequence of direct hybridization and homologous recombination with breeding protocols contributing different back-crossing and selection strategies. Some cross combinations among primary gene pool members may require embryo rescue, but cytogenetic manipulation procedures are not necessary.

The secondary gene pool is composed of the polyploid *Triticum* and *Aegilops* species, which share one genome with the three genomes of wheat. The diploid species of the "Sitopsis" sections are included in this pool, and hybrid products within this gene pool demonstrate reduced chromosome pairing. Gene transfers occur as a consequence of direct crosses, breeding protocols, and homologous exchange between the related genome or through use of special manipulation strategies among the non-homologous genome. Embryo rescue is a complementary aid for obtaining hybrids.

Diploid and polyploid species are members of the tertiary gene pool. Their genomes are non-homologous. Homologous exchanges cannot affect genetic transfers, but genomic homoeology of these species does permit the transfer of genes by somewhat complex protocols, facilitated by irradiation or callus culture-mediated translocation induction.

4 Varietal Groups/Classes

The type of cultivar of a crop used for commercial production is depended on the mode of reproduction, economics of cost of seed production and the crop returns, the nature, feasibility, and ease of seed production and distribution system and the environment under which crop production occurs. Wheat is a monoecious plant with perfect flowers. It reproduces sexually as an autogamous or self-pollinated crop although limited (3%) cross pollination is possible. Wheat cultivars are developed mainly by three means: introduction, selection, and hybridization. Introduction of wheat cultivars grown elsewhere has been successful in the early establishment and development of wheat production areas. The "Green Revolution" in the 1960s was as a result of the introduction of cultivars developed at Centro Internacional de Mejoramiento de Maíz y Trigo (CIMMYT) into more than 25 countries where the introduced cultivars were very productive and successful. CIMMYT germplasm adapted to high input environments resistant to lodging as a result of the semi-dwarf gene (*Rht1*) from the Japanese source "Norin 10" was often introduced as cultivars/lines and selected for local adaptability and desirability and the resulting selected genotype was then released as a cultivar. However, most modern cultivars are developed by artificial hybridization or crossbreeding followed by rigorous selection for desired traits.

In addition to landraces and local populations still grown in some areas, three types of wheat cultivars are commercially grown: hybrids, multi-lines, and pure-lines. Hybrid wheat cultivars are produced by the use of chemical gametocides and cytoplasmic-genetic male sterility. However, hybrid wheats are grown in a limited acreage due to the lack of heterotic groups for economic traits and the expense involved in producing them. Two types of multi-line wheat cultivars are produced (i) mixture of near-isogenic lines differing in the resistance genes to one or a few wheat diseases and (ii) mixture of distinct cultivars. Multi-line wheat cultivars are also grown commercially on a limited basis in few countries. The majority of present day wheat cultivars are pure-lines and occupy the majority of wheat acreage worldwide.

Based on genomic constitution, commercially grown wheat is of two main classes (i) durum wheat, a tetraploid with AABB genomes and (ii) common wheat, a hexaploid with AABBDD genomes. Common wheat is further classified into two categories (winter vs. spring wheat) based on the distinct growing seasons. Winter wheat, which normally accounts for 70–80% of US production, is sown in the fall and harvested in the spring or summer while spring wheat is planted in the spring and harvested in late summer or early fall. Outside of the USA, spring wheat is grown in most countries except in northern Europe where winter wheat is dominant.

Based on kernel color, endosperm hardness, and other quality characteristics spring wheat is classified into three distinct classes:

Hard red spring (HRS) wheat: Contains the highest percentage of protein of the wheat classes, making it excellent for bread wheat with superior milling and baking characteristics. The majority of the crop is grown in the Northern Great Plains of USA (Montana, North Dakota, South Dakota, and Minnesota) and Canada. In USA, North Dakota is the leading producer of HRS wheat accounting for 50% of US production. USA produced spring wheat is largely exported to Central America, Asia (Japan and Philippines), and parts of Europe.

Soft white (SW) wheat: Contains low protein content but is high yielding. It produces flour for baking cakes, crackers, cookies, pastries, quick breads, muffins, and snack foods. SW wheat is grown mainly in the Pacific Northwest and to a lesser extent in California, Michigan, Wisconsin, and New York. SW is also grown in significant amounts in Canada. It is exported to Egypt, Morocco, and Far East Asian region.

Hard white spring (HWS) wheat: The newest class of wheat to be grown in the USA. Closely related to red wheats (except for color genes), this class of wheat has a milder, sweeter flavor, equal fiber and similar milling and baking properties. Flour from HWS wheat is used mainly in the production of yeast breads, hard rolls, bulgur, tortillas, and oriental noodles. HWS wheat is used primarily in domestic markets, although it is also exported in limited quantities.

5 Current Goals of Wheat Breeding

With the ever-increasing human population the demand for wheat globally is bound to increase considerably. Global wheat production must increase at 2% annually to meet future demands. The potential of increasing the global arable land is limited; hence, future increases in wheat production must be achieved by enhancing the productivity to the land already in use. Developing cultivars with increased grain yield potential, superior end-use quality, tolerance to abiotic stresses, and enhanced resistance to diseases and pests would be essential (Singh et al., 2007). The objectives of a wheat breeding program determine largely the parents used in the development of breeding populations, methods of selection performed in the breeding population, and the resource allocation. The major objectives of most present day wheat breeding programs include the following:

5.1 Grain Yield

The most important objective of any wheat breeding program is to enhance grain yield. Grain yield is a complexly inherited trait of low to moderate heritability and is strongly influenced by environmental conditions. Higher grain yields are usually associated with lower protein content and delayed maturity. Wheat breeders attempt to enhance grain yield without adversely affecting the protein content, which is detrimental to its marketability, and the time to maturity, that may increase the chances of production failure. Yield enhancement is often achieved by not only selecting for greater genetic potential for yield *per se* but also by selecting for resistance to biotic and abiotic stresses that may limit the expression of the cultivar's maximum yield potential. Breeding for enhanced yield depends on (i) the average yield potential of the selected population, (ii) selection intensity, (iii) genotypic variation for yield potential, and (iv) the degree to which genotypic differences in yield potential are expressed in the selection nursery. Visual selection and single plant evaluation for yield are not effective; hence, most breeders tend to select for maturity, photoperiod response, shattering resistance, short stature, harvest index, and numerous disease reactions in early generation selections. These traits directly or indirectly affect the yield potential of selected genotypes.

Wheat breeders use different approaches for yield evaluation due to the genetic complexity of the trait and the considerable interaction between yield and environment. Tests across several years and locations are conducted to identify lines which are genetically superior and stable for yield. Yield trials are conducted using various experimental designs that include checks, randomization, and replications to enhance the precision and accuracy of the selection process.

Rajaram (2001) emphasized that future yield increases will be based on (i) the development and deployment of yield-enhancing, genetically based technologies to produce future wheat cultivars, (ii) research and development of germplasm with polygenically conferred multiple disease resistance targeted toward emerging cropping systems such as zero tillage, (iii) solidifying research targeting marginal environments both in terms of developing adapted improved germplasm and conserving the natural resource base, especially in drought-prone areas, and (iv) using biotechnology to aid conventional plant breeding by developing transgenics and exploiting the application of molecular markers.

Management practices that optimize the genetic yield potential in available cultivars need to be developed. Furthermore, the development of genotypes that interact positively with new crop management practices may be one means of increasing productivity, especially in more favorable environments (Ransom et al., 2007).

5.2 Grain Quality

Wheat grain quality relates to how successfully wheat and flour perform in consumer products and industrial processes. Enhancing wheat quality improves processing efficiencies, makes more desirable and more diverse consumer products and ensures the competitiveness of farmers, grain merchandisers, millers, and end processors. Wheat quality criteria may vary drastically depending on the end-use. Similarly, wheat cultivars may show large differences in their processing and end-use quality attributes. Therefore, in setting breeding priorities and strategies, one must determine: the cultivar's intended end-uses and/or the demands of the targeted market, specific quality traits to breed for, and genotype × environment × management interactions that may influence the quality of the resulting cultivar.

Once parental lines are characterized and the crossing plan defined, the probability of selecting desirable lines depends on the intensity and effectiveness of the quality-selection pressure applied; the best results are obtained by breeding for the targeted environment (warm, dry, wet, or erratic) and screening F_3-F_5 lines (using in some cases Marker Assisted Selection (MAS) for some traits) for desirable genes and allelic variations controlling grain-compositional traits (Arbelbide and Bernardo, 2006), complemented by rapid, high-throughput conventional small-scale tests such as flour sedimentation and Near Infrared Reflectance Spectroscopy (NIRS), which are related to end-use processing quality (Peña et al., 2002; Souza et al., 2002). Because Marker Assisted Selection (MAS) and conventional small-scale quality tests explain end-use quality only partially, in advanced breeding stages (F_6–F_8), quality screening should be based on more specific food-processing (dough viscoelasticity and mixing properties, starch pasting properties, baking performance) and end-product quality attributes (Peña et al., 2002; Souza et al., 2002). Finally, multi-location yield trials exposing advanced elite lines to environmental variation and farmer's crop management practices are necessary to identify the few genotypes combining stable yield and quality attributes across locations and years.

5.3 Resistance to Biotic Stresses

Globally the major fungal diseases of wheat, caused by biotrophs, include the three rusts, powdery mildew, and the bunts and smuts; whereas, those caused by hemibiotrophs include Septoria tritici blotch, Septoria nodorum blotch, spot blotch, tan spot, Fusarium head blight (FHB) (Fig. 1), etc. The biotrophs are highly specialized and significant variation exists in the pathogen population for virulence to specific resistance genes. Evolution of new virulence through migration, mutation, recombination of existing virulence genes and their selection is more frequent in rust and powdery mildew fungi. Therefore, breeding for resistance to these diseases needs a critical analysis to enhance the durability of resistance. Physiological races are known to occur for most bunts and smuts, however, evolution and selection of new races is less frequent. Changes in pathogen races are also less frequent for diseases caused by hemibiotrophs; however, the importance of some of these pathogens has increased dramatically in those countries where residue retention has become a common practice of conservation agriculture.

Among the major challenging insects to wheat production include Hessian fly, stem sawfly, cereal leaf beetle, greenbug, grasshoppers, midge, and wheat curl mite. The insects and mites that damage wheat production have complex biology, varied reproductive behaviors, diverse food and survival habits, and powers of dispersal. This makes breeding for pest tolerance or resistance very challenging. As with the

Fig. 1 Fusarium Head Blight (FHB) known as "scab" symptoms on spikes (**a**) and damage on wheat kernels ("Scabby Kernels" or "Tombstones") (**b**). (*Photo: Mergoum Mohamed*)

wheat diseases, incorporation of both vertical and horizontal resistance to insects is being attempted when developing resistant cultivars. MAS in combination of traditional disease/insect damage control strategies are being followed to develop cultivars with enhanced resistance to the major diseases and pests. Currently, a more integrated approach to pest management is being employed that includes cultural practices, genetic resistance, biological control, and chemical protection.

5.4 Tolerance to Abiotic Stresses

Various environmental stresses and weather-induced losses affect wheat production. Important among these factors are drought causing poor seedling emergence/establishment and stress during the life cycle, flooding, pre-harvest sprouting, extreme temperatures (heat and freezing), wind (lodging or grain shatter), and mineral stress (deficiency or toxicity). Breeding for resistance/tolerance to abiotic stresses is generally more challenging than most other stresses due to their complex, inconsistent, and elusive nature.

When breeding for abiotic stresses both direct and indirect selection strategies are followed. In direct selection, breeders intentionally place experimental plots in areas of wheat cultivation where the stress exists consistently and uniformly. However, in the indirect selection strategy, traits which affect directly or indirectly the targeted trait are selected. Both direct and indirect selection strategies were responsible for the successful development of cold hardiness in cultivars in the Northern Great Plains of North America. This has resulted in the production of winter wheat in areas where historically spring wheat was grown. Molecular techniques in combination with traditional abiotic stress assessment strategies are being followed to develop cultivars with enhanced resistance to abiotic stresses.

6 Breeding Methods and Techniques

In modern wheat breeding programs different breeding procedures and techniques are followed depending upon a number of factors including (i) genetics of the trait to be developed, (ii) resources available, and (iii) personal preference of the breeder. Several different methods can be used to breed superior cultivars and most present day wheat breeders follow a combination of several procedures, each based specifically on the programs objectives and the resources available, to him or her. The major breeding methods that are followed during development of superior wheat cultivars include the following:

6.1 Backcrossing Selection

Backcrossing is a breeding method used to transfer one or few genes from one parent (donor parent) that may be un-adapted or exotic to another parent (recipient parent) that is an adapted cultivar in a given region. Backcross derived cultivars possess the additional genes for resistances to diseases and pests that were lacking in the parent (recipient) cultivar. Backcross breeding methodology is simple and predictable and can be further hastened by the incorporation of off-season, greenhouse nurseries or use of double haploids. However, this methodology of breeding is not very successful in handling complex traits like yield; hence, breeders rarely rely on this methodology alone. Backcross breeding method is used mainly to improve parental lines or eliminate defects in useful genotypes.

6.2 Pedigree Selection

The pedigree selection method involves alternate parent–progeny evaluation starting from early generations through to the advanced generations. Pedigree selection, involving combination and transgressive breeding, is most often used by wheat breeders in the development of superior cultivars. Selection for easily identifiable characters with high heritability is effective in early generations while complex traits like yield that have low heritabilites are selected in later generations. Shuttle breeding using off-season nurseries and greenhouse facilities are utilized to fasten the breeding process in most modern wheat breeding programs. Following the pedigree method in the strict sense is slow, labor intensive, tedious, and requires significant resources when dealing with large number of crosses and their progenies. At North Dakota State University (NDSU), Fargo, the HRS wheat breeding program handles large number of segregating populations in a combination of the pedigree and bulk methods. Spikes of F_2 plants are selected and grown as head rows the following season. Selection for desired traits is done on a row basis from the F_3

to the F_5 generations. Since the selected rows are harvested as bulks there is sufficient seed to evaluate quality, adaptation, diseases, and yield earlier than with the conventional pedigree method. Additionally, the pedigree method at NDSU, Fargo, is combined with shuttle breeding using New Zealand and Arizona, USA nurseries to speed up the selection process.

6.3 Bulk Selection

In traditional seed selection systems, natural selection was the key feature of bulk selection that resulted in improvement, as surviving individuals tended to be the fittest. However, in modern wheat breeding programs both natural and artificial selection are performed in the bulk method also referred as selected-bulk method. In bulk populations artificial selection may be performed for disease and pest resistance, spike type, height, plant architecture, and awn development. Often the bulk method is followed when stresses selected for are absent or when dealing with crosses between different wheat types. Most wheat breeders perform bulk selection on traits that are important but inconsistent in occurrence like diseases or environmental stresses and if the genetic information of the trait in consideration is not clearly understood. In these situations, bulk populations are advanced until appropriate conditions occur for expression of the traits and both natural and artificial selection favor the desired genotypes. At the end of the bulking period, this is when a high degree of homozygosity is achieved, individual plants are selected and their progenies are evaluated as in the pedigree method.

6.4 Single Seed Descent

In the single seed descent (SSD) method of breeding the objective is to advance the early generation rapidly with the expenditure of minimal resources. The SSD method capitalizes on advancing two to three generations per year using greenhouse/growth chamber and field/off season nurseries. Generally, SSD is used in developing populations for genetic studies or just rapid advance of early generation breeding material and in later stages the breeding material is evaluated as in the pedigree selection method. However, modification of SSD involving single-spike descent or single-hill descent methods are more commonly used. In these modifications wheat breeders are utilizing both the greenhouse and field facility and try to hasten the breeding process. The single-spike descent method requires more space and resources than SSD but gives breeder more accurate and reliable information and helps ensure that progeny of each genotype will survive the selection generation.

6.5 Recurrent Selection

Recurrent selection, more often used in cross-pollinated crops, is being adopted in various wheat breeding programs with the long-term objective of developing germplasm with a broad genetic base. Recurrent selection, which is a cyclic process alternating between selection and hybridization, is also gaining more interest in present day wheat breeding programs specially when dealing with complex traits. Wheat breeders using recurrent selection do not make all possible inter-matings among selected genotypes/parents of a given cycle. Additionally, they tend to delay selection to advanced generations to get sufficient seed/homozygosity to accurately evaluate breeding lines for the trait in consideration. In the HRS wheat breeding program at NDSU, Fargo, one of the major challenges is to breed for FHB resistance. The genetics of FHB resistance/tolerance is complex and disease assessment is difficult. In this program the advance generation or double haploid breeding material is evaluated and selected germplasm is re-hybridized in order to pyramid the genes for FHB resistance. Good success, achieved at NDSU in tackling FHB is attributed to the combined use of the pedigree method, double haploidy, and recurrent selection breeding methods.

6.6 Double Haploidy

Haploid production followed by chromosome doubling results in the creation of genetically pure lines (double haploids) within a short period of time. In most wheat breeding programs, double haploid production is done by anther culture or wheat/maize wide hybridization. Double haploidy enables wheat breeders to achieve completely homozygous lines in one generation from early generation (F_1 or F_2) breeding material. This procedure eliminates several generations of selfing normally required before uniform lines can be evaluated in yield trials. Not only does double haploidy fasten the breeding process, but it also saves resources/costs required in the advancement and evaluation of segregating breeding material through traditional breeding approaches. This is an important breeding procedure that shortens the release of improved cultivars to 6–7 years as opposed to the conventional 10–12 years. This technique, however, induces less genetic variation than is generated by segregation and recombination in traditional breeding methods.

6.7 Hybrid Wheat

Hybrid wheat production has been carried out mainly by private companies. These private companies use either cytoplasmic male sterility (CMS) or chemical hybridizing agents (CHA) for commercial hybrid production. The CMS hybrid production involves three different lines called the A-line, B-line, and R-line. The A-line is the female parent and is male sterile due to gene(s) present in the cytoplasm. The R-line or "Restorer line" is the male line and is used to cross-pollinate with the A-line to produce hybrid seed. The R-line has one or more nuclear genes which can override the male sterility trait of the A-line resulting in the production of fertile hybrid seed. The B-line or "Maintainer line" possesses similar genetic make-up to the A-line except it does not have the genes for male-sterility in the cytoplasm and is used to maintain seed. The CHA hybrid production involves spray of chemicals, prior to anthesis on the female parents which results in male sterility. Later the male-sterile plants are pollinated by wind-borne pollen from untreated male-parents. Seeds produced on the female parents which is harvested carefully and marketed as hybrid seed.

Technical capabilities for controlled pollination continue to enhance resulting in increasing the efficiency and number of hybrid wheat produced and evaluated. At this stage there is limited hybrid wheat development and production worldwide. Furthermore, there is relatively limited research and resources allocated to the hybrid wheat so there is currently few "heterotic" groups identified. However, the availability of CMS and CHA systems enhances the breeder's ability to discover, manipulate, exploit, and use genetic variability to develop superior hybrid wheat.

6.8 Mutation Breeding

Ionizing radiation (X-rays, gamma rays, and neutrons), ultraviolet light, and chemical mutagens (ethyl methane sulfonate and diethyl sulfate) have been successfully used to create mutations. Years of intensive efforts in mutation breeding, however, have resulted in few successful accomplishments. Hence, mutation breeding is presently mainly used to compliment other wheat breeding approaches. Evaluation of mutagenized populations is performed in the M_2 and M_3 generations following the mutation and subsequently the selected population is handled using traditional breeding methods. The main traits for which useful mutations have been secured include changes in morphology, physiology, reproductively, chemical composition, and disease and more recently resistance to "Imazamox" (Imidazolinone) herbicide.

6.9 Shuttle Breeding

Shuttle breeding, pioneered at CIMMYT, Mexico, was originally used to speed up the breeding process by advancing and testing breeding material at contrasting environments resulting into more than a single generation per year. Results reveal higher success in shuttle breeding due to the exposure of the breeding material to contrasting disease spectra, soil types, photoperiod length, and diverse environmental constraints. The success of shuttle breeding resulted in part in the "Green

Revolution" in the 1960s to 1970s wherein semi-dwarf, high-yielding, photoperiod insensitive, and disease-resistant cultivars were developed that were highly responsive to fertilization and other inputs.

Wheat breeders in the Prairie regions of the Great Plains of North America have developed their own shuttle breeding concept which is highly effective in their achieving the breeding objectives. At the HRS wheat breeding program, NDSU, Fargo, the main breeding evaluation is conducted in North Dakota. However, evaluation of the breeding material is also done in contrasting environments at Arizona, USA, China, and New Zealand (Fig. 2). This shuttle breeding concept also provides breeders the opportunity to tackle specific traits in a given site. In the NDSU breeding program evaluation in China is mainly done to evaluate resistance to FHB, a major constraints in wheat production in the USA. Since FHB infection is highly influenced by environmental conditions, it is very important that the resistance is expressed in several different environments and crop growing conditions worldwide. The high success achieved in developing FHB resistant cultivars however is not attributed to evaluation carried out in China only. The creation of FHB nurseries using artificial inoculation coupled with mist irrigation in different parts of the wheat growing regions in the USA was the key behind releasing FHB resistant cultivars.

6.10 Marker-Assisted Selection

Molecular markers can help breeders select for particular genes. Three broad, practical criteria must be satisfied before MAS can be effectively implemented in

Fig. 2 An overview of the "Off-season" winter nurseries of spring wheat from NDSU and other spring wheat breeding programs grown at a location near Christchurch, New Zealand. (*Photo: Steve Inwood*)

a breeding program: (i) efficiency/gain compared to phenotypic selection, (ii) usefulness of markers in breeding-relevant populations, and (iii) the cost, throughput, and expertise required. Markers for end-use quality (Gale, 2005) and disease resistance (Liu and Anderson, 2003) are most likely to meet the first two criteria. Inconsistent QTL effects across breeding populations are the main reason that there are few examples of their use in MAS programs in wheat and other crops. Theoretical considerations aside, it should be noted that any MAS activity is competing against or complimenting, and in most cases, not replacing a well-established evaluation system based on phenotype. Most markers in use today are PCR-based simple sequence repeats (SSR) or sequence-tagged-site (STS) markers. Backcrossing with markers and parental characterization for key genes are cost effective ways of utilizing this technology. A recently completed USDA-funded project, MAS wheat, utilized backcrossing with DNA markers to introgress 43 genes into 75 genetic backgrounds (Dubcovsky, 2004).

Efficient implementation of MAS demands the use of high-throughput equipment and trained personnel. Although MAS is becoming a new capability in many wheat breeding programs, its implementation is limited by the cost to support trained personnel and purchase equipment and reagents. Establishment of the USDA-ARS Regional Small Grains Genotyping Centers in the USA has dramatically increased the capabilities of breeders to apply MAS by providing access to high throughput DNA extraction and genotyping equipment. With such facilities, MAS activities have expanded to include BC_1, F_2, and F_3 populations. However, only a fraction of genotypes potentially segregating for important genes can be accommodated, even with this equipment and technology (Bonnett et al., 2005). More efficient DNA extraction technologies and marker platforms [e.g. single nucleotide polymorphisms (SNPs)] will allow more complete implementation of MAS in wheat breeding programs in the future.

7 Major Breeding Achievements

To enhance production wheat cultivars need to be high yielding, resistant to biotic and abiotic stresses, and possess high end-use quality. The main reason for the Green Revolution in the 1960s was the development of wheat cultivars that were semi-dwarf, lodging resistant, fertilizer responsive, high yielding with resistance to major diseases and pests. The lodging resistance was attributed to the short stature controlled by genes originating from the Japanese cultivar "Norin 10". Wheat breeding is an ongoing process and there are several successes which include the following:

7.1 Grain Yield

Wheat breeding has enabled dramatic increase in grain yield. The last 50 years saw major success in wheat breeding largely due to the use of shuttle breeding (Fig. 2), wide adaptation, incorporation of durable resistance, and international multi-location testing, resulting in gains in stability of cultivars performance and the wise use of genetic variability enhanced yield gains in the released cultivars (Rajaram, 2001).

The genetic gain in yield potential in high output wheat systems worldwide since the mid-1960s has been ~1% per year (Sayre et al., 1997). A better understanding of the various genetic and non-genetic factors contributing to yield increase has resulted in the development of superior cultivars. Genetic improvements have made a substantial increase in the yields of the wheat worldwide (Fig. 3). Grains in yield potential are attributed mainly to increased kernel weight per spike, reduced plant height, lodging resistance (Fig. 4), and increased harvest index. Additional gains in the genetic potential for yield have been attributed to resistance to diseases and pests and tolerance to adverse environmental conditions.

7.2 Grain Quality

The processing quality of wheat, and our understanding of the factors that determine quality, has improved greatly in the last two decades resulting in the development of cultivars with superior grain quality. Color, kernel hardness, gluten strength, and grain protein concentration are key traits that influence end-use quality of wheat. The improvements in quality largely resulted from intercrossing existing cultivars and elite lines possessing contrasting quality characters and then selecting individuals possessing all desired quality traits. Efficiency of breeding has

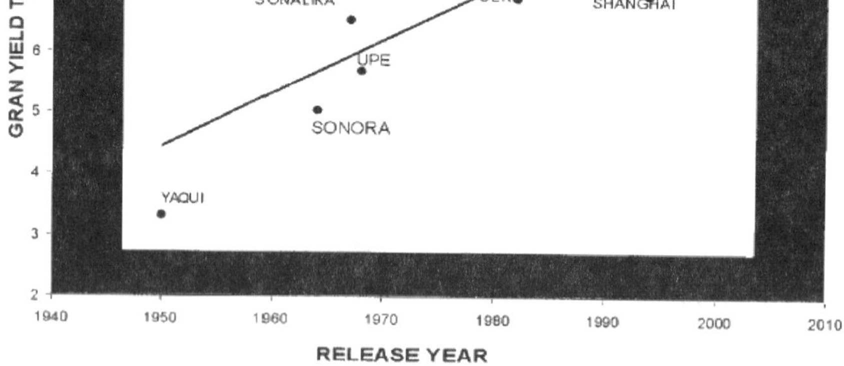

Fig. 3 Yield advance of CIMMYT varieties and lines over 50 years period (Rajaram 2001)

Fig. 4 Severe lodging observed on a susceptible HRSW cultivar grown under irrigated and high inputs environments in Arizona, USA. (*Photo: Mohamed Mergoum*)

Table 2 Main wheat quality traits and their genetic control

S. No.	Quality trait	Gene	Chromosomal location
1	Flour color		7A and 7B
2	Yellow pigmentation	Various additive genes	A and B chromosomes
3	Alpha amylase	*Amy-1, Amy-2*	Groups 6 and 7
4	Polyphenoloxidase		2AL and 2D
5	Grain hardness (puroindolins)	*Pina-D1, Pinb-D1*	Short arm of 5D
6	Protein content	*Pro-1, Pro-2*	5D
7	Glutenins	*Glu-1, Glu-3*	1A, 1B, and 1D
8	Gliadins	*Gli-1, Gli-2*	Groups 1 and 6
9	Starch granule-bound synthase	*Wx-1*	7AS, 4AL, and 7DS

been improved through new technologies such as NIRS for measuring protein concentration and color factors that has increased the potential to predict milling yield and protein quality. MAS for protein subunits are used to screen early generation populations for quality potential where appropriate subunit differences are present. The use of these novel technologies, in association with the development of doubled haploid techniques, may facilitate further genetic analysis of both novel and traditional quality traits and result in more success in enhancing wheat quality.

Success in breeding for end-use quality is due to gains in genetically controlled physical (grain size, color, hardness), compositional (mainly storage proteins and starch), and biochemical (enzymatic activity, gluten visco-elasticity, starch pasting properties) grain traits (Table 2) that are also influenced by genotype × environment (heat and moisture stresses) interactions (Spiertz et al., 2006, Zhang et al., 2006) and nutrient availability, particularly at the grain-filling stage (Dupont et al., 2006; Otteson et al., 2007).

The main factor defining wheat end-uses is gluten, which is formed by insoluble endosperm proteins that give water-flour dough its visco-elastic properties. Gluten is composed of monomeric gliadins and polymeric glutenins. Gliadin subunit groups (α-; β-; γ-; ω-gliadins) are controlled by genes in the complex *Gli-1* and *Gli-2* loci, while glutenins, high- (HMW-GS) and low- (LMW-GS) molecular weight subunits are controlled by genes in the complex *Glu-1* and *Glu-3* loci, respectively. Allelic variations, mainly at *Glu-1*, *Glu-3*, and *Gli-1*, are responsible for most variation in dough properties (strength and extensibility), dough-mixing properties, and bread- and Chinese noodle-making quality (Békés et al., 2006; Branlard et al., 2001). HMW-GS contribute mainly to dough strength, while LMW-GS and ω-gliadins contribute mainly to dough extensibility and viscosity (Békés et al., 2006; Branlard et al., 2001; He et al., 2005).

Knowledge of grain quality-related characteristics and desirable/undesirable quality-related genes or allelic combinations has allowed breeders to plan crosses that have given more success in generating desirable wheat quality types (Békés et al., 2006; Souza et al., 2002). Electrophoresis (sodium dodecyl sulfate polyacrylamide gel electrophoresis, SDS-PAGE) is commonly used to identify allelic variations at *Glu-1*, *Glu-3* and *Gli-1* and characterize parental lines. Information on glutenin subunit and gliadin composition helps breeders design crosses aimed at achieving allelic combinations known to contribute positively to dough properties required for producing leavened and flat breads, flour noodles, cookies, and pasta. Rapid, small-scale, high-throughput tools are essential in quality improvement; efficiency in breeding for improved quality can be enhanced by using MAS and NIRS in screening. MAS offers a better option for specific traits such as grain color, hardness, and proteins, because it is performed on leaf tissue before seed-setting; this allows eliminating lines with undesirable traits before harvesting.

7.3 Resistance to Diseases and Pests

Although a major emphasis in the past was given to use race-specific major genes to control rust diseases with limited success due to their fast breakdown, utilization of durable or slow rusting resistance (Johnson, 1988) has been more effective. Durable resistance to leaf and stripe rusts and powdery mildew involves slow rusting genes that have small to intermediate but additive effects. Although, the best characterized genes with pleiotropic effects in conferring resistance to the above three diseases are *Lr34/Yr18/Pm38* and *Lr46/Yr29/Pm39* (Spielmeyer et al., 2005), various other genomic regions are now known to harbour additional slow rusting resistance genes (Singh et al., 2004). Presence of a single or a couple of slow rusting genes in a cultivar is often not sufficient for satisfactory control under high disease pressure. However, cultivars with "near immune" levels of resistance were developed by combining three to five slow rusting genes (Lillemo et al., 2005). Slow-rusting genes based resistance to leaf and stripe rusts is common in modern, high-yielding spring wheat germplasm originating from CIMMYT (Singh et al., 2007) and is used in further wheat improvement.

Stem rust, historically known to cause severe losses to wheat production, has been controlled effectively through the use of genetic resistance in semi-dwarf cultivars. Resistance gene *Sr31*, located on wheat-rye 1B/1R translocation contributed to a high level of resistance in several wheat cultivars developed worldwide in recent years. Consequently stem rust disease is often not considered important in wheat breeding in many countries. However, detection of *Puccinia graminis tritici* race Ug99 in 1999 in Uganda with its broad virulence, including the virulence for *Sr31*, its migration to Kenya, Ethiopia and its detection in 2006 in Yemen and Sudan, and evolution of its variant with virulence to resistance gene *Sr24* were recognized as highly significant events (Singh et al., 2006b). One of the major challenge to wheat breeding is identifying or developing and diffusing adapted resistant cultivars effective against Ug99 race. Sources of durable resistance to Ug99, including some modern wheats have been identified (Singh et al., 2006b) in combination with additional resistance genes, mostly of alien origin, can provide effective control. If race-specific genes are used, they must be deployed in combination to enhance their longevity and MAS will prove to be very useful for success.

The Yangtze river basin of China with about 7 million ha has traditionally been known to be highly prone to the FHB or scab epidemics. FHB disease incidences leading to epidemics are now frequent in North America, Europe, and South America. In addition to crop losses, the fungus also produces mycotoxins, such as deoxynivalenol (DON) that accumulates in the grain and renders grain unsuitable for human and livestock consumption (Fig. 1). Although several genomic regions are now known to contribute quantitative resistance (Buerstmayr et al., 2002; Anderson et al., 2001), a gene from a Chinese cultivar, "Sumai 3", in the short arm of chromosome 3B has shown the largest and consistent effect in reducing disease severity and mycotoxin accumulation (Anderson et al., 2001). At NDSU, Fargo, the cultivar "Alsen" was released in 2000 and it possesses the 3BS resistance gene (Mergoum et al., 2005b). Subsequently "Steele-ND" and "Howard", released in 2004, possesses the resistance to FHB derived from wheat relative *Triticum dicoccoides* (Mergoum et al., 2005a, 2006a). The latest and the most effective FHB resistant cultivar, "Glenn", was released in 2005 and possesses resistance of both Sumai 3 and *Triticum dicoccoides* (Mergoum et al., 2006b). However, the Sumai 3 source of resistance to FHB was reported to possess grain shattering (Zhang and Mergoum, 2007a,b), an undesirable trait for modern wheat cultivars (Fig. 5). In the recent cultivar, "Glenn", this linkage is broken and Glenn is reported to be both FHB and shattering resistant (Mergoum et al., 2006b).

Further progress in enhancing the level of FHB resistance beyond the current level can come from a breeding strategy that would favor the accumulation of multiple major and minor genes from various sources into a single genotype. This is being attempted at the NDSU, HRS wheat breeding program by combining the

Fig. 5 Grain shattering symptoms observed on a HRSW susceptible genotype grown under field in Prosper, ND. USA. (*Photo: Mohamed Mergoum*)

different breeding methodologies involving traditional methods like the modified pedigree method and recurrent selection and modern breeding tools like double haploidy and MAS.

Diverse sources of resistance to leaf spotting diseases which include Septoria tritici blotch, Septoria nodorum blotch, spot blotch, and tan spot are now available in improved semi-dwarf wheats (Singh and Rajaram, 2002) and are utilized in breeding wheats for areas where these diseases are important (Duveiller et al., 2007). A high proportion of elite germplasm developed and evaluated in the Northern Plains of North America has resistance to leaf spotting diseases (Singh et al., 2006a; Mergoum et al., 2007). This is mainly due to the integration of field screening with greenhouse evaluation and the use of novel techniques like toxins, culture filtrates, and MAS in germplasm screening.

The permanence and effect of cultivars developed for resistance to insects and pests is a very challenging component of wheat breeding. Resistant cultivars in association with other integrated pest management strategies are used successfully to counter Hessian fly, stem sawfly, and leaf beetle. Breeding for Hessian fly is the most successful and 31 major resistance genes have been used (Williams et al., 2003). Most resistant cultivars possess multiple major resistance genes to have effective and durable resistance. Resistance to stem sawfly is associated primarily with breeding for stem solidness and several high-yielding cultivars with a high level resistance to stem sawfly have been released. Genetic resistances to other important insects including greenbug, leaf beetle, and curl mite have been identified and presently the development of resistant cultivars is in progress. Higher success in breeding for resistance to disease and pests has been achieved in recent years due to the integrated approach involving traditional and molecular breeding strategies.

7.4 Tolerance to Abiotic Stresses

Direct and indirect selection approaches, in recent years, have resulted in significant success in developing cultivars effective against abiotic stresses. Drought regularly limits wheat production in almost 50% of the wheat cropped area (Pfeiffer et al., 2005). Wheat breeders have successfully developed cultivars better adapted to moisture stress conditions resulting in 0.4–1.3% grain yield increase per annum in many drier wheat producing areas (Byerlee and Traxler, 1995). Major success achieved in wheat production under drought is due to the development of superior cultivars and their production under improved agronomic practices. Different breeding strategies including better understanding of genotype × environment interactions to make reliable selection, development of reliable and repeatable drought screening methods, and exploiting physiological basis of drought resistance have contributed to the success. The identification of molecular markers associated in drought resistance and their use in MAS in breeding programs may improve selection for abiotic stress resistance in the future.

Considerable success has been achieved in breeding for lodging resistance by developing semi-dwarf cultivars, resulting partly in the Green Revolution. The improved lodging resistance conferred by reducing culm length and increasing harvest index has further allowed exploitation of yield promoting factors like response to irrigation and fertilization.

In recent years major emphasis, with considerable success, has been put on breeding for pre-harvest sprouting resistance especially in white wheat, a trait that is believed to be linked to sprouting. Genetic control of resistance to pre-harvest sprouting has been identified and used in cultivar development with success. Although pre-harvest sprouting is environmentally sensitive, selection for high dormancy, desired plant morphology, and reduced rate of water uptake by spikes have resulted in the development of resistant cultivars. Additionally, the selection of genotypes with reduced levels of α-amylase activity in the grain obtained through laboratory testing and MAS for pre-harvest sprouting have significantly contributed to the development of resistant cultivars.

Considerable success has been obtained in selecting wheat tolerant to acidic soils and aluminum toxicity as laboratory tests, MAS, and selection in fields can be performed. However, limited success has been made in developing genotypes tolerant to heat, flooding, salt and alkaline soils and wheat breeders are looking for new avenues to address these challenges.

8 Integration of Novel Technologies in Breeding Programs

The past 15 years have witnessed substantial progress in the development and advancement of numerous genomics tools in wheat including mapping, sequencing, expressed sequence tags (ESTs), large insert libraries, gene cloning, bioinformatics, and analysis of gene expression. These technologies have contributed to our understanding of the wheat genome and its genetic improvement. Compared to other major crop plants, genomic analysis of wheat is more difficult because it is polyploid and has a relatively large genome (Arumuganathan and Earle, 1991), more than 6 and 30 times larger than maize and rice, respectively. This is due to large amounts of repetitive DNA (~90%). Genetic mapping is further hampered by low levels of polymorphism (Anderson et al., 1993).

Genetic and physical maps are the foundation for genomics studies. The genetic maps of wheat are well-populated with microsatellite markers that are useful for MAS by breeders and anchoring other types of markers (Somers et al., 2004). Physical maps have been constructed based on deletion lines (Endo and Gill, 1996) that take advantage of wheat's ability to tolerate the absence of chromosomal segments due to compensation by homoeologues. BAC libraries have been produced for several *Triticum* species (see http://agronomy.ucdavis.edu/Dubcovsky/BAC-library/ITMIbac/ITMIBAC.htm). These libraries are essential tools for physical mapping and also facilitate sequencing and map-based cloning efforts.

Agronomicaly important genes have been cloned in wheat, including the cereal cyst nematode resistance gene, *Cre3* (Lagudah et al., 1997); semi-dwarfing gene *Rht-B1* (syn. *Rht1*) (Peng et al., 1999); leaf rust resistance genes *Lr10* (Feuillet et al., 2003) and *Lr21* (Huang et al., 2003); the vernalization genes *Vrn1* (Yan et al., 2003), *Vrn2* (Yan et al., 2004), and *Vrn3* (Yan et al., 2006); aluminum tolerance gene *ALMT1* (Sasaki et al., 2004); powdery mildew resistance gene *Pm3* (Srichumpa et al., 2005); and the domestication gene *Q* (Simons et al., 2006).

Because of its large amounts of repetitive DNA, sequencing the entire wheat genome using existing technology is not practical. However, there is increasing evidence that the gene space of wheat is concentrated (Yan et al., 2003) and this should facilitate future sequencing strategies that focus on gene-rich regions. Although estimates of base pairs per cM, one estimate of gene density, vary dramatically in wheat depending upon the chromosomal location, it is encouraging that the maximum gene density is similar to that observed in *Arabidopsis* and rice. Most of the publicly available wheat sequence data are ESTs [more than 800,000 as of this writing (http://www.ncbi.nlm.nih.gov/dbEST/dbESTsummary.html)]. In addition to being potential markers and landmarks, ESTs serve as important connecting points among species. For example, a wheat EST may have sequence homology with other organisms in which the function of the corresponding gene has been determined, thus leading to the prediction of function for the wheat homolog.

The synteny of cultivated wheat with its diploid relatives and partial synteny with other grasses, including rice, greatly facilitates genomics activities related to gene discovery (Devos, 2005). At the very least, the synteny with rice can be exploited as a means of providing markers to more fully saturate the syntenous region in wheat (Liu and Anderson, 2003). If a trait or biochemical pathway exists in both rice and wheat, rice may serve as the source or intermediary to find the homologous gene in wheat. However, many important targets for selection (e.g. bread-making properties and resistance to several important diseases) will be unique to wheat. Another pitfall in the use of rice sequence is that a number of rearrangements exist, even at the sequence level.

The International Wheat Genome Sequencing Consortium (http://www.wheat-genome.org/index.html) was recently organized to coordinate the development of DNA-based tools and resources that result from the complete sequence of the common (hexaploid) wheat genome. Further advances in genomic technologies such as SNP markers, map-based cloning, and array-based analysis of gene expression will contribute to our understanding of the wheat genome and the genetic improvement of this staple crop.

9 Foundation Seed Production and Intellectual Property Issues

During the multiplication of cultivars for use as seed, it is essential that the genetic purity of the cultivar is maintained. To ensure purity and good heath of seed of wheat cultivars, elaborate seed production programs exit. Production of pedigreed seed can be separated into four different levels: Breeder seed, Foundation seed, Registered seed, and Certified seed. Breeder seed is produced in isolation under the direct supervision of the wheat breeder who develops the cultivar and is the purest form of a cultivar (Fig. 6). Selection is performed to eliminate off-types and extra care is taken to prevent out-crossing or natural hybridization and mechanical mixture. Foundation seed is derived from breeder seed. Its production is carefully

Fig. 6 A field of seed increase of a new spring wheat cultivar grown under raised bed irrigation system in the Northern State of Sonora, Mexico (*Photo: Mohamed Mergoum*)

supervised or approved by seed producers of the parent seed agency. This agency is generally an agricultural experiment station or a state foundation seed program with trained experts to handle seed production. Foundation seed is produced in the area of adaptation of the concerned cultivar and has to be certified by a seed certifying agency for purity and other characteristics. Registered seed is derived from foundation seed and is produced under the supervision of a seed certifying agency. Production of registered seed is under control of a registered seed producer. Registered seed is planted by seed growers to produce the certified seed. To be certified, seed must meet the prescribed requirements regarding purity and quality and certified seed is available to general distribution to farmers for commercial commodity production.

Plant Variety Protection (PVP) and plant breeder's rights (PBR) are the rights granted by the government to a plant breeder, originator, or owner of a cultivar to exclude others from producing or commercializing the seed of a cultivar for a minimum period. To qualify for PBR protection, the cultivar has to be novel, distinct from existing cultivars, and uniform and stable in its essential characteristics. The PVP/PBR protects the released cultivars but not the genetic components and the breeding procedures used in the development of the cultivar. Wheat breeders strive to develop cultivars with novel traits which can be patented for these novel traits. The PBR protections are generally valid for 20–25 years; however, there is no restriction on the use of these cultivars as genetic material in the development of future cultivars. One drawback of PVP and patents is that it limits sharing of germplasm and the use of cultivars with novel traits may be restricted.

10 Future Prospects

Globally by the year 2020 one billion metric tons of wheat will be required compared to the current production of nearly 626 million metric tons (Rajaram, 2001). This challenging projection for wheat needs by 2020 can be achieved provided there is continuous support and investment in agricultural research and development, especially in conventional wheat breeding, integrated pest management, improved seed multiplication and distribution systems, and optimum and efficient use of inputs. Future wheat breeding will need to focus on redesigning the wheat plant, discovery and assembly of hybrid vigor, efficient management of water and drought, genetically superior systems of uptake and translocation of nutrients, suitable germplasm adapted to conservation tillage practices, durable and multiple disease resistance that would ensure superior germplasm for future cultivars with high and stable yield potential available for commercial production (Rajaram, 2001). Additionally, both traditional plant breeding in association with modern plant breeding techniques like double haploid and MAS need to be integrated to have a successful and productive wheat breeding program.

References

Anderson, J.A., Churchill G.A., Autrique J.E., Tanksley S.D. and Sorrells M.E. (1993). Optimizing parental selection for genetic linkage maps. Genome 36, 181–186.

Anderson, J.A., Stack R.W., Liu S., Waldron B.L., Fjeld A.D., Coyne C., Moreno-Sevilla B., Mitchell F.J., Song Q.J., Cregan P.B. and Frohberg R.C. (2001). DNA markers for Fusarium head blight resistance QTLs in two wheat populations. Theor. Appl. Genet. 102, 1164–1168.

Arbelbide, M. and Bernardo R. (2006). Mixed-model QTL mapping for kernel hardness and dough strength in bread wheat. Theor. Appl. Genet. 112, 885–890.

Arumuganathan, K. and Earle E.D. (1991). Nuclear DNA content of some important plant species. Plant Mol. Biol. Rep. 9, 208–218.

Békés, F., Kémeny S. and Morell M. (2006). An integrated approach to predicting end-product quality of wheat. Eur. J. Agron. 25, 155–162.

Bergthaller, W.J. (1997). New uses of wheat gluten and non-starch wheat components. In: J.L. Steele and O.K. Chung (eds.). Proceedings of the First International Wheat Quality Conference. Grain Industry Alliance, Kansas, USA. pp. 285–301.

Bonnett, D.G., Rebetzke G.J. and Spielmeyer W. (2005). Strategies for efficient implementation of molecular markers in wheat breeding. Mol. Breed. 15, 75–85.

Branlard, G., Dardevet M., Saccomano R., Lagoutte F. and Gourdon J. (2001). Genetic diversity of wheat storage proteins and bread wheat quality. In: Z. Bedo and L. Lang (eds.). Wheat in a Global Environment. Kluger Academic Publishers, The Netherlands. pp. 157–169.

Buerstmayr, H., Lemmens M., Hartl L., Doldi L., Steiner B., Stierschneider M. and Ruckenbauer P. (2002). Molecular mapping of QTLs for Fusarium head blight resistance in spring wheat. I. Resistance to fungal spread (type II resistance). Theor. Appl. Genet. 104, 84–91.

Byerlee, D. and Traxler G. (1995). National and international wheat improvement research in the post green revolution period. Evolution and impacts. Am. J. Agric. Econ. 77, 268–278.

Culshaw, D. (1997). Straw as fuel. In: G.M. Campbell, C. Webb and S.L. McKee (eds.). Cereals, Novel Uses, and Processes. Plenum Press, New York, pp. 153–158.

Devos, K.M. (2005). Updating the 'Crop circle. Curr. Opin. Plant Biol. 8, 155–162.

Dubcovsky, J. (2004). Marker assisted selection in public breeding programs: The wheat experience. Crop Sci. 44, 1895–1898.

Dupont, F.M., Hurkman W.J., Vensel W.H., Tanaka C., Kothari K.M., Chung O.K. and Altenbach S.B. (2006). Protein accumulation and composition in wheat grains: Effects of mineral nutrients and high temperature. Eur. J. Agron. 25, 96–107.

Duveiller, E., Singh R.P. and Nicol J.M. (2007). The challenges of maintaining wheat productivity: Pests, diseases, and potential epidemics. Euphytica 157, 417–450.

Dvořák, J. (1998). Genome analysis in the *Triticum-Aegilops* alliance, In: A.E. Slinkard (eds.), Proceedings of the 9th International Wheat Genetics Symposium, Vol. 1, Saskatoon, Saskatchewan, Canada, University Extension Press, University of Saskatchewan, Saskatoon, Saskatchewan. pp. 8–11.

Endo, T.R. and Gill B.S. (1996). The deletion stocks of common wheat. J. Hered. 87, 295–307.

Faridi, H. (1988). Flat breads. In: Y. Pomeranz (ed.). Wheat Chemistry and Technology. American Association of Cereal Chemists, St. Paul, MN. pp 457–506.

Feuillet, C., Travella S., Stein N., Albar L. and Keller B. (2003). Map-based isolation of the leaf rust disease resistance gene *Lr10* from the hexaploid wheat (*Triticum aestivum* L.) genome. Proc. Natl. Acad. Sci. USA 100, 15253–15258.

Gale, K.R. (2005). Diagnostic DNA markers for quality traits in wheat. J. Cereal Sci. 41, 181–192.

Gill, B.S., Appels R., Botha-Oberholster A.M., Buell C.R., Bennetzen J.L., Chalhoub B., Chumley F., Dvořák J., Iwanaga M., Keller B., Li W., McCombie W.R., Ogihara Y., Quetier F. and Sasaki T. (2004). A workshop report on wheat genome sequencing: International Genome Research on Wheat Consortium. Genetics 168, 1087–1096.

He, Z.H., Liu A.H., Peña R.J. and Rajaram S. (2003). Suitability of Chinese wheat cultivars for production of northern style Chinese steamed bread. Euphytica 131, 155–163.

He, Z.H., Liu L., Xia X.C., Liu J.J. and Peña R.J. (2005). Composition of HMW and LMW glutenin subunits and their effects on dough properties, pan bread, and noodle quality of Chinese bread wheats. Cereal Chem. 82, 345–350.

Huang, L., Brooks S.A., Li W., Fellers J.P., Trick H.N. and Gill B.S. (2003). Map-based cloning of leaf rust resistance gene *Lr21* from the large and polyploidy genome of bread wheat. Genetics 164, 655–664.

Janda, J., Bartoš J., Šafář J., Kubaláková M., Valárik M., Číhalíková J., Šimková H., Caboche M., Sourdille P., Bernard M., Chalhoub B. and Doležel J. (2004). Construction of a subgenomic BAC library specific for chromosomes 1D, 4D, and 6D of hexaploid wheat. Theor. Appl. Genet. 109, 1337–1345.

Johnson, R. (1988). Durable resistance to yellow (stripe) rust in wheat and its implications in plant breeding. In: N.W. Simmonds and S. Rajaram (eds.). Breeding strategies for resistance to the rusts of wheat, CIMMYT, Mexico. pp. 63–75.

Kahlon, T.S. (1989). Nutritional implications and uses of wheat and oat kernel oil. Cereal Foods World 34, 872–875.

Lagudah, E.S., Moullet O. and Appels R. (1997). Map-based cloning of a gene sequence encoding a nucleotide binding domain and a leucine-rich region at the *Cre3* nematode resistance locus of wheat. Genome 40, 659–665.

Li, W., Zhang P., Fellers J.P., Friebe B. and Gill B.S (2004). Sequence composition, organization and evolution of the core Triticeae genome. Plant J. 40, 500–511.

Lillemo, M., Chen X.M., He Z.H. and Singh R.P. (2005). Leaf rust resistance gene *Lr34* is involved in powdery mildew resistance of CIMMYT bread wheat line Saar. In: H.T. Buck, J.E. Nisi and N. Solomon (eds.). Wheat Production in Stressed Environments. Mar del Plata, Argentina. pp. 17.

Liu, J.J., He Z.H., Zhao Z.D., Peña R.J. and Rajaram S. (2003). Wheat quality traits and quality parameters of cooked dry white Chinese noodles. Euphytica 131, 147–154.

Liu, S. and Anderson J.A. (2003). Marker assisted evaluation of Fusarium head blight resistant wheat germplasm. Crop Sci. 43, 760–766.

Maningat, C.C. and Seib P.A. (1997). Update on wheat starch and its uses. In: J.L. Steele and O.K. Chung (eds.). Proceedings of the First International Wheat Quality Conference. Grain Industry Alliance, Kansas, USA. pp. 261–284.

McKee, D. (2006). Focus on Thailand. World Grain 24(8), 16–20.

Mergoum, M., Frohberg R.C., Miller J.D. and Stack R.W. (2005a). Registration of 'Steele-ND' wheat. Crop Sci. 45, 1163–1164.

Mergoum, M., Frohberg R.C. and Stack R.W. (2005b). Breeding hard red spring wheat for Fusarium head blight resistance: Successes and challenges. In: H.T. Buck, J.E. Nisi and N. Solomon (eds.). Wheat Production in Stressed Environments. Mar del Plata, Argentina. pp. 161–167.

Mergoum, M., Frohberg R.C., Olson, T., Friesen T.L., Rasmussen J.B. and Stack R.W. (2006a). Registration of 'Howard' wheat. Crop Sci. 46, 2702–2703.

Mergoum, M., Frohberg R.C., Olson T., Friesen T.L., Rasmussen J.B. and Stack R.W. (2006b). Registration of 'Glenn' wheat. Crop Sci. 46, 473–474.

Mergoum, M., Singh P.K., Ali S., Elias E.M., Anderson J.A., Glover K.D. and Adhikari T.B. (2007). Reaction of elite wheat genotypes from the northern Great Plains of North America to Septoria diseases. Plant Dis. 91, 1310–1315.

Mujeeb-Kazi, A. and Rajaram S. (2002). Transferring alien genes from related species and genera for wheat improvement. Bread Wheat: Improvement and Production. In: B.C. Curtis, S. Rajaram and H. Gómez Macpherson (eds.). FAO Plant Production and Protection Series No 30. Food and Agriculture Organization of United Nations, Rome.

Otteson, B.N., Mergoum M. and Ransom J.K. (2007). Response of spring wheat to varying seeding rate and nitrogen management. Agron. J. (in press: A07–0002).

Palazzolo, G. (2003). Cereal bars: they're not just for breakfast anymore. Cereal Foods World 48, 70–72.

Peña, R.J. and Pfeiffer W.H. (2005). Breeding methodologies and strategies for durum wheat quality improvement. In: C. Royo, M.N. Nachit, N. Di Fonzo, J.L. Araus, W.H. Pfeiffer and G.A. Slafer (eds.). Durum Wheat Breeding: Current Approaches and Future Strategies. Food Product Press. The Haworth Press Inc., New York. pp. 663–772.

Peña, R.J., Trethowan R.M., Pfeiffer W.H. and van Ginkel M. (2002). Quality (end-use) improvement in wheat. Compositional, genetic, and environmental factors. In: A.S. Basra and L.S. Randhawa (eds.). Quality Improvement in Field Crops. Food Product Press, Haworth Press, Inc., New York. pp. 1–37.

Peng, J.R., Richards D.E., Hartley N.M., Murphy G.P., Devos K.M., Flintham J.E., Beales J., Fish L.J., Worland A.J., Pelica F., Sudhakar D., Christou P., Snape J.W., Gale M.D. and Harberd N.P. (1999). 'Green revolution' genes encode mutant gibberellin response modulators. Nature 400, 256–261.

Pfeiffer, W.H, Trethowan R.M., van Ginkel M., Ortiz M.I. and Rajaram S. (2005). Breeding for abiotic stress tolerance in wheat. In: M. Ashraf and P.J.C. Harris (eds.). Abiotic Stresses: Plant Resistance through Breeding and Molecular Approaches. The Haworth Press, Inc. NY. pp. 401–489.

Popineau, Y., Huchet B., Larré C. and Bérot S. (2002). Foaming and emulsifying properties of fractions of gluten peptides obtained by limited enzymatic hydrolysis and ultrafiltration. J. Cereal Sci. 35, 327–335.

Prior, D. (1997). The cradle of civilization: A snapshot of the flour milling industry Feed & Grain, October 1997, pp. 15–17.

Rajaram, S. (2001). Prospects and promise of wheat breeding in the 21st century. Euphytica 119, 3–15.

Ransom, J.K., Endres G.J. and Schatz B.G. (2007). Sustainable improvement of wheat yield potential: The role of crop management. J. Agric. Sci. 14, 55–61.

Riley, R. and Chapman V. (1958). Genetic control of the cytological diploid behavior of hexaploid wheat. Nature 182, 713–715.

Šafář, J., Bartos J., Janda J., Bellec A., Kubaláková M., Valárik M., Pateyron, S., Weiserová J., Tušková J., Číhalíková J., Vrána J., Šimková H., Faivre-Rampant P., Sourdille P., Caboche M., Bernard M., Doležel J. and Chalhoub B. (2004). Dissecting large and complex genomes: flow sorting and BAC cloning of individual chromosome from bread wheat. Plant J. 39, 960–968.

Sasaki, T., Yamamoto Y., Ezaki B., Katsuhara M, Ahn S.J., Ryan P.R., Delhaize E. and Matsumoto H. (2004). A wheat gene encoding an aluminum activated malate transporter. Plant J. 37, 645–653.

Sayre, K.D., Rajaram S. and Fisher R.A. (1997). Yield potential progress in short bread wheat in the northwest Mexico. Crop Sci. 37, 36–42.

Sears, E.R. (1954). The aneuploids of common wheat. Mol. Agr. Exp. Sta. Res. Bull. 572, 1–59.

Sears, E.R. (1966). Nullisomic-tetrasomic combinations in hexaploid wheat. In: R. Riley and K.R. Lewis (eds.). Chromosome Manipulation and Plant Genetics. Oliver & Boyd, Edinburgh. pp. 29–45.

Simons, K.J., Fellers J.P., Trick H.N., Zhang Z., Tai Y.S., Gill B.S. and Faris J.D. (2006). Molecular characterization of the major wheat domestication gene *Q*. Genetics 172, 547–555.

Singh, P.K., Mergoum M., Ali S., Adhikari T.B., Elias E.M., Anderson J.A., Glover K.D. and Berzonsky W.A. (2006a). Evaluation of elite wheat germplasm for resistance to tan spot. Plant Dis. 90, 1320–1325.

Singh, R.P., Hodson D.P., Jin Y., Huerta-Espino J., Kinyua M.G., Wanyera R., Njau P. and Ward R.W. (2006b). Current status, likely migration and strategies to mitigate the threat to wheat production from race Ug99 (TTKS) of stem rust pathogen. CAB Reviews: Perspectives in Agriculture, Veterinary Science, Nutrition and Natural Resources 1 (No. 054), 13pp.

Singh, R.P., Huerta-Espino J., Sharma R., Joshi A.K. and Trethowan R.M. (2007). High yielding spring bread wheat germplasm for irrigated agro-ecosystems. Euphytica 157, 351–363.

Singh, R.P. and Rajaram S. (2002). Breeding for disease resistance in wheat. In: B.C. Curtis, S. Rajaram and H. Gomez Macpherson (eds.). Bread Wheat Improvement and Production. FAO, Rome. pp. 141–156.

Singh, R.P., William H.M., Huerta-Espino J. and Rosewarne G. (2004). Wheat Rust in Asia: Meeting the Challenges with Old and New Technologies. New directions for a diverse planet: Proceedings of the 4th International Crop Science Congress Brisbane, Australia, 26 Sep–1 Oct 2004 | ISBN 1 920842 20 9 | www.cropscience.org.au

Srichumpa, P., Brunner S., Keller B. and Yahiaoui N. (2005). Allelic series of four powdery mildew resistance genes at the *Pm3* locus in hexaploid bread wheat. Plant Physiol. 139, 885–895.

Somers, D.J., Isaac P. and Edwards K. (2004). A high-density microsatellite consensus map for bread wheat (*Triticum aestivum* L.). Theor. Appl. Genet. 109, 1105–1114.

Souza, E.J., Graybosch R.A. and Guttieri M.J. (2002). Breeding wheat for improved milling and baking quality. In: A.S. Basra and L.S. Randhawa (eds.). Quality Improvement in Field Crops. Food Product Press, Haworth Press, Inc., New York. pp. 39–74.

Spielmeyer, W., McIntosh R.A., Kolmer J. and Lagudah E.S. (2005). Powdery mildew resistance and *Lr34/Yr18* genes for durable resistance to leaf and stripe rust cosegregate at a locus on the short arm of chromosome 7D of wheat. Theor. Appl. Genet. 111, 731–735.

Spiertz, J.H.J., Hamer, R.J., Xu, H.Y., Primo-Martin, C., Don C. and van der Putten, P.E.L. 2006. Heat stress in wheat; effects on grain weight and quality within genotypes. Europ. J. Agr. 25, 89–95.

Varady, K.A., Wang Y. and Jones P.J.H. (2003). Role of policosanols in the prevention and treatment of cardiovascular disease. Nutr. Rev. 61, 376–383.

Williams, C.E., Collier N., Sardesai C.C. and Ohm H.W. (2003). Phenotypic assessment and mapped markers for *H31*, a new wheat gene conferring resistance to Hessian fly (Diptera: Cecidomyiidae). Theor. Appl. Genet. 107, 1516–1523.

Yan, L., Fu D., Li C., Blechl A., Tranquilli G., Bonafede M., Sanchez A., Valarik M., Yasuda S. and Dubcovsky J. (2006). The wheat and barley vernalization gene *VRN3* is an orthologue of FT. Proc. Natl. Acad. Sci. USA 103, 19581–19586.

Yan, L., Loukoianov A., Blechl A., Tranquilli G., Ramakrishna W., SanMiguel P., Bennetzen J.L., Echenique V. and Dubcovsky J. (2004). The wheat *VRN2* gene is a flowering repressor down-regulated by vernalization. Science 303, 1640–1644.

Yan, L., Loukoianov A., Tranquilli G., Helguera M., Fahima T. and Dubcovsky J. (2003). Positional cloning of the wheat vernalization gene *Vrn1*. Proc. Natl. Acad. Sci. USA 100, 6263–6268.

Zhang, Y., He Z., Zhang A., van Ginkel M., Peña R.J. and Ye G. (2006). Pattern analysis on protein properties of Chinese and CIMMYT spring wheat cultivars sown in China and CIMMYT. Aus. J. Agric. Res. 57, 811–822.

Zhang, G. and Mergoum M. (2007a). Developing evaluation methods for kernel shattering in spring wheat. Crop Sci. 47, 1841–1850.

Zhang, G. and Mergoum M. (2007b). Molecular mapping of kernel shattering and its association with Fusarium head blight resistance in a Sumai3 derived population. Theor. Appl. Genet. 115, 757–766.

Rye Breeding

H.H. Geiger and T. Miedaner

Abstract Rye (*Secale cereale* L.) is mainly a European cereal with about 75% of the global production growing in Russia, Belarus, Poland, Germany, and Ukraine. It has the best overwintering ability, and the highest tolerance to drought, salt, or aluminium stress from all small-grain cereals. Harvest is used for bread making, feed, and in growing demands for ethanol and biomethane production as a renewable energy source. Hybrid rye is competitive to triticale and wheat also on better soils and grown in Germany on about 70% of the total rye acreage. Rye developed in the Middle East as a secondary crop, cultivated rye has its greatest diversity in landraces and populations from Central and East Europe. Their utility for breeding has considerably increased by progress in marker-based introgression of donor chromosome segments. Resistance breeding is presently focused on leaf and stem rust (*Puccinia recondita*, *P. graminis* f.sp. *secalis*), ergot (*Claviceps purpurea*), and *Fusarium* diseases. Leaf blotch (*Rhynchosporium secalis*) and soilborne viruses might gain more attention in the future. Main breeding goals are grain yield, straw shortness, lodging resistance, high kernel weight, tolerances to pre-harvest sprouting and abiotic stresses. Population varieties comprise open-pollinated and synthetic varieties. Both are derived from self-incompatible breeding populations which are steadily improved by recurrent half- or full-sib selection. Open pollinated varieties (OPVs) constitute selected fractions of those populations whereas synthetic varieties are composed of specifically selected parents from which they can identically be reconstituted. Most modern population varieties contain germplasm from two or more genetically distant gene pools. Hybrid breeding is based on self-fertile gene pools and cytoplasmic genic male sterility (CMS) is used as hybridizing mechanism. Long-lasting breeding cycles are needed for the development of seed parent lines since testcrossing is only possible after the inbred lines have been converted to CMS analogues by repeated backcrossing. Options to speed up this process are discussed. Development of restorer lines is straightforward once

H.H. Geiger(✉)
University of Hohenheim, Institute of Plant Breeding, Seed Science, and Population Genetics, D-70593 Stuttgart, Germany, e-mail: geigerhh@uni-hohenheim.de

M.J. Carena (ed.), *Cereals*,
DOI: 10.1007/978-0-387-72297-9, © Springer Science + Business Media, LLC 2009

effective restorer genes have been introduced to the respective breeding populations. Recurrent improvement of fertility restoration is most efficiently accomplished by recombining selected inbred lines after the first or second testcrossing stage. Commercial hybrid seed production requires well-skilled farmers, careful seed processing, and deliberate logistics since rye produces huge amounts of pollen which may be transported over long distances. Even the slightest genetic contamination of the CMS pre-basis and basis seed production may render the respective seed lots worthless for subsequent multiplication. To reduce the cost of the final step of seed production, the CMS seed parent and the pollinator parent are grown as a mixture in a 95:5 ratio. Thus, only about 95% of the certified seed consists of true hybrid seed. Whereas the remainder 5% are randomly intermated plants of the pollinator. However, the latter generally are poor competitors and therefore do not impair the yielding performance of the 'hybrid' stand. In the last decades, population and hybrid breeding led to substantial progress in grain yield and other traits.

1 Introduction

Rye (*Secale cereale* L.) is a diploid ($2n = 2x = 14$) annual, cross-pollinated cereal with an effective gametophytic self-incompatibility system (Lundqvist, 1956). Similar to many crops of the Old World, *S. cereale* evolved in the Fertile Crescent of the Near East. Main regions of diversity are Turkey, Libanon, Syria, Iran, Iraq, and Afghanistan. Rye was, however, never cultivated as a crop there but grew and still grows as a weed within the stands of barley and wheat. Annual rye forms evolved in this agricultural context by natural selection leading to semi- to non-shattering ears, larger kernels, and dormancy ('primitive rye'). Populations growing at higher altitude show an excellent cold hardiness. The first cultivation of rye took place in the region around the Caspian Sea at about 3000–4000 BC. Rye came to Eastern Europe by Slavic people. During their migration to the West at about 500 BC, they brought the knowledge of rye growing to the Germanic, Celtic, and Finnish peoples. During the whole Middle Ages and modern times till the 1960s, rye was the major cereal crop from Germany to Eastern Siberia. Large breeding progress in the self-pollinated crops, wheat and barley, lead to a decrease of the rye acreage in regions where stress tolerance is less important. On a world-wide basis, rye acreage was nearly halved in the last decade (Table 1). Main rye producing countries presently are Russia, Belarus, Poland, Germany, and Ukraine. By far the highest grain yields are obtained in Germany illustrating the high potential of the crop under intensive growing conditions.

Rye is mostly grown as a winter cereal. Spring rye is superior in extremely cold areas or where the snow cover lasts longer than 3 months. Compared to other cereals, rye has the best overwintering ability and the highest tolerance to drought, salt, or aluminium stress among all small-grain cereals. Rye excels in considerable growth during late fall and resumes growth very quickly in early spring. Rye is

Table 1 Main rye growing countries and acreages in 1995 and 2005 and average grain yield in 2005 (FAO, 2006)

Continent/Country	Acreage (1,000 ha)		Yield (t ha^{-1})
	1995	2005	2005
Europe	8,613	5,620	2.46
Russian Federation	3,250	1,900	1.91
Poland	2,452	1,410	2.38
Belarus	969	580	2.16
Germany	861	555	5.07
Ukraine	552	610	1.95
Asia	1,126	587	1.87
USA	374	337	1.77
World	10,206	6,598	2.37

more productive than other cereals on infertile, sandy, or acid soils, as well as on poorly prepared land. Hybrid rye is competitive to triticale and wheat even on better soils due to its higher yield potential.

About 50–75% of the harvest is used for bread making. Traditional rye bread is the dark sour bread known in North Germany, Finland, the Baltics, Poland, Belarus, and the Russian Federation. In Sweden, Denmark, and parts of Germany, rye flour is commonly mixed with 25–50% wheat flour for bread making. The remainder rye harvest is used for feeding, the production of alcohol (Schnaps, Vodka), and as a resource for renewable bioenergy (bioethanole, biomethane, and combustion).

2 Germplasm and Use of Genetic Resources

Cultivated rye displays a broad range of genetic diversity reflecting the great ecological differences among the various growing areas. As expected from a cross-pollinated crop, a higher amount of genetic diversity can be found within than among populations (Persson and von Bothmer, 2000). A recent marker-based diversity study with landraces and varieties from Nordic countries, Germany, and Poland revealed eight clusters differing in origin (Persson et al., 2006). Large gene bank collections exist in various countries (Table 2) comprising East European cultivars, landraces from Europe, Asia, and South America, primitive populations from the Near East, and wild *Secale* species. In European gene banks, 9,901 accessions are stored, one-third of which are likely to be duplicated (Podyma, 2003). Additionally, 236 accessions are available from *S. silvestre*, *S. iranicum*, and *S. montanum*.

In practical rye breeding, genetic resources have not been intensively utilized for a number of reasons: (1) exotic germplasm generally lacks adaptation to the targeted growing area, (2) substantial difference in performance between elite and exotic germplasm for polygenic traits, (3) exotic germplasm is lacking inbreeding tolerance, (4) little is known about their genetic distance to established heterotic groups, and (5) genetic phenomena such as pleiotropy, epistasis, and coupling phase linkage between desired and undesired alleles may hinder a direct utilization

Table 2 Large collections of *Secale* accessions (Podyma, 2003)

Institution	Country	No. of accessions
N.I. Valivov Institute of Plant Industry	Russia	2,685
Botanical Garden of the Polish Academy of Science	Poland	1,630
Plant Breeding and Acclimatization Institute, Radzikow	Poland	1,354
Leibniz Institute of Plant Genetics and Crop Plant Research (IPK), Gatersleben	Germany	1,207
Research Institute of Crop Production	Czech Rep.	659
Aegean Agricultural Research Institute	Turkey	512
Instituto Nacional de Investigacion y Tecnologia Agraria y Alimentaria	Spain	428
Others	Various	1,426

(Haussmann et al., 2004). Despite these handicaps, exotic germplasm may contain genomic segments that can improve oligo- and polygenically inherited traits in highly selected breeding populations (de Vicente and Tanksley, 1993).

The targeted exploitation of genetic resources is possible by advanced backcross analysis of quantitative trait loci (AB-QTL; Tanksley and Nelson, 1996) or via introgression libraries (Eshed et al., 1992). An introgression library consists of a set of lines, each carrying a single marker-defined donor chromosome (DC) segment introgressed from a genetic resource into the background of an elite recipient line (Zamir, 2001). Ideally, the introgressed DC segments are evenly distributed over the whole recipient genome and the total genome of the 'exotic' donor is represented in the set of near-isogenic lines. In rye, two introgression libraries were established with the Iranian primitive rye accession 'Altevogt 14160' as donor (Sušić, 2005; Falke et al., 2008a) by marker-assisted selection using amplified fragment length polymorphisms (ATLP) and simple sequence repeat markers (Hackauf and Wehling, 2002). The libraries comprise 38 and 40 BC_2S_3 lines, respectively, jointly covering approximately 70% of the total donor genome. Most of the introgression lines harbour one to three homozygous DC segments with a mean length of about 12 cm. A comprehensive phenotypic evaluation of the libraries revealed considerable genetic variation for quantitatively inherited baking quality traits (Falke et al., 2008) and pollen-fertility restoration (Falke et al., 2009). Thus, these results demonstrate that introgression libraries can serve as a valuable tool for broadening the genetic base of rye breeding as well as for detecting and validating QTL (Zamir, 2001).

3 Disease Resistance

Important diseases of rye in Central and East Europe are snow mold (*Microdochium nivale*), foot rot caused by a complex of *Helgardia herpotrichoides*, *H. acuformis* (syn. *Pseudocercosporella herpotrichoides* var. *herpotrichoides*, var. *acuformis*),

M. nivale, and *Fusarium* spp., powdery mildew (*Blumeria graminis* f.sp. *secalis*), leaf rust (*Puccinia recondita*), stem rust (*P. graminis* f.sp. *secalis*), leaf blotch (*Rhynchosporium secalis* and other fungi), ergot (*Claviceps purpurea*), and Fusarium head blight (FHB) caused by *M. nivale* and various *Fusarium* species. Soilborne cereal mosaic virus (SBCMV), soilborne wheat mosaic virus (SBWMV), and wheat spindle streak mosaic virus (WSSMV), transferred by the soilborne fungus *Polymyxa graminis*, cause new diseases in some regions of Germany (Kastirr et al., 2006). For powdery mildew and leaf rust, both qualitative and quantitative resistances were reported, whereas quantitative variation was found for foot rot, head blight, and ergot.

In population varieties, the spread of diseases is reduced by collective buffering. Compared with line varieties in self-pollinated crops, rye populations are more likely to harbour resistances accumulated by natural selection (e.g., Geiger et al., 1988; Mirdita, 2006). They can be improved for their disease resistance by recurrent selection (RS) methods based on half- or full-sib families (HSF or FSF) (*cf.* Sect. 5.1). In hybrid rye, resistance breeding is simplified by the availability of inbred lines for reproducible testing, high genotypic variance between these lines, and the possibility to introgress major genes by backcrossing.

For efficient resistance selection, it is crucial to know the most important population parameters (Table 3). Genotypic variance was found to be very large for powdery mildew and leaf-rust resistance caused by both quantitative and qualitative resistances jointly segregating in a population (Wilde et al., 2006). In this case, quantitative resistances can only be detected when the masking effect of the race-specific resistance genes is eliminated by using appropriate pathogen races being virulent to all qualitative resistances. For foot-rot resistance, genotypic variance is significant, but generally low in self-fertile (SF) materials (Miedaner et al., 1995). Selection for lodging resistance does not necessarily lead to a correlated response for foot-rot resistance.

Table 3 Survey of population parameters determining the gain from selection for resistance to five rye diseases based on experimental results

Parameter	Mildew, leaf rust	Foot rot	FHB	Ergot
Variance components				
Inbred lines *per se* (L)	very large	small	large	moderate
L × environment	small-moderate	moderate	large	very large
GCA	++	++	++	++
SCA	ns	ns	(++)	ns
Heritability	high	high	moderate	small-mod
Heterosis	inconsistent	very small	moderate	mod negative
Genetic correlation				
Inbreds – hybrids	high	high	ns	moderate
Trait assessment	fast, easy	laborous	moderate	laborous
Expected selection gain	high	mod-small	moderate	mod-small

For details, see text; *mod* = moderate, *ns* = non-significant, ++ = very important, () = not in all years

For FHB resistance, genotypic variance between selfed lines is large and thus can easily be exploited. Variation for ergot resistance is available in all materials tested so far including cytoplasmic genic male sterile (CMS) materials and self-incompatible (SI) populations (Mirdita, 2006). Environmental effects plays a moderate to large role in all pathosystems. Even artificial inoculation with highly aggressive isolates may result in strongly deviating infection levels in different environments and cause significant genotype × environment interactions. This is especially valid for FHB and ergot resistance. Heritability estimates are generally lower for the latter two resistances compared with resistance to powdery mildew, leaf rust, and foot rot.

General combining ability (GCA) variance is much more important than specific combining ability (SCA) variance in all pathosystems (Table 3). The only exception is FHB resistance where significant SCA was found, however, not consistently across years (Miedaner and Geiger, 1996). No correlation was found between inbred lines and their hybrids in this pathosystem (Miedaner et al., 2003), in sharp contrast to the other pathosystems. Both findings necessitate multi-environmental evaluation of testcross progeny. The use of moderately susceptible testers is crucial to gain a maximum genetic differentiation. Moderate heterosis for resistance was found in FHB resistance. For the foot and leaf diseases, heterosis is practically absent. For ergot resistance, crosses normally display a higher disease severity in terms of weight of sclerotia than their inbred lines (Mirdita, 2006). Selection intensity is mainly restricted by the cost of inoculation and trait assessment. The wind-borne diseases can be provocated rather easily and scoring at one date on one leaf or even a single whole-plant rating will suffice (Miedaner et al., 2002). The other extremes are foot-rot resistance requiring individual scoring of at least ten stems per plot (Miedaner et al., 1995), and ergot resistance necessitating separation of sclerotia from the grain and determining their weight proportion (Mirdita et al., 2008).

Taking all parameters together, the expected selection gain is highest for mildew and leaf-rust resistance and lowest for foot-rot resistance. For FHB resistance, the necessity to test at the hybrid level substantially reduces the selection gain. These characteristics of the above pathosystems have implications on the method and generation in which selection for disease resistance should be practiced.

In hybrid breeding, mildew and leaf-rust resistance can easily be implemented in the regular line development scheme (*cf.* Sect. 5.2.3) by selecting already in early selfing generations. For all other diseases, specific pre-breeding procedures are required aiming at an increased frequency of resistance alleles in the elite materials. For improving foot-rot and ergot resistance, the introgression of positive alleles from SI elite populations or genetic resources is recommended to increase the genetic variance. Agreement between line and GCA effects allows selection among lines *per se*, that is, without prior testcrossing. Only for FHB resistance, selection should predominantly be carried out at the non-inbred level. A correlated reduction can then be expected for low deoxynivalenol content.

Since no substantial heterosis for resistance was found in any of the pathosystems, selection is necessary in both the seed- and the pollinator-line gene pools. In the future, experiments designed to estimate population parameters should be combined with QTL analyses. Marker-assisted selection may considerably add to progress in resistance breeding as recently demonstrated for FHB resistance in wheat (Wilde et al., 2007).

4 Use and Breeding Goals

Most goals in winter rye breeding are similar to those in other small-grain cereals. Generally, hybrid breeding is more flexible than population breeding in creating varieties with specific characteristics.

Grain yield is by far the most important trait for rye growers. The average grain yield in Europe presently (2004–2005) varies from 1.9 t ha^{-1} in Russia to 5.1 t ha^{-1} in Germany (Table 1). Generally, hybrids yield 15–20% more than population varieties (*cf.* Sect. 6).

Straw shortness and lodging resistance are important breeding goals. However, since the culm is the major assimilation organ in rye (Nalborczyk et al., 1981), extremely short-strawed rye varieties do not reach high grain yield levels, in particular under severe stress. Therefore, dwarf or semi-dwarf varieties never gained large acreages.

Tolerance to drought and nutrient stress are important components of yield stability because rye is widely grown on poor, sandy soils. Compared to wheat, rye has a much higher tolerance to abiotic stresses, such as drought, nitrogen deficiency, and high concentrations of aluminium, zinc, sodium, and acidity.

Baking quality: For milling and baking, mainly a high kernel weight and resistance to pre-harvest sprouting is demanded. Because of its low dormancy, rye kernels may start germinating already before harvest if the weather is warm and moist. This leads to a deterioration of the starch and considerably reduces baking quality. Indirect selection for low α-amylase activity by the falling number method is effective in hybrid materials without negative influence on yield and other agronomic traits (Wehmann et al., 1991). Indeed, some modern German hybrid varieties combine both high yielding performance and high falling number. Another important quality component is a high pentosane content that can indirectly be measured by near-infrared spectroscopy (Rode et al., 2005).

Feeding quality: Rye is increasingly used as an animal feed either on farm or by compound feed producers. The grain is rich in energy and contains more digestible protein and total digestible nutrients than oats or barley and a higher starch content than barley. Rye is not recommended in the diet for weaning pigs, growing chicken, and turkeys and should be restricted to 40–50% of the diet for other animals because of its high concentration of pentosanes (Boros, 2007). They negatively affect feed intake, feed conversion efficacy, and growth rate in animals. Thus, pentosanes have reverse effects on feeding and baking quality which means that contrasting genetic materials are needed for the two usages. At present, only in Poland, low-pentosane inbred lines are being developed by selecting for low extract viscosity (Madej et al., 1990).

Ethanol production: Rye is well suited for ethanol production as a renewable energy source. In the European Union, adding biofuels into gasoline is obligatory. Distiller's residues are marketed as protein concentrate for pigs, where rye with its favourable protein composition has an advantage over wheat and barley. The grain may not contain mycotoxins or alkaloids caused by *Fusarium* or ergot infections. Specific breeding goals for high ethanol production are high starch content, high starch yield, and high enzyme activity (Rode et al., 2005). For selecting starch content, the thousand kernel mass can be used as an indirect trait. Protein and pentosan content should be low.

Biomethane production: The use of rye as biomass substrate is rapidly growing in Germany. In humid regions, starting from about 800 mm of rain, a sequence of forage rye and maize will reach higher biomass yields than maize alone. In spring, rye is the fastest growing winter cereal and excels in a low specific water consumption. On the typical dry-warm maize sites, maize alone is safer because soil water may not suffice for growing two crops per year. On dry or cold sites, silage rye may be more economic than maize. Harvest is carried out in the milky ripe stage of grain development when the highest methane yield is obtained. Main breeding goals are high biomass yield and lodging resistance. The methane yield per kilogram dry matter did not differ among rye genotypes.

Forage and pasture: Rye is an excellent forage crop especially when combined with clover and ryegrass. It generally provides more forage than other small grains in late fall and early spring because of its rapid growth and its adaptation to low temperatures. For the same reasons, rye fits well into erosion control programs. For best forage quality, rye should be cut between early heading and the milky ripe stage. Main breeding goals are an early start of growth in spring and high rust resistance.

5 Breeding Methods and Techniques

5.1 Population Breeding

Population breeding comprises the development of open-pollinated varieties (OPVs) and synthetic varieties (Schnell, 1982). In cross-fertilized species like rye, such varieties constitute panmictic populations and can be regrown by the farmer without noteworthy yield reduction. Open-pollinated and synthetic populations differ in their genetic buildup. Whereas the former constitute selected fractions of one (or several) breeding population(s), the latter are established by intermating highly selected parental units with subsequent multiplication under open-pollination. The parental units may originate from more than one gene pool.

Fig. 1 Pollination control by foliar walls (*left*) or cabins (*right*)

5.1.1 Breeding Open-pollinated Varieties

Over decades, rye breeders used various modifications of a half-sib family (HSF) recurrent selection (RS) scheme for population improvement aiming at OPVs. A typical procedure includes four steps requiring 4 years:

1. Selection among equally spaced (e.g., 25×25 cm^2) so-called mother plants mainly for disease resistance, productive tillering, straw stiffness, spike characteristics, disease resistance, and general appearance.
2. Progenies of selected mother plants, that is HSFs, are evaluated in unreplicated drilled observations plots at two to three locations. Beside the above traits, lodging resistance, sprouting resistance, and grain quality are important breeding goals at this stage of selection.
3. Multiplication of remnant seed of the best HSFs by open-pollination in plots separated from each other by spatial isolation or by foliar isolation walls or cabins (Fig. 1).
4. Multi-environment yield trials of the advanced HSFs, here designated by (HSF)^{2}s, on six- to eight-rowed, 5–10 m^2 plots with one to two replicates per environment. Grain yield, stress tolerance, and lodging resistance are the most important breeding goals at this final selection stage.

To shorten the RS cycle from 4 to 2 years without renouncing yield trials, the mother plant genotypes can be multiplied by two to three cloning steps before planting. This way, about 20 plants per clone can be produced. The clones are transplanted into an isolated field to which the respective breeding population had been drilled some weeks earlier with gaps for the clones such that the latter will completely be surrounded by population plants. This furnishes enough seed of each HSF for unreplicated yield trials on 5 m^2 plots at three to four locations. A major disadvantage of this otherwise highly effective RS scheme is the great labour demand for cloning, particularly since it has to be accomplished during the most burdening labour peak in autumn between harvest and planting.

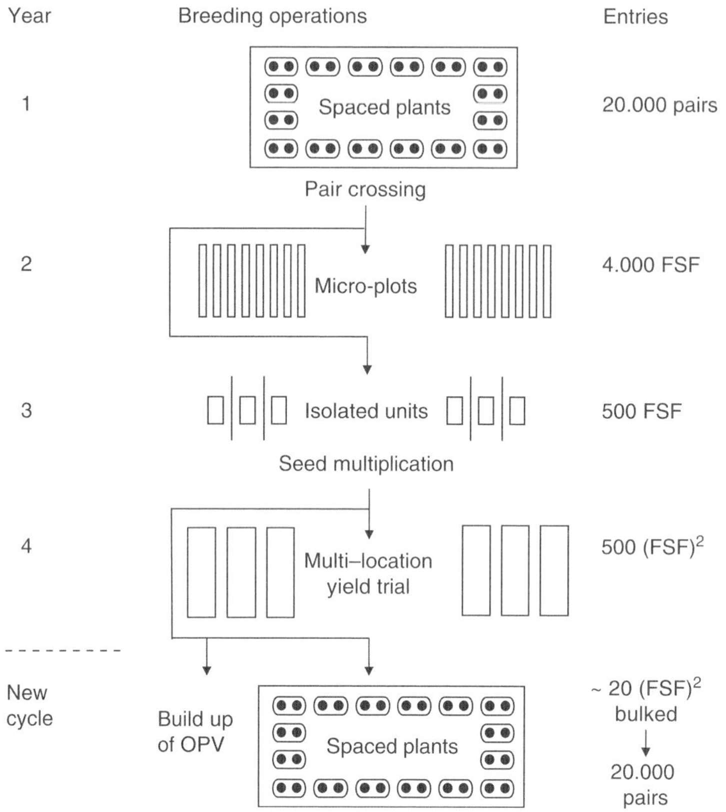

Fig. 2 Flow diagram and dimensioning example (number of entries) of a full-sib family (*FSF*) recurrent selection scheme for intrapool population improvement in self-incompatible rye; $(FSF)^2$ = progeny of an FSF advanced under open pollination in a pollen isolation device

In the course of time, many breeders changed from HSF to full-sib family (FSF) selection (Fig. 2). The latter requires the production of pair crosses under pollination bags in the first year. Because of the self-incompatibility of rye this is possible without emasculation. In the second year, FSFs instead of HSFs are grown in observation plots, and best FSFs are multiplied under pollen isolation in the third year. Finally, in the fourth year, $(FSF)^2$s are evaluated in multi-location yield trials.

The expected response to selection is greater for the FS than for the HS scheme because of complete parental control and greater genetic variance between the test units [FSF vs HSF and $(FSF)^2$ vs $(HSF)^2$, respectively]. However, producing the pair crosses requires a considerably higher experimental input than the corresponding steps in the HS scheme. Furthermore, only a weak selection pressure is possible among the plant pairs in year 1, and more $(FSF)^2$s than $(HSF)^2$s have to be saved after each RS cycle to comply with a minimum effective population size (Walsh, 2004).

As in the HS scheme, cloning the parent plants would allow to half the cycle length of the FS scheme, and the same pros and cons would apply. This high-input short-cycle scheme is well suited as a 'crash' procedure for rapidly increasing the yield level of an otherwise satisfactory breeding population.

Most modern OPVs are built up by merging selected fractions of two or even more genetically distant populations representing a set of heterotic pools. This way, part of the 'panmictic-midparent heterosis' (Lamkey and Edwards, 1999) can be used for enhancing the varietal performance. Experimental data (Hepting, 1978) indicate that crossing two genetically distant rye populations may increase the yield level of the population cross by 10–20% above that of the parent populations. However, at Hardy–Weinberg equilibrium which is rapidly attained during seed multiplication, about half of this increase is lost due to a corresponding drop in heterozygosity.

5.1.2 Breeding Synthetic Varieties

The parental components, in short 'parents', of a synthetic variety can be HS or FS families, clones, inbred lines, or other materials which can be preserved so that the variety can identically be re-composed. However, long-term preservation of rye clones, for example, by in vitro culture, is very difficult and expensive. Clones, therefore, have not gained practical relevance in synthetic breeding so far. Selfing SI genotypes (rather than cloning) is possible if the plants are cultivated at high temperature (30–35 °C) shortly before and during anthesis (Wricke, 1978). However, only few seeds per spike can be obtained this way and many genotypes don't respond to the treatment at all. Indeed most synthetics are composed of HS or FS families, and thus OPVs and synthetics basically have the same genetic structure and breadth.

Although SF and SI materials do not differ in flowering characteristics, SF plants display selfing rates of about 20–50% under open-pollination (Geiger and Schnell, 1970; Wricke, 1979). The resulting inbreeding was shown to drastically lower the performance level of SF synthetics below that of SI synthetics (Singh et al., 1984). Therefore, high-combining SF inbred lines as developed in hybrid breeding are not suited as parents of synthetic varieties.

Selection of parents is mostly based on their intra-pool breeding values as ascertained in RS programs. If parents from two or more heterotic groups are to be combined, it would be desirable to additionally consider their inter-pool breeding values. Production of testcross seed could be accomplished growing the candidates in isolated plots with excessive tester pollen. However, such selection procedures require much higher efforts or take more time to complete a selection cycle than the HS and FS schemes described above. It therefore appears questionable whether inter-pool testing is worthwhile in breeding synthetic rye varieties. Diallel or factorial crosses for estimating SCA effects are not rewarding since SCA variance determines only a negligible part of the genetic differentiation among synthetics.

The optimum number of parents principally depends on the same factors as the number of recombination units in RS. The effective population size should range between 25 and 50 to comply with an upper limit of 0.01 of the inbreeding coefficient (Crow and Kimura, 1970; Walsh, 2004). This corresponds to about 7–13 HSFs, 13–25 FSFs, or 25–50 clones.

5.2 Hybrid Breeding

Hybrid breeding allows (1) to fully exploit the 'panmictic-midparent heterosis' (Lamkey and Edwards, 1999) of crosses between genetically distant populations (heterotic groups) and (2) to capitalize not only on GCA but also on SCA effects for varietal improvement.

5.2.1 Base Materials

In Germany, where hybrid rye breeding started around 1970 (Geiger and Schnell, 1970), all former leading population varieties belonged to either the 'Petkus' or the 'Carsten' gene pool. In diallel variety crosses (Hepting, 1978), combinations between Petkus and Carsten varieties generally displayed the highest panmictic-midparent heterosis and the highest hybrid performance. The two gene pools were therefore chosen as starting material for the development of seed- and pollinator-line development, respectively. Later on, East European gene pools were found which proved to be heterotic counterparts to both the Petkus and the Carsten pool.

The natural self-incompatibility of rye can be overcome by self-fertility genes which were detected in various European materials (Ossent, 1938; Wricke, 1969; Wolski, 1970). These genes could easily be transferred to the above SI materials by repeated backcrossing since self-fertility is dominant over self-incompatibility.

Several sources of CMS were found in rye (Geiger and Schnell, 1970; Kobyljanskij, 1971; Łapiński, 1972; Klyuchko and Belousov, 1972; Adolf and Winkel, 1985). They can be classified into two major groups, the 'P' and the 'V' type (*P*ampa and *V*avilov, respectively; Łapiński and Stojałowski, 2001). Maintainers of the P-type CMS were found at rather high frequency in all rye cultivars studied so far (Geiger et al., 1995), whereas the V type is very difficult to maintain (Madej, 1976; Winkel et al., 1979). Almost all hybrid varieties listed so far are produced by means of the P cytoplasm. Fertility restorer genes, on the other hand, are rare for the P and abundant for the V type. The first restorer for P CMS was found in a European inbred line (Geiger, 1972). Restoring ability in European rye materials, however, is often unsatisfactory leading to hybrids with a reduced pollen shedding (Geiger et al., 1995) and a higher incidence of ergot infection under adverse weather conditions. More effective and environmentally stable restorer genes were recently detected in gene bank accessions of Iranian and South American primitive ryes on chromosome 4RL (Miedaner et al., 2000, 2005). Narrow-linked PCR-based markers are available for two of these sources: Pico Gentario and IRAN IX (*Rfp1* and *Rfp2*, respectively; Stracke et al., 2003). These new restorers were transferred to elite pollinator lines by marker-assisted backcrossing and have substantially improved pollen shedding of newly released hybrids (Wilde, personal communication).

5.2.2 Genetic Structure of Hybrids

All presently listed hybrids are crosses between a CMS single cross as seed parent and a restorer synthetic as pollinator (Geiger, 1982):

$$(A_{CMS} \cdot B) \cdot Syn_{Rf}$$

The parent lines A and B of the CMS single cross may be derived from the same or two different heterotic groups, both being unrelated to the pollinator gene pool.

The seed parent has to be absolutely male sterile under a wide range of environmental conditions and to furnish adequate amounts of high-quality seed. To meet the first requirement, the seed parent has to be genetically uniform to facilitate visual male sterility control during pre-basis and basis seed production (*cf.* Section 5.3). To comply with the second requirement, the seed parent needs to be sufficiently vigorous. Taken together, the two requirements can best be fulfilled by a single cross between unrelated homozygous lines.

The restorer synthetic usually is composed of two inbred lines. The genetic structure of the hybrid then corresponds to that of a double cross. As the vigour of inbred lines has considerably been improved over selection cycles, breeders are recently attempting to narrow down the genetic breadth of the pollinator parent by combining two sister lines or by using just one partially inbred line. This allows to more efficiently exploit SCA effects and to minimize the risk of negative epistatic effects due to a disruption of coadapted parental gene arrangements caused by segregation (Geiger, 1988). On the other hand, a genetically broader synthetic is a more secure pollinator due to a longer pollen shedding period.

Limited data is available on the yielding stability of different hybrid types. Becker et al. (1982) compared balanced sets of three-way and double crosses. The authors found no consistent stability trend in favour of one or the other type. However, the three highest yielding three-way crosses significantly surpassed the three best double crosses.

5.2.3 Parent Line Development

Parent line development in commercial hybrid breeding programs is exclusively done by selfing. No routinely applicable doubled-haploid technology exists so far (Flehinghaus-Roux et al., 1995; Tenhola-Roininen et al., 2006; *for review see*

Fig. 3 Production of inbred lines by bagging (*left*) and by multiplication in a foliar greenhouse (*right*)

Altpeter and Korzun, 2007). During the selfing phase, extensive testing for line and testcross performance is practised. Production of testcross seed requires a reliable male-sterility system since manual emasculation would not furnish enough seed for multi-environment yield trials. The P-CMS mechanism detected by Geiger and Schnell (1970) proved to be best suited for this purpose, whereas gametocides and transgenic approaches did not yet reach practical applicability. Thus, both testcross and commercial hybrid seed production (see below) are exclusively based on CMS.

Multiple cycles of line development have resulted in highly SF breeding materials. Inbred lines can easily be maintained by bagging or multiplicated in pollen isolation cabins or other devices (Fig. 3).

Yet, considerable inbreeding depression for grain yield and yield components occur in the course of inbreeding (Geiger and Miedaner, 1999). Seed yields of selected homozygous lines range between 30% and 50% relative to the performance level of non-inbred materials.

Development of CMS and Maintainer Lines

Parent line development generally comprises one or two stages of early-generation selection for line *per se* performance and two subsequent stages of testcross selection (Fig. 4). Transfer of lines into the CMS-inducing cytoplasm by backcrossing, usually starts after the S_2 line *per se* test and is continued throughout the testcrossing phase. Seed production for the first stage of testcross selection is produced on S_4-line analogues in BC_1 and for the second stage on S_6-line analogues in BC_2. Testers are the most elite restorer synthetics from the pollinator gene pool.

In this standard scheme of seed-parent development, testcrossing is considerably delayed by the preceding backcrossing steps. To avoid this delay, Geiger (2006) suggested to develop representative male-sterile testers from the 'opposite'

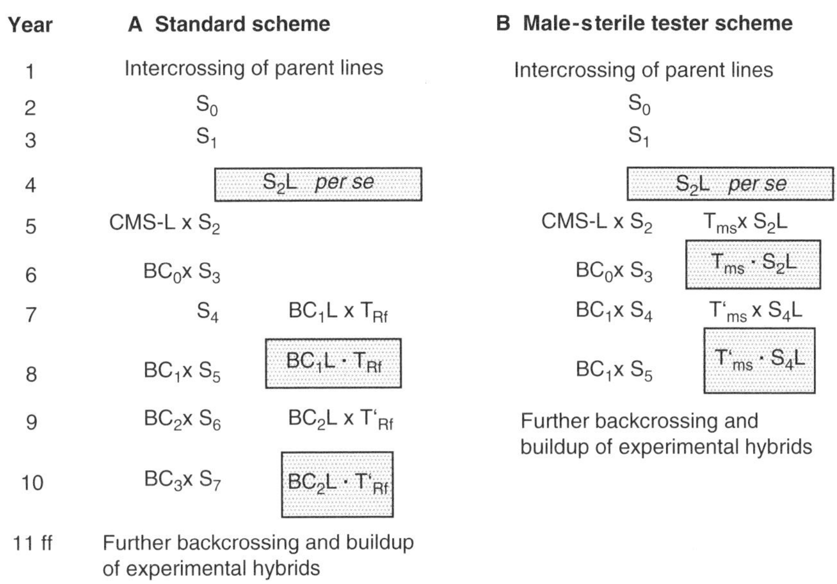

Fig. 4 Flow diagram of seed-parent line development. **A** Standard scheme, **B** male-ste rile tester scheme; S_x = selfing generation *x*, *L* = line, *CMS* = cytoplasmic genic male sterility, BC_x = backcross generation *x*, *T*, *T′* = different testers, *Rf* = restorer factor, *ms* = male sterile; A×B = production of cross A times B (parent generation), A · B = F_1 generation of that cross

pollinator gene pool. Then the selected S_2 lines could be testcrossed immediately after the line evaluation phase, and S_4 rather than S_6 lines could be used for the second stage of testcross selection. This would save 2 years, and the S_2 and S_4 lines would furnish more accurate combining ability data than the BC_1 and BC_2 CMS analogues of S_4 and S_6 lines, respectively. However, developing male-sterile testers representing the pollinator gene pool is very difficult since sterility maintenance/ fertility restoration is under complex genetic control [*for review see*, Stojałowski et al. (2004)]. Marker-based backcrossing may eventually overcome this difficulty. In a transitional period, CMS singles from an unrelated seed parent gene pool may serve as substitutes for CMS ‘opposite’ testers. A preferable alternative would be to employ gametocides for creating male-sterile testers. This would allow to use testers with normal cytoplasm and, in contrast to using CMS testers, would furnish male-fertile testcrosses and growing pollinator stripes in the yield trials would become dispensable. Unfortunately no rye gametocide is available at present.

Development of Pollinator Lines

On the pollen-parent side, line development (Fig. 5) is less complicated and laborious than on the seed parent side, since no parallel backcrossing is required. Moreover, inbreeding needs not necessarily be continued to complete homozygosity since the pollinator synthetic will be heterogeneous anyway.

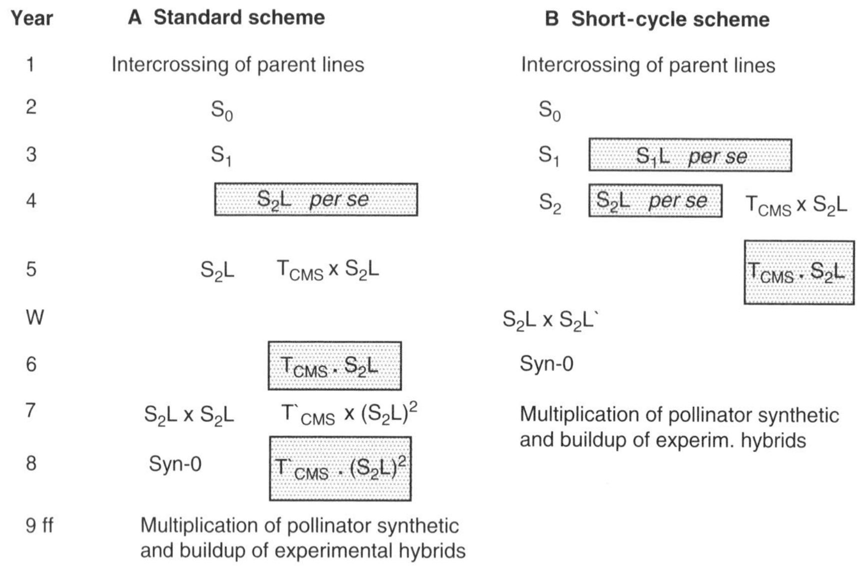

Fig. 5 Flow diagram of pollinator line development. **A** Standard scheme, **B** short-cycle scheme; *W* = winter season, *Syn-1* = synthetic generation 1, for further abbreviations, see Fig. 4

Partially inbred lines may have excellent fertility restoring ability and according to quantitative genetic theory (Wricke and Weber, 1986), the GCA variance among S_2 or S_3 lines is not much lower than that of fully inbred lines.

In order to take full advantage of the easier testing procedure on the pollinator side, Geiger (2006) suggested to further shorten the breeding cycle by evaluating the lines *per se* already in generation S_1 and restrict the testcrossing phase to one stage only. This would of course impair the accuracy and precision of line and testcross evaluation but would reduce the breeding cycle by two generations and thus may considerably speed up the *annual* selection gain. Model calculations in maize (Gordillo and Geiger, 2008a, b) showed that optimized one-stage selection schemes generally lead to the fastest breeding progress.

5.2.4 Recurrent Improvement of Combining Ability

RS for GCA is pivotal for medium- to long-term breeding progress. This requirement is most efficiently met if RS and line development are fully integrated such that after each line development cycle, a fraction of superior new lines is intermated to establish the S_0 population of the next breeding cycle. The number of selected lines has to be large enough to avoid a too fast decline of the genetic variance and thus to ensure selection response over many cycles. Periodically including best lines from related RS programs also contributes to counterbalance the loss of genetic variance (Gordillo and Geiger, 2008b).

In two-stage testcross selection schemes, the question arises whether the recombination units (= lines to be intermated) should be selected after the first or the second evaluation stage. The advantage of the former option would be a shorter cycle length whereas the latter furnishes more accurate and precise performance data. Model calculations for optimizing line development schemes (Tomerius, 2001; Tomerius et al., 2008) indicated that the second option will be superior since the relative increase in selection gain per cycle is larger than the relative reduction in cycle length. The number of lines to be recombined depends on the testing intensity (years, locations, and replicates), the inbreeding coefficient of the lines, and the cycle length, and should not fall below 20–30 in the aforementioned breeding schemes.

5.2.5 Buildup of Experimental Hybrids

Once the very best lines have been identified, they are combined to CMS single crosses on the seed-parent side and to restorer synthetics on the pollinator side. Promising hybrids are then built up based on the estimated GCA of the parents for the various performance traits. SCA effects are only considered in the final evaluation of the experimental hybrids. There are two reasons for neglecting SCA before building up these hybrids: First, estimation of SCA would require additional 2 years and second, only one quarter of the SCA variance among factorial single crosses contributes to the variance between double-cross hybrids. Furthermore, a compilation of GCA and SCA variance component estimates over many rye experiments showed that even for grain yield, GCA is far more important than SCA, particularly in interpool crosses (Tomerius, 2001).

5.3 Commercial Hybrid Seed Production

Hybrid seed production is a multi-stage procedure (Fig. 6) requiring well-skilled farmers, careful seed processing, and deliberate logistics.

Since rye as a wind-pollinated species produces huge amounts of pollen, which may be transported over great distances [*for review see* Feil and Schmid (2002)], careful isolation is necessary to avoid genetic contamination of seed production fields. Reproduction coefficients range between 60 and 100 in spaced-planted stands of inbred lines and between 50 and 80 in stands of non-inbred materials drilled at regular seeding rate (Wilde, 2007, personal communication).

Multiplying the CMS A line and its maintainer version generally requires 2–3 years. The CMS and male-fertile version are grown stripe-wise in a 2:1 to 3:1 ratio depending on the field size. The fields should be located in non-rye-growing regions to minimize the risk of contamination by alien pollen or by mechanical mixtures during harvest or seed processing. Repeated careful checks for off-types are indispensable before, during, and after anthesis. Even extremely few off-types (especially

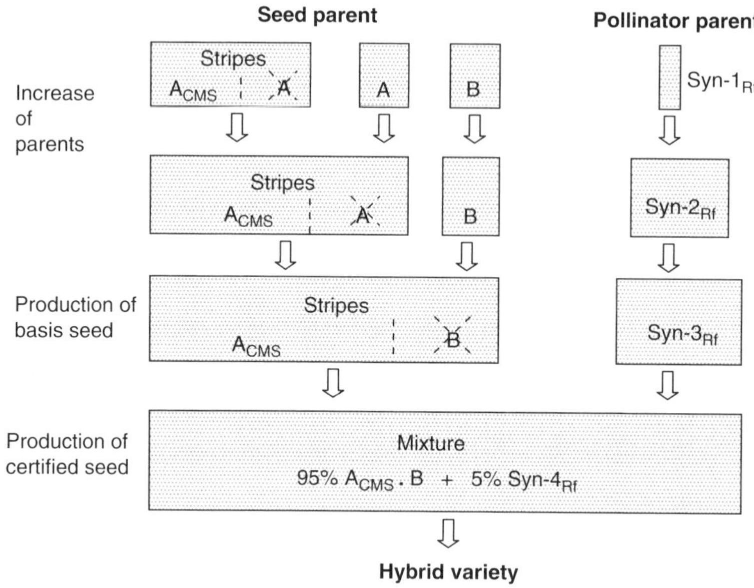

Fig. 6 Flow diagram of hybrid seed production; for abbreviations, see Figs. 4 and 5; broken-line crosses designate maintainer-line stripes that are eliminated after flowering

restorers) reaching pollen shedding in a CMS stand will render the respective field worthless for further seed multiplication or basis seed production. Furthermore, the growing region should feature a mild climate since inbred lines are sensitive to severe abiotic stress. In Europe, such regions exist, for example, in South France and North Italy.

Well-isolated fields are also required for multiplying the non-restorer B line and the male-fertile maintainer version of the CMS A line. However, the risk of mispollination is lower than for a CMS stand since male-fertile stands produce abundant pollen themselves and thus abate the fertilization rate of alien pollen.

Production of the CMS single cross follows the same rules as described for the CMS A line. On large enough fields, the female:male ratio may be raised up to 4:1 with 6–8 m broad CMS stripes (Fig. 7).

Quality management should include molecular fingerprinting of adequate seed samples from all pre-basis and basis seed production fields. This should also extend to the plasmotype (CMS vs normal) since male-fertile A line plants are very difficult to detect in drilled stands of the CMS A line. If they remain undetected they will produce male-fertile versions of the seed-parent single cross and may cause high proportions of selfings in the certified seed production fields. This would of course seriously lower the performance of the final ‘hybrid’. To minimize seed admixtures from the maintainer stripes to the CMS stripes during harvest, the former are removed shortly after flowering.

Fig. 7 Seed- and pollinator-parent stripes in a basis seed production field in North Italy (*Source:* P. Wilde)

Restorer synthetics are built up and multiplied in the same way as regular synthetic varieties. However, it is advisable to multiply and conserve restorer synthetics under stricter spatial isolation than prescribed for population varieties, since outcrosses generally are more vigorous than the partially inbred synthetic and thus may cause a high proportion of mispollinations.

The final step of seed production leading to certified seed for the farmer is accomplished in a mixed stand of the CMS seed parent and the restorer synthetic in a ratio of about 95:5. This procedure considerably reduces the cost of hybrid seed production compared to drilling the female and male parent in alternating stripes. However, since the seed grown on the pollinator plants cannot economically be separated from the true hybrid seed grown on the CMS single cross, a 'hybrid' variety may contain up to 5% pollinator parent plants. But although the restorer synthetic is considerably lower yielding than the hybrid, the 5%-admixture is too small to significantly affect the performance of the variety. To protect the seed production field from mispollination, it is surrounded by a 5- to 10-m-broad 'mantle' of the restorer synthetic. The mantle is removed before harvest. Isolation distances to other fields follow the regulations for population varieties.

5.4 *Integrating Population and Hybrid Breeding*

Both population and hybrid varieties are requested on the seed market since the relative economic merits of the two categories of variety depend on the farming system, the market conditions, the intended usage, and other factors. An integrated breeding strategy is therefore desirable to make optimum use of the genetic, human, and technical resources of a breeder. Unfortunately, this is difficult to achieve mainly for three reasons:

1. Self-incompatibility is indispensable in population varieties (*cf.* Sect. 5.1), whereas self-fertility is a precondition for hybrid breeding. Thus, improvement of SI population varieties by introducing superior SF hybrid breeding materials is not directly possible. Eliminating SF genes from SF × SI crosses is very cumbersome since the SI parents are genetically heterogeneous and several cycles of selfing, selection, and intercrossing are needed to purge the introduced SF genes. Diagnostic molecular markers may substantially reduce these difficulties in the future.
2. Introducing superior SI germplasm into the SF hybrid breeding materials, on the other hand, does not cause reproduction problems since SF × SI crosses show regular seed setting and in subsequent selfing generations, SF is 'automatically' fixed. However, a major problem arises from the high mutational load of deleterious recessives carried by SI germplasm. This leads to high frequencies of defective segregants during the selfing period and drastically reduces the recovery of acceptable inbred lines. Marker-based introgression of individual DC segments allows to largely circumvent this problem but is extremely expensive (*cf.* Sect. 2).
3. In hybrid breeding, materials need to be grouped into genetically distant gene pools and selection is practiced within those pools (*cf.* Sect. 5.2.1). A corresponding heterotic grouping of the SI materials allows to directly use them for broadening the genetic base of the SF gene pools. However, for maximizing progress in population breeding, the most promising strategy is to merge all high-yielding heterotic groups, since the expected gain from selection is higher in one genetically broad-based population than averaged across two or more narrow-based gene pools. Thus, the population breeder has to decide whether backing-up hybrid breeding or boosting population breeding should be given highest priority.

In conclusion, integrating hybrid and population breeding in one comprehensive breeding program is a major challenge requiring manifold further research efforts.

6 Major Achievements of Breeding

Intensive population as well as hybrid breeding is practised in all main rye-growing countries. In Germany, hybrids cover about 70% of the total rye acreage ranging from 50% to 90% depending on the region (Anonymous, 2006). Some of these hybrids are also widely distributed in neighbouring countries. According to the statutory German variety trials ('Wertprüfungen'), grain yields of the best hybrids surpass those of the best population varieties by 15–20% (Fig. 8). From 1982 to

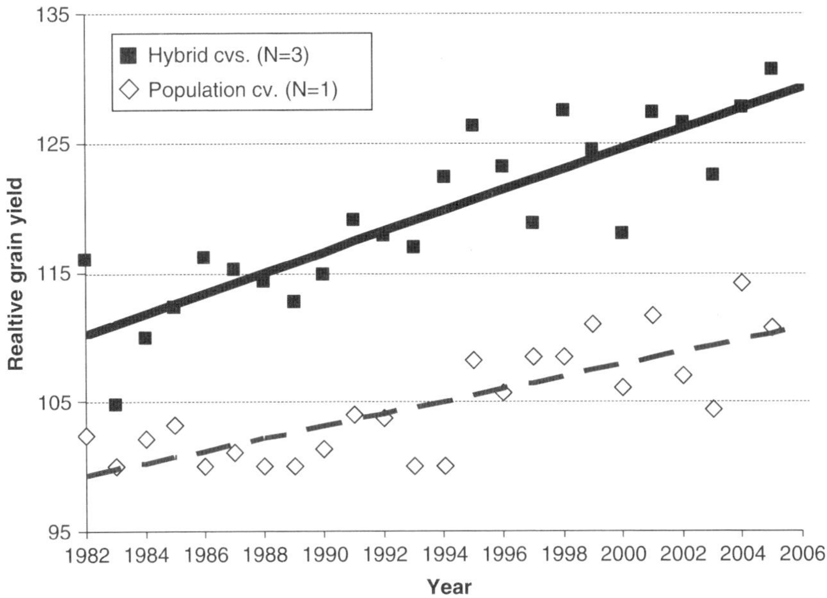

Fig. 8 Relative grain yields (% of long-term standard cultivar, 6.65 t ha^{-1} = 100) of the highest yielding three hybrids and the best population cultivar (*cv.*) in the statutory German variety trials 1982–2005

2005, the average annual genetic gain from selection amounted to 51 kg ha^{-1} for the hybrids and 30 kg ha^{-1} for the populations. In estimating these figures, the influence of non-genetic factors was eliminated by relating all data to the performance of a long-term standard variety ('Halo').

The superiority of hybrids over population varieties does not only pertain to grain yield but also to a shorter plant stature, lodging resistance, and bread-making quality (Fig. 9). Furthermore, for most traits, a greater genetic range exists among hybrids than among population varieties. From both, hybrid and population breeding, varieties combining superior lodging resistance, leaf-rust resistance, and thousand-kernel weight were developed. Yet the most favourable trait combinations are found in the hybrids as exemplified by contrasting the modern hybrid variety 'Visello' with the best population variety 'Conduct' in Fig. 9. Hybrids, therefore, became highly attractive to farmers.

In conclusion, significant progress has been achieved in both population and hybrid breeding. Improvements in breeding methodology have greatly contributed to this success. Hybrid varieties presently predominate in most West and Central European rye-growing areas but population varieties still have their merits.

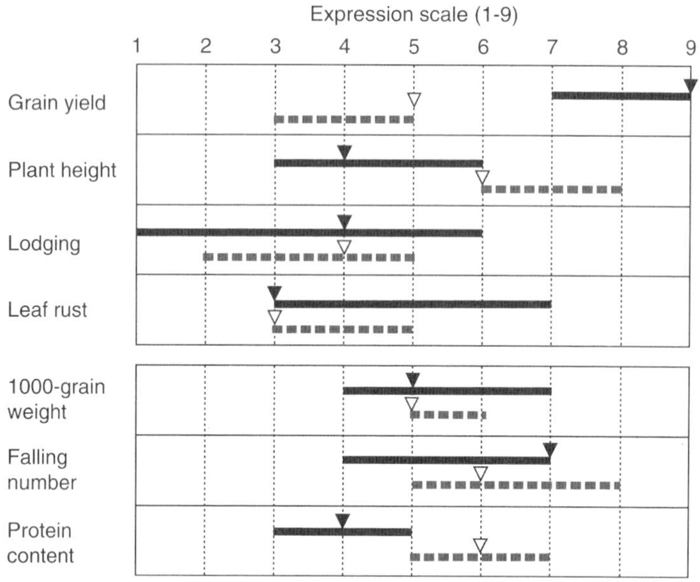

Fig. 9 Range of important traits (on a 1–9 scale) of 17 hybrid varieties (*bold line*) and 11 open-pollinated varieties (*broken line*) contained in the German Recommended List 2006. Filled triangles refer to the best hybrid variety ('Visello') and open triangles to the best population variety ('Conduct')

References

Adolf, K. and Winkel, A. (1985) A new source of spontaneous sterility in winter rye – preliminary results. In: Proceedings of Eucarpia Meeting of the Cereal Section on Rye. 11–13 June 1985, Svalöv, Sweden, pp. 293–306.

Altpeter, F. and Korzun, V. (2007) Rye. In: E.C. Pua and M.R. Davey (Eds.), Biotechnology in Agriculture and Forestry, Vol. 59, Transgenic Crops IV. Springer Berlin, Heidelberg, Germany, pp. 107–117.

Anonymous (2006) Erntebericht zur Qualität des deutschen Roggens. Bundesforschungsanstalt für Ernährung und Lebensmittel (BfEL), Institut für Getreide, Kartoffel- und Stärketechnologie, Detmold, Germany.

Becker, H.C., Geiger, H.H. and Morgenstern, K. (1982) Performance and phenotypic stability of different hybrid types in winter rye. Crop Sci. 22, 340–344.

Boros, D. (2007) Quality aspects of rye for feed purposes. In: Proceedings of International Symposium on Rye Breeding & Genetics. 28–30 June 2006, Groß Lüsewitz, Germany, Vortr. Pflanzenzuechtg. 71, 80–85.

Crow, J.F. and Kimura, M. (1970) An Introduction to Population Genetics Theory. Harper and Row, New York.

de Vicente, M.C. and Tanksley S.D. (1993) QTL analysis of transgressive segregation in an interspecific tomato cross. Genetics 134, 585–596.

Eshed, Y., Abu-Abied, M., Saranga, Y. and Zamir, D. (1992) A genome-wide search for wild-species alleles that increase horticultural yield of processing tomatoes. Theor. Appl. Genet. 93, 877–886.

Falke, K.C., Susic, Z., Hackauf, B., Korzun, V., Schondelmaier, J., Wilde, P., Wehling, P., Wortmann, H., Mank, J., Rouppe van der Voort, J., Maurer, H., Miedaner, T. and Geiger, H. H. (2008) Establishment of introgression libraries in hybrid rye (*Secale cereale* L.) from an Iranian primitive rye accession as a new tool for rye breeding and genomics. Theor Appl Genet 117, 641–652.

Falke, K.C., Wilde, P. and Miedaner, T. (2009) Rye introgression lines as source of alleles for pollen-fertility restoration in Pampa CMS. Plant Breed (in press).

FAO (2006) FAOSTAT database. Internet resource: http://faostat.fao.org/site/567/default.aspx (verified Jan 18, 2007).

Flehinghaus-Roux, T., Deimling, S. and Geiger, H.H. (1995) Anther-culture ability in *Secale cerale* L. Plant Breed. 114, 259–261.

Feil, B. and Schmid, J.E. (2002) Dispersal of maize, wheat and rye pollen: A contribution to determining the necessary isolation distances for cultivation of transgenic crops. Shaker, Aachen.

Geiger, H.H. (1972) Wiederherstellung der Pollenfertilität in cytoplasmatisch männlich sterilem Roggen. Theor. Appl. Genet. 42, 32–33.

Geiger, H.H. (1982) Züchtung synthetischer Sorten. III. Einfluß der Vermehrungsgeneration und des Selbstungsanteils. Vortr. Pflanzenzüchtung 1, 41–72.

Geiger, H.H. (1988) Epistasis and heterosis. In: B.S. Weir (Ed.), Proceeding of Second International Conference on Quantitative Genetics. 31 May – 5 June 1987, Raleigh, NC, Sinauer Assoc. Inc., Sunderland, MA, USA, pp. 395–399.

Geiger, H.H., Schumacher, A.E. and Billenkamp, N. (1988) Häufigkeiten von vertikalen Resistenzen und Virulenzen im Roggen-Mehltau-Pathosystem. Vortr. Pflanzenzüchtung 13, 5–17.

Geiger, H.H. (2006) Strategies of hybrid rye breeding. In: Proceedings of International Symposium on Rye Breeding & Genetics. 28–30 June 2006, Groß Lüsewitz, Germany, p. 16.

Geiger, H.H. and Schnell, F.W. (1970) Cytoplasmic male sterility in rye (*Secale cereale* L.). Crop Sci. 10, 590–593.

Geiger, H.H., Yuan, Y., Miedaner, T. and Wilde, P. (1995) Environmental sensitivity of cytoplasmic genic male sterility (CMS) in *Secale cereale* L. In: U. Kück and G. Wricke (Eds.), Genetic Mechanisms for Hybrid Breeding. Adv. Plant Breed. 18, 7–17.

Geiger, H.H. and Miedaner, T. (1999) Hybrid rye and heterosis. In: J.G. Coors and S. Pandey (Eds.), Genetics and Exploitation of Heterosis in Crops. Crop Science Society. America, Madison, Wisconsin, USA, pp. 439–450.

Gordillo, G.A. and Geiger, H.H. (2008a) Optimization of DH-line based recurrent selection procedures in maize under a restricted annual loss of genetic variance. Euphytica, 161, 141–154.

Gordillo, A., and Geiger, H.H. (2008b) Alternative recurrent selection strategies using doubled haploid lines in maize breeding. Crop Sci. 48: 911–922.

Hackauf, B. and Wehling, P. (2002) Identification of microsatellite polymorphisms in an expressed portion of the rye genome. Plant Breed. 121, 17–25.

Haussmann, B.I.G., Parzies, H.K., Presterl, T., Sušić, Z. and Miedaner, T. (2004) Plant genetic resources in crop improvement (Review). Plant Genet. Resour. 2, 3–21.

Hepting, L. (1978) Analyse eines 7 × 7-Sortendialles zur Ermittlung geeigneten Ausgangsmaterials für die Hybridzüchtung bei Roggen. Z. Pflanzenzüchtg. 80, 188–197.

Kastirr, U., Wortmann, H., Schmiedchen, B., Rabenstein, F. and Kühne, T. (2006) Occurrence of soil-borne viruses in rye and prospects for improving resistance to these viruses. In: Proceedings of International Symposium on Rye Breeding & Genetics. 28–30 June 2006, Groß Lüsewitz, Germany, p. 34.

Klyuchko, P.F. and Belousov, A.A. (1972) Results of a genetic study of cytoplasmic male sterility in winter rye (in Russian). Genetika 8(7), 9–15.

Kobyljanskij, V.D. (1971) The production of sterile analogues of winter rye, sterility, maintainers and fertility restorers (in Russian). Tr. Prikl. Bot. Genet. Sel. 44, 76–85.

Lamkey, K.R. and Edwards, J.W. (1999) Quantitative genetics of heterosis. In: J.G. Coors and S. Pandey (Eds.), The Genetics and Exploitation of Heterosis in Crops. ASA, CSSA, SSSA, Madison, WI, pp. 31–48.

Łapiński, M. (1972) Cytoplasmic-genic type of male sterility in *Secale montanum* Guss. Wheat Inf. Serv. 35, 25–28.

Łapiński, M. and Stojałowski, S. (2001) Occurrence of male sterility-inducing cytoplasm in rye (*Secale* spp.) populations. Pl. Breed. Seed Sci. 48(2), 7–23.

Lundqvist, A. (1956) Self-incompatibility in rye. I. Genetic control in the diploid. Hereditas 42, 293–348.

Madej, L. (1976) The genetical characteristics of three sources of male sterility in rye (*Secale cereale* L.). Hod. Rósl. Aklim. i Nasienn., Warszawa 20, 157–174.

Madej, L., Raczynska-Bojanowska, K. and Rybka, K. (1990) Variability of the content of soluble non-digestible polysaccharides in rye inbred lines. Plant Breed. 104, 334–339.

Miedaner, T. and Geiger, H.H. (1996) Estimates of combining ability for resistance of winter rye to *Fusarium culmorum* head blight. Euphytica 89, 339–344.

Miedaner, T., Ludwig, W.F. and Geiger, H.H. (1995) Inheritance of foot rot resistance in winter rye. Crop Sci. 35, 388–393.

Miedaner, T., Glass, C., Dreyer, F., Wilde, P., Wortmann, H. and Geiger, H.H. (2000) Mapping of genes for male-fertility restoration in 'Pampa' CMS winter rye (*Secale cereale* L.). Theor. Appl. Genet. 101, 1226–1233.

Miedaner, T., Gey, A.-K., Sperling, U. and Geiger, H.H. (2002) Quantitative-genetic analysis of leaf-rust resistance in seedling and adult-plant stages of inbred lines and their testcrosses in winter rye. Plant Breed. 121, 475–479.

Miedaner, T., Wortmann, H. and Geiger, H.H. (2003) Genetics of deoxynivalenol (DON) contamination caused by Fusarium head blight in hybrid rye. Pl. Breed. Seed Sci. 48, 69–78.

Miedaner, T., Wilde, P. and Wortmann, H. (2005) Combining ability of non-adapted sources for male-fertility restoration in Pampa CMS of hybrid rye. Plant Breed. 124, 39–43.

Miedaner, T., Sušić, Z., Seggl, A., Hackauf, B., Korzun, V., Schondelmaier, J., Wilde, P., Wortmann, H., Rouppe van der Voort, J., Wehling, P. and Geiger, H.H. (2006) Entwicklung und Evaluierung einer Introgressionsbibliothek bei Roggen. Ber. Tag. Vereinig. Pflanzenzüchter und Saatgutkaufleute Österreichs, Gumpenstein/Austria 57: Im Druck.

Mirdita, V. (2006). Genetische Variation für Resistenz gegen Mutterkorn (*Claviceps purpurea* [Fr.] Tul.) bei selbstinkompatiblen und selbstfertilen Roggenpopulationen. Diss. Univ. Hohenheim. Internet resource: http://www.uni-hohenheim.de/ub/opus/volltexte/2006/148/ (verified Jan 08, 2007).

Mirdita, V., Dhillon, B.S., Geiger, H.H. and Miedaner, T. (2008) Genetic variation for resistance to ergot (*Claviceps purpurea* [Fr.] Tul.) among full-sib families of winter rye (*Secale cereale* L.). Theor. Appl. Genet. 118, 85–90.

Nalborczyk, E., Nalborczyk, T. and Wawrzonowska, B. (1981) Models of photosynthetic activity in cereals. In: G. Akoyunoglou (Ed.), Photosynthesis VI. Photosynthesis and Productivity, Photosynthesis and Environment, pp. 97–105. Balaban International Science Services, Philadelphia, Pa.

Ossent, H.P. (1938) Zehn Jahre Roggenzüchtung in Müncheberg. Züchter 10, 255–261.

Persson, K. and von Bothmer, R. (2000) Assessing the allozyme variation in cultivars and Swedish landraces of rye (*Secale cereale* L.). Hereditas 132, 7–17.

Persson, K., von Bothmer, R., Gullord, M. and Gunnarsson, E. (2006) Phenotypic variation and relationships in landraces and improved varieties of rye (*Secale cereale* L.) from northern Europe. Genet. Resour. Crop Evol. 53 (4), 857–866.

Podyma, W. (2003) Rye genetic resources in Europe. Pl. Breed. Seed Sci. 48 (2/2), 37–44.

Rode, J., Schumann, E., Weber, E., Wilde, P. and Wortmann, H. (2005) Rye breeding for bioethanole production. Proceedings of 11th International Conference On Renewable Resources and Plant Biotechnology. NAROSSA, 06–07 June 2005, Poznan, Polen.

Roux, S.R., Miedaner, T., Geiger, H.H., Knopf, E., Wilde, P. and Wortmann, H. (2003) Combining ability *vs.* population performance of genetic resources in rye. Pl. Breed. Seed Sci. 48, 45–48.
Schnell, F.W. (1982) A synoptic study of the methods and categories of plant breeding. Z. Pflanzenzüchtg. 89, 1–18.
Singh, R.K., Geiger, H.H., Diener, C. and Morgenstern, K. (1984) Effect of number of parents and synthetic generation on the performance of self-incompatible and self-fertile rye populations. Crop Sci. 24, 306–309.
Stojałowski, S., Łapiński, M. and Masojć, P. (2004) RAPD markers linked with restorer genes for the C-source of cytoplasmic male sterility in rye (*Secale cereale* L.). Plant Breed. 123, 428–433.
Stracke, S., Schilling, A.G., Förster, J., Weiss, C., Glass, C., Miedaner, T. and Geiger, H.H. (2003) Development of PCR-based markers linked to dominant genes for male-fertility restoration in Pampa CMS of rye (*Secale cereale* L.). Theor. Appl. Genet. 106, 1184–1190.
Sušić, Z. (2005). Experimental and simulation studies on introgressing genomic segments from exotic into elite germplasm of rye (*Secale cereale* L.) by marker-assisted backcrossing. Ph.D. thesis, University of Hohenheim, Germany.
Tanksley, S.D. and Nelson, J.C. (1996) Advanced backcross QTL analysis: a method for the simultaneous discovery and transfer of valuable QTL from unadapted germplasm into elite breeding lines. Theor. Appl. Genet. 92, 191–203.
Tenhola-Roininen, T., Immonen, S. and Tanhuanpää, P. (2006) Rye doubled haploids as a research and breeding tool – a practical point of view. Plant Breed. 125, 584–590.
Tomerius, A. (2001) Optimizing the development of seed-parent lines in hybrid rye breeding. Diss. Univ. Hohenheim, Stuttgart. Internet resource: http://www.uni-hohenheim.de/ub/opus/volltexte/2001/10/ (verified Jan 08, 2007).
Tomerius, A., Miedaner, T. and Geiger, H.H. (2008) A model calculation approach towords the optimization of a standard scheme of seed-parent line development in hybrid rye breeding. Plant Breed. 127, 433–440.
Walsh, B. (2004) Population- and quantitative-genetic models of selection limits. Plant Breed. Rev. 24, 177–225.
Wehmann, F., Geiger, H.H. and Loock, A. (1991) Quantitative-genetic basis of sprouting resistance in rye. Plant Breed. 106, 196–203.
Wilde, K., Geiger, H.H. and Miedaner, T. (2006) Significance of host complexity and diversity for race-specific leaf-rust resistance in self-fertile synthetic rye populations. Plant Breed. 125, 225–230.
Wilde, F., Korzun, V., Ebmeyer, E., Geiger, H.H. and Miedaner, T. (2007) Comparison of phenotypic and marker-based selection for Fusarium head blight resistance and DON content in spring wheat. Mol. Breed. 19, 357–370.
Winkel, A., Müller, H.W. and Gabow, G. (1979) Über die gegenwärtigen Möglichkeiten zur Schaffung von Hybridsorten und zur Nutzung von Sterilitätssystemen in der Roggenzüchtung. Tag.-Ber., Akad. Landwirtsch.-Wiss. DDR, Berlin 168, 171–177.
Wolski, T. (1970) Studies on breeding of rye. Genetica Pol., Warszawa/Poznań 11, 1–26.
Wricke, G. (1969) Untersuchungen zur Vererbung der Selbstfertilität beim Roggen (*Secale cereale*). Theor. Appl. Genet. 39, 371–378.
Wricke, G. (1978) Pseudo-Selbstkompatibilität beim Roggen und ihre Ausnutzung in der Züchtung. Z. Pflanzenzüchtg. 81, 140–148.
Wricke, G. (1979) Degree of self-fertilization under free pollination in rye populations containing self-fertility gene. Z. Pflanzenzüchtg. 82, 281–285.
Wricke, G. and Weber, W.E. (1986) Quantitative Genetics and Selection in Plant Breeding. deGruyter, Berlin.
Zamir, D. (2001) Improving plant breeding with exotic genetic libraries. Nat. Rev. Genet. 2, 983–989.

Grain Sorghum Breeding

Robert G. Henzell and David R. Jordan

Abstract Grain sorghum [*Sorghum bicolor* (L.) Moench] is a relatively drought- and heat-resistant crop. World wide it is used as feed and food grain. In Australia, it is used as a feed grain and is grown under rain-fed conditions. Water availability to the plant is the major constraint to production. This chapter describes aspects of the Department of Primary Industries and Fisheries sorghum breeding program. The overall aim of this program is the development of germplasm which is licensed to the private sector for the development of hybrid cultivars and use in their breeding programs. This latter aspect has ensured program focus and the ready adoption by industry of program products. The specific objectives of the program are the development of germplasm with resistance to the sorghum midge, drought resistance (stay-green) and yield. High levels of midge resistance have been developed and combined with significant levels of stay-green and improved yield under Australian conditions.

1 Introduction

Sorghum bicolor (L.) Moench is a relatively drought- and heat-resistant cereal crop. In developed countries, it is used as a feed grain and for food and feed in the less developed countries such as Africa and Asia. World wide its importance is likely to increase because of the pressure on reliable food and animal feed supply, the latter due in part, to a world population becoming more reliant on animal protein. Sorghum's relative drought and heat resistance may also increase its importance world wide if the predicted effects of global warming come to pass. Additionally, sorghum has a wide variety of other uses including beverages, building materials, fuel (particularly in regions denuded of wood), ethanol production, broom production and many others.

Cultivars are either inbreds or hybrids in the underdeveloped countries and F1 hybrids in the more developed countries. The cytoplasmic-nuclear male sterility

R.G. Henzell(✉)
Department of Primary Industries and Fisheries, Hermitage Research Station, MS 508, Warwic Q4370, Australia, e-mail: bob.henzell@dpi.qld.gov.au

M.J. Carena (ed.), *Cereals*,
DOI: 10.1007/978-0-387-72297-9,

system used to facilitate the production of hybrid cultivars was reported by Stephens and Holland (1964) and resulted from the interaction between the A1 cytoplasm from the durra race (see later Taxonomy section) and the non-restorer genes from the kafir race.

The nuclear genetics of this system was reviewed by Rooney (2000). Other cytoplasmic-nuclear systems were reported by Schertz (1983).

In Australia, grain sorghum is grown on approximately 0.8 Mha with an average yield of 2.6 t/ha. It is grown under dryland conditions in an environment that is characterised by water stress particularly during grain fill.

2 Origin of *Sorghum bicolor* (L.) Moench

Harlan and De Wet (1971) have an excellent discussion on the origin and domestication of *Sorghum bicolor* (L.) Moench. It is generally agreed that *S. bicolor* originated and was domesticated in the Sub-Saharan region of Africa and spread to India and China. It is likely that the Sub-Saharan and north east regions of Africa (particularly Ethiopia and Sudan) are the primary centres of origin and diversity and that India and China are secondary centres. A tertiary pool of diversity is considered to be the 19 wild species indigenous primarily to Australia, but also to South East Asia and Africa (Lazarides et al., 1991). It is also generally agreed that *S. bicolor* (L.) Moench spp *bicolor* [later and above referred to as *Sorghum bicolor* (L.) Moench] was derived from the wild species *S. bicolor spp verticilliflorum* and *S. bicolor spp drummondii* in Africa and from *S.halepense* and *S. propinquum* in Asia.

3 Taxonomy of the Genus *Sorghum*

Doggett (1988) and Dahlberg (2000) have comprehensive discussions on the classification of sorghum. Their papers form the basis of the following discussion along with some key publications for further reading. It is common that there are variable interpretations of taxonomic literature and that for sorghum is no exception.

Sorghum (described by Linnaeus in 1773 and named by Moench in 1794) belongs to the Family *Poaceae*, Tribe Andropogoneae, which consists of 16 sub-tribes, one of which is *Sorghastrae* (Stapf, 1917; Garber, 1950). Garber (1950) considered this sub-tribe comprised two main genera, *Cleistachne* and *Sorghum*. Snowden (1935, 1936) and Garber (1950) suggested that the genus *Sorghum* comprises six sub-genera.

This discussion relates only to the species *Sorghum bicolor* (L.) Moench sub-genus *Sorghum*. All grain sorghums are in this species.

The genetic diversity within *S. bicolor* (L.) Moench forms the basis of the thousands of years' natural and farmer/user selection and the grain and forage sorghum breeding programs that have occurred internationally during the last century. The cultivated taxa were first grouped into 28 species by Snowden (1936). Classification schemes since then have all been based on his historic work. These 28 species were grouped into *S. bicolor* (L.) Moench for logistical reasons and because there are no reproductive barriers between any of the Snowden species.

Later all named *S. bicolor* (L.) Moench accessions were first grouped into working groups (WG) by Murty and Govil (1967) and later into five races and ten intermediate groups by Harlan and de Wet (1972). This grouping was based on plant phenotype and has proved to capture a major part of the genetic diversity in the species. Hence, this grouping, particularly grouping on the basis of race, proved very useful for breeders. The five races are bicolor, kafir, caudatum, durra and guinea.

4 Cytogenetics and Genetics of *S. bicolor* (L.) Moench

The cytogenetics and nuclear genetics of *S. bicolor* has recently been reviewed by Rooney (2000). Conclusions are that it is a diploidised tetraploid with a basic chromosome number of 5. However, more recent molecular evidence (Sprangler et al., 1999) suggests it may be a diploid with $n = 10$ chromosomes. From the point of view of an applied breeder, sorghum is treated as a diploid.

Sorghum has extensive structural genomic resources such as dense genetic marker maps, back libraries, physical maps and a large quantity of sequence information. The full sequence of the sorghum genome was made available to researchers in soon.

Numerous QTL studies have been carried out to elucidate the genetic architecture of important traits. However, internationally, the use of molecular marker technology in applied breeding programs has been limited.

5 Sources of Genetic Diversity

There is a large amount of genetic diversity in this species, many accessions of which have been classified, characterised and evaluated although there are major gaps, especially for multigenic traits (Rosenow and Dahlberg, 2000). Most of this diversity is represented in land races from Africa, India and China. The large number of accessions in the 122 collections reported in the IPGRI data base in 2006 presents a challenge for sorghum breeders who can work on only a portion of these. A core collection (10%) has been developed at ICRISAT on the basis of country of origin and phenotypic. Deu et al. (2006) and de Oliveira et al. (1996) have written two papers where molecular markers have been used in an attempt to measure genetic diversity in landraces. In general, markers are effective in describing locality of collection diversity but also to a lesser degree discriminating amongst races. Also, the USDA-ARS/Texas A&M University's Sorghum Conversion Program (Stephens et al., 1967) has been a fruitful source of much of the germplasm used in breeding programs in Australia and elsewhere. The sorghum conversion program is a pre-breeding exercise where dwarfing and photoperiod insensitivity genes are backcrossed into exotic lines which are tall, photoperiod sensitive. Substitution of the major genes controlling these traits greatly enhances the ease with which this genetic diversity can be used by breeding programs targeting subtropical and temperate environments where mechanical harvesting is used.

6 Sorghum Breeding in Australia

In this chapter, we describe the approaches used and results achieved by a sorghum breeding program conducted by the Department of Primary Industries and Fisheries (DPI&F) in Queensland. The objective of most plant breeding programs is to produce cultivars that are grown commercially by farmers. In contrast, germplasm enhancement programs aim to develop lines or populations which have been improved for particular characteristics that can then be used in breeding programs as parent material.

In Australia, there are four private sector programs which provide all of the commercially used cultivars (F1 hybrids). The DPI&F breeding program is the only current public sector program in Australia and it operates as a germplasm enhancement program providing strategic breeding services to the sorghum industry. The public sector and the private sector programs have complementary roles and work closely together. The DPI&F program's role is the development of germplasm which is licensed to the private sector for the development and sale of hybrid cultivars.

The DPI&F program focuses on the assemblage of genes for important traits in a relatively adapted genetic background with respect to grain yield. The latter is achieved by testing the general combining ability for yield in a yield testing program less rigorous than that for the identification of cultivars but sufficient to eliminate poor combiners. The specific major objectives of the public program include resistance to the sorghum midge, drought resistance (stay-green) and grain yield. These objectives obviously reflect the major constraints to profitable grain sorghum production in Australia.

6.1 Resistance to the Sorghum Midge (Stenodiplosis sorghicola [Coquillette])

The sorghum midge was a major constraint to the profitable production of grain sorghum in Australia. The female midge deposits eggs in the spikelet at anthesis and the larvae destroy the grain. Before host plant resistance, chemical control was used resulting in an annual cost to industry of $A10 M (estimated in 1980). Our midge resistance breeding program commenced in 1975 following the reporting of germplasm with midge resistance (Johnson et al., 1973).

6.1.1 Mechanisms of Midge Resistance

Two major mechanisms of resistance have been reported – oviposition antixenosis (Franzmann, 1993) and antibiosis (Sharma et al., 1993; Hardy et al., 2001; Tao et al., 2003). The former mechanism results in fewer eggs being laid and the latter results in death of the developing larvae before they damage the grain. The reasons for these mechanisms are unknown, although it is suspected that spikelet structure is implicated in the oviposition antixenosis mechanism, midge laying fewer eggs in genotypes with small tightly appressed glumes. The reason for the antibiosis mechanism is unknown. Oviposition antixenosis is the major mechanism used and only recently has the antibiosis mechanism been incorporated in the breeding program. It is likely that midge resistance will be durable because of the mechanism of resistance and because the selection pressure on the insect will be reduced because of the susceptible forage and weedy sorghums in the system. Resistance also has an effect on the population dynamics (particularly population density) of the sorghum midge.

6.1.2 Genetics of Midge Resistance

The genetics of the oviposition antixenosis mechanism is complex, reviewed by Henzell et al. (1996). It is a polygenic trait that varies, amongst crosses, from recessive to partially dominant and that both specific and general combining ability are significant (Page, 1979) suggesting some diversity of resistance genes. This has been supported by evidence of gene pyramiding in the breeding program.

A QTL mapping study by Tao et al. (2003) identified two QTLs for oviposition antixenosis explaining a modest percentage of the genetic variation in a recombinant inbred line population. One of these QTLs corresponds to a region under selection from SC 165-14E, a source of resistance. Retrospective analysis of pedigrees in our breeding program using molecular markers (Jordan et al., 2004) shows that this chromosome region from the line SC 165-14E has been selected through a number of cycles of phenotypic selection and crossing.

In contrast, the genetic architecture of the antibiosis mechanism may be relatively simple with a single QTL explaining a large percentage of the phenotypic variance (Tao et al., 2003).

6.1.3 Sources of Midge Resistance

The major sources of resistance used in the program are from the USDA-ARS/ Texas A&M University sorghum conversion program (Henzell et al., 1996). The key lines from the conversion program include TAM2566, SC 170-6, SC 173C, SC 165-14E, SC 108, each being race caudatum. Advanced lines from the Texas A&M University sorghum breeding program were also used as were lines from ICRISAT in India.

The ICRISAT line, ICSV745, has been used as the source of the antibiosis gene.

6.1.4 Breeding Methods for Midge Resistance

The pedigree and limited backcross methods have been employed to maintain local adaptation while increasing resistance. The multigenic nature of midge resistance, necessitated multiple cycles of crossing, evaluation, and selection of parents and additional crossing to commence a new cycle with the occasional infusion of new germplasm. Some parents were included on the basis of other traits, for example grain yield and stay-green. The use of such midge susceptible germplasm obviously slows progress in increasing midge resistance, but our approach has been to include characters for local adaptation rather than to concentrate on midge resistance alone.

The biology of the sorghum midge creates considerable challenges for the plant breeder. The life cycle of the midge is two weeks in summer with most of that time being spent in juvenile form (egg, larvae or pupae) with adult midge living for only a single day. Sorghum midge emerge from infected spikelets early in the morning, mate and then the females lay their eggs in flowering sorghum spikelets. Only florets that are actually flowering are vulnerable to infection by the midge. The short lifespan, high reproductive rate and the short viability of the adult midge results in large day-to-day variation in midge pressure. This is further complicated by environmental factors such as wind, rainfall and temperature that have large impacts on midge activity. Plants that vary in flowering time by one or two days can experience very different midge pressures. As a result selection using conventional field trials would result in low heritability and response to selection.

To counter this difficulty, a managed environment approach was used to increase heritability in our selection environments by reducing day-to-day and seasonal variation in midge pressure (Henzell et al., 1994). This approach involved pre-planting of infector rows at regular temporal and spatial intervals throughout the nursery increase the potential high midge populations and reduce day-to-day variation. Individual panicles were marked with spray paint to indicate the day they flowered. Midge damage is visually accessed on plants that flowered on a particular day and compared with a set of check genotypes. This method allows for the removal of temporal effects due to day-to-day midge pressure. This method does not eliminate the possible confounding effect of visiting non-preference where midge show preference for particular genotypes given choice but this preference is not associated with resistance in a no choice environment. While preference can be significant in some genotypes, the correlation between the field test where preference is possible and the cage test, where it is not, is high (D. Butler and B.A. Franzmann, personal communication).

Table 1 Midge tested rating, % seed loss due to midge damage and yield loss (g) in four hybrids varying in level of midge resistance – 2002 (A. Hardy, personal communication)

Hybrid	Midge tested rating	Seed loss (%)	Yield loss (g)[a]
ATx623/RQL12	1	86[b]	82
Tullock	4	52	38
AQL39/RQL36	7	31	23
A23277/R40386	8+	4	16

[a] Yield loss in grams per panicle over a standard range of midge densities
[b] Some panicles were periodically protected from midge exposure to prevent 100% damage

6.1.5 Results of Breeding for Midge Resistance

The project has been successful in increasing the level of resistance to a point that results in field immunity to the sorghum midge (Table 1). These results were obtained in a cage test where visiting non-preference was eliminated. The 'Midge Tested Rating' in the table is a measure of resistance determined by the DPI&F and the seed industry which can be used by growers to determine Economic Injury Levels and hence a chemical control program if needed (Franzmann et al., 1996).

These results show that the hybrid ATx623/RQL12 loses at least eight times the weight of grain than does our experimental hybrid A23296/R40386. Rarely would it be economic for graingrowers in Australia to chemically control midge on a hybrid such as A23296/R40386 (i.e. 'field immunity'). Germplasm with this level of resistance has been licensed to the seed industry breeders who are either using it as direct hybrid parents or incorporating it into their breeding programs.

This highly resistant material has only the oviposition antixenosis mechanism of resistance. We are in the process of adding the antibiosis mechanism to this material that will not only further raise the level of resistance but should also add to the durability of resistance.

6.1.6 Adoption of Midge-Resistant Hybrids in Australia

The adoption of this technology has been rapid and to a high level. Now, all commercial hybrids have a level of midge resistance. This reflects the commercial importance of the trait but also to farmers being provided with a management information package which varies with the level of resistance for particular hybrids. Industry and DPI&F have determined a standard method of testing and of describing levels of resistance, called the Midge Tested Rating Scheme referred to above and in Table 1.

6.1.7 Additional Benefits of Midge Resistance

The benefits of midge-resistant hybrids extend beyond the successful control of midge alone to the development of a stable integrated pest management (IPM) strategy to control other major insect pests of sorghum, Helicoverpa armigera (heliothis) and aphids. Additionally, a biological virus spray (NPV) is being used in place of chemicals to control heliothis, now the most damaging insect pest of grain sorghum in Australia (Murray et al., 2001). The value of beneficial insects in controlling a wide range of insect pests is, in many cases, sufficient to reduce infestations of heliothis and other insect pests such as aphids below economically damaging levels (Murray et al., 2001). Reduced chemical usage also lessens the threat of the development of insecticide resistance. The benefits of IPM-managed midge-resistant hybrids are being exploited in regional or area-wide management (AWM) of heliothis on the Darling Downs and elsewhere in Australia (Murray et al., 2000). It is estimated that ongoing increased economic control of heliothis and other insect pests in IPM-managed midge-resistant hybrids equates to $20/ha (Hardy, unpublished data), while the additional environmental and marketing benefits are significant.

7 Drought Resistance Breeding

Two distinctly different types of drought stress response have been identified and described in sorghum depending on their incidence relative to anthesis (Rosenow and Clark, 1981). The pre-anthesis response occurs between floral initiation and anthesis while the post-anthesis response occurs during the grain filling period. Much of Australia's sorghum is grown in clay soils with a farming system that utilises stored soil moisture. As a result post-anthesis drought stress is the predominant type of drought experienced by Australian sorghum crops. It has been this form of drought resistance that has been targeted by our program.

Two strategies have been employed in the DPI&F sorghum breeding program attempting to improve drought resistance: indirect selection for traits known to be associated with yield under drought and direct selection for yield in drought environments.

7.1 Indirect Selection for Drought Resistance

In contrast to most other cereals sorghum is a perennial crop grown as an annual and if seasonal conditions permit the leaves and stem will remain green even as the grains mature. Drought stress during the grain filling period leads to senescence of the plant, reduced grain size, the invasion of the stem by a complex of fungi and stalk lodging. The latter has a major impact on harvestable yield. Sorghum plants which tend to retain more green leaf under this type of stress are said to have the stay-green trait. The stay-green trait is a clear example of the successful use of indirect selection for post-anthesis drought resistance breeding.

7.1.1 Stay-Green

Genotypes with stay-green were first reported in sorghum by Rosenow (1977). Stay-green has been associated with higher grain yields under stress (Henzell et al., 1992; Borrell et al., 1999; Jordan et al., 2003). It has also been associated with lodging resistance (Rosenow et al., 1983) and larger grain size.

The strongest evidence for the value of the trait comes from a retrospective analysis of 15 years of the DPI&F sorghum breeding trials that sampled 48 environments. This analysis indicated that stay-green was positively associated with a best linear unbiased estimate of hybrid grain yield (Jordan et al., 2003). Further analysis of this dataset showed that these environments could be classified into two groups on the basis of G $\times$ E interactions. Stay-green was positively associated with yield in one of these groups that contained environments where post-anthesis drought stress was likely to have occurred. However, stay-green was not associated either positively or negatively with grain yield in the other group containing environments where post-anthesis drought is less frequent (unpublished data). Importantly, this suggests that stay-green is not associated with low grain yield in high yield environments.

7.1.2 Sources of Stay-Green

Rosenow et al. (1983) reported a number of stay-green genotypes. Testing in Australia of these indicated B35 to be the best source of the trait. B35 is a partially converted derivative of IS12655 a line collected in Ethiopia. QL12 is another source of stay-green that has been used in our program. It is a derivative of KS19 which in turn is a derivative of the Nigerian line Short Kaura, the probable source of QL12's stay-green. Its stay-green is less strongly expressed but when combined in a hybrid with B35 exhibits exceptional post-anthesis drought resistance (Borrell et al., 2000).

7.1.3 Inheritance of and Molecular Markers for Stay-Green

Early studies on the inheritance of B35 stay-green were reviewed by Tao et al. (2000) These 'traditional' studies suggest the trait is relatively simply inherited and partially dominant. However, the marker work, internationally, also reviewed and reported by Tao et al. (2000), suggested there are at least five regions associated with the trait. However, Jordan et al. (2004), in a pedigree analysis study using RFLP markers of the DPI&F program, found that there is one region associated with stay-green that has been under strong phenotypic selection in the program. This region was not identified in any of the QTLs found in the literature reviewed by Tao et al. (2000).

Selection will be much more efficient when validated markers are identified and used because of its relatively low heritability. This is due to the expression of the trait requiring specific environmental conditions that result in large genotype by environment interactions.

Current work involves large-scale validation in a range of genetic backgrounds to enable marker-assisted selection in our breeding program. At this stage all selection for stay-green is based on phenotype.

7.1.4 Stay-Green Breeding Methodology

The pedigree and limited backcross methods have been used. As for midge resistance, the approach has been to combine a number of traits including stay-green, midge resistance, appropriate agronomic type (phenology and height) and grain yield. The latter two traits being critical as a 'stay-green' like phenotype can be generated by plants with low harvest index and variation in stay-green expression can result from variation in phenology.

There are numerous genes involved, necessitating multiple cycles of crossing, evaluation, and selection of parents and additional crossing to commence a new cycle with the occasional infusion of new germplasm.

Only phenotypic selection has been used to date. This has proved to be reasonably effective because the trait is expressed in hybrid yield tests when post-anthesis drought stress occurs.

7.1.5 Stay-Green Results

Progress has been made in transferring a moderate level of the trait from B35 to otherwise locally adapted genotypes and combining this with a high level of midge resistance. B35's level of stay-green has not been recovered. There is a need to increase the level of stay-green. Part of future stay-green breeding will involve marker-assisted backcrossing of the validated markers for the region.

7.1.6 Adoption by Growers

At this time, the adoption of this technology is low because there are few 'stay-green' commercial hybrids available and these have only a low level of stay-green. Material with much higher levels of stay-green combined with high levels of midge resistance are now in private breeding programs so it is reasonable to expect that the real benefits of the trait will soon be available to graingrowers.

The annual value of this trait to industry when fully developed and adopted has been estimated to be about $A30 M.

7.2 *Direct Selection for Drought Resistance*

The breeding methods used for breeding for midge resistance and stay-green resulted in the narrowing of the genetic base in the DPI&F program (Jordan et al., 1998). This, potentially, has at least two detrimental effects. One is a reduction in the genetic variance and hence heritability for example for yield in the breeding program gene pool. The other effect is a reduction in heterosis which, in general, increases with the genetic diversity of the hybrid parents.

7.2.1 Increasing Genetic Diversity

A program to increase the genetic diversity in the gene pool was implemented. It is anticipated that this will increase heritability. Initially, this was achieved by introgressing putative genetic diversity using the pedigree breeding method. More recently, a limited backcross method (BC1) has been used where the non-recurrent parent introduces diversity and the recurrent parent is a proven source of midge resistance and stay-green. Both methods have proven to be effective in increasing yield while maintaining significant levels of midge resistance and stay-green. In the case of the limited backcross method, hybrids have been produced, higher in yield than that of the recurrent parent (Jordan et al., 2006).

At this stage, the non-recurrent parents are chosen not only on their proven agronomic worth, but also on the basis of their suspected (e.g. on the basis of region of collection and/or race) or known genetic diversity. Deu et al. (2006) and de Oliveira et al. (1996) have shown that DNA markers allow a measure of genetic diversity on the basis of their regional origin and to a lesser extent diversity associated with taxonomic races. It has yet to be determined if diversity determined with the marker systems used to date has any relation with diversity with respect to important agronomic traits, many of which are multigenic in inheritance.

7.2.2 Defining the Target Population of Environments

In addition to increasing genetic diversity, other newer, technologies are being tested in the program of directly selecting for drought resistance in Australia (Borrell et al., 2006). In Australia, grain sorghum commonly experiences water deficits even though it is normally grown in soils with relatively high water holding capacity. The phenological timing, severity and duration of deficits varies spatially (within and between tests) and temporally resulting in significant genotype by environment interactions, in turn resulting in low heritability for yield. Differences in phenology and reaction to stress amongst the test genotypes add to this complexity.

Methods of reducing the effects of genotype by environment interactions are being employed. One such method involves defining the target population of environments – TPEs (Comstock, 1977; Borrell et al., 2006). On this basis and to simplify TPEs for sorghum in Australia, Chapman et al. (2000) used a sorghum crop growth model (Hammer and Muchow, 1994) and long-term weather and soils data, to calculate a water stress index and grouped these by pattern analysis. This approach not only identifies particular environment types but also indicates how frequently these environment types occur in the TPE. Podlich and Cooper (1998) suggested that genetic gain would be enhanced by using this information to weight the data from individual tests depending on how frequently the environment of the test occurred in the TPE. This method could also be used for selection for specific adaptation to a particular pattern of water stress.

7.2.3 Statistical Methodology

Another tool is the use of newer statistical methodologies for spatial design and analysis. It is common, particularly under stress conditions, that there are strong spatial trends in the stress index within a test. This results in large error variances. Newer methods have been developed and used to significantly improve this situation (Chan and Eccleston, 2003; Cullis and Gleeson, 1991; Gilmour et al., 1997). Smith et al. (2001) reported a 6–46% increase in heritability for yield using these techniques and Kelly (personal communication) reported an increase of 3–22% in the genetic gain from DPI&F yield tests.

8 Conclusions

In this chapter, we have provided a description of a successful germplasm enhancement program in sorghum. There is a diversity of sorghum breeding programs world wide and we have presented only this program as a case study of a successful germplasm enhancement exercise. The reasons for the success of the program in making genetic progress have been the development of a good understanding of the genetic architecture and biology of the traits in question. This allowed the development of selection systems which maximised heritability either through increasing genetic variability or by decreasing environmental variability or both. While the program focused on particular traits, namely, stay-green and midge resistance, considerable effort was made to ensure that adaptation to the target environment was maintained via selection for acceptable performance in multi-environment trials.

Future breeding sorghum breeding programs, including our own, will alter markedly as new technologies such as whole genome marker screening, simulation modelling and advances in statistical methods are adopted. However, some fundamental issues will remain such as the need to understand the genetic architecture and biology of traits. This understanding allows appropriate and efficient breeding methodologies to be developed.

Acknowledgements To the Department of Primary Industries and Fisheries, Queensland, and the Grains Research and Development Corporation for financial support of the breeding program. Also to Dr R.L. Brengman for the significant contribution he made to the program during 1979 to 1996.

References

Borrell, A.K., Bidinger, F.R., and Sunitha, K. (1999) Stay-green associated with yield in recombinant inbred sorghum lines varying in rate of leaf senescence. In: *International Sorghum and Millets Newsletter*. 40, 31–34.

Borrell, A.K., Hammer, G.L., and Henzell, R.G. (2000) Does maintaining green leaf area in sorghum improve yield under drought? II. Dry matter partitioning and yield. *Crop Science*. 40, 1037–1048.

Borrell, A.K., Jordan, D.R., Mullet, J., Henzell, R.G., and Hammer, G.L. (2006) Drought Adaptation in Sorghum. In: *Drought Adaptation in Cereals*. The Hawath Press Inc. 335–399.

Chapman, S.C., Cooper, M.C., Hammer, G.L., and Butler, D.G. (2000) Genotype by environment interactions affecting grain sorghum. II. Frequencies of different seasonal patterns of drought stress are related to location effects on hybrid yields. *Australian Journal of Agricultural Research*. 57, 209–222.

Chan, B.S.P., and Eccleston, J.A. (2003) On the construction of nearest-neighbour balanced row-column designs. *Biometrics*. 45, 97–106.

Comstock, R.E. (1977) Quantitative genetics and the design of breeding programs. In: *Proceedings of the International Conference on Quantitative Genetics*. 705–718. Ames, IA, Iowa State University Press.

Cullis, B.R., and Gleeson, A.C. (1991) Spatial analysis of field experiments – an extension to two dimensions. *Biometrics*. 47, 1449–1460.

Dahlberg, J.A. (2000) Classification and characterization of Sorghum. In: C.W. Smith and R.A. Frederiksen (Eds). Sorghum: origin, history, technology, and production. John Wiley & Sons, Inc. New York, NY.

Deu, M., Rattunde, F., and Chantereau, J. (2006) A global view of genetic diversity in cultivated sorghums using a core collection. *Genome*. 49, 168–180.

de Oliveira, A.C., Richter, T., and Bennetzen, J.L. (1996) Regional and racial specificities in sorghum germplasm assessed with DNA markers. *Genome*. 39, 579–587.

de Wet, J.M.J., and Haran, J.R. (1971) The origin and domestication of Sorghum bicolor. *Econ. Bot*. 25, 128–135.

Doggett, H. (1988) Sorghum. 2nd. Ed. John Wiley & Sons, New York.

Franzmann, B.A. (1993) Ovipositional antixenosis to Contarinia sorghicola (Coquillett) (Diptera: Cecidomyiidae). *Journal of Australian Entomology Society*. 32, 59–64.

Franzmann, B.A., Butler, D.G., Henzell, R.G., Fletcher, D., and Cutler, J.H. (1996) A sorghum industry scheme assigning midge resistance levels to commercial hybrids. In: M.A. Foale, R.G. Henzell, and J.F. Kneipp (Eds). *Proceedings of the Third Australian Sorghum Conference*, Tamworth, 20–22 February 1996. Australian Institute of Agricultural Science, Melbourne, Occasional Publication No. 93. pp. 359–363.

Garber, E.D. (1950) Cytotaxonomic studies in the genus Sorghum. University of California Publications in Botany. 23, 283–361.

Gilmour, A.R., Cullis, B.R., and Verbyla, A.P. (1997) Accounting for natural and extraneous variation in the analysis of field experiments. *Journal of Agricultural Biological and Environmental Statistics*. 2, 269–293.

Hardy, A.T., Tao, Y.Z., and Franzmann, B.A. (2001) New Host-Plant Resistance to the Sorghum Midge in Grain Sorghum and the use of Molecular Marker Mediated Selection. In: A.K.

Borrell, R.G. Henzell, and J.F. Kneipp (Eds). '*Proceedings of 4th Australian Sorghum Conference*', Kooralbyn, Gold Coast Hinterland, 5–8 February (CD Rom).

Hammer, G.L., and Muchow, R.C. (1994) Assessing climatic risk to sorghum production in water-limited subtropical environment. I. Development and testing of a simulation model. *Field Crops Research*. 36, 221–234.

Harlan, J.R., and de Wet, M.J. (1971) The origin and domestication of Sorghum bicolor. *Econ. Bot.* 25, 128–135.

Harlan, J.R., and de Wet, M.J. (1972) A simplified classification of cultivated sorghum. *Crop Science*. 12, 172–176.

Henzell, R.G., Brengman, R.L., Fletcher, D.S., and McCosker, A.N. (1992) Relationship between yield and non-senescence (stay-green) in some grain sorghum hybrids grown under terminal drought stress. In: M.A. Foale, R.G. Henzell, and P.N. Vance. (Eds), *Proceedings of the Second Australian Sorghum Conference*. pp. 355–359. Gatton, February 4–6, 1992. Occasional Publication No.68, Melbourne: Australian Institute of Agricultural Science.

Henzell, R.G., Franzmann, B.A., and Brengman, R.L. (1994) Sorghum midge resistance research in Australia. *International Sorghum and Millets Newsletter*. 35, 41–47.

Henzell, R.G., Peterson, G.C., Teetes, G.L., Franzmann, B.A., Sharma, H.C., Youm, O., Ratnadass, A., Toure, A., Raab, J., and Ajayi, O. (1996) Breeding for resistance to panicle pests of sorghum and pearl millet. In: *Proceedings of the International Conference on Genetic Improvement of Sorghum and Pearl Millet*. September 23–27, Lubbock, Texas. pp. 481–489.

Johnson, J.W., Rosenow, D.T., and Teetes, G.L. (1973) Resistance to the sorghum midge in converted exotic sorghum cultivars. *Crop Science*. 13, 174–175.

Jordan, D.R., Tao, Y.Z., Godwin, I.D., Henzell, R.G., Cooper, M., and McIntyre, C.L. (1998) Loss of genetic diversity associated with selection for resistance to sorghum midge in Australian sorghum. *Euphytica*. 102, 1–7.

Jordan, D.R., Tao, Y.Z., Godwin, I.D., Henzell, R.G., Cooper, M., and McIntyre, C.L. (2003) Prediction of hybrid performance in grain sorghum using RFLP markers. *Theoretical and Applied Genetics*. 106, 559–567.

Jordan, D.R., Tao, Y.Z., Godwin, I.D., Henzell, R.G., Cooper, M., and McIntyre, C.L. (2004) Comparison of identity by descent and identity by state for detecting genetic regions under selection in a sorghum pedigree breeding program. *Molecular Breeding*. 14, 441–454.

Jordan, D.R., Hammer, G.L., and Henzell, R.G. (2006) Breeding for yield in the DPI&F breeding program. In: A.K. Borrell, R.G. Henzell, and D.R. Jordan. (Eds). *Proceedings from the Fifth Australian Sorghum Conference*. Gold Coast, Australia, 30-Jan to 2 Feb., 2006, CDROM format, Range Media Pty. Ltd. ISGN; 0-6464673-0-1.

Lazarides, M., Hacker, J.B., and Andrew, M.H. (1991) Taxonomy, Cytology and Ecology of Indigenous Australian Sorghums. (Sorghum Moench: Andropogoneae: Poaceae *Australian Systematic Botany*. 4, 591–635.

Murty, B.R., and Govil, J.N. (1967) Description of 70 groups in genus sorghum based on a modified Snowden classification. *Indian Journal of Genetics*. 27, 75–91.

Murray, D., Miles, M., and Ferguson, J. (2000) Area-wide management of heliothis – results of current studies. In: *Proceedings of the Tenth Australian Cotton Conference*, Brisbane. pp. 11–17.

Murray, D., Ferguson, J., Lloyd, R., Hopkinson, J., Maclean, S., and Powell, R. (2001) Advances in heliothis management on grain sorghum in Australia. In: A.K. Borrell, R.G. Henzell, and J.F. Kneipp (Eds). '*Proceedings of 4th Australian Sorghum Conference*', Kooralbyn, Gold Coast Hinterland, 5–8 February (CD Rom).

Page, F.D. (1979) Resistance to sorghum midge (*Contarinia sorghicola* Coquillet) in grain sorghum. *Australian Journal of Experimental Agriculture and Animal Husbandry*. 19, 97–101.

Podlich, D.W., and Cooper, M. (1998) QU-GENE: A simulation platform for quantitative analysis of genetic models. *Bioinformatics*. 14, 623–653.

Rooney, W.L. (2000) Genetics and cytogenetics. In: C.W. Smith and R.A. Frederiksen (Eds). Sorghum: origin, history, technology, and production. John Wiley & Sons, Inc. New York, NY.

Rosenow, D.T. (1977) Breeding for lodging resistance in sorghum. In: *Proceedings of the 32rd Corn and Sorghum Research Conference* 4, 171–185. DC: American Seed Trade Association.

Rosenow, D.T., and Clark, L.E. (1981) Drought tolerance in sorghum. In: *Proceedings Annual Corn and sorghum research Conference*, December 9–11, 1981, Chicago, IL.

Rosenow, D.T., and Dahlberg, J.A. (2000) Collection, Conversion and Utilization of Sorghum. In: C. Wayne Smith and A. Richard (Eds). Sorghum: origin, history, technology and production. Frederiksen ISBN 0-471-24237-3 John Wiley and Son, Inc.

Rosenow, D.T., Quisenberry, J.E., Wendt, C.W., and Clark, L.E. (1983) Drought tolerant sorghum and cotton germplasm. *Agricultural Water Management*. 7, 207–222.

Schertz, K.F. (1983) Potentials with new cytoplasmic male sterility systems in sorghum. *Proc. Genet. Soc. Am.* 38, 1–10.

Sharma, H.C., Vidyasagar, P., and Subramanian, V. (1993) Antibiosis component of resistance in sorghum to sorghum midge, *Contarinia sorghicola. Annals Applied Biology*. 123, 469–483.

Smith, A.B., Cullis, B.R., and Thompson, R. (2001) Analysing variety by environment data using multiplicative mixed models and adjustments for spatial field trend. *Biometrics*. 57, 1138–1147.

Snowden, J.D. (1935) The classification of the cultivated Sorghums. Bull. Misc. Information, No. 5. Royal Botanic Gardens, Kew, England. pp. 221–255.

Snowden, J.D. (1936) The cultivated races of Sorghum. Allard and Co., London.

Sprangler, R. Zaithik, B., Russo, E., and Kellogg, E. (1999) Andropogoneae Evolution and generic Limits in Sorghum (Poaceae). *Systematic Botany*. 16, 279–299.

Stapf, O. (1917) Flora of Tropical Africa. I. Reeve and Co. London.

Stephens, J.C., and Holland, R.F. (1964) Cytoplasmic male-sterility for hybrid seed production. *Agronomy Journal*. 46, 20–25.

Stephens, J.C., Miller, F.R., and Rosenow, D.T. (1967) Conversion of alien sorghums to early combine genotypes. *Crop Science*. 7, 396.

Tao, Y.Z., Henzell, R.G., Jordan, D.R., Butler, D.G., Kelly, A.M., and McIntyre, C.L. (2000) Identification of genomic regions associated with stay-green in sorghum by testing RILs in multiple environments. *Theoretical and Applied Genetics*. 100, 1225–1232.

Tao, Y.Z., Hardy, A., Drenth, J., Henzell, R.G., Franzmann, B.A., Jordan, D.R., Butler, D.G., and McIntyre, C.L. (2003) Identifications of two different mechanisms for sorghum midge resistance through QTL mapping. *Theoretical and Applied Genetics*. 107, 116–122.

Durum Wheat Breeding

Conxita Royo, Elias M. Elias, and Frank A. Manthey

Abstract This chapter summarizes the scientific and technical knowledge for durum wheat breeding, giving some examples of the methods applied in national programs. Section 1 refers to the importance of durum wheat in the world. Sections 2 and 3 give technical details on genetic diversity and the choice of germplasm, while the main varietal groups are explained in Section 4. Information about the major breeding achievements, current goals of breeding and breeding methods and techniques are covered by Sections 5, 6 and 7 respectively. The integration of new biotechnologies, particularly marker assisted selection, into breeding programs is described on Section 8, while information about foundation seed production and intellectual property rights are given on Section 9.

1 Introduction

Durum wheat [*Triticum turgidum* ssp. *turgidum* convar. *durum* (Desf.) MacKey] is one of the oldest cultivated cereal species in the world. It is of great importance in cereal areas of the Mediterranean Basin and North America, where the great bulk of world production of this crop and land under cultivation with it are concentrated (Table 1). The area annually planted with durum wheat worldwide is estimated to be about 13.5 million ha, though it has shown a decreasing tendency since the 1970s when it was close to 18 million ha (Belaid, 2000). The European Union devotes around 3.5 million ha to its cultivation, with a production of around 9.2 million metric tons. Canada is the second largest producer in the world and the greatest exporter. Average global yields have increased from 1.4 t ha^{-1} during the 1970s to more than 2 t ha^{-1} in recent years, leading to a great increase in total production. However, a reduction in global production occurred in 2005 due to lower plantings in the major EU durum-producing countries (Italy and Spain), combined with a severe drought affecting growing areas in the Mediterranean Basin.

C. Royo(✉)
IRTA (Institute for Food and Agricultural research and Technology) Generalitat de Catalunya, Spain, Cereal Breeding, e-mail: conxita.royo@irta.es

M.J. Carena (ed.), *Cereals*,
DOI: 10.1007/978/-0-387-72297-9,

Table 1 Area, yields, and production of durum wheat in the world in 2004 and 2005

Country	Area (000 ha)		Yield (t/ha)		Production (000 t)	
Year	2004	2005	2004	2005	2004	2005
Algeria	1,369	1,000	1.33	1.00	1,816	1,000
Argentina	57	54	3.16	2.96	180	160
Australia	200	200	2.00	2.00	400	400
Austria	15	15	4.00	4.00	60	60
Canada	2,141	2,200	2.32	2.16	4,962	4,750
France	406	415	5.05	3.98	2,050	1,650
Germany	8	8	6.25	5.75	50	46
Greece	500	500	2.00	2.00	1,000	1,000
India	450	450	2.67	2.67	1,200	1,200
Italy	1,870	1,450	3.05	2.41	5,700	3,500
Kazakhstan	100	100	1.00	1.00	100	100
Mexico	230	240	5.22	5.00	1,200	1,200
Morocco	1,111	1,050	1.82	0.71	2,025	750
Portugal	145	120	1.14	0.58	165	70
Russia	1,000	1,000	1.00	1.20	1,000	1,200
Spain	910	850	3.10	1.18	2,825	1,000
Syria	830	830	2.53	2.53	2,100	2,100
Tunisia	830	750	1.69	1.53	1,400	1,150
Turkey	1,100	1,100	2.18	2.09	2,400	2,300
UK	1	1	6.00	6.00	6	6
USA	956	993	2.56	2.58	2,450	2,560
World	14,229	13,326	2.33	1.97	33,089	26,202

Source: USDA (http://www.fas.usda.gov/pecad/highlights/2005/07/durum2005/)

In the SEWANA region (South Europe, West Asia, and North Africa), durum wheat is mainly grown under rainfed conditions, characterized by unpredictable rainfall and a large incidence of abiotic and biotic stresses. Drought and heat during the grain filling period, nutrient deficiencies, soil problems, diseases, and pests are the main yield constraints. The Mediterranean Basin is also the largest consumer of durum wheat products (spaghetti, macaroni, couscous, bulgur, frekeh, etc.), and the most significant import market.

2 Genetic Diversity

Durum wheat originated and became diversified in the Middle and Near East and in North Africa (MacKey, 2005). On the basis of the geographic origin and ecophysiological characterization of a number of Mediterranean and West Asian durum wheat landraces, the species *T. turgidum* L. ssp. *durum* (Desf.) Husn was subdivided during the last century into three botanical sections, namely *mediterranea*, *syriaca*, and *europea* (Grignac, 1965).

Recent studies indicate that the genetic diversity in durum wheat seems to be structured, at least in part, according to a geographical pattern (Moragues et al., 2006, 2007; Maccaferri et al., 2005). In a study including durum wheat accessions from 22 countries, the widest genetic diversity within countries was found in the Indian germplasm, while the lowest corresponded to the Bulgarian one (Pecetti et al., 1992). A high level of diversity has also been found in germplasm from the Middle East region (Yang et al., 1991), particularly Jordan (Rawashdeh et al., 2007), Turkey (Akar and Ozgen, 2007), Iran, Egypt, Afghanistan, and Ethiopia (Asins and Carbonell, 1989). The Ethiopian durum wheat seems to constitute a different pool from that of other geographical regions (Pecetti et al., 1992). The Mediterranean genetic pool appears to be different from the Southwest Asian (Moragues et al., 2006) and the North American ones (Maccaferri et al., 2005), while gene pools from Syria and Jordan seem to be closely related (Pecetti et al., 1992).

Durum wheat landraces, which were widely cultivated in the early twentieth century, were later increasingly replaced by improved varieties. The introduction of productive semi-dwarf cultivars resulted in the abandon of the genetically diverse, locally well-adapted but unimproved landraces, and the extinction of on-farm genetic variability. It has been suggested that the level of genetic diversity underlying the successful modern varieties may have fallen due to the limited number of ancestors, the relative uniformity of the pursued ideotype (Autrique et al., 1996; Pecetti and Annicchiarico, 1998), the high selection pressure applied in breeding programs, and the relatively small number of varieties currently in cultivation (Skovmand et al., 2005). Pedigree analysis has revealed that, in some cases, the genetic background of the modern pool of elite durum wheat varieties is narrow (Maccaferri et al., 2005). A study including 51 cultivars derived from the CIMMYT/ICARDA breeding program found that 15 ancestors were present in the pedigree of at least 80% of the cultivars, 5 of them being present in all of them (Autrique et al., 1996). Similarly, nine ancestors are present in the pedigree of more than half the cultivars developed in Russia within the framework of various breeding programs (Martynov et al., 2005). Moreover, Tunisian cultivars seem to be genetically more similar than the old ones (Medini et al., 2005). The main risk of a narrowing of the genetic background of the modern genetic pool is one of increased vulnerability to diseases and pests (Frankel et al., 1995) and a fall in the abiotic stress tolerance, particularly to the drought and high temperatures that are typical of many regions growing durum wheat.

However, the results of several studies not only do not evidence an overall decrease in the genetic diversity of durum wheat due to past breeding activities (Martos et al., 2005) but even reveal that it is increasing over time as a result of the introgression of genetic variability (Autrique et al., 1996; Maccaferri et al., 2003; Martynov et al., 2005). CIMMYT and ICARDA, the two international centres operating with durum wheat, have largely helped to widen the genetic pool of current cultivars; shuttle breeding and germplasm exchange all around the world have been key factors in creating the current overall variation in durum wheat. Molecular analyses have revealed that the most recent CIMMYT-derived founders are genetically distant from the old Mediterranean ones (Maccaferri et al., 2003).

Around 79,000 tetraploid and 253,000 unspecified *Triticum* accessions are currently available in gene bank collections around the world (Skovmand et al., 2005). The characterization of germplasm maintained in gene banks is crucial for exploiting the existing genetic variability for traits of economic importance such as yield, yield stability, grain quality, and tolerance to biotic and abiotic stresses. Molecular genetic markers have proven to be a valuable tool for identifying accessions containing genes and alleles of interest, frequently masked in undesirable phenotypes.

3 Choice of Germplasm

The first step in a breeding program consists in the creation of variability, usually by hybridization, with the aim of accumulating enough genetic variation and providing useful gene recombinations for the target traits in the progeny. The choice of parents for crosses requires a priorization of the goals to be achieved by breeding, and the collection and characterization of genetic sources carrying favourable alleles for the target traits.

The parents to be used are usually chosen for their performance in terms of the main breeding objectives: yield potential, end-use quality, and resistance to abiotic and biotic stresses (Table 2). Crosses of elite × elite durum wheat adapted parents are usually the choice when there is sufficient genetic variability within the species for the target traits. However, when diversity within durum wheat is limited, other sources of variation such as crosses with *Triticeae* relative species must be explored. Crosses with bread wheat have been extensively used in durum wheat breeding to introgress favourable alleles for many traits such as plant vigour, winter hardiness, and fusarium head blight (FHB) tolerance. The Chinese bread wheat cultivar 'Sumai 3' has been widely used to incorporate tolerance to FHB into durum wheat.

Because of the low intraspecific genetic variability for drought tolerance in durum wheat, some wild relatives such as *T. turgidum* ssp. *dicoccoides* and *T. urartu* have been proposed as donors for drought resistance in prebreeding programs (Kara et al., 2000; Valkoun, 2001). Other wild relatives such as *Hordeum chilense* (Forster et al., 1990) and *Thinopyrum bessarabicum* (King et al., 1997) have been used to introgress tolerance to salinity in durum wheat (Mano and Takeda, 1998; Colmer et al., 2006).

Most of the characteristics relevant for the agronomic performance of cultivars are complex quantitative traits regulated by several genes. Genetic gains of grain yield – the most common example of this kind of traits – have usually been achieved after the Green Revolution by crossing a good number of high-yielding parents with good combining ability. However, a large spectrum of genetic variation for yield components has been reported (Elias et al., 1996b). Genotypes with extreme expression for the number of grains per spike, grains per spikelet, and spike length have been developed from *T. polonicum* and other alien donors (Al-Hakimi et al., 1997). An excellent review of wide crosses for durum wheat improvement may be found in Mujeeb-Kazi (2005).

Table 2 Durum wheat cultivars and lines identified as resistant or tolerant to diseases affecting this species

Disease	Pathogen	Resistant/tolerant cultivars and lines	References
Leaf rust	*Puccinia triticina*	Golden Ball, Lloyd, Medora, Pelissier, Quilafen, Stewart 63, Wakomma	Zhang and Knott, 1990
		Hualita, Jupare 2001, Llareta INIA, Pohowera, Somateria	Singh et al., 2005
			Herrera-Foessel et al., 2005
		Ardente, Aronde, Creso, Colosseo	Martinez et al., 2007
Stem rust	*Puccinia graminis* f. sp. *tritici*	Golden Ball, Maruccos 623, Peliss, Petterson ML68-14	Roelfs et al., 1992
			McIntosh et al., 1995
Powdery mildew	*Blumeria graminis* f. sp. *tritici*	Valnova, Valforte, Valgerardo	Vallega and Zitelli, 1973
Fusarium head blight	*Fusarium graminearum*	Rugby	Elias et al., 1995
			Elias et al., 1996a
		Creso, Crispiero, Enduro, Fenix, Ixos, Neodur, Primadur, Tresor, Vento	Balmas et al., 1999
		Ajaia 3/Silver 16, Eupoda 3, Ghaz 1, Kitza 12, Lhne/Akaki//Dipper, Netta 1/Gan, Nokikana 23, Srn 2/Ru/Duilio, Wizza 1	Singh et al., 2005
Bunt	*Tilletia tritici*	Russello SG7, Cappelli, Azizia, Garigliano, Capeiti 8	Grasso, 1968 Bozzini, 1971
Leaf blotch	*Septoria tritici*	Haurani, Kunduru 1149, Senatore Capelli	ICARDA, 1980
		Aus1/5/Cndo/4/Bry*2/Tace//II27655/3/Time/Zb/2*2 W, Cali/ship 2//Fillo 7, BD 2337, BD 2338, BD 2339	Singh et al., 2005
		CMH82A. 1062/3/GdoVZ394//Sba81/Plc/4/Aaz 1/Crex/5/Hui//Cit71/CII	
		Gdfl/*T. dicoccoides*–SY20013//Bcr, Plata 6 / Green 17, Porron 1, Zeina 2, Zeina 4	
		Azizia, Garigliano, Cappelli, Capeiti 8	Rosielle, 1972
Tan spot	*Pyrenophora*	Joda, Oscar, Pabellon, Sham-3, WL5023, S3-6	Singh et al., 2005

4 Varietal Groups

The genetic pools of durum wheat may be classified according to a pattern of adaptability and geographical distribution. About 86% of the durum wheat cultivated in the world has a spring growth habit, whereas the remaining 14% are winter and facultative types, mostly restricted to the areas around the Mediterranean, Black, and Caspian Seas (Palamarchuk, 2005). The following are some of the most representative varietal groups:

4.1 The Italian Pool

The importance of durum wheat in Italy and the noteworthy breeding efforts devoted from the beginning of the twentieth century to improving this species make the Italian pool one of the most, if not the most, representative within the Mediterranean Basin. Although other countries conducted breeding programs in the early decades of the last century, Italy may be considered as a pioneer in durum wheat improvement.

Numerous selections were obtained by Italian breeders during the early decades of the twentieth century from a very large pool of Mediterranean landraces (Di Fonzo et al., 2005). One of the most widely spread was the variety 'Senatore Capelli', an 'africanum' type selected from the population 'Jean Retifah' from Algeria that was released in 1915, used in further crosses and is still grown in some areas (Di Fonzo et al., 2005). However, only small-yield increases were achieved at the early times since most landraces were tall and very sensitive to lodging. Breeding efforts resulted in the release of a number of improved varieties from 1950 to 1975. 'Appulo' and 'Trinakria' were two of the most outstanding varities, due to their yielding ability, quality, and good adaptation to drought (Grifoni, 1964; Ballatore, 1973). The varieties 'Viscardo Montarani', 'Carlo Jucci', and 'Giovani Raineri' were obtained from crosses with hexaploid wheat aiming to enhance the number of fertile florets per spikelet. Research on mutation breeding as a way to induce shorter plants with strong straw resulted in the selection of lines with higher yielding ability and short straw such as 'Castelporziano' and 'Castelfusano', derived from the cultivar 'Capelli' (Scarascia-Mugnozza et al., 1972). The gap between durum and bread wheat yield potential was filled by the release in 1974 of 'Creso', a high-quality variety derived from a cross between a CIMMYT dwarf line and cv. 'Castelfusano'.

During the last few decades, the Italian pool has been enriched with the incorporation of new gene pools, mainly from CIMMYT germplasm. The varieties 'Simeto', 'Duilio', 'Arcangelo', 'Creso', 'Colosseo', 'Ciccuio', 'Ofanto', 'Grazia', 'Appulo', 'Rusticano', 'Radioso', 'Appio', 'Svevo', 'Neodur', 'Zenit', and 'Meridiano' are among the ones most cultivated by Italian farmers (Di Fonzo et al., 2005).

4.2 The CIMMYT Pool

Durum wheat germplasm developed by CIMMYT has been the most widely used by national programs worldwide. Variety release statistics reveal that more than 90% of the durum wheat varieties released in developing countries from 1991 to 1997 were introduced or derived from germplasm developed at CIMMYT (Pfeiffer and Payne, 2005).

Early breeding efforts during the 1960s and 1970s were devoted to the introgression of dwarfing genes and alleles for photoperiod insensitivity, improvement of floral fertility, and enhanced biotic stress resistance (Pfeiffer et al., 2000). Further efforts focused on agronomic components associated with high genetic yield potential and wide adaptation in combination with acceptable quality attributes. Varieties such as 'Jori 69' (released in 1969), 'Cocorit 71', 'Mexicali 75', and 'Yavaros 79' were widely adopted and some of them are still grown in many countries. During the 1980s, a new generation of durum wheat varieties arose ('Altar 84', 'Aconchi 89') from the development of the ideotype concept, with balanced increase in all yield components. Later efforts concentrated on the stabilization of grain yield potential and quality improvement.

Because of the appearance in Mexico in 2001 of a new leaf rust (*Puccinia triticina* Eriks) race –designated as BBG/BR – the majority of the CIMMYT germplasm, including some extensively grown cultivars such as 'Mexicali 75', 'Yavaros 79', and 'Altar 84', became susceptible. Since 2001, the main objective of the CIMMYT durum wheat breeding program has concentrated on developing leaf rust-resistant germplasm and widening the genetic basis of leaf rust resistance in this widely adapted gene pool. The high-yielding, rust-resistant cultivar 'Jupare C2001' was released as an immediate emergency measure in Northern Mexico in 2001 and grown commercially in 2002. Two other cultivars, 'Samayoa C2004' and 'Banamichi C2004', were released shortly thereafter to provide genetic resistance based on different major genes and better quality attributes. Internationally, most leaf rust-resistant germplasms have been distributed since 2003, with more recently a considerable improvement in the overall levels of industrial quality attributes in the material distributed. Whereas wide adaptation combined with durable resistance to rusts – including the new threat of stem rust – is still the main focus of the CIMMYT program, substantial emphasis is placed on quality and drought tolerance-related traits.

4.3 The North American Pool

Durum introduction to the USA began in 1850 when the US Department of Agriculture (USDA) introduced to the farmers the varieties 'Algerian Flint', 'Turkish Flint', 'Syrian Spring', and 'Arnautka'. Soon farmers found these varieties not to be adapted to their region and were hard to mill and therefore were used for feed.

In 1870, durum wheat from Nicaragua was introduced to Texas. Again, millers found it hard to mill and by 1890 durum had almost disappeared from production (Carleton, 1901). In 1894, the USDA collected 1,000 spring and winter durum varieties from all parts of the world and evaluated them for agronomic traits, diseases, and production. The varieties 'Nicaragua' from Nicaragua, 'Missogen' and 'Volvo' from Greece, 'Medeah', 'Pelissier', and 'El Safra' from Algeria, 'Candial' from Argentina and Chili, and 'Arnautka' and 'Kubanka' from the former USSR were selected for production (Joppa and Williams, 1988). 'Kubanka' became the leading cultivar from 1910 to 1920. Pure line selections were made from these collected varieties. As a result, 'Acme' was selected from 'Kubanka' in 1909, 'Mindum' was selected from common wheat and released in 1917, and 'Pentad' was selected from a former USSR collection and released as red durum in 1903 (Joppa & Williams, 1988). 'Pentad' gained popularity because of its resistance to stem rust (caused by *Puccinia graminis* Per.:Pers. f. sp. *tritici* Eriks. & E. Henn), but it was given a separate class because of its red kernel colour which made it unsuitable for pasta manufacturing. Over the years, most of these cultivars became susceptible to the stem rust race 56, with the exception of 'Pentad'. The newly established durum breeding program attempted to transfer resistance from 'Pentad' but it proved to be difficult to transfer and the resulted progenies had very poor quality (Joppa et al., 1988). Crosses for stem rust resistance with 'Vernal' emmer were more successful. 'Mindum' had a moderate level of resistance to stem rust and was the leading cultivar from 1920 to 1940.

Because of the importance of durum wheat for the state of North Dakota, the durum plant breeding and genetic program was initiated in 1929 (Joppa et al., 1988) and is the only public research project that develops and releases durum wheat cultivars in USA. A year earlier, a similar breeding program was established in Canada (Knott, 1995). 'Carleton' and 'Stewart' were the first cultivars released by the North Dakota breeding program to the farmers in 1943 as result of crossing 'Mindum' to 'Vernal' emmer and back crossing twice to 'Mindum' (Joppa et al., 1988). By 1949, these two cultivars became the most popular cultivars replacing 'Mindum'. In 1953, 'Stewart 221' was developed in Canada by backcrossing 'Stewart' to 'Mindum'. In 1950, the stem rust race 15B attacked all the durum grown in USA and Canada and caused epidemics in 1953 and 1954. 'Khapli' emmer and PI94701 were found to be resistant to race 15B. In 1955, 'Langdon' and 'Yuma' with stem rust resistance from 'Khapli' and 'Ramsey' and 'Towner' with resistance from PI94701 were released to the farmers as resistant durum cultivars (Joppa et al., 1988). 'Langdon' had good resistance to stem rust, but in 1957, it was overcome by race 15B-2. 'Langdon' was crossed to derivatives of crosses of 'Khapli' emmer and as a result, the two cultivars 'Wells' and 'Lakota' were released to the farmers in 1960 as resistant to all races and have yield similar to 'Langdon'. However, both had small kernels and low milling yield. Both were crossed to other cultivars for stem rust diversity and kernel size. In 1966, 'Leeds' was released as a durum cultivar with stem rust resistant, large kernels, and good milling yield. In 1963, also 'Stewart 63' was released by the University of Saskatchewan as a stem rust-resistant cultivar. Prior to 'Stewart 63', Canada was growing US cultivars such as 'Mindum', 'Carlton', 'Stewart', and 'Ramsey' (Clarke, 2005). Because of the epidemic of stem rust that lasted 30 years, the breeding program had little time to accomplish other objectives. However, in 1950, the cultivar 'Nugget' was released because of its yellow pigment and reduced plant height.

In 1956, the semi-dwarf genes were introduced to the North American germplasm to increase yield. However, progenies from crosses of semi-dwarf sources with other cultivars did not result in higher yield (Joppa, 1973). Since then, several semi-dwarf cultivars have been released from North Dakota such as 'Cando' (1975), 'Calvin' (1978), 'Lloyd (1983)', and 'Plaza' (1999) and from Canada, such as AC 'Morse' (1996) and AC 'Navigator' (1999). In the 1970s, several medium height to tall cultivars were released from North Dakota that had higher yields than 'Leeds', such as 'Rollete' (1971), 'Ward' (1972), 'Rugby' (1973), 'Crosby' (1973), and 'Botno' (1973). All these cultivars had high yields, disease resistance, and good colour but lacked gluten strength. The good colour in these cultivars was incorporated from the Australian durum wheat 'Heiti' (Joppa et al., 1988). In the 1970s, Canada released the cultivars 'Wascana' (1971), 'Wakooma' (1972), and 'Coulter' (1978). In 1970, there was a shift in the objectives of the breeding program, and emphases were placed on developing strong gluten durum cultivars in order to compete in the international expert market. 'Cappelli' a strong gluten durum cultivar from Italy, was used to introduce gluten strength to North Dakota germplasm. In 1976, 'Edmore' was released as the first strong gluten durum cultivar followed by 'Vic' (1979), Lloyd, 'Monroe' (1985), and 'Renville' (1988). At the same time period, Canada released the cultivars 'Medora' (1981), 'Arcola' (1984), and 'Sceptre' (1985). The current objective of the breeding program in North Dakota is to release cultivars that have good agronomic traits and disease resistance, and possess excellent quality for the domestic industry and the international market. For genetic diversity, some germplasm has been introduced from Europe and CIMMYT. However, most of the cultivars that have been developed in the last 15 years in North Dakota have over 95% of North American germplasm in their background. These cultivars are 'Munich' (1995), 'Ben' (1996), 'Belzer' (1997), 'Mountrail' (1998), 'Maier' (1998), 'Lebsock' (1999), 'Plaza' (1999), 'Pierce' (2001), 'Dilse' (2002), 'Divide' (2005), 'Alkabo' (2005), and 'Grenora' (2005). The highest yielding cultivars are 'Alkabo', 'Divide', 'Grenora', 'Lebsock', and 'Mountrail'. 'Dilse', 'Divide', and 'Maier' have the highest grain protein concentration. 'Alkabo', 'Belzer', 'Divide', 'Grenora', and 'Pierce' have very strong gluten. Several cultivars also have been released by Canada in the last 15 years such as 'Plenty' (1990), AC 'Melita' (1994), AC Morse, AC 'Avonlea' (1997), AC Navigator, AC 'Pathfinder' (1999), AC 'Napoleon' (1999), 'Strongfield' (2003), and 'Commander' (2004). AC 'Napoleon' and 'Strongfield' have low cadmium content and 'Commander' has high pigment content and very strong gluten properties.

4.4 The Winter Pool

Local winter and facultative durum wheat populations were formed around 7000 BP in Anatolia, Turkey (Nesbitt and Samuel, 1996). Facultative and winter varieties were derived from selections from old local collections during the early twentieth century in countries from the Caucasian region (Armenia, Azerbaijan, Russian Federation, and Georgia), the Near East region (Iran and Turkey), the Balkan region (Bulgaria, Macedonia, and Romania), and the North Black Sea region (Ukraine). Some of them, such as the local varieties 'Ak-Bugda' and 'Sara Bugda', have been cultivated until recently (Palamarchuk, 2005). In the late nineteenth century, the local facultative varieties 'Agili', 'Sbei', and 'Hamira' were grown in Tunisia and 'Souri', 'Caid de Souef', and 'Kahla' were grown in Algeria (Flaksberger, 1935). The West Asian facultative landraces 'Horani', 'Hourani', 'Gaza', and 'Haiti' and the Transylvanian variety 'Arnaut' have been widely used by many breeders (Palamarchuk, 2005). Varieties 'Selcuelu 97', 'Yilmar 98', and 'Ankara 98' from Turkey, 'Pandur' and 'Condur' from Romania, 'Martondur 1' and 'Martondur 2' from Hungary, 'Leukurum 21' from the Russian Federation, and 'Dnepryana' and 'Leukurum 21' from the Ukraine are examples of cold-tolerant varieties released in the last decade (Palamarchuk, 2005).

5 Major Breeding Achievements

The first breeding efforts devoted to durum wheat were made in areas close to its centre of origin and diversification, and consisted in selection within local landraces and crosses between them (Tesemma and Bechere, 1998; Royo and Briceño-Félix, in press). However, two of the most important milestones in the history of durum wheat breeding were reached after the establishment of CIMMYT in 1965: the introgression of dwarfing genes from genetically distant sources into adapted germplasm and the incorporation of photoperiod insensitivity.

The Japanese hexaploid cultivar 'Norin 10' was the original donor of the dwarfing gene most widely used in durum wheat breeding: the *Rht-B1* (formerly *Rht1*). This gene, located on the short arm of chromosome 4B (Gale et al., 1975; Gale and Marshall, 1976; McVittie et al., 1978), confers insensitivity to exogenous application of low concentrations of gibberellic acid (GA) solution (Gale and Gregory, 1977). Alternative sources of dwarfism have been some GA-responsive genes also derived from Japanese varieties (e.g., Akagomughi, Saitama 27), and probably other minor genes first introduced by the Italian breeder Nazareno Strampelli (Borghi, 2001). Dwarfing genes conferred tolerance to lodging and hence adaptation to high rates of fertilizer application. During the second half of the twentieth century, semi-dwarf wheats replaced old tall varieties in the irrigated and high-yielding regions of the world.

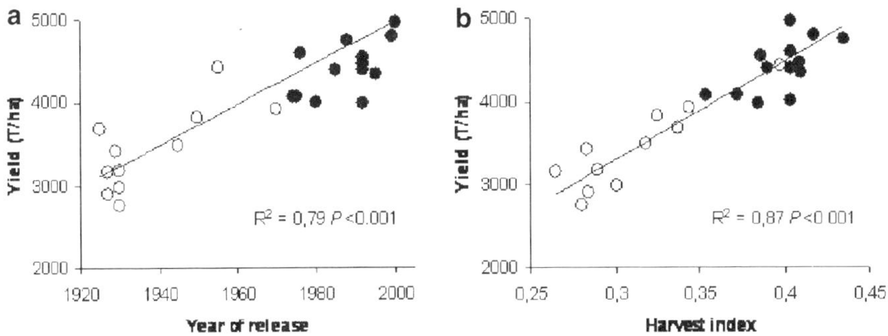

Fig. 1 **(a)** Yield of 24 durum wheat cultivars from Italy and Spain versus their year of release. **(b)** Relationship between harvest index and yield of the same set of cultivars. Data are means across 12 environments from experiments conducted in Spain. Closed circles refer to varieties carrying dwarfing genes

Photoperiod-insensitive durum wheat lines resulted from a shuttle program conducted in Mexico from 1945 onward. Segregating populations were shuttled between two Mexican environments that elicited contrasting responses from the plant types: Cd. Obregon and Toluca (Rajaram and van Ginkel, 2001). The incorporation of photoperiod insensitivity allowed durum wheat, a long-day species, to be grown under short winter days, permitting the wide spread of Mexican semi-dwarf wheats (Borlaug, 1995).

Few studies have evaluated yield progress in durum wheat. Gains reached by CIMMYT's durum wheat breeding program were estimated by Waddington et al. (1987) to be almost 3% per year over the 1960–1984 period, and by Pfeiffer et al. (1996) to be 1.7% per year between 1967 and 1994. In Canada, McCaig and Clarke (1995) reported yield improvements in durum wheat of about 0.81% per year, while a global genetic gain of 0.61% per year has been reported for Italian and Spanish durum wheats grown from 1930 to the present (Fig. 1a) (Royo et al., 2008). Discrepancies in the values of genetic gain may reflect the larger investment in international centres than in national programs and the different yield potential of the environments in which experiments were conducted. The yield advantage of semi-dwarf varieties carrying the *Rht-B1* gene becomes evident in environments producing more than 3.5 t ha^{-1}, and is maximized in the top-yielding sites (Worland and Snape, 2001). Grain yield increases have been mostly based on increases in harvest index (Fig. 1b) and number of grains per unit area (McCaig and Clarke, 1995; De Vita et al., 2007; Royo et al., 2007), via an enhanced number of spikes and grain set (Royo et al., 2007), while the mean weight of the grains has remained virtually unchanged (Waddington et al., 1987; Pfeiffer et al., 2000; Álvaro et al., 2008). Most of these changes have been attributed to a pleiotropic effect of the *Rht-B1* dwarfing gene (Worland and Snape, 2001).

Durum wheat is milled to produce a coarsely ground endosperm called semolina, which is used to make pasta and couscous. Durum flour (a by-product of milling) is used to make bread, particularly in the Mediterranean region. The quality requirements for these products are similar and include semolina rich in yellow (carotenoid) pigments and high in protein and gluten strength. Thus, most durum breeding programs are focused on improving the semolina yellowness, protein content, and protein strength.

Semolina colour is evaluated at the F_5 and subsequent generations in the NDSU durum breeding program. Semolina colour (carotenoid pigment concentration) is a highly heritable trait and is controlled by additive gene effects (Johnston et al., 1983; Joppa et al., 1988). The high heritability indicates that only a few genes with several alleles control the characteristic. Using Psy1–1 and Psy2–1 allele-specific markers and chromosome mapping, the *Psy1* and *Psy2* genes were located on chromosomes 7 and 5, respectively. Four quantitative trait loci (QTL) underlying phenotypic variation in endosperm colour were identified on chromosomes 2A, 4B, 6B, and 7B. Psy1–1 locus co-segregated with the 7B QTL (Pozniak et al., 2007).

Yellowness of pasta depends on the carotenoid pigment concentration and oxidative enzyme activity in the semolina. Oxidative enzymes, lipoxygenase, polyphenol oxidase, and peroxidase, are found in wheat kernel. Lipoxygenase is associated with loss of pigment content, particularly during pasta processing. Lipoxygenase enzymes catalyze the oxidation of polyunsaturated fatty acids containing a *cis*, *cis*-1, 4-pentadiene system, producing conjugated *cis*, *trans*-diene hydroperoxides (Siedow, 1991). Radicals produced during the intermediate steps of this reaction are responsible for oxidative degradation of carotenoid pigments. Lipid oxidation by lipoxygenase can occur during the processing and drying of pasta, which results in a loss of yellow colour in pasta products (Icard-Verniere and Feillet, 1999; Irvine and Winkler, 1950). Borrelli et al. (1999) reported a 16.3% loss of carotenoid pigments during pasta processing. They reported that isoenzymic forms LOX-2 and LOX-3, active at pH of dough, were responsible for the loss of colour in pasta.

Variation exists in the expression of lipoxygenase genes (Manna et al., 1998). Use of QTL may eventually allow screening of germplasm for oxidative enzymes earlier in the breeding cycle. Carrera et al. (2007) reported the existence of a duplication at the *Lpx-B1* locus and an allelic variation for a deletion of the *Lpx-B1*/*1* copy resulted in 4.5-fold reduction in lipoxygenase activity and improved pasta colour but not semolina colour. They reported a molecular marker for the deletion on chromosome 4B. A second lipoxygenase locus *Lpx-A3* was mapped on the homoeologuous region on chromosome 4A and was associated with semolina and pasta colour but not with lipoxygenase activity in the mature grain. *Lpx-B1* locus has been mapped on the short arm of chromosome 4B whereas the *Lpx-A3* locus was mapped on the long arm of chromosome 4A.

Grain protein concentration and gluten strength are important quality traits in durum wheat. Commercial pasta typically contains 7 g of protein per serving (56 g). To meet protein requirement for labelling, the semolina would need 12.5% protein (12% mb) or 12.2% protein at 14% mb. Durum cultivars with high-protein content produce pasta products with greater cooked firmness and increased tolerance to overcooking (Dexter and Matsuo, 1977; Grzybowski and Donnelly, 1979). Protein concentration is evaluated at the F_5 and subsequent generations in the NDSU durum breeding program. Whole grain protein is evaluated using a Technicon InfraAlyzer, where the targeted whole grain protein concentration is 13.5% or higher, at 12% mb.

Emmer wheat (*T. turgidum* L. var. *dicoccoides*) represents a source of genetic variability for grain protein concentration and can provide useful sources of high-protein genes for introgression into durum breeding germplasms. Joppa and Cantrell (1990) substituted individually all 14 chromosomes of a high-protein *T. dicoccoides* accession FA-15-3 into the genetic background of the durum cultivar 'Langdon'. This and the subsequent studies showed that 5 of the 14 *T. dicoccoides* substitution lines (2A, 3A, 6A, 5B, and 6B) have a higher grain protein concentration than the 'Langdon' parent, which indicates that one or more genes that control protein concentration are located on these chromosomes (Joppa and Cantrell, 1990, 1991; Cantrell and Joppa, 1991). Joppa et al. (1997) later mapped a QTL for protein concentration in the LDN(Dic-6B) substitution line. They used a recombinant inbred chromosome line population where the 6B chromosome of *T. dicoccoides* and 'Langdon' were recombined in an otherwise 'Langdon' background. This high-protein locus (*QGpc.ndsu.6Bb*) was located on the short arm and near the centromere of chromosome 6B, flanked by restriction fragment length polymorphism (RFLP) loci *Xmwg79* and *Xabg387*. They also showed that the *QGpc.ndsu.6Bb* locus was inherited as a single genetic factor (i.e., segregating in a 1:1 ratio in a recombinant inbred population) and explained 66% of the total phenotypic variation in protein concentration.

Chee et al. (2001) studied the introgression of this high grain protein genetic factor(s) into adapted durum germplasm and molecular mapping to verify the 6B chromosomal region associated with the high-protein concentration. Analysis of this character using simple regression and interval mapping procedures identified a locus near *Xcdo365* and *Xmwg79* on chromosome 6B that has a major effect on grain protein concentration. This high-protein locus, which explained up to 72% of the phenotypic variance, accounted for a 15 g kg^{-1} increase in average protein concentration and accounted for all the protein content differences between the two parents. The 6B source of high-protein concentration is being used in several durum breeding programs.

Storage proteins of durum wheat are composed of gliadins and glutenins. Gluten is the protein matrix that is formed when gliadins and glutenins are hydrated and mixed together. There is a market for the three gluten strengths. Weak gluten is desirable for crimped or stamped pasta products, strong gluten is desirable for long goods, and very strong gluten is desirable for blending with lower quality semolina and may be advantageous in bread products. Visual selection for gluten strength can be practiced starting at the F_2 generation (McClung and Cantrell, 1986). Strong gluten is linked with the gene *Rg1* for glume colour in durum wheat (Liesle et al., 1981; Hare et al., 1986). White glume colour is associated with strong gluten, while buff or brown colour is associated with weak gluten. In the F_2 population, only white glume colour plants are selected in the NDSU durum breeding program. Gluten strength at the F_3 and subsequent generations is evaluated by the SDS micro-sedimentation test developed by Dick and Quick (1983). A sedimentation value below 30 mm indicates weak gluten, and a sedimentation value of 35 or higher indicates strong gluten. SDS micro-sedimentation test is effective in distinguishing weak from strong gluten, but is less effective in distinguishing strong from very strong gluten. In later generations, gluten/dough strength can be evaluated using SDS micro-sedimentation, mixograph, wet gluten/gluten index tests, alveograph, and glutograph.

Gluten strength in durum wheat is highly heritable and its inheritance primarily additive (Braaten et al., 1962). Gluten strength has been found to be related to allelic variation at the γ-gliadin (*Gli-B1*) locus. Bushuk and Zillman (1979) reported that durum cultivars possessing the γ-gliadin band 45 are of strong gluten type and yield pasta products with greater firmness and other desirable cooking qualities. Pasta products made from cultivars possessing the γ-gliadin band 42, on the other hand, have poor cooking quality. This allelic variation at the *Gli-B1* locus is now believed to be only an association (marker) and not a causal agent for pasta quality. Gliadin proteins provide cohesiveness to the gluten protein matrix and do not contribute greatly to gluten strength due to their non-aggregating properties. It is now generally accepted that genes coding for most γ- and ω-gliadins are located on the short arm of homoeologous group 1 chromosomes (the *Gli-1* locus) and most genes coding for α- and β-gliadins are on the short arm of homoeologous group 6 chromosomes (the *Gli-3* locus). Both the *Gli-1* and *Gli-3* are complex, multigenic loci. Each locus contains a tightly linked genes family clustered in a discrete order and inherited as a block (Bietz, 1987).

Glutenin proteins can form polymeric complexes and are responsible for strength and elasticity of gluten. Like the gliadins, the glutenins are found to be inherited in blocks. Glutenins are divided into poorly soluble HMW and the ethanol-soluble LMW subunits (Payne et al., 1984). Payne et al. (1984) showed that HMW glutenins are synthesized by homoeologous loci on the long arms of chromosome 1A, 1B, and 1D. These HMW glutenins gene loci are denoted *Glu-A1*, *Glu-B1*, and *Glu-D1* for homoeologous chromosomes 1A, 1B, and 1D, respectively. The LMW glutenin subunits are coded by genes at the *Glu-3* locus located on the short arm of homoeologous group 1 chromosomes (Jackson et al., 1983). The *Glu-B3* locus on chromosome 1B is of particular importance to durum wheat quality because of the LMW-1 and LMW-2 subunits coded to this locus. Pogna et al. (1988, 1990) determined that the LMW-2 subunits can strongly influence pasta-making quality of durum wheat. The γ-gliadin band 45 at the *Gli-B1* locus is only a marker for gluten strength, and LMW glutenin subunits at the *Glu-B3* locus are responsible for cooking quality of pasta. It has been shown that the *Gli-B1* locus is tightly linked (2 recombinant units) to the *Glu-B3* locus on the short arm of chromosome 1B (Pogna et al., 1990). Furthermore, in most durum cultivars, γ-gliadin 42 is linked in the coupling phase to the LMW-1 subunits, and γ-gliadin 45 is linked in the coupling phase to the LMW-2 subunits. According to Feillet (1988), the LMW glutenin proteins tend to form a strongly aggregated protein matrix through heat treatments. A high content of LMW glutenin results in the formation of a large-aggregated protein matrix, which contributes to pasta firmness (Feillet, 1988).

Dick, J.W. and Quick, J.S. (1983) A modified screening test for rapid estimation of gluten strength in early-generation durum wheat breeding lines. Cereal Chem. 60, 315–318.

Di Fonzo, N., Ravaglia, S., DeAmbrogio, E., Blanco, A. and Troccoli, A. (2005) Durum wheat improvement in Italy. In: C. Royo, M.N. Nachit, N. Di Fonzo, J.L. Araus, W.H. Pfeiffer and G.A. Slafer (Eds.), *Durum Wheat Breeding: Current Approaches and Future Strategies*. Food Products Press, New York, pp. 825–881.

Donmez, E., Sears, R.G., Shroyer, J.P. and Paulsen, G.M. (2001) Genetic gain in yield attributes of winter wheat in the Great Plains. Crop Sci. 41, 1412–1419.

Duggan, B.L., Richards, R.A., van Herwaarden, A.F. and Fettell, N.A. (2005) Agronomic evaluation of a tiller inhibition gene (*tin*) in wheat. I. Effect on yield, yield components, and grain protein. Aust. J. Agric. Res. 56, 169–178.

Elias, E.M., Stack, R.W. and Mitchell-Fetch, J. (1995) To search for sources of resistance to fusarium head blight in durum wheat. *Current Research on Fusarium Head Blight of Small Grains*, North Dakota State University, USA.

Elias, E.M., Stack, R.W. and Joppa, L.R. (1996a) Identification and development of resistant durum germplasm to fusarium head blight. *Current Research on Fusarium Head Blight of Small Grains*. North Dakota State University, USA.

Elias, E.M., Steiger, D.K. and Cantrell, R.G. (1996b) Evaluation of lines derived from wild emmer chromosome substitutions: II. Agronomic traits. Crop Sci. 36, 228–233.

Elouafi, I., Nachit, M.M. and Martin, L.M. (2001) Identification of a microsatellite on chromosome 7B showing a strong linkage with yellow pigment in durum wheat (*Triticum turgidum* L. var. *durum*). Hereditas 135, 255–261.

Fedak, G., Cao, W., Xue, A., Savard, M., Clarke, J. and Somers, D.J. (2007) Enhancement of Fusarium head blight resistance in bread wheat and durum by means of wide crosses. In: H.T. Buck, J.E. Nisi and N. Salomón (Eds.), *Wheat Production in Stressed Environments*. Springer, The Netherlands, pp. 91–95.

Fehr, W.R. (1987) (Eds.) *Principles of Cultivar Development*. McGraw-Hill, Inc. New York.

Flaksberger, C.A. (1935) Cereals Wheat. *State Agricultural Publishing Company*. ML., pp. 1–434.

Forster, B.P., Phillips, M.S., Miller, T.E., Baird, E. and Powell, W. (1990) Chromosome location of genes controlling tolerance to salt (NaCl) and vigour in *Hordeum vulgare* and *H. chilense*. Heredity 65, 99–107.

Feillet, P. (1988) Protein and enzyme composition of durum wheat. P. 93–119. In: G. Fabriani and C. Lintas (Eds.) *Durum Wheat: Chemistry and Technology*. American Society of Cereal Chemists, Inc., St. Paul, MN, pp. 47–68.

Frankel, O.H., Brown, A.H.D. and Burdon, J.J. (1995) *The Conservation of Plant Biodiversity*. Cambridge: Cambridge University Press.

Gale, M.D. and Gregory, R.S. (1977) A rapid method for early generation selection of dwarf genotypes in wheat. Euphytica 126, 733–739.

Gale, M.D. and Marshall, G.A. (1976) The chromosomal location of *Gai1* and *Rht1* genes for gibberellin insensitivity and semi-dwarfism in a derivative of Norin 10 wheat. Heredity 37, 283–289.

Gale, M.D., Law, C.N. and Worland, A.J. (1975) The chromosomal location of a major dwarfing gene from Norin 10 in new British semi dwarf wheats. Heredity 35, 417–421.

Gelin, J.R., Elias, E.M. and Kianian, S.F. (2005) Evaluation of two durum wheat (*Triticum turgidum* L. var. *durum*) crosses for pre-harvest sprouting resistance. Field Crops Res. 97, 188–196.

Grasso, V. (1968) Gli ustilaginali dei cereali in Italia (1938–1968). *Abstract od Annali della Facoltà di Agraria*, Bari, Italy, Vol. XII, 239–308.

Grzybowski, R.A. and Donnelly, B.J. (1979) Cooking properties of spaghetti: factors affecting cooking quality. J. Agric. Food Chem. 27, 380–384.

Grifoni, R. (1964) Appulo: nuovo grano duro precoce. Terra Pugliese, 13, 1.

Grignac, P. (1965) Contribution a l'étude du *Triticum durum* Desf. Paris: Thèse Fac. De Toulouse.

Hare, R.A. (1997) Characterization and inheritance of adult plant stem rust resistance in durum wheat. Crop Sci. 37, 1094–1098.

Hare, R.A., Du Cros, D.L. and Barnes, W.C. (1986) Genetic linkage between glume color and gliadin proteins in durum wheat. Crop Sci. 26, 831–833.

Herrera-Foessel, S.A., Singh, R.P., Huerta-Espino, J., Yuen, J. and Djurle, A. (2005) New genes for leaf rust resistance in CIMMYT durum wheats. Plant Dis. 89, 809–814.

Herrera-Foessel, S.A., Singh, R.P., Huerta-Espino, J., Crossa, J., Djurle, A. and Yuen, J. (2007) Evaluation of slow rusting resistance components to leaf rust in CIMMYT durum wheats. Euphytica 155, 361–369.

Icard-Verniere, C. and Feillet, P. (1999) Effects of mixing conditions on pasta dough development and biochemical changes. Cereal Chem. 76, 558–565.

ICARDA (1980). Annual Report for RDTN, Aleppo, Syria.

Irvine, G.N. and Winkler, C.A. (1950) Factors affecting the color of macaroni. II. Kinetic studies of pigment destruction during mixing. Cereal Chem. 27, 205–218.

Jackson, E.A., Holt, L.M. and Payne, P.I. (1983) Characterization of high-molecular-weight gliadin and low-molecular-weight glutenin subunits of wheat endosperm by two-dimensional electrophoresis and the chromosomal location of their controlling genes. Theor. Appl. Genet. 66, 29–37.

Johnston, R.A., Quick, J.S. and Hammond, J.J. (1983) Inheritance of semolina color in six durum wheat crosses. Crop Sci. 23, 607–610.

Joppa, L.R. (1973) Agronomic characteristics of near-isogenic tall and semidwarf lines of durum wheat. Crop Sci. 13, 743–746.

Joppa, L.R. and Cantrell, R.G. (1990) Chromosomal location of genes for grain protein content in wild tetraploid wheat. Crop Sci. 30, 1059–1064.

Joppa, L.R. and Williams, N.D. (1988) Genetics and beeding of durum wheat in the United States. In: G. Fabriani and C. Lintas (Ed.), *Durum Chemistry and Technology*. American Association of Cereal Chemists, St. Paul, pp. 47–68.

Joppa, L.R., Hareland, G.A. and Cantrell, R.G. (1991) Quality characteristics of the Langdon durum-*dicoccoides* chromosome substitution lines. Crop Sci. 31, 1513–1517.

Joppa, L.R., Du, C., Hart, G.E. and Hareland, G.A. (1997) Mapping gene(s) for grain protein in tetraploid wheat (*Triticum turgidum* L.) using a population of recombinant inbred chromosome lines. Crop Sci. 37, 1586–1589.

Kara, Y., Martin, A., Souyris, I., Rekika, D. and Monneveux, P. (2000) Root characteristics in durum wheat (*T. turgidum* conv. *durum*) and some wild *Triticeae* species. Genetic variation and relationship with plant architecture. Cereal Res. Commun. 28, 247–254.

King, I.P., Forster, B.P., Law, C.C., Cant, K.A., Orford, S.E., Gorham, J., Reader, S. and Miller, T.E. (1997) Introgression of salt-tolerance genes from *Thinopyrum bessarabicum* into wheat. New Phytologist 137, 75–81.

Knott, D.R. (1995) Durum wheat. In: A.E. Slinkard and D.R. Knott (Eds.), *Harvest of Gold: The History of Field Crop Breeding in Canada*. University Extension Press, University of Saskatchewan, Saskatoon, Canada. pp. 67–81.

Liesle, D.F., Kosmolack, F.G. and Kovacs, M. (1981) Association of glume color with gluten strength and gliadin proteins in durum wheat. Can. J. Plant Sci. 61, 149–151.

Maccaferri, M., Sanguineti, M.C., Donini, P. and Tuberosa, R. (2003) Microsatellite analysis reveals a progressive widening of the genetic basis in the elite durum wheat germplasm. Theor. Appl. Genet. 107, 783–797.

Maccaferri, M., Sanguineti, M.C., Donini, P., Porcedu, E. and Tuberosa, R. (2005) A retrospective analysis of genetic diversity in durum wheat elite germplasm based on microsatellite analysis: a case study. In: C. Royo, M.N. Nachit, N. Di Fonzo, J.L. Araus, W.H. Pfeiffer and G.A. Slafer (Eds.), *Durum Wheat Breeding: Current Approaches and Future Strategies*. Food Products Press, New York, pp. 99–142.

MacKey, J. (2005) Wheat: Its concept, evolution and taxonomy. In: C. Royo, M. Nachit, N. Di Fonzo, J.L. Araus, W.H. Pfeiffer and G.A. Slafer (Eds.), *Durum Wheat Breeding: Current Approaches and Future Strategies*. Food Products Press, New York, pp. 3–61.

Manna, F., Borrelli, G.M., Massardo, D.R., Wolf, K., Alifano, P., Del Giudice, L. and DiFonzo, N. (1998) Differential expression of lipoxygenase genes among durum wheat cultivars. Cereal Res. Commun. 26, 23–30.

Mano, Y. and Takeda, K. (1998) Genetic resources of salt tolerance in wild *Hordeum* species. Euphytica 103, 137–141.

Martinez, F., Sillero, J.C. and Rubiales, D. (2007) Resistance to leaf rust in cultivars of bread wheat and durum wheat grown in Spain. Plant Breeding 126, 13–18.

Martos, V., Royo, C., Rharrabti, Y. and García del Moral, L.F. (2005) Using AFLPs to determine phylogenetic relationships and genetic erosion in durum wheat cultivars released in Italy and Spain throughout the 20th century. Field Crops Res. 91, 107–116.

Martynov, S.P., Dobrotvorskaya, T.V. and Pukhalskiy, V.A. (2005) Analysis of genetic diversity of spring durum wheat (*Triticum durum* Desf.) cultivars released in Russia in 1929–2004. Russ. J. Genet. 41, 1113–1122.

Maskus, K.E. (2006) Intellectual property rights in agriculture and the interests of Asian-Pacific economies. The World Economy 29, 715–742.

McCaig, T.N. and Clarke, J.M. (1995) Breeding durum wheat in western Canada. Historical trends in yield and related variables. Can. J. Plant Sci. 75, 55–60.

McIntosh, R.A., Wellings, C.R. and Park, R.F. (1995) *Wheat Rusts: An Atlas of Resistance Genes*. CSIRO, Australia.

McClung, A.M. and Cantrell, R.G. (1986) Inheritance of gluten, color, and gluten strength in durum wheat. Euphytica 35, 885–890.

McVittie, J.A., Gale, M.D., Marshall, G.A. and Westcott, B. (1978) The intrachromosomal mapping of the Norin 10 and Tom Thumb dwarfing genes. Heredity 40, 67–70.

Medini, M. Hamza, S., Rebai, A. and Baum, M. (2005) Analysis of genetic diversity in Tunisian durum wheat cultivars and related wild species by SSR and AFLP markers. Genet. Resour. Crop Evol. 52, 21–31.

Monneveux, P., Reynolds, M.P. and Ribaut, J.M. (2005) Physiological basis of yield potential in durum wheat. In: C. Royo, M. Nachit, N. Di Fonzo, J.L. Araus, W.H. Pfeiffer and G.A. Slafer (Eds.), *Durum Wheat Breeding: Current Approaches and Future Strategies*. Food Products Press, New York, pp. 425–460.

Moragues, M., Zarco-Hernández, J., Moralejo, M.A. and Royo, C. (2006) Genetic diversity of glutenin protein subunits composition in durum wheat landraces [*Triticum turgidum* ssp. *turgidum* convar. *durum* (Desf.) MacKey] from the Mediterranean basin. Genet. Resour. Crop Evol. 53, 993–1002.

Moragues, M., Moralejo, M., Sorrells, M.E. and Royo, C. (2007) Dispersal of durum wheat [*Triticum turgidum* L. ssp. *turgidum* convar. *durum* (Desf.) MacKey] landraces across the Mediterranean basin assessed by AFLPs and microsatellites. Genet. Resour. Crop Evol. 54, 1133–1144.

Mujeeb-Kazi, A. (2005) Wide crosses for durum wheat improvement. In: C. Royo, M. Nachit, N. Di Fonzo, J.L. Araus, W.H. Pfeiffer and G.A. Slafer (Eds.), *Durum Wheat Breeding: Current Approaches and Future Strategies*. Food Products Press, New York, pp. 703–743.

Nalam, V.J., Vales, M.I, Watson, C.J., Kianian, S.F. and Reera-Lizarazu, O. (2006) Map-based analysis of genes affecting the brittle rachis character in tetraploid wheat (*Triticum turgidum* L.). Theor. Appl. Genet. 112, 373–381.

Nesbitt, M. and Samuel, D. (1996) From staple crop to extinction? The archaeology and history of hulled wheat. In: S. Padulosi, K. Hammer and J. Heller (Eds.), *Hulled Wheats*. Rome, International Plant Genetics Resources Institute, pp. 40–99.

Nicholson, P., Gosman, N., Draeger, R., Thomsett, M., Chandler, E. and Steed, A. (2007) The Fusarium head blight pathosystem: status and knowledge of its components. In: H.T. Buck,

J.E. Nisi and N. Salomón (Eds.), *Wheat Production in Stressed Environments*. Springer, The Netherlands, pp. 23–36.

Otto, C.D., Kianian, S.F., Elias, E.M., Stack, R.W. and Joppa, L.R. (2002) Genetic dissection of a major Fusarium head blight QTL in tetraploid wheat. Plant Mol. Biol. 48, 625–632.

Palamarchuk, A. (2005) Selection strategies for traits relevant for winter and facultative durum wheat. In: C. Royo, M. Nachit, N. Di Fonzo, J.L. Araus, W.H. Pfeiffer and G.A. Slafer (Eds.), *Durum Wheat Breeding: Current Approaches and Future Strategies*. Food Products Press, New York, pp. 599–644.

Pardey, P., Koo, B. and Nottenburg1, K. (2003) Creating, protecting, and using crop biotechnologies worldwide in an era of intellectual property. In WIPO-UPOV Symposium on intellectual property rights in plant biotechnology. Geneva. pp. 2–39.

Payne, P.I., Corfield, K.G., Jackson, E.R. and Law, C.N. (1984) Wheat storage proteins: Their genetics and their potential for manipulation by plant breeding. Philos. R. Soc. Lond. Ser. B. 304, 359–371.

Pecetti, L. and Annicchiarico, P. (1998) Agronomic value and plant type of Italian durum wheat cultivars from different eras of breeding. Euphytica 99, 9–15.

Pecetti, L., Annicchiarico, P. and Damania, A.B. (1992) Biodiversity in a germplasm collection of durum wheat. Euphytica 60, 229–238.

Pfeiffer, W.H. and Payne, T.S. (2005) CIMMYT durum wheat improvement program. In: C. Royo, M. Nachit, N. Di Fonzo, J.L. Araus, W.H. Pfeiffer and G.A. Slafer (Eds.), *Durum Wheat Breeding: Current Approaches and Future Strategies*. Food Products Press, New York, pp. 1031–1048.

Pfeiffer, W.H., Sayre, K.D. and Mergoum, M. (1996) Enhancing genetic grain yield potential in durum wheat and triticale. In: M.P. Reynolds, S. Rajaram and A. McNab (Eds.), *Increasing Yield Potential in Wheat: Breaking the Barriers*. CIMMYT, Mexico, pp. 90–100.

Pfeiffer, W.H., Sayre, K.D. and Reynolds, M.P. (2000) Enhancing genetic gain yield potential and yield stability in durum wheat. In: C. Royo, M.M. Nachit, N. Di Fonzo, and J.L. Araus (Eds.), *Durum Wheat Improvement in the Mediterranean Region: New Challenges*. Options Méditerranéennes, Series A. 40, 88–93.

Pogna, N.E., Lafiandra, D., Feillet, P. and Autran, J.C. (1988) Evidence of a direct causal effect of low molecular weight subunits of glutenins of on gluten viscoelasticity in durum wheat. J. Cereal Sci. 7, 211–214.

Pogna, N.E. Autran, J.C., Mellini, F. Lafiandra, D. and Feillet, P. (1990) Chromosome 1B-encoded gliadins and glutenin subunits in durum wheat: genetics and relationship of gluten strength. J. Cereal Sci. 11, 15–34.

Pozniak, C.J., Knox, R.E., Clarke, F.R. and Clarke, J.M. (2007) Identification of QTL and association of a phytoene synthase gene with endosperm colour in durum wheat. Theor. Appl. Genet. 114, 525–537.

Rajaram, S. (2001) Prospects and promise of wheat breeding in the 21st century. Euphytica 119, 3–15.

Rajaram, S.D. and van Ginkel, M. (2001) Mexico: 50 years of international wheat breeding. In: A.P. Bonjean and W.J. Angus (Eds.), *The World Wheat Book, a History of Wheat Breeding*. Lavoisier Publishing, Paris, pp. 579–608.

Rawashdeh, N.K., Haddad, N.I., Al-Ajlouni, M.M. and Turk, M.A. (2007) Phenotypic diversity of durum wheat (*Triticum durum* Desf.) from Jordan. Genet. Resour. Crop Evol. 54, 129–138.

Rebetzke, G.J., Richards, R.A., Sirault, X.R.R. and Morrison, A.D. (2004) Genetic analysis of coleoptile length and diameter in wheat. Aust. J. Agric. Res. 55, 733–743.

Rebetzke, G.J., Richards, R.A., Fettell, N.A., Long, M., Condon, A.G., Forrester, R.I. and Botwright, T.L. (2007) Genotypic increases in coleoptile length improves stand establishment, vigour and grain yield of deep-sown wheat. Field Crops Res. 100, 10–23.

Reynolds, M.P, Van Beem, J., Van Ginkel, M. and Hoisington, D. (1996) Breaking the yield barriers in wheat: a brief summary of the outcomes of an international consultation. In: M.P.

Reynolds, S. Rajaram and A. McNab (Eds.), *Increasing Yield Potential in Wheat: Breaking the Barriers*. CIMMYT, Mexico, pp. 1–10.
Reynolds, M.P., Pellegrineschi, A. and Skovmand, B. (2005) Sink-limitation to yield and biomass: a summary of some investigations in spring wheat. Ann. Appl. Biol. 146, 39–49.
Roelfs, A.P., Singh, R.P. and Saari, E.E. (1992) *Rust Diseases of Wheat: Concepts and Methods of Disease Management*. CIMMYT, Mexico, D.F.
Royo, C., Nachit, M., Di Fonzo, N., Araus, J.L., Pfeiffer, W.H. and. Slafer, G.A. (Eds.) (2005) *Durum Wheat Breeding: Current Approaches and Future Strategies*. Food Products Press, New York, 1084 pp.
Royo, C., Alvaro, F., Martos, V., Ramdani, A., Isidro, J., Villegas, D. and García del Moral, L.F. (2007) Genetic changes in durum wheat yield components and associated traits in Italian and Spanish varieties during the 20th century. Euphytica 155, 259–270.
Royo, C. and Briceño-Félix, G.A. Spanish Wheat Pool. In: A.P. Bonjean, W.J. Angus and M. van Ginkel (Eds.), *The World Wheat Book 2, a History of Wheat Breeding*. Groupe Limagrain. Lavoisier Publishing, Paris (in press).
Royo, C., Martos, V., Ramdani, A., Villegas, D., Rharrabti, Y. and García del Moral, L.F. (2008) Changes in yield and carbon isotope discrimination of Italian and Spanish durum wheat during the 20th century. Agron J. 100, 352–360.
Rosielle, A.A. (1972) Source of resistance in wheat to speckle leaf blotch caused by *Septoria tritici*. Euphytica 21, 152–161.
Ruuska, S.A., Rebetzke, G.J., van Herwaarden, A.F., Richards, R.A., Fettell, N.A., Tabe, L. and Jenkins, C.L.D. (2006) Genotypic variation in water-soluble carbohydrate accumulation in wheat. Funct. Plant Biol. 33, 799–809.
Scarascia-Mugnozza, G.T., Bagnara, D. and Bozzini, A. (1972) Mutagenesis applied to durum wheat: results and perspective. In: *Induced Mutations and Plant Improvement*, IAEA Publ, pp. 183–198.
Siedow, J.N. (1991) Plant lipoxygenase: Structure and function. Annu. Rev. Plant Physiol. Plant Mol. Biol. 4, 145–188.
Singh, R.P., Huerta-Espino, J., Pfeiffer, W. and Figueroa-Lopez, P. (2004) Occurrence and impact of a new leaf rust race on durum wheat in Northwestern Mexico from 2001 to 2003. Plant Dis. 88, 703–708.
Singh, R.P., Huerta-Espino, J., Fuente, G., Duveiller, E., Gilchrist, L., Henry, M. and Nicol, J.M. (2005) Resistance to diseases. In: C. Royo, M. Nachit, N. Di Fonzo, J.L. Araus, W.H. Pfeiffer and G.A. Slafer (Eds.), *Durum Wheat Breeding: Current Approaches and Future Strategies*. Food Products Press, New York, pp. 291–327.
Singh, R.P., Kinyua, M.G., Wanyera, R., Njau, P., Jin, Y. and Huerta-Espino, J. (2007) Spread of a highly virulent race of *Puccinia graminis tritici* in Eastern Africa: Challenges and opportunities. In: H.T. Buck, J.E. Nisi and N. Salomón (Eds.), *Wheat Production in Stressed Environments*. Springer, The Netherlands, pp. 51–57.
Skovmand, B., Warburton, M.L., Sullivan, S.N. and Lage, J. (2005) Managing and collecting genetic resources. In: C. Royo, M. Nachit, N. Di Fonzo, J.L. Araus, W.H. Pfeiffer and G.A. Slafer (Eds.), *Durum Wheat Breeding: Current Approaches and Future Strategies*. Food Products Press, New York, pp. 143–163.
Snape, J.W. and Moore, G. (2007) Reflections and opportunities: gene discovery in the complex wheat genome. In: H.T. Buck, J.E. Nisi and N. Salomón (Eds.), *Wheat Production in Stressed Environments*. Springer, The Netherlands, pp. 677–684.
Solomon, K.F., Labuschagne, M.T. and Viljoen, C.D. (2007) Estimates of heterosis and association of genetic distance with heterosis in durum wheat under different moisture regimes. J. Agric. Sci. 145, 239–248.
Tesemma, T. and Bechere, E. (1998) Developing elite durum wheat landrace selections (composites) for Ethiopian peasant farm use: raising productivity while keeping diversity alive. Euphytica 102, 323–328.

Vallega, J., and Zitelli, A, (1973) New high yielding Italian durum wheat varieties. In Proc. of the Symposium on Genetics and Breeding of Durum Wheat, ed. G.T. Scarascia Mugnozza, University of Bari, Italy, pp 373–399.

Valkoun, A.J. (2001) Wheat pre-breeding using wild progenitors. Euphytica 119, 17–23.

Waddington, S.R., Osmazai, M., Yoshida, M. and Ranson, J.K. (1987) The yield of durum wheats released in Mexico between 1960 and 1984. J. Agric. Sci. Camb. 108, 469–477.

Watanabe, N., Akond, A.S.M.G.M. and Nachit, M.M. (2006) Genetic mapping of the gene affecting polyphenol oxidase activity in tetraploid durum wheat. J. Appl. Genet. 47, 201–205.

Worland, T. and Snape, J.W. (2001) Genetic basis of worldwide wheat varietal improvement. In: A.P. Bonjean and W.J. Angus (Eds.), *The World Wheat Book, a History of Wheat Breeding*. Lavoisier Publishing, Paris, pp. 59–100.

Yang, R.C., Jana, S. and Clarke, J.M. (1991) Phenotypic diversity and associations of some potentially drought-responsive characters in durum-wheat. Crop Sci. 31, 1484–1491.

Zhang, H. and Knott, D.R. (1990) Inheritance of leaf rust resistance in durum wheat. Crop Sci. 30, 1218–1222.

Barley

R.D. Horsley, J.D. Franckowiak, and P.B. Schwarz

Abstract Barley (*Hordeum vulgare* L.) is the cereal crop with the widest range of production areas in the world. Compared to other cereals, barley is fourth in world production behind maize (*Zea mays* L.), wheat (*Triticum aestivum* L.), and rice (*Oryza sativa* L.). Barley has many uses, including livestock feed and forage, human food, and malt beverages. Barley to be used for malting must meet specifications for germination, kernel size and weight, grain protein, activity of several enzymes, and many other traits. Barley for livestock and human food uses has much fewer restrictions, but they are also critical in cultivar utilization. Likewise, quality traits for barley used as forage are less well-defined, but they are important in cultivar acceptance. This chapter outlines the different types of barley (e.g. six- vs. two-rowed and malting vs. feed), items to consider when choosing parents for crossing, current goals barley breeders, major breeding achievements, an example of a breeding scheme for developing malting barley cultivars, examples of integration of biotechnology methods into breeding programs, and issues related to cultivar release and intellectual property protection.

1 Introduction

Barley (*Hordeum vulgare* L.) is the cereal crop with the widest range of production areas in the world. It is often the last cereal crop grown at the highest altitudes in the Andes and Himalaya mountains; adjacent to the deserts of Africa, the Middle East, and China; and near the artic circle in the northern reaches of Asia, Europe, and North America. Because of its wide adaptation, barley often is grown where maize (*Zea mays* L.) is unadapted or is not competitive with barley. Barley is frequently considered by farmers the safest and easiest annual cool-season crop to grow for grain. Compared to other cereals, barley is a distant fourth in world production

R.D. Horsley(✉)
North Dakota State University, Department of Plant Sciences, NDSU Dept. 7060, Po Box 6050, Fargo, ND 58108-6050, USA, e-mail: richard.horsley@ndsu.edu

M.J. Carena (ed.), *Cereals*,
DOI: 10.1007/978-0-387-72297-9,

behind maize, wheat (*Triticum aestivum* L.), and rice (*Oryza sativa* L.) (FAO, http://faostat.fao.org/).

Barley has many uses, including livestock feed and forage, human food, and malt beverages. Barley to be used for malting must meet specifications for germination, kernel size and weight, grain protein, activity of several enzymes, and many other traits. Barley for livestock and human food uses has much fewer restrictions, but they are also critical in cultivar utilization. Likewise, quality traits for barley used as forage are less well-defined, but they are important in cultivar acceptance.

2 Genetic Diversity

Like many cereal crops, the center of origin of barley is the Fertile Crescent area of southwest Asia (Bothmer et al., 2003, for review and references). The genus *Hordeum* consists of approximately 32 species (Bothmer et al., 2003), of which *H. vulgare* subsp. *vulgare* ($2n = 2x = 14$) is referred to as cultivated barley and the original progenitor of cultivated barley, *H. vulgare* subsp. *spontaneum* [(C. Koch) Thell.] ($2n = 2x = 14$) is often referred to as "wild barley". *H. vulgare* subsp. *spontaneum* is found growing today from Morocco in northern Africa through southwest Asia and into the highlands of central Asia. *H. vulgare* subsp. *spontaneum* is an important, but underutilized source of genes for resistances to multiple diseases (Fetch et al., 2003), agronomic performance (Pillen and Leon, 2003), and possibly malt quality (Matus et al., 2003; Erkkilä et al., 1998).

The primary genepool of *H. vulgare* consists of cultivars of cultivated barley, landraces, and *H. vulgare* subsp. *spontaneum* (Harlan and de Wet, 1971). A major advantage of using *H. vulgare* subsp. *spontaneum* as a source of genes is that it readily crosses with cultivated barley. The secondary genepool consists only of the perennial grass *H. bulbosum* L. Genes from *H. bulbosum* can be transferred to cultivated barley via crossing, but embryo rescue must be used to generate viable plants. In addition, the success of transferring genes is dependent on growth conditions, especially temperature, and genotypes of the cultivated barley and *H. bulbosum* used as parents (Pickering, 1989). An example of the successful transfer of a gene from *H. bulbosum* to cultivated barley is the transfer of a gene for powdery mildew resistance, incited by *Erysiphe graminis* DC. f. sp. *hordei* Ém. Marchal (Pickering et al., 1995). The remaining *Hordeum* species belong to the tertiary genepool and obtaining successful crosses between them and cultivated barley is very difficult (Bothmer et al., 1983).

3 Types of Barley

The floral structure of the barley plant is a spike, or, as it is sometimes referred to, a head or ear (Bergal and Clemencet, 1962). Barley can be divided into two groups based on spike morphology, six-rowed and two-rowed barley (Fig. 1).

Fig. 1 Comparison of the spike morphology of two-rowed barley (*left*) and six-rowed barley (*right*)

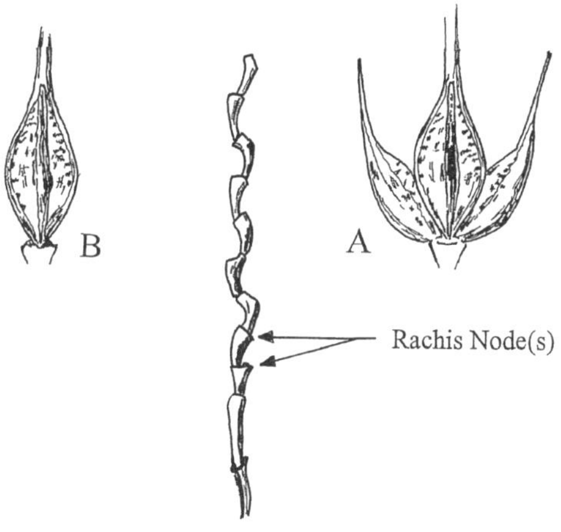

Fig. 2 Diagram showing fertile kernels present at a rachis node in (**a**) six-rowed barley and (**b**) two-rowed barley

To appreciate the difference between these two groups, an understanding of spike morphology is needed. The central axis of the spike, the rachis, is composed of nodes and internodes. Attached at the rachis nodes are spikelets that can later develop into kernels. A barley spikelet comprises an individual floret with surrounding bracts, the lemma and palea and two subtending (outer) glumes. In both two- and six-rowed barley, each rachis node has three spikelets, but the fertility (or sterility) of the spikelets differs in each type. In six-rowed barley, all three spikelets of the rachis node have a fertile spikelet that can develop into kernel. Thus, each rachis node in the mature spike of six-rowed barley has three kernels (Fig. 2). When the rachis is viewed from one side, there appears to be three rows of kernels. However, when the spike is viewed from the top, there appears to be six rows of kernels. In two-rowed barley, only the central spikelet is fertile and will develop into a kernel. The two lateral spikelets are sterile. When the spike is viewed from above, there appears to be two rows of kernels.

The control of the fertility of the lateral spikelets is controlled by genes at the *vrs1* locus on chromosome 2H (Franckowiak and Lundqvist, 1997; Komatsuda et al., 2007). The two-rowed character is dominant; hence, six-rowed is homozygous recessive (*vrs1 vrs1*) for the *vrs1* locus. The size of the lateral spikelets in barley is controlled by the *Int-c* locus on chromosome 4H (Lundquist and Franckowiak). Large lateral spikelet size is dominant; thus, the genotype of a six-rowed spike is *vrs1 vrs1 Int c Int-c* and a two-rowed spike has the genotype *Vrs1 Vrs1 int-c int-c*. Because the row number and size of lateral spikelets are under control of single genes, there are no problems making crosses between six-rowed and two-rowed barley.

Barley also can be grouped based on its intended use, malting or non-malting, and the two types are usually segregated for storage. The level of segregation is dependent on the requirements of the brewer. The malt blend used by a brewer may be a general list of what cultivars may be included, or it may be specified that the blend contain specific percentages of particular cultivars. In the latter case, each cultivar must be stored and malted separately and then the cultivars are blended before the malt shipment is sent to the brewer.

Barley malt is produced in a process where the barley is steeped in water to bring the moisture to $\approx$43% moisture, germinated for 4–5 days, and dried or kilned. These three processes are performed under closely monitored conditions. During the malting process, the protein-cell wall matrix of the endosperm is modified or broken down by enzymes that are produced and/or released during germination. This process, called malt modification, exposes starch granules in the endosperm that can be acted upon by amylytic enzymes during brewing to produce fermentable sugars and releases nutrients used by yeast during fermentation. Both two-rowed and six-rowed barley can be used for malting, but two-rowed barley is more commonly used worldwide due to its larger kernels, higher malt extract levels, and tradition. Large amounts of six-rowed barley are used in North America because of its higher enzyme content and other characteristics that are useful when beer is made with malt plus adjuncts (e.g., rice, maize, or sugar) (Schwarz and Horsley, http://brewingtechniques.com/bmg/schwarz.html).

Other categories used to group barley include growth habit (i.e., spring, winter, or facultative) and hull adherence (i.e., hulled or hulless). In many countries where barley is a traditional crop, desirable attributes are associated with specific cultivars or groups of cultivars for example soft straw, black seed, smooth awns, waxy starch, or specific photoperiod responses. Malting barley cultivars commonly have the spring growth habit; however, some winter six-rowed cultivars, such as Plaisant in France, have been developed and used for malting. In the United States, breeding programs at Oregon State University and the United States Department of Agriculture – Agricultural Research Service (USDA-ARS) program at Aberdeen, Idaho have as a priority development of winter malting barley. In Europe, non-malting barley cultivars typically have the winter growth habit. Cultivars with this growth habit often have a higher yield potential than spring barley because grain fill occurs during the cooler part of summer as compared to spring barley. Emphasis in developing winter malting barley in Europe is increasing due to farmers' demands and the need to ensure there is sufficient quantities of barley suitable for malting and brewing. In areas where winter temperatures are milder, production of fall-sown spring barley is common. The worldwide increase in demand for fuels derived from plants, biofuels, will result in more competition among the crops that growers produce. Land sown to crops used for production of biofuels, such as maize, sorghum (*Sorghum bicolor* L.), rapeseed (*Brasica napus* L. and *B. campestris* L.), and wheat will likely increase at the expense of crops such as barley and pulses.

To date, almost all malting barley for brewing beer is hulled because the hulls serve as a filter bed during the lautering process. However, there is some discussion on use of hulless barley for brewing when mash filters are used instead of lautering tuns. The largest use of hulless barley is for human food consumption. Hulless barley is a food staple in the Andean and Himalayan regions, and is gaining popularity in many other regions of the world due to its health benefits (Behall and Hallfrisch, 2006, for review and references). Hulless barley also has been used for some classes of livestock feed and has potential for the production of ethanol.

4 Choice of Germplasm

The choice of germplasm to use for crossing is most critical for developing locally acceptable barley cultivars. Barley cultivars need to be adapted to specific production areas and designed for specific uses. Malting barley, for example, is unique from most other crops in that it is usually marketed and stored on an identity preserved basis. As described earlier, malting barley cultivars are kept segregated from non-malting barley cultivars, and often from each other, when they are sold and stored. This segregation is done because cultivars used for malting and brewing must meet a long list of specific criteria before they will be accepted for use by maltsters and brewers. Guidelines for these characteristics, which include measurements on barley and malt, often are provided to the breeders by organizations in the countries where they work. Table 1 presents target requirements specified by the American Malting Barley Association, Inc. (AMBA, Milwaukee, WI) in the United States for two-rowed and six-rowed barley to be used by its member companies. The parameters that receive the most attention in breeding programs are grain protein, kernel plumpness, malt extract, enzymatic activity (α-amylase and diastatic power), and measures of modification (β-glucan content, viscosity, or friability). A detailed description of these malt parameters and others can be found in Kunze (2004).

Because of the long list of parameters that must be met, development of malting barley cultivars is frequently done using crosses between parents that have acceptable quality. Any limitations in malt quality of either parent usually appear in the progeny. This need to make crosses between parents with acceptable quality has led to very narrow germplasm bases because variability already has been severely

Table 1 Barley and malt quality specifications provided to barley breeder in the United States by the American Malting Barley Association, Inc. (Milwaukee, WI)

	Two-rowed barley	Six-rowed barley
Barley factors		
Plump kernels†	>90%	>80%
Thin kernels‡	<3%	<3%
Germination§	>98%	>98%
Protein	11.0–13.0%	11.5–13.5%
Skinned and broken kernels	<5%	<5%
Malt factors		
Total protein	10.8–12.8%	11.3–13.3%
On 7/64″ sieve	>70%	>60%
Measures of malt modification		
Beta-glucan (ppm)	<100	<120
Fine-coarse extract difference	<1.2%	<1.2%
Kohlbach index	40–47%	42–47%
Turbity (NTU)	<10	<10
Viscosity (absolute cP)	<1.50	1.50
Congress wort		
Soluble protein	4.4–5.6%	5.2–5.7%
Extract (fine grind db)	>81.0%	>79.0%
Color (°ASBC)	1.6–2.2	1.8–2.2
Free amino nitrogen	>180	>190
Malt enzymes		
Diastatic power (°ASBC)	>120	>140
Alpha-amylase (20° DU)	>45	>45

† Percent of kernels retained on a sieve with 0.24 × 1.9-cm slotted openings
‡ Percent of kernels passing through a sieve with 0.20 × 1.9-cm slotted openings
§ Percent of kernels germinated after 72 h in a petri-dish with 4 mL of water

restricted by adaptation to specific production areas. For example, Horsley et al. (1995) stated that in the early 1990s, all cultivars developed by six-rowed barley breeding programs in the Midwest of the United States and the eastern Prairie Provinces of Canada could be traced back to 15 accessions obtained in the early 1890s. However, even within the narrow germplasm bases of malt barley programs, gains still are being made. Rasmusson and Phillips (1997) theorize that gains are made in the narrow germplasm bases due to de novo variation and elevated epistasis.

In developing cultivars designed for livestock feed, the choice of parents to choose is of less importance because growers are usually not paid premiums for specific quality parameters as is done for malting barley. Breeding for high yield per se and minimum production inputs receive the most attention even though studies suggest that barley quality is important in feeding livestock (Juskiw et al., 2005). Incorporating disease resistance genes becomes especially important because application of expensive fungicides to feed barley can be cost prohibitive. Breeders developing hulless barley for the human consumption look to develop cultivars that are high in soluble fiber to reduce the risk of cardiovascular disease (Behall et al., 2004). The trait that receives the most attention is breeding for increased barley β-glucan content.

5 Major Breeding Achievements

A unique feature of malting barley is that the life of cultivars can easily be 10 years or more. In fact, some malting barley cultivars remain in the marketplace so long that their name almost becomes a generic name for barley. For example, in the 1980s and 1990s, when international buyers purchased barley from Canada or the United States, they requested Harrington. Buyers purchasing barley from Australia at one time requested Clipper or Schooner.

A feature of the narrow germplasm base of malting barley is that the contributions of older cultivars, especially for disease resistance, are well known. In all Midwest United States malting barley cultivars developed in North America since the mid-1950s, the *Rpg1* gene that confers resistance to wheat stem rust [incited by *Puccinia graminis* f. sp. *tritici* (Eriks. & E. Henn.) D.M. Henderson] traces back to the cultivar Peatland (Steffenson, 1992). Peatland is a pure line selection made at the University of Minnesota from a Swiss landrace. The breeding line NDB112, developed at North Dakota State University in the 1950s, has been used worldwide as a source of resistance to the net and spot forms of net blotch [incited by *Drechslera teres* (Sacc.) Shoemaker f. *teres* and *Drechslera teres f. maculata* Smedeg, respectively] and spot blotch [incited by *Cochliobolus sativus* (Ito & Kuribayashi) Drechs. ex Dastur]. NDB112 is the source of durable resistance to spot blotch found six-rowed cultivars developed in the Midwest of the United States and the Western Prairie Provinces of Canada since the mid-1960s (Steffenson et al., 1996; Wilcoxson et al., 1990).

Another example of a high percentage of cultivars having common genes is *mlo* alleles that confer stable resistance to powdery mildew [incited by *Blumeria graminis* (DC.) E.O. Speer f. sp. *hordei* Ém. Marchal]. It has been estimated that in over 70% of the European barley cultivars developed since the 1970s powdery mildew is controlled by *mlo* alleles (Jorgenson, 1992). The *mlo* gene was first identified in a barley mutant in 1942, but it was later found in an Ethiopian landrace collected in the 1930s (Jorgenson, 1992).

The cultivar Diamant, developed using mutagenesis of the cultivar Valticky in the former Czechoslovakia, is an important source of the semidwarf (*sdw1*) character in European barley cultivars. Triumph, developed from Diamant in the 1960s in the former East Germany, is another important cultivar that can be found in the pedigree of a high percentage of European malting barley cultivars (Russell et al., 2000). Triumph also contributed a gene for resistance to leaf rust (incited by *Puccinia hordei* Otth); however, this gene (*Rph9.z*) and a number of other leaf rust resistance genes have been overcome by changes in pathogen races in

P. hordei. Stable leaf rust resistance in Europe is now provided by adult plant resistance genes from the cultivar Vada or combinations of *Rph* genes.

These examples illustrate that the introduction of new genes into adapted germplasm can be problematic and time-consuming. Testing the effects of new genes before they are introduced into elite germplasm can be done using various approaches that include development of near-isogenic lines (Wiebe, 1968). For example, backcross-derived lines are used to compare genes in a common genetic background and to minimize linkage effects. The largest group of backcross-derived lines (over 750) developed is in the "Bowman" genetic background (Davis et al., 1997).

Extensive quality parameters as are found for malting barley do not exist for feed barley cultivars. However, feed quality attributes are being developed for several classes of livestock. Using marker-assisted selection (MAS), the cultivarValier was developed to incorporate excellent agronomic performance and improved feed characters for beef cattle (Blake et al., 2002).

6 Current Goals of Breeding

The goal for breeding is dependent on whether the primary target is malting or non-malting barley. For malting barley, maltsters and brewers specify quality parameters that must be met. Failure to meet all of their criteria usually results in the cultivar not being recommended for production as a malting barley cultivar, regardless of the other advantageous characteristics it may have. Cultivars also must be accepted by growers because they might have more profitable alternatives. The malting and brewing industry is conservative and is very particular when accepting new cultivars. It is not unusual for a single cultivar to dominate the market for 20 years or more. Examples of cultivars dominating a market for 10 years or more are the six-rowed cultivar Larker and Robust from the United States, Conquest and Bonanza from Canada, and Plaisant from France; and the two-rowed cultivars Harrington from Canada, Clipper and Schooner from Australia, Optic from England, and Scarlett from Germany.

A high priority in many breeding programs is development of cultivars with disease resistance to one or more pathogens or pests. As stated earlier, barley is considered a low input crop; thus, it is often not financially expedient to use chemical control. The easiest and most cost effective method of disease control is growing resistant cultivars. The list of barley diseases caused by bacteria, fungi, and viruses is plentiful (Mathre, 1997). Table 2 provides a partial list of important bacterial, fungal, and viral diseases of barley. Barley diseases exist that can infect all parts of the plant. A series of different root and crown rots severely limit root development and plant survival. The effects of this group of diseases are especially noticed during dry years or in dry areas of the field such as hill tops because roots are unable to uptake sufficient water. Diseases that affect the foliage of the plants such as the rusts, blotches, and leaf blights can severely reduce the photosynthetic

Table 2 A partial list of common diseases of barley, causal organism, and most common growth stage of infection

Disease	Causal organism or vector[a]	Plant parts infected
Bacterial		
Bacterial leaf blight	*Pseudomonas syringae*	Leaves
Bacterial kernel blight	*Pseudomonas syringae*	Kernels
Bacterial blight	*Xanthomonas translucens*	Leaves
Fungal		
Common root rot	*Bipolaris sorokiniana*	Roots
Pythium root rot	*Pythium* ssp.	Roots
Rhizoctonia root rot	*Rhizoctonia solani*	Roots
Spot blotch	*Bipolaris sorokiniana*	Leaves
Net blotch	*Pyrenophora teres*	Leaves
Scald	*Rynchosporium secalis*	Leaves
Powdery mildew	*Erysiphe graminis*	All aerial plant parts
Leaf rust	*Puccinia hordei*	Leaves
Stripe rust	*Puccinia striiformis* f. sp. *hordei*	Leaves
Stem rust	*Puccinia graminis* f. sp. *tritici*	Leaves
Covered smut	*Ustilago hordei*	Kernels
Loose smut	*Ustilago nuda*	Kernels
Head blight (scab)	*Fusarium* ssp.	Kernels
Viral		
Barley stripe mosaic	Seedborne	Leaves
Barley yellow dwarf	Aphids	Leaves
Barley yellow mosaic	Soilborne fungus *Polymyxa graminis*	Leaves

[a]Causal organism for bacteria and fungi, and vector for viruses

capacity of the plants and may produce toxins. These diseases often result in yield losses and excessive grain protein for malting due to reduced grain size or thin grain. Finally, diseases of the spike such as kernel blights and smuts can reduce yield, but more importantly they can limit the marketability of the grain. Grain that is contaminated with smut or kernel blight can be severely discounted or rejected at the point of sale. Some pathogens such as *Fusarium* spp. produce blighted kernels and also mycotoxins. In the Midwest of the United States in early 2000s, barley with greater than 0.5 ppm of the mycotoxin deoxynivalenol (DON) was discounted nearly \$20 per tonne. Grain with DON levels greater than 3.0 μg/g was rejected entirely for malting.

Breeding for resistance to insects or nematodes often is another priority of barley breeders. Insects that cause damage due to viruses or toxins transferred during feeding are of the most concern. Four aphid species are known to transfer the *Barley Yellow Dwarf Virus* (BYDV) during feeding. These aphids are the bird cherry-oat aphid, *Rhopalosiphum padi* (L.); corn leaf aphid, *R. maidis* (Fitch); English grain aphid, *Sitobion avenae* (Fabricius); and green bug, *Schizaphis graminum* (Rondani). *Barley Yellow Dwarf* Virus is the most important viral disease of barley and is found worldwide. Use of resistant cultivars has provided effective control for this disease. Russian wheat aphid, *Diuraphis noxia* (Kurdjumov), can cause severe losses in some barley production areas by injecting a toxin into the plant while feeding. As with the aphids that transmit BYDV, development of resistant cultivars is the best control against damage by this pest.

Three nematode genera are known to cause economic damage to barley (Mathre, 1997). Cereal cyst nematodes are a complex of several species of nematode that belong to the *Heterodera avenea* group. In seriously damaged plants, the roots are stunted and produce many lateral roots. Control of this pest has been accomplished using resistant cultivars among other methods. The cereal root-knot nematode (*Meloidogyne naasi* Franklin) is known to cause economic yield losses in barley. Damage caused by this complex of nematodes is similar to that of the cereal cyst nematodes. Control of damage due to cereal root-knot nematodes is usually accomplished using crop rotation, not genetic resistance. The root-lesion nematodes, *Pratylenchus* spp., can cause damage in arid environments when water stress occurs.

Breeding for abiotic stresses are also a high priority for many barley breeders. A non-exhaustive list of these stresses includes drought and flooding, high and low temperatures, mineral deficiencies and toxicities, poor soil tilth (structure), or pre-harvest sprouting. A complicating factor of abiotic stresses is that plants often are exposed to multiple stresses at the same time and common responses to the stresses often are elicited in plants (Langridge et al., 2006). This fact makes breeding for abiotic stresses much more difficult than breeding for biotic stresses. Functional genomic technologies are being used to gain a better understanding of how plants respond to abiotic stresses (Langridge et al., 2006). However, establishment of effective breeding strategies based on this information will be challenging.

Cultivars and accessions, including unadapted genotypes, land races, and wild barley accessions that are resistant or tolerant to most economically important biotic and abiotic stresses, usually can be found after sometimes exhaustive searches (Ullrich et al., 1995). However, sources of the resistance genes are often poorly adapted to the target environment and desirable genes may show linkage drag to undesirable traits. When a new problem (e.g., a disease or pest) arises or resistance breaks down in currently grown cultivars, breeders can develop improved cultivars in a short time if they have been working on the problem. When a loss of resistance is unanticipated, it can easily take more than 10 years to incorporate new resistance into acceptable cultivars. In the case of malting barley, this time frame is grossly under-estimated because stringent quality parameters must be maintained.

To maintain barley as profitable part of a cropping system, both productivity and market access are important issues. Where both malting and non-malting barley cultivars are grown, risk assessments by growers are based on marketability and often determine which types of barley and cultivars are sown. Central Europe is a production area where the balance shifted toward feed barley cultivars that have a winter growth habit and a higher yield potential because grain-fill occurs during a cooler portion of the summer. In areas where production of arable crops is a high-risk undertaking, barley cultivars having multiple resistances should be advantageous.

This challenge has been undertaken by the ICARDA barley breeding programs in Mexico and Syria. The ICARDA-Mexico program has demonstrated that use of shuttle breeding can result in the development of cultivars that are high yielding in both short-day and long-day environments (Vivar, 2001). This germplasm also has many disease resistances and abiotic stress tolerances in improved genetic backgrounds, but this resource is difficult to use when malting barley is targeted.

A major challenge in barley breeding is the adaptation of introduced germplasm to new production areas. The reasons for this can range from the occurrence of new disease pathotypes to altered malt quality requirements for export markets. Such production areas are often marginal for arable crops or have unique combinations of production constraints. Using molecular tools to develop cultivars by design, Eglinton et al. (2006) demonstrated that breeding of cultivars for a specific production area can be successful. However, the use of similar techniques in other programs may be limited by the number of desirable genes for which closely linked markers have been identified (Collard et al., 2005).

Many barley breeders are expected to develop improved cultivars using limited financial and physical resources. The two-rowed barley breeding program at North Dakota State University (NDSU) in the United States is an example of where the existing barley genetic resources were poorly adapted to a production area when the program was started in the early 1970s. One advantage the two-rowed introductions had was that they were more productive than the six-rowed cultivars developed for the region during dry seasons in western North Dakota. To improve two-rowed lines, crosses were made between two-rowed introductions and adapted six-rowed cultivars. The most important selection criteria in early generations were low grain protein (Foster et al., 1967) and kernel plumpness. The first cultivar released from this program was Bowman (Franckowiak et al., 1985). Using a North Dakota material, Choo et al. (2005) demonstrated that large grain size and low protein are associated, and Emebiri et al. (2007) showed that low grain protein and high yield are associated. The unanticipated result of the later study was recovery of doubled-haploid (DH) lines from a single cross with yields equal to those of the best check cultivar.

7 Breeding Methods and Techniques

To illustrate the various breeding methods and techniques used in combination to develop barley, a breeding scheme used at NDSU for developing malting barley cultivars is described. This scheme uses a modified-pedigree breeding method and off-season nurseries to reduce the length of time needed to develop new cultivars by up to three years. Applications of alternative methods of breeding, such as recurrent selection and use of DHs or breeding for non-malting barley cultivars will be described later. The possible substitution of these methods for those used in the NDSU scheme will be described.

The breeding scheme used at NDSU is a collaborative effort between the breeding team of a breeder, plant pathologist, and cereal chemist; regional agronomists; and the malting and brewing industries. The organization representing the malting and brewing industries is the American Malting Barley Association, Inc. (AMBA). As described earlier, this organization provides breeders in the United States with baseline criteria that must be met by new cultivars before they will be recommended for use by their members.

7.1 NDSU Breeding Scheme

Each year, over 150 crosses are made with the hope that a new cultivar will be developed. From the time of crossing to cultivar uptake, 10–12 years of extensive testing are done on experimental lines to evaluate their agronomic, malting, and brewing quality traits. A description of the testing a barley line must go through before it is released as a named cultivar is listed below. The key features are rapid generation advance and maximization of selection opportunities in early generations.

7.2 Year 1

Crosses of selected parents are made in the fall greenhouse nursery to incorporate such characters as high yield, disease resistance, straw strength, drought tolerance, and malt quality. In the winter greenhouse nursery, hybrid seeds (F_1) from the crosses are sown in the greenhouse to increase seed for summer evaluation.

F_2 populations are grown in the field during summer in North Dakota. The number of plants grown for each cross depends on the purpose of the cross. For example, in crosses made to incorporate Fusarium head blight (FHB) resistance, up to 18,000 F_2 plants may be grown. In a typical cross made for malting quality improvements, the F_2 population size is approximately 2,000 to 3,000 plants. From the better crosses, individual spikes are selected from plants that express favorable plant characteristics for maturity, kernel plumpness, straw strength, spike fertility, and plant height. After harvest, all spikes from a cross are bulk threshed, the grain is cleaned, and the seed is sized to remove kernels that pass through a sieve with 0.2×1.9-cm slotted openings.

7.3 Year 2

A sample of the cleaned-sized F_3 seed is sown in October in an off-season nursery near Christchurch, New Zealand, for generation advancement. In February at maturity, about 200 spikes are harvested from each cross. The harvested spikes are then threshed and sown in progeny rows (1.5 m long) during the spring at up to two locations to avoid weather-caused loss of specific nurseries. From the best crosses, approximately 1,500 progeny rows are selected at the end of the growing season based on appearance. These F_4 rows are evaluated for straw strength, maturity, plant height, and disease resistance. From within each selected row, one spike is selected from each of the three "best" plants. Selection of spikes is based on kernel plumpness, spike length, seed set, and kernel color. After the spikes are selected, the remaining portion of the row is harvested in bulk, threshed, and sent to the Barley Quality Laboratory under the direction of Dr. Paul Schwarz, Department of Plant Sciences.

The spikes selected from each row are individually threshed and the seed from the three spikes are compared to identify the "best" two. Selection among grain from spikes is based on kernel color, blighted kernels, kernel plumpness, and hull retention. Seed from the two selected spikes is sown in late October as individual progeny rows (2 m long) in the off-season nursery near Yuma, Arizona.

During the winter, quality prediction tests are conducted in Dr. Schwarz's laboratory on the F_4 rows that were bulk harvested. Quality tests performed are kernel assortment (i.e., kernel plumpness) test weight, barley protein using near infra-red reflectance spectrometry, and barley diastatic power using an assay that utilizes papain to release bound β-amylase.

7.4 Year 3

On the basis of quality data, rows in Arizona having favorable quality are considered for harvest in late March. Among the rows with acceptable quality, rows are selected for harvest based on maturity, plant height, uniformity, and straw strength. From the rows harvested, approximately 675 lines are advanced to replicated preliminary yield trials (PYTs) that are randomized using a lattice design and grown at two locations in North Dakota. The lines in the PYTs are assigned an "ND number" and seed increase is started to provide seed for the following year's trials. Lines in the PYTs are evaluated for grain yield, maturity, plant height, foliar diseases, lodging, straw breakage, and test weight. Seed from one location of lines having good agronomic performance are sent to the USDA-ARS Cereal Crops Research Unit (USDA-ARS-CCRU) at Madison, Wisconsin, for malt quality analysis and to Dr. Stephen Neate in the Department of Plant Pathology, NDSU. Malt quality characters studied on each line are barley protein, kernel plumpness, kernel color, soluble wort protein, malt extract, fine-coarse difference, wort viscosity, diastatic power, and alpha amylase activity. Dr. Neate evaluates the lines for reaction to the net and spot blotch pathogens on both seedlings and adult plants in the greenhouse.

7.5 Year 4

On the basis of agronomic and malting quality data, about 75 ND lines are advanced to the intermediate yield trial (IYT) at seven locations in North Dakota and one location in eastern Montana. The lines are evaluated for the same agronomic, disease, and malt quality traits studied in the PYTs. In addition, larger scale pure seed increases of the IYT lines are produced. Reactions to the net blotch and spot blotch pathogens are evaluated in field nurseries by Dr. Neate. Seed of promising ND lines from two to three locations is sent to the USDA-ARS-CCRU for malt quality analyses.

7.6 Year 5

About 20 promising ND lines are advanced to the advance yield trial (AYT) based on agronomic and malt quality data collected over the previous two summers. The AYT is grown at the same location as the IYT. Lines are evaluated for the same agronomic, disease, and malt quality traits studied in the previous yield trials. Seed of promising lines from two to three locations is forwarded to the USDA-ARS-CCRU for malt quality analyses. On the basis of the agronomic and malt quality characteristics, up to four lines are sent to the AMBA for the first year of Pilot Scale Evaluation. Seed for this evaluation is obtained from two of the best six locations of large increase plots that are grown specifically for the AMBA's tests. For this evaluation, 17 kg of each entry is divided among the malting and brewing members of AMBA and malt quality is compared to the check cultivars.

7.7 Year 6

ND lines submitted for pilot scale evaluation are entered in varietal yield trial (VYT) at the same locations as the AYT, in North Dakota Agricultural Experiment Station (NDAES) Varietal Trials at six locations across North Dakota, and in Cooperative Regional Yield Trials grown at 8 to 15 locations across the barley growing regions in the United States. Lines are evaluated for the same agronomic, disease, and malt quality traits studied in the previous yield trials. Grain samples of lines passing first-year pilot scale evaluation and having good agronomic potential are submitted to AMBA for second-year pilot scale evaluation.

7.8 Year 7

ND lines passing first-year pilot scale evaluation are entered in VYT, Cooperative Yield Trials, and NDAES Varietal Trials. Lines are evaluated for the same agronomic, disease, and malt quality traits studied in the previous yield trials. Seed of promising ND lines is increased in the Arizona off-season nursery incase permission is granted by the AMBA for an ND line to be advanced to plant scale evaluation. Seed sufficient to sow about 250 hectares is needed from the off-season increase. In general, the AMBA allows only one line per breeding program to be entered for plant scale evaluation.

The most promising ND line passing its second year of pilot scale evaluation is submitted to the AMBA for plant scale evaluation. In this evaluation, the malting and brewing quality of large grain samples ($\approx$435–655 MT) are used for evaluation by one or more maltsters and brewers.

In preparation for possible release, seed purification of the ND-line begins. Approximately 1,000 single spike selections are sown as single progeny rows. Rows containing off-type plants are removed and the remaining rows are harvested in bulk.

7.9 Year 8

The ND-line submitted for plant scale evaluation is tested in the same yield trials mentioned for year 7 and evaluated for the same agronomic, disease, and malt quality traits. If the ND line passes its first year of plant scale testing, it is entered for the second year of plant scale testing. Seed from the bulked head rows is sown by the NDAES Foundation Seedstocks Program to produce Breeder seed in preparation for possible cultivar release.

Occasionally, extremely promising lines may be released as named cultivars at this time. This is done if growers and NDSU scientists, regional agronomists, and growers see an economic benefit from immediate distribution of the cultivar.

7.10 Year 9

The ND line submitted for its second year of plant scale testing is entered in the same yield trials as the previous two years and evaluated for the same agronomic, disease, and malt quality traits.

On the basis of acceptance of an ND line by the AMBA, it is assigned a cultivar name and released as a recommended malting barley cultivar. Breeder seed is sown by the NDAES Foundation Seedstocks Program to produce Foundation seed.

7.11 Year 10

Foundation seed is sold to growers that are members of the North Dakota Crop Improvement Association so they can produce Registered Seed.

No two breeding programs can use the same breeding scheme. Variations to the scheme above can be made at any stage to accommodate specific resources and breeding goals. For example, DH techniques or MAS based on DNA markers could be used to shorten the length of time needed to develop new cultivars by several years. These methods will be described in greater detail in the next section. Off-season or glasshouse nurseries also can reduce the length of time needed to develop new cultivars by providing for two or more field generations per year. Locations used for off-season nurseries by breeders in the northern hemisphere include the southwestern United States, Argentina, and New Zealand. Finally, non-malting or feed barley cultivars can be developed much quicker than malting barley cultivars because extensive quality testing is not needed. Feed barley cultivars often can be developed three to four years quicker than malting barley cultivars.

8 Integration of New Biotechnologies in Breeding Programs

Utilization of DH technology to extract lines from F_1 plants produces homozygous lines in about one year. This can reduce the time needed to develop new cultivars by two to three years because early generation testing of material while selfing is not needed. Additionally, the DH method may be advantageous for breeders developing winter barley cultivars because they do not usually use off-season nurseries for generation advancement. Two methods of DH production are used for barley, the *Hordeum bulbosum* L. method (Chen and Hayes, 1989) and the microspore culture method (Ziuddin et al., 1992). In the *H. bulbosum* method, F_1 plants are typically used as the female and crossed with pollen from selected *H. bulbosum* clones. Following pollination, the seven chromosomes from *H. bulbosum* are gradually eliminated in most embryos, leaving the growing embryo with only seven chromosomes from the female plant. Because the endosperm collapses, the growing embryo has to be rescued before two weeks under sterile conditions and cultured on growth media, resulting in production of mostly haploid plants ($n = 7$). A very small proportion of the embryos will undergo spontaneous doubling to $2n = 14$. To double the chromosome number to $2n = 14$, the crown of the plant is soaked in colchicine. In our experience in using this method, less than 15% of the pollinated florets result in DH plants.

The production of DH plants by culturing microspores from immature anthers was used on a limited basis prior to the mid-1990s because success was genotype specific and larger numbers of albino plants were obtained. Ziuddin et al. (1992) developed a system of DH production from microspores that greatly reduced these problems and several breeding programs adopted this method. Successful utilization of this method is dependent on having high-quality growth chambers and facilities to nurse the DH plants produced.

Selection of F_1 plants from complex crosses or plants in early generations using molecular marker-assisted selection (MAS) has been employed in barley with limited success. In theory, selection of plants based on their DNA composition early in the breeding program should increase efficiency because only plants with the genes of interest would be evaluated. Microsattelite or simple sequence repeat (SSR) markers are most commonly used; however, success is often limited to simply inherited traits for which closely linked markers have been identified. Traits that were selected with some success using MAS include resistance to cereal cyst nematode and boron tolerance (Ogbonnaya et al., 1998) and covered smut, incited by *Ustilago hordei* (Pers.) Lagerh (Ardiel et al., 2002). The development of markers for MAS of lines with resistance to FHB is receiving a lot of attention. Microsatellite markers linked to FHB-resistance quantitative trait loci (QTL) have been mapped in several resistance sources (Horsley et al., 2006; Dahleen et al., 2003; Mesfin et al., 2003; Ma et al., 2000; de la Pena et al., 1999; Zhu et al., 1999). However, successful use of these markers for MAS has been limited. Recently, research on utilizing single nucleotide polymorphism (SNP) markers has increased significantly. It is hoped that these markers will permit successful utilization of MAS over a wider range of crosses and traits.

Transformation of barley offers another method for introducing new traits into barley. However, many industries, including maltsters, brewers, and food producers do not accept transformed barley for processing. Stable transformation of barley using microparticle bombardment was first reported by Wan and Lemaux (1994). A method of using *Agrobacterium tumefaciens* for transforming barley was developed shortly afterward (Tingay et al. 1997). For both methods, the cultivar Golden Promise has been found the easiest to transform; thus, it is widely used. An inadequacy of Golden Promise is that is a non-malting barley cultivar which is not adapted for production in stressed environments. An alternative to transforming Golden Promise is the two-rowed malting barley cultivar Conlon (Manoharan et al., 2006).

9 Cultivar Release and Intellectual Property Issues

Depending on where the cultivar is being developed or utilized, different rules for regional testing and registration apply. Similarly, the rules for sales of seed and protection of intellectual property rights are dependent on the country where the cultivars were developed and/or will be sold. The general procedures for registering and protecting a new cultivar are listed below for Australia, Canada, the European Union, and the United States.

9.1 Australia

Barley cultivar development in Australia is primarily done by public breeding programs funded by the federal government through the Grains Research and Development Corporation (GRDC) and state governments. Since the start of this century, the breeding effort has changed from ones conducted by each state to one program (Barley Breeding Australia, BBA) with a national focus. Barley cultivars and lines nearing release are evaluated in the National Variety Trial (NVT) system established by GRDC and managed by the Australia Crop Accreditation System, Ltd. Cultivars recommended for release by BBA are put out for public tender and bids by commercialization parties are evaluated by a panel including the breeder or the barley industry development officer, the Department of Primary Industries' business manager, and GRDC. Before a cultivar can receive a recommendation or accreditation as a malting barley cultivar, it must undergo an evaluation overseen by Barley Australia. Barley Australia was formed out of the Malting Barley

Development Council in 2005. Similar to the AMBA in the United States, Barley Australia is composed of members in the malting and brewing industries and facilitates collaboration between their members and public barley researchers. The accreditation process requires two years of evaluation that includes plant scale malting evaluation by a maltster that is a member of the Malting and Brewing Industry Barley Technical Committee and pilot brewing conducted by Pilot Brewing Australia in Melbourne. On the basis of results of these two years of evaluations, Barley Australia meets to decide if the cultivar should be accredited as a malting barley cultivar. Breeder seed is maintained by the breeding organization. Maintenance of other classes of seed and seed quality assurance are responsibilities of the commercialization partner.

9.2 Canada

In Canada, the number of public barley breeding programs outnumbers the number of private programs. A formal cultivar registration system is mandated by the federal Seeds Act and Regulations. The Canadian Food Inspection Agency (CFIA) is responsible for implementing the Act and Regulations. There are various registration recommending committees across the country for most commercial crops. Essentially, all malting barley cultivars are tested and recommended for registration in western Canada, the area where most barley is grown. The Prairie Recommending Committee for Oat and Barley (PRCOB) provides recommendations to the CFIA on which candidates should become registered cultivars. Prior to registration, prospective cultivars are evaluated in Co-operative ("Co-op") and Collaborative ("Collab") trials for a minimum of two years at each stage, with the second year of Co-op trials and the first year of Collab trials typically taking place in the same year. At the Co-op stage, agronomic performance, disease resistance, and malting quality are all evaluated, while at the Collab stage only malting quality is evaluated. Malting quality evaluation during the Co-op trials is coordinated by the Grain Research Laboratory (GRL), Canadian Grain Commission (CGC), and during the Collab trials by the Brewing and Malting Barley Research Institute (BMBRI). BMBRI industry members and the GRL carry out the malting tests at both stages, sharing the results with the PRCOB, which then makes recommendations to the CFIA Variety Registration Office on whether a candidate should be registered. The PRCOB also evaluates and makes recommendations on non-malting lines. Post-registration, malting and brewing companies carry out plant scale evaluation of registered malting barley cultivars and decide which cultivars they will use. The Canadian Malting and Brewing Technical Center (CMBTC) publishes an annual list of "recommended malting barley varieties" as one of the signals to barley growers on the anticipated demand for specific cultivars for both the domestic and export markets. Production and sale of pedigreed seed of registered cultivars is carried out under license from the breeding institution, typically by one or more seed companies. The CFIA's Web site lists all registered cultivars and their license holders. Intellectual property rights can be protected by obtaining a Plant Breeder Right certificate under Canada's Plant Breeder's Rights legislation.

9.3 European Union

Regulations and laws for cultivar registration in the European Union (EU) can appear very complex for someone working outside the system. Because of these complexities and differences in the regulations between countries, it is impossible in this chapter to provide an in-depth review on the processes for cultivar registration and protection of intellectual property rights. The reader is encouraged to contact the official cultivar registration office in each country for precise details. In general, development of cultivars in Europe is dominated by large multinational companies such as Limagrain and Syngenta, and many small family or regional/sub-regional companies. The number of public breeding programs in Europe is much fewer than in North America and Australia. Cultivars developed by public programs and family and regional/sub-regional companies often market their cultivars through larger multipurpose agricultural companies or conglomerates. European barley breeding companies are dependent on royalties; thus, developing a cultivar that is adapted to more than one county is an important goal because this can significantly increase the royalty stream.

The information presented below represents the rules and procedures for the 27 countries that make up the EU. The process of getting approval to increase and sell seed of a new cultivar requires two separate procedures that are typically done concurrently. The first procedure is called Distinct Uniform and Stable (DUS) testing. This procedure is more commonly called the International Union for the Protection of New Varieties of Plants (UPOV) testing. The second procedure is the national listing/national registration trials. These DUS and national listing tests are carried out separately and the results are intended for separate audiences. The DUS testing can be done in one EU country chosen by the breeder, as long as they are a citizen of an EU country. After two years of DUS testing, the breeder receives what is called the *UPOV Report on Technical Examination* and protection for the cultivar. Protection is granted for 25 years, but without a national listing, the breeder does not have the right to produce and sell seed. To receive a national listing, the breeder can choose one preferred country to submit the cultivar for testing, and this country can be different from the one where DUS testing is conducted. The country that performs the national listing trial will purchase the UPOV report from the country that did the DUS testing.

Rules for the national listing trials are based on the laws of the country where the trials are conducted. The length of the trial is dependent on the crop being tested and the country where the tests occur. Typically, the trials are conducted for two or three years and the new cultivars are compared to reference cultivars. Each country has different reference cultivars and these can change in each year of testing. Permission to advance to a second year of national listing testing is dependent on the rules of the country where the tests are done. Some countries, such as Germany, require that a new cultivar must be above a specified index based on multiple traits before it can be advanced to a second year of testing. Other countries have no such rule, so every cultivar can go through the entire two to three years of testing. For a cultivar to receive a national listing, it must be superior to the reference cultivar(s) for most traits. If a cultivar successfully receives a national listing, the listing is good for 10 years. After 10 years, the breeder needs to reapply to keep the cultivar listing current.

Upon completion of DUS and national listing testing in one country, the breeder has one year to apply for cultivar protection across the entire EU. The application is a formal act that does not require additional field testing. The EU protection is important because it allows the breeder to recuperate royalties in each EU country where the cultivar is grown. The EU protection is good for 25 years as long as the breeder keeps their national listing current. If the breeder does not reapply for national listing after 10 years, the EU protection ends. Finally, EU protection or a national protection does not prevent other breeders within the EU from using protected cultivars as parents for crossing.

The testing described above is done regardless if the cultivar has potential use for malting and brewing. To determine the utility of a cultivar, the breeder must have their own quality laboratory or make arrangements for an outside laboratory to evaluate the malting potential of a line before it enters DUS and national listing tests. Depending on the country where national listing tests are done, evaluation of malt quality may be part of the process. For example, malt quality is evaluated in Germany and France as part of their national listing testing. Nevertheless, none of the quality evaluations done before or during national listing testing give any indication if a cultivar will be used by the malting and brewing industries.

Determination of a cultivar's suitability for malting and brewing is typically determined after DUS and national listing tests are completed. In some countries, quality evaluation is organized by allied members of the malting and brewing industries, which may include breeders, brewers, maltsters, national or regional institutes, or universities. For example, the evaluation program in France is overseen by the Comité Biére Malt Orge (CBMO), in Germany by the Berlin Program, and in Great Britain by the Cereal Technical Advisory Committee (CTAC). However, because many EU countries do not have such national organizations, they look to the European Brewing Convention (EBC) for evaluating new cultivars that are nationally listed. The EBC coordinates the testing of new cultivars in field trials under variable growing conditions and the countries where the tests are conducted are assigned to one of four regions (Table 3). The EBC produces reports that summarize data from both national trials and their trials so members of the European malting and brewing industries can identify new cultivars that may be of interest, especially those that have acceptable agronomic performance and malt quality when produced in multiple regions. As is done across the world, large international brewers will perform their own plant scale malting and brewing evaluations to determine if they will use a new cultivar.

Table 3 Countries conducting malting barley evaluation trials sponsored by the European Brewing Congress

Region	Country
North	Denmark
	Sweden
	Finland
	Estonia
West	United Kingdom
	France
	Belgium
	The Netherlands
Central	Germany
	Czech Republic
	Hungary
	Austria
	Croatia
	Slovakia
South	Portugal
	Spain
	Italy
	Bulgaria

9.4 United States

In the United States, publicly funded breeding programs at universities and the USDA-ARS outnumber private companies. No formal national system of crop registration is done in the United States. Instead, each institution or company has their own criteria for determining if a cultivar should be released. The determination of whether a cultivar will be added to the AMBA list of recommended malting barley varieties is described previously and is dependent on a minimum of two years of pilot scale and two years of plant scale evaluation by the AMBA's member malting and brewing companies. In general, most cultivars from the public programs are released to grower organizations (e.g., Crop Improvement Associations) in their states. Foundation seed of these releases is handled by the university or state agricultural experiment station that released the cultivar. Protection of intellectual rights is done by obtaining a Plant Variety Protection (PVP) certificate; however, the decision to obtain a PVP certificate is not consistent across states or even cultivars.

Acknowledgments The authors express their appreciation to Mr. Jan Hartmann of BayWa AG in Munich, Germany; Mr. Scott Heisel of the American Malting Barley Association in Milwaukee, Wisconsin, USA; and Ms. Erin Armstrong of the Brewing and Malting Barley Research Institute in Winnipeg, Canada, for supplying and reviewing information utilized in the "Cultivar Release and Intellectual Property Issues" section of this chapter.

References

Ardiel, G.S., T.S. Grewal, P. Deberdt, B.G. Rossnagel, and G.J. Scoles. 2002. Inheritance of resistance to covered smut in barley and development of a tightly linked SCAR. Theor. Appl. Genet. 104:457–464.

Behall, K.M., and J.G. Hallfrisch. 2006. Effects of barley consumption in CVD risk factors. Cereal Foods World 51:12–15.

Behall, K.M., D.J. Scholfield, J.G. Hallfrisch. 2004. Lipids significantly reduced by diet containing barley compared to whole wheat and brown rice in moderately hypercholesterolemic men. J. Am. Med. Assoc. 23:55–62.

Bergal, P., and M. Clemencet. 1962. The botany of the barley plant. *In* A.H. Cook (ed.) Barley and malt: biology, biochemistry and technology, pp. 1–23. Academic Press, New York, 1962.

Blake, T., J.G.P. Bowman, P. Hensleigh, G. Kushna, G. Carlson, L. Welty, J. Eckhoff, K. Kephart, D. Wichman and P.M. Hayes. 2002. Registration of "Valier" barley. Crop Sci. 42:1748–1749.

Bothmer, R. von, J. Flink, N. Jacobsen, M. Katimäki, and T. Landström. 1983. Interspecific hybridization with cultivated barley. Hereditas 99:219–244.

Bothmer, R. von, K. Sato, T. Komatsuda, S. Yasuda, and G. Fischbeck. 2003. The domestication of cultivated barley. *In* R. von Bothmer, Th. van Hintum, H. Knüpffer, and K. Sato (eds.), Diversity in Barley (*Hordeum vulgare*), pp. 9–27. Elsevier Science B.V., Amsterdam, The Netherlands.

Chen, F., and P.M. Hayes. 1989. A comparison of *Hordeum bulbosum* mediated haploid production efficiency in barley using in vitro floret and tiller culture. Theor. Appl. Genet. 77:701–704.

Choo, T.M., B. Vigier, K.H. Ho, S. Ceccarelli, S. Grando, J.D. Franckowiak. 2005. Comparison of black, purple, and yellow barleys. Genet. Resour. Crop Evol. 52:121–126.

Collard, B.C.Y., M.Z.Z. Jahufer, J.B. Brouwer, and E.C.K. Pang. 2005. An introduction to markers, quantitative trait loci (QTL) mapping and marker-assisted selection for crop improvement: The basic concepts. Euphytica 142:169–196.

Dahleen, L., H. Agrama, R. Horsley, B. Steffenson, P. Schwarz, A. Mesfin, and J. Franckowiak. 2003. Identification of QTLs associated with Fusarium head blight resistance in Zhedar 2 barley. Theor. Appl. Genet. 108:95–104.

Davis, M.P., J.D. Franckowiak, T. Konishi, and U. Lundqvist (eds.). 1997. Barley Genetics Newsletter, Volume 26. American Malting Barley Assoc., Inc. Milwaukee, WI.

Eglinton, J., S. Coventry, and K. Chalmers. 2006. Breeding outcomes from molecular genetics. p 743–749. *In* C.F. Mercer (ed.) Proc. of the 13th Australasian Plant Breeding Conference, Christchurch, New Zealand 18–21 April 2006.

Erkkilä, M.J., R. Leah, H. Ahokas, and V. Cameron-Mills. 1998. Allele-dependent barley grain β-amylase activity. Plant Physiol. 117:679–685.

Emebiri, L., D.B. Moody, C. Black, M. van Ginkel, and E. Hernandez. 2007. Improvements in malting barley grain yield by manipulation of genes influencing grain protein content. Euphytica 155:185–194.

Fetch, T. G., Jr., B.J. Steffenson, and E. Nevo. 2003. Diversity and sources of multiple disease resistance in *Hordeum spontaneum*. Plant Dis. 87:1439–1448.

Foster, A.E., G.A. Peterson, and U.J. Banasick. 1967. Heritability of factors affecting malting quality of barley, *Hordeum vulgare* L. Emend. Lam. Crop Sci. 7:611–613.

Franckowiak, J.D., and U. Lundqvist. 1997. BGS 6, Six-rowed spike 1, *vrs1*. Barley Genet. Newsl. 26:49–50.

Franckowiak, J.D., A.E. Foster, V.D. Pederson, and R.E. Pyler. 1985. Registration of Bowman barley. Crop Sci. 25:883.

Horsley, R., D. Schmierer, C. Maier, C. Urrea, B. Steffenson, P. Schwarz, J. Franckowiak, M. Green, B. Zhang, and A. Kleinhofs. 2006. Identification of QTL associated with Fusarium head blight resistance in barley accession CIho 4196. Crop Sci. 46:145–156.

Horsley, R.D., P.B. Schwarz, and J.J. Hammond. 1995. Genetic diversity in malt quality of North American six-rowed spring barley. Crop Sci. 35:113–118.

Jorgensen, J.H., 1992. Discovery, characterization and exploitation of Mlo powdery mildew resistance in barley. Euphytica 63:141–152.

Juskiw, P., J. Helm, L. Oatway, and J. Nyachiro. 2005. Breeding for feed barley quality in hulled two-row barley. 12th Australian Barley Technical Symposium (ABTS) http://www.cdesign.com.au/proceedings_abts2005/papers%20(pdf)/tues_1620.pdf.

Komatsuda, T., M. Pourkheirandish, C. He, P. Azhaguvel, H. Kanamori, D. Perovic, N. Stein, A. Graner, T. Wicker, A. Tagiri, U. Lundqvist, T. Fujimura, M. Matsuoka, T. Matsumoto, and M. Yano. 2007. Six-rowed barley originated from a mutation in a homeodomain-leucine zipper I-class homeobox gene. PNAS 104:1424–1429.

Kunze, W. 2004. Technology brewing and malting; 3rd International edition – in English. VLB, Berlin.

Langridge, P., N. Paltridge, and G. Fincher. 2006. Functional genomics of abiotic stress tolerance in cereals. Brief. Funct. Genomic Proteomic 4:343–354.

Lundquist and Franckowiak, http://wheat.pw.usda.gov/ggpages/bgn/26/

Ma, Z., B. Steffenson, L. Prom, and L. Lapitan. 2000. Mapping of quantitative trait loci for Fusarium head blight resistance in barley. Phytopathology 90:1079–1088.

Manoharan, M., L.S. Dahleen, T.M. Hohn, S.M. Neate, X. Yu, N.J. Alexander, S.P. McCormick, P. Bregitzer, P.B. Schwarz, and R.D. Horsley. 2006. Expression of 3-Oh-Triothecene acetyltransferase in barley and effects on Fusarium head blight. Plant Sci. 171:699–706.

Mathre, D.E. (ed.) 1997. Compendium of Barley Diseases, second edition. Amer. Phytopath. Soc., St. Paul, MN.

Matus, I., A. Corey, T. Filchkin, P.M. Hayes, M.I. Vales, J. Kling, O. Riera-Lizarazu, K. Sato, W. Powell, and R. Waugh. 2003. Development and characterization of recombinant chromosome substitution lines (RCSLs) using *Hordeum vulgare* subsp. *spontaneum* as a source of donor alleles in a *Hordeum vulgare* subsp. *vulgare* background. Genome 46:1010–1023.

Mesfin, K., K. Smith, R. Dill-Macky, C. Evans, R. Waugh, C. Gustus, and G. Muehlbauer. 2003. Quantitative traits loci for Fusarium head blight resistance in barley detected in a two-rowed by six rowed population. Crop Sci. 43:307–318.

Ogbonnaya, F.C., D.B. Moody, A. Rehman, J.M. Kollmorgen, and H.A. Eagles. 1998. Marker assisted selection for resistance to cereal cyst nematode and boron tolerance in barley. p. 189. *In* Plant & Animal Genome VI Conference Abstracts.

de la Pena, R., K. Smith, F. Capettini, G. Muehlbauer, M. Gallo-Meagher, R. Dill-Macky, D. Somers and D. Rasmusson.1999. Quantitative trait loci associated with resistance to Fusarium head blight and kernel discoloration in barley. Theor. Appl. Genet. 99:561–569.

Pickering, R.A. 1989. Plant regeneration and variants from calli derived from immature embryos of diploid barley (*Hordeum vulgare* L.) and *H. bulbosum* L. crosses. Theor. Appl. Genet. 78:105–112.

Pickering, R., A. Hill, M. Michel, and G. Timmerman-Vaughan. 1995. The transfer of a powdery mildew resistance gene from *Hordeum bulbosum* L. to barley (*Hordeum vulgare* L.) chromosome 2 (2I). Theor. Appl. Genet. 91:1288–1292.

Pillen, K., and J. Leon. 2003. Advanced backcross QTL analysis in barley (*Hordeum vulgare* L.). Theor. Appl. Genet. 107:340–352.

Rasmusson, D.C. and R.L. Phillips. 1997. Plant breeding progress and genetic diversity from de novo variation and elevated epistasis. Crop Sci. 37:303–310.

Russell, J.R., R.P. Ellis, W.T.B. Thomas, R. Waugh, J. Provan, A. Booth, J. Fuller, P. Lawrence, G. Young, and W. Powell. 2000. A retrospective analysis of spring barley germplasm development from "foundation genotypes" to currently successful cultivars. Mol. Breed. 6:533–568.

Steffenson, B.J. 1992. Analysis of durable resistance to stem rust in barley. Euphytica 63:153–167.

Steffenson, B.J., P.M. Hayes, and A. Kleinhofs. 1996. Genetics of seedling and adult plant resistant to net blotch (*Pyrenophora teres* f. *teres*) and spot blotch (*Cochliobolus sativus*) in barley. Theor. Appl. Genet. 92:552–558.

Tingay, S., D. McElroy, R. Kalla, S. Fleg, M. Wang, S. Thornton, and R.I.S. Brettell. 1997. *Agrobacterium tumefaciens*-mediated barley transformation. Plant J. 11:1369–1376.

Ullrich, S.E., D.M. Wesenberg, H.E. Blockelman, and J.D. Franckowiak. 1995. International cooperation in barley germplasm activities. *In* R.R. Duncan (ed.) International germplasm transfer: past and present, pp. 157–170. CSSA Special Publ. 23. CSSA and ASA, Madison, WI.

Vivar, H.E. 2001. Two decades of barley breeding. *In* H.E. Vivar and A. McNab (eds.). Breeding Barley in the New Millennium: Proceedings of an International Symposium, pp. 77–82. CIMMYT, Mexico., D.F.

Wan, Y., and P. Lemaux. 1994. Generation of large numbers of independent transformed fertile barley plants. Plant Physiol. 104:37–48.

Wiebe, G.A. 1968. Breeding. *In* Barley: Origin, botany, culture, winterhardiness, genetics, utilization, pests. pp.96–104. (Agric. Handbook No. 338, Agric. Res. Service, U.S. Dept. Agric., Washington, D.C. or USDA Agric. Handb. No. 338.).

Wilcoxson, R.D., D.C. Rasmusson, and M.R. Miles. 1990. Development of barley resistant to spot blotch and genetics of resistance. Plant Dis. 74:207–210.

Zhu, H., L. Gilchrist, P. Hayes, A. Kleinhofs, D. Kudrna, Z. Liu, L, Prom, B. Steffenson, T. Toojinda, and H. Vivar. 1999. Does function follow form? Principal QTLs for *Fusarium* head blight (FHB) resistance are coincident with QTLs for inflorescence traits and plant height in a doubled-haploid population of barley. Theor. Appl. Genet. 99:1221–1232.

Ziuddin, A., A. Marsolais, E. Simion, and K.J. Kasha. 1992. Improved plant regeneration from wheat anther and barley microspore culture using phenylacetic acid (PAA). Plant Cell Rep. 11:489–498.

Winter and Specialty Wheat

P. Baenziger, R. Graybosch, D. Van Sanford, and W. Berzonsky

Abstract Wheat is the most widely grown crop in the world. Winter wheat is primarily common wheat ($2n = 6x = 42$) which has extensive germplasm resources that are used in breeding, often for disease and insect resistance. Though wheat can be used as a forage crop and its grain for animal feed, the primary uses of common wheat are to make products used for human consumption; hence end-use quality is also a major breeding objective. The quality characteristics of these products are often associated with kernel hardness which affects milling, kernel color, and specific climatic zones or regions. The soft red and white wheat cultivars of the Eastern and Southeastern U.S. are generally used to make breakfast cereals, cookies, cakes, and crackers. The hard red and white wheat cultivars of the Great Plains are used predominantly for leavened products such as bread. The soft white wheat cultivars of the Pacific Northwest are often exported and used to make noodles or steam breads. These end-uses and production (adaptation) regions determine the germplasm pools used by wheat breeders. All of the common self pollinated breeding methods are used to breed new wheat cultivars. The choice of breeding method is usually based upon breeding objective and program resources. Breeding methods and objectives are evolving with new technology and market changes.

1 Introduction

The two main commercial types of wheat are durum (*Triticum durum* L., $2n = 4x = 28$) and common (*Triticum aestivum* L., $2n = 6x = 42$) wheat, the latter being the more widely grown. Wheat has three growth habits, namely winter (wheat grown primarily during the winter months, that requires vernalization to flower, and can withstand prolonged periods of below freezing temperatures), facultative (wheat grown primarily during the winter months in mild climates, that may or may not require vernalization to flower, and cannot withstand prolonged periods of below

P. Baenziger(✉)
362D Plant Science Building, Department of Agronomy and Horticulture, University of Nebraska, Lincoln, NE 68583-0915 USA

M.J. Carena (ed.), *Cereals*,
DOI: 10.1007/978-0-387-72297-9, © Springer Science + Business Media, LLC 2009

freezing temperatures), and spring (wheats grown primarily during the spring and summer months, that normally do not require vernalization to flower, and cannot withstand moderate periods of below freezing temperatures). Growth habit should be viewed as a continuum from winter wheat to facultative wheat to spring wheat. Because wheat can be grown through the winter or summer, is very drought tolerant, and is used primarily for human consumption, it is the most widely grown crop in the world (218,283,000 ha; http://www.nass.usda.gov/Publications/Ag_Statistics/2007/chap01.pdf, verified May 19, 2008; note Web address will change annually, so the primary address is http://www.nass.usda.gov/Publications/Ag_Statistics, which will have the most recent Agricultural Statistics). In this chapter, we will concentrate on winter wheat breeding using examples of breeding which target cultivar development in the United States, but examples relating to global winter wheat breeding will be discussed as appropriate. It should be noted that as a winter annual, wheat is often grown in rotation with summer annual crops if the growing season is long enough and moisture is sufficient.

2 Genetic Diversity and Germplasm Selection

In North America, all cultivated winter wheat species are forms of common wheat. The desert durum wheats of the southwest United States are facultative wheats. An allohexaploid, *T. aestivum* has three genomes (A, B, and D), each composed of seven chromosome pairs. These genomes derive from related ancestral species. Chromosomes of each genome are numbered 1–7, with each number designating a homoeologous set. Homoeologous chromosomes are similar both in structure and gene content, and thus, they can compensate for each other as was demonstrated in the production of nullisomic–tetrasomic lines (Sears, 1953). Chromosomes derived from related species of the grass tribe Triticeae also may substitute for wheat chromosomes (Friebe et al., 1996). Wild and domestic relatives of wheat routinely are utilized as valuable sources of genes for wheat improvement. Wide or interspecific hybridization, when followed by radiation treatment or the induction of homoeologous recombination, can result in the successful transfer of chromosomes arms, or smaller chromosome segments, resulting in the formation of wheat-alien translocation lines (Friebe et al., 1996). Manipulating homoeologous chromosome recombination to incorporate alien chromatin is known as chromosome engineering, and it has been a mainstay of wheat improvement programs for the past 50 years.

The majority of alien chromosome segments conferring disease and pest resistance to wheat have been derived from species of the genus *Aegilops*, two species of perennial wheat grasses, *Thinopyrum elongatum* (Host) Dewey and *Elytrigia intermedia* (Host) Nevski, and from rye (*Secale cereale* L.) (Friebe et al., 1996). An examination of pedigrees of recent entrants in the US Department of Agriculture-Agricultural Research Service (USDA-ARS) Hard Winter Wheat Regional Nursery Trials (http://www.ars.usda.gov/Research/docs.htm?docid=11932, verified May19,

2008) indicates that US hard winter wheat breeders are utilizing various germplasm lines as donors of genes from wild and cultivated relatives. Examples include KS96WGRC39, which carries a leaf rust (incited by *Puccinia triticina* Eriks) resistance gene (*Lr41*) and tan spot [incited by *Pyrenophora tritici-repentis* (Died.) Drechs.] resistance from *Aegilops tauschii* (Brown-Guedira et al., 1999); KS90WGRC10, also carrying *Lr41* from *A. tauschii*; KS91WGRC11, providing *Lr42*, again from *A. tauschii* (Cox et al., 1994); and KS93WGRC27, which carries *Wsm-1*, a gene conferring resistance to wheat streak mosaic virus. *Wsm-1* is present on a T4DL·4Ai#2S, wheat-*A. intermedium* chromosome (Gill et al., 1995). North American breeders of hard winter wheat long have used the *Sr24 and Lr24* resistance genes, which are in the cultivar Agent but which are derived from *T. elongatum* (McIntosh et al., 1976), as well as a greenbug [*Schizaphis graminum* (Rondani)] resistance gene (*Gb3*) derived from *A. tauschii* and present in the wheat cultivar Largo (Hollenhorst and Joppa, 1983). Winter wheat breeders in the US Pacific Northwest (PNW) have been incorporating resistance to eyespot (straw-breaker foot rot incited by *Tapesia yallundae* Wallwork and Spooner) derived from *A. ventricosa* via the wheat VPM-1. VPM-1 also has served as a donor of genes for resistance to cereal cyst nematode (CCN; *Heterodera avenae*), stem rust (incited by *P. graminis Pers.: Pers. f. sp. tritici* Eriks and E. Henn), stripe rust (incited by *P. striiformis* Westendorp f. sp. *tritici*), and leaf rust (Seah et al., 2000).

Both soft and hard winter wheat breeders have long utilized genes for resistance to pathogens and insect pests from rye (*Secale cereale* L.), especially those on chromosome arm 1RS. Both 1BL.1RS (originally transferred from the Russian cultivar Kavkaz) and 1AL.1RS (from the cultivar Amigo) have been introduced into the soft and hard winter wheat gene pools. These chromosome arms originally served as sources of leaf and stem rust resistance, resistance to powdery mildew, and, in the case of 1AL.1RS, resistance to greenbug. In addition, both chromosomes must confer, at least in some genetic backgrounds, broad agronomic adaptation, via stress resistance (Graybosch, 2001). The 1AL.1RS chromosome translocation is more prevalent in the US advanced hard winter wheat breeding lines, while 1BL.1RS is more prevalent in US soft winter wheat breeding lines (http://www.ars.usda.gov/Research/docs.htm?docid=11932).

Reduced end-use quality is rarely attributed to transfers from nonrye sources. For example, Divis et al. (2006) found no significant or consistent quality problems associated with the presence of *Wsm-1*. However, wheats carrying 1RS often exhibit undesirable baking characteristics. Desirable genes from rye are linked to traits that alter the grain composition of the recipient wheat cultivars. These effects have been the subject of several review articles (see Graybosch, 2001).

While the majority of wheat cultivars are developed through matings of elite germplasm, unadapted wheat landraces and accessions from the USDA-ARS National Small Grains Collection (Aberdeen, ID) frequently serve as donors for important traits. Examples include Russian wheat aphid [Diuraphis noxia (Mordvilko)] resistance, derived from several landraces (Souza, 1998; Souza et al., 1991), the null granule-bound starch synthase gene mutants used to develop waxy (amylose-free) wheats (Morris and Konzak, 2001), and novel high molecular weight glutenin subunits derived from *A. tauschii* and *T. turgidum* var. *dicoccoides*.

3 Varietal Groups

Winter wheat in the United States may be divided into four varietal groups or gene pools as follows: (1) Eastern and Southeastern soft wheats, (2) Southern Great Plains wheats, (3) Northern Great Plains wheats, and (4) Pacific Northwest (PNW) wheats. Each represents distinct groups in terms of both agroecological adaptations and end-use properties (soft wheat for uses ranging from cookies to cakes to cereals or hard wheats primarily used for bread making). On a global scale, winter wheat can also be divided into varietal groups or gene pools based upon agroecological adaptation and end-use properties. It should be noted that the nomenclature can be confusing. Hard wheat in Europe can refer to durum wheat, whereas, soft wheat can refer to common wheat, regardless of its kernel hardness. Also, in countries where warm season crops such as maize (*Zea mays* L.) or sorghum (*Sorghum vulgare* L.) are not grown or expensive to import, wheat is often grown as an animal feed.

Breeding efforts within each group largely consist of matings of elite types from the same gene pool, with matings between wheats from alternative gene pools primarily made to transfer genes for pest and pathogen resistance. Wheat cultivars produced in the eastern region of the United States comprise soft red winter (SRW) and soft white winter (SWW) market classes. SWW wheat is grown predominately in Michigan and New York. In Michigan, historically, SWW wheats have been grown for their use in breakfast cereals. Millers and end users in southern states have shown interest in expanding SWW acreage because of their potential use in health food products, but concerns about sprouting and the ability to produce a consistently high quality product have largely stalled these efforts. Thus, most of the wheat produced east of the Mississippi river is SRW wheat. Much of the SRW wheat crop is milled domestically, while approximately one third is exported (http://www.ers.usda.gov/Data/Wheat/WheatYearbook.aspx#Trade, verified May 19, 2008).

Eastern wheat breeders have always contended with numerous fungal diseases that flourish in their typically high rainfall, humid target environments. Recently, however, this region has been dominated by Fusarium head blight or head scab (incited by *Fusarium* spp.) Since the mid 1990s, when the Northern US Corn Belt experienced a severe head scab epidemic, resistance has become a key breeding objective for many, if not all of the breeding programs in the eastern wheat region. Much of this breeding research has been facilitated by the US Wheat and Barley Scab Initiative (http://www.scabusa.org, verified May 19, 2008). Breeders have made much progress in developing resistant varieties since this initiative was undertaken. This has required use of scab-resistant Chinese spring wheats such as "Sumai 3" (Yu, 1982), "native" SRW resistance such as "Truman," and a combination of the two. The component of this disease that differentiates it from many other diseases in its potential to cause losses is the array of mycotoxins (primarily deoxynivalenol) produced by the fungus upon invasion of the plant (Bai and Shaner, 1994). Food and feed safety concerns have resulted in very low tolerances of this toxin in finished products, and flour millers have come to regard absence of deoxynivalenol as a primary quality target goal (Mesterházy et al., 1999). While millers and end users have always been concerned about having resistance to Fusarium head blight, the production of mycotoxins in scabby wheat has elevated this concern.

The eastern wheat region consists of the area from Louisiana to New York and eastern Kansas to the Atlantic coastal plain. In this region, cultivars are often licensed to companies with broad multistate sales targeting wide adaptation. Breeders identify broadly adapted wheat lines by entering candidate cultivars in two USDA-ARS regional nurseries grown annually at 30 locations each. Advanced and preliminary breeding lines are often entered in a host of collaborative nurseries that are organized by the breeders themselves.

Winter wheats of the PNW region typically have the longest grain-fill periods, and subsequently often are low in grain protein content. Because of periodic snow cover which insulates and protects the crowns, winter hardiness typically is less than that of Northern Great Plains wheats. Various quality types are produced, including soft white, club, hard white, and hard red. PNW wheats typically incorporate resistance to stripe and leaf rusts, eyespot (strawbreaker foot rot), powdery mildew [incited by *Blumeria graminis* (DC.) E.O. Speer], Cephalosporium stripe (incited by *Hymenula cerealis* Ellis and Everh.), and dwarf bunt (incited by *Tilletia controversa* Kühn in Rabenh).

As previously noted, VPM-1 resistance genes are heavily utilized in this region, primarily originating from the winter wheat cultivar Madsen (Allan et al., 1989). Precipitation is high and irrigation is widely practiced, although PNW wheats cultivated further from the Pacific coast, in regions such as Idaho, Utah, and Montana, receive less moisture and hence, have more drought tolerance.

Winter wheat is cultivated in the Great Plains of North America across a wide area, stretching from southern Texas north to South Dakota, and from the Missouri River Valley to the eastern slopes of the Rocky Mountains. Winter wheats also are grown along with spring wheats from South Dakota north to the prairies of Canada. Such a large geographic area represents an array of ecological habitats, and hence, the need for cultivars with diverse agronomic properties. Wheats of this region may be considered members of two broad gene pools, the Southern Great Plains (Texas, Oklahoma, Kansas, and Colorado) and the Northern Great Plains (Nebraska, South Dakota, North Dakota, Wyoming, and Montana). The Great Plains region generally is semiarid, and irrigation, while common, typically is used to produce higher value crops such as maize (*Zea mays* L.). Hence, the majority of the wheat in this region is cultivated under dryland (rainfed) environments, and drought tolerance is a trait of paramount importance. From the earliest cultivation of wheat in this area (~1870) until 1980, hard red winter wheat was the predominant type. In the early 1980s, breeders, commencing in Kansas, initiated the development of hard white winter wheats. As of this writing, breeders have released the hard white winter wheat cultivars Rio Blanco, Arlin, Trego, Betty, Heine, Lakin, Danby, Avalanche, Intrada,

Platte, Nuplains, NuFrontier, Antelope, and Arrowsmith, but hard red winter cultivars still predominate in the region.

Wheats of the Southern Great Plains gene pool often are bred for dual purpose, namely, grazing and grain production. Winters, especially in Texas and Oklahoma, are mild, and livestock producers typically will feed cattle on wheat pasture that grows continuously during the winter months. Depending on spring rainfall amounts and the price of grain, cattle may be removed in spring, and a grain crop harvested. Hence, wheats of this region typically are selected for regrowth potential after grazing. Other characteristics of Southern Great Plains wheats include resistance to leaf rust, tan spot, greenbug, Russian wheat aphid, and soil-borne mosaic virus. Southern Great Plains wheats also are early to break spring dormancy, and are early heading. They typically lack sufficient winter hardiness for cultivation north of the Kansas/Nebraska border. They also are nearly all semidwarf types. Since 2000, periodic outbreaks of stripe rust have plagued the Southern Great Plains, whereas, before, stripe rust rarely was seen in this region, as the rust pathogen typically did not multiply under the warm, arid conditions. However, the evolution of new races (Chen, 2005), not slowed by typical Great Plains environmental conditions, has led to periodic outbreaks in areas rarely exposed to this disease in the past. Breeding for resistance in the Southern Great Plains has become a priority for all breeding programs, since the majority of wheats adapted to this region are susceptible to new race(s).

While wheat is cultivated across a vast area in the Northern Great Plains, at present, there are five active breeding programs, three university based and two private. Mandatory traits of Northern Great Plains wheats include winter hardiness and resistance to stem rust. Snow cover in this region often is short lived; hence, winter wheats must be capable of surviving severe winter conditions. Other important traits include resistance to Hessian fly (*Mayetiola destructor* Say), leaf and stripe rust, and soil-borne mosaic virus. Wheat streak mosaic virus probably is the most commonly encountered disease in this region. To date, cultivars with effective resistance are just beginning to be deployed (e.g., RonL). However, breeding lines with both the *Wsm-1* resistance gene and acceptable agronomic performance have been identified (Divis et al., 2006), and the recent release of Mace indicates good progress is being made in developing cultivars with resistance to this pathogen. Two notable differences in the Northern Great Plains gene pool, as compared to the Southern Great Plains one, are the more frequent occurrence of tall or so-called "tall-semidwarf" wheat cultivars and of photoperiod sensitive wheat cultivars. In the drier western regions of the Northern Great Plains, tall wheats are established early, can be planted to moisture due to their longer coleoptile, perform well, and provide sufficient height for effective and efficient harvesting by combine (Budak et al., 1995). They also leave adequate straw for winter grazing, soil moisture retention, and erosion control. Examples of recently released tall-semidwarf wheats for this region include Millennium and Husker Genetics Brand Overland. There still remain considerable hectares of conventional tall wheats such as Buckskin and Scout 66, released several decades ago; there are also more recently released tall wheats, such as Pronghorn and Goodstreak. Photoperiod sensitivity is needed because of its epistatic interactions with vernalization genes to provide greater low temperature tolerance for a longer period of time that is needed in regions with longer winters (Mahfoozi et al., 2001).

4 Breeding for End Use Quality

Hard wheats traditionally have been used for the production of leavened bakery products, such as breads, rolls, bagels, and baguettes, and for many different types of Asian noodles. Great Plains breeding programs generally are designed to develop wheats for large-scale mechanized bakeries using sponge-and-dough technology and, hence, they aim to produce cultivars with strong tolerance to over-mixing. For use in US commercial bread baking, hard wheat cultivars should possess grain protein concentrations of ~130 g/kg, good mixing tolerance, adequate bread loaf volume, and should produce loaves with a good internal appearance. Flour water absorption should be high enough to meet product specifications. Tolerance to over-mixing is defined in commercial settings as the range of mix times above and below peak dough development, at which a given lot of flour will function properly in a baking procedure. This is an important variable, since in mechanized plants, under- or over-mixed doughs can result in poor final product or in a stoppage of production ("down-time"). Breeding programs are unable to precisely replicate the commercial definition of "tolerance," due to the larger sample sizes, long times, and bake evaluation resources necessary for its estimation. Wheat breeding programs have instead relied on small-scale measures of gluten strength, either via the sodium dodecyl sulfate (SDS) sedimentation test or via some type of physical dough testing procedure, most commonly the mixograph. In later generations, small-scale (~100 g flour) "pup-loaf" bake procedures are used. High small-scale loaf volumes, which likely represent the combined effects of high protein concentration and high gluten content, are the best predictors of overall commercial quality potential. Small-scale testing typically starts at the F_4 or F_5 generations, with early culling based on mixograph properties and grain protein concentrations. Baking tests generally are conducted each year on breeding lines at the F_6 through F_8 generations (Baenziger et al., 2001).

While numerous schemes have been proposed for the selection of wheats based either on high molecular weight (HMW) glutenin subunit composition (Payne, 1987) or on the direct measurement of glutenin content (Singh et al., 1990), such procedures have not gained favor in US hard wheat breeding programs. This situation partially derives from the relative high frequency of "optimal" subunits already present in US hard wheat gene pools (Graybosch, 1992), and the conclusion that essentially the same type of information is likely derived from simpler and less expensive tests.

The bread making quality of nearly all hard wheats improves with increased flour protein concentration. Increased protein concentration is somewhat amenable to improvement by breeding based on safe, accurate, and simple tests, such as near-infrared reflectance spectroscopy (NIR). Expression of protein concentration is, however, highly variable, with much of the overall variation being due to environment (Peterson et al., 1992). Genes conditioning increased protein concentration have been discovered, and, more importantly, there seem to be a number of independent loci. Thus, combining genes could be used to increase protein concentration. Two notable sources of higher flour protein are Atlas 66 and Plainsman V. These two sources seem to differ in their effects on dough strength. Atlas 66 types, such as Lancota, have shorter mix times, but good mixing tolerance and good loaf volume potential. Plainsman V types have long mixing times, strong tolerance to over-mixing, and high loaf volume potential. The Plainsman V source has been used to develop superior baking quality cultivars (Peterson et al., 1993), such as Jagger, Jagalene, Karl, Karl 92, Nuplains, and Wesley.

Interest in developing hard white wheat cultivars has required that additional quality screening procedures be added to breeding programs. Asian wet noodles require lower protein content and a more mellow gluten type than that used for bread production. Breeders now typically decide whether they should retain only those wheats with gluten types and protein contents suitable for making bread products or whether they should retain "dual purpose" white wheats with gluten types and protein contents suitable for both bread and noodle products. Fortunately, US hard white winter wheat cultivars developed and intended for bread end-use products are similar in protein content and quality to hard red winter wheat cultivars (Pike and MacRitchie, 2004). Product discoloration also is a concern, especially for fresh Asian noodles which are produced and sold within 48 h, but small-scale tests exist that can address this issue. Enzyme polyphenol oxidase (PPO) levels in the flour have been associated with the undesirable discoloration of many food products derived from white wheats. Thus, newly developed white wheats should have low or, if possible, negligible levels of PPO. Simple and nondestructive assays to screen seed for PPO levels have been developed (Bernier and Howes, 1994), and these are being applied in early generation selection. Low PPO hard white winter wheat cultivars released to date include Lakin and Platte. Flour starch swelling capacity is also important when developing white wheats for certain Asian noodles, particularly Japanese udon noodles. Cultivars lacking at least one gene for the production of granule-bound starch synthase (GBSS), called "partial waxy" cultivars, exhibit higher starch swelling and are desirable for the production of udon-type noodles (Graybosch, 1998; Hung et al., 2006). Breeders can select for this trait by testing for starch pasting properties using a Rapid ViscoAnalyser, the official designation or by testing for the absence of a GBSS gene using molecular markers (Zhang et al., 2008).

In general, hard red wheats are more tolerant to preharvest seed sprouting than hard white wheats (Mares and Ellison, 1989). Preharvest sprouting results in a release of degradative enzymes, especially amylases, with a concomitant loss in test weight, seed germination, and flour baking quality. Genetic variation does exist among white wheats for this trait, and white wheats with genes conditioning tolerance have been identified (Mares, 1992). In breeding programs, the falling number test, which provides an indirect measure of α-amylase activity, can be used in early generations to identify white wheats with preharvest sprouting tolerance. However, the falling number test may be used to segregate lines only if sprouting has already occurred. In years in which environmental conditions are not "favorable" for preharvest sprouting, irrigation systems might be needed to induce trait expression. Alternatively, germination tests on grain harvested at physiological maturity, and subsequently dried, have been used to select lines for resistance to preharvest sprouting (Wu and Carver, 1999). Rio Blanco, Danby, and Nuplains are hard white winter wheat cultivars which possess genes conditioning tolerance to preharvest sprouting.

Grain hardness probably is the single most important quality factor, as it ultimately governs the type of end-use quality applications for which a wheat cultivar may be used. Because of partitioning of gene pools based on grain characteristics in the United States, many programs rarely screen for hardness in early generations. In addition, the level of desired hardness is easily detected during milling for small-scale quality tests. The availability of the single-kernel hardness tester (Perten) makes hardness screening during early generations relatively easy in a breeding program. In addition, this instrument can determine seed weight and size, and more recent models have been outfitted with NIR technology for determining protein concentration. This instrument also calculates the deviation within a sample for each trait, which is useful information when one is attempting to develop more uniform cultivars. Uniformly hard or soft kernels are very important for milling and for the production of flour where traits such as starch damage impact end-use functionality.

For soft wheat, end products include cookies, crackers, cakes, other pastries and waffles, pretzels, soup thickeners, and biscuits. The diverse properties of these products require a wheat type which has soft kernel texture, does not exhibit preharvest sprouting, resists insect infestation, resists disease infection and mycotoxin development, and exhibits high milling yield and high test weight. For acceptable quality pastry products, medium protein content and strength, and low alkaline water retention are also important. For production of high quality crackers, pretzels, and flat breads, a wheat type would be required to have medium kernel texture, higher protein content and strength, and medium alkaline water retention. As markets continue to emerge for new products developed from SWW and SRW wheat types, the traditional need for weak gluten to assure large cookie diameter has been expanded to include the need for stronger gluten soft wheat types for the production of various crackers. Because it is often difficult to source wheat types with diverse end-use quality characteristics, SWW and SRW wheat cultivars developed for emerging products markets often are most successfully marketed strictly on an identity-preserved basis.

5 Breeding Methods

There are numerous self-pollinated breeding methods (e.g., pedigree, bulk, backcross, double-haploid, and single seed descent) that can be applied to winter wheat improvement, and these methods have been described in the necessary detail in various plant breeding textbooks. Suffice it to say, all of these methods have been used successfully by winter wheat breeders, and the breeding method choice will often depend upon a breeder's individual situation. For example, in countries with high labor costs (e.g., Europe, Japan, parts of North and South America), the bulk breeding method is often employed because it is very labor efficient and suited to the use of mechanical drills and harvesters. In countries with lower labor costs and with high equipment costs, the pedigree method is often used because it provides much more information on each selected line than does the bulk breeding method. The backcross breeding method is likely used when the objective is to add a single important trait to an existing adapted cultivar, such as when a new pathogen or pest strain arises or when a pathogen or pest expands its area of adaptation (e.g., the Russian wheat aphid entering the United States in 1985). Single-seed descent, though commonly used in spring wheat breeding because of its short generation time (does not require a vernalization period), tends to be used less frequently in winter wheat breeding. The double-haploid method is preferred for rapidly developing inbred lines in winter wheat. Either anther/microspore culture or the wheat × maize haploid induction system can be used. The wheat × maize haploid induction system is probably more commonly used because dependence on the wheat genotype used is less of a factor in creating double-haploids. However, all double-haploid breeding methods are relatively labor intensive and, hence, it is costly to use in many countries.

Though the common breeding methods have been described individually and in detail in textbooks (e.g., Acquaah, 2007), in practice, few are used on a strictly individual basis and without being combined with other methods. Most breeders use a "mix and match" approach where they might use one breeding method for an early generation objective and then change to a different method for a later generation objective. Common examples of this approach would be using a pedigree method to select plants with excellent disease or insect resistance (highly heritable traits in the early generations) and with good plant phenotype followed by using double-haploid or single seed descent to produce pure lines. This process eliminates many of the obviously unwanted types early, thus reducing the needed number of lines, before the labor intensive production and testing of pure lines is required. Similarly, bulk breeding could be used to eliminate spring growth habit types (again a highly heritable trait) and followed by practicing pedigree selection using the remaining winter hardy progeny.

Though often not discussed in detail, the generation in which the final selection is practiced in winter wheat breeding has very important ramifications on the uniformity of the resultant cultivar (Baenziger et al., 2006). The earlier the generation of final selection, the more likely it is that the line will have heterozygous loci which will lead to heterogeneity (a mixture of homozygous lines) in later generations. With each generation of self-pollination, heterozygosity is reduced and, hence, the amount of heterogeneity in later generations is also reduced. In countries, such as the United States, lines derived from early generations are tolerated because cultivar heterogeneity can be described as a cultivar characteristic for intellectual property protection (part of Plant Breeders' Rights). Consequently, many cultivars are F_3- or F_4-derived lines. However, in many other countries (particularly Europe) where cultivar uniformity (e.g., little heterogeneity) is required to obtain Plant Breeders' Rights, most final selections are made in later generations. The double-haploid breeding method, the use of which results in a completely homozygous line, has a distinct advantage in these countries.

As an outgrowth of the US Wheat and Barley Scab Initiative, the concept of regional genotyping laboratories was proposed and labs were established to foster marker-assisted selection in plant breeding (Van Sanford et al., 2001). The three USDA-ARS labs that serve winter wheat breeding programs in the United States are located in Manhattan, KS; Raleigh, NC; and Pullman, WA. These labs have provided public breeding programs the opportunity to carry out marker-assisted breeding in association with accelerated backcrossing and forward crossing of parents. Some of this research was ongoing in individual breeding programs, but the high throughput sequencers that are used for allele fragment analysis in these labs allow breeding programs to undertake large-scale projects. These projects often involve the collaboration of several breeding programs, and the size of these projects would otherwise preclude individual programs from undertaking them. An excellent example of a very large-scale collaborative project that involves the genotyping labs is the wheat Coordinated Agricultural Projects (CAP) Initiative (http://maswheat.ucdavis.edu/, verified May 19, 2008). In terms of winter wheat participation this involves at least ten breeding programs in as many different states.

6 Transgenic Wheats

Because of a lack of acceptance by consumers, no genetically modified (GM) or transgenic wheat cultivars have been deployed anywhere in the world. Nonetheless, experiments with transgenic wheat can be, and are, conducted. USDA-APHIS maintains a public database of all field trials with transgenic wheat (http://www.aphis.usda.gov/brs/database.html, verified May 19, 2008). According to this database, over 400 such trials have been completed in the United States. Transgenic phenotypes evaluated include those presumably expressing traits with enhanced gluten strength, resistance to *Fusarium* head blight, glyphosate resistance, starch modifications, increased grain lysine content, tolerance to heavy metals, resistance to take-all [incited by *Gaeumannomyces graminis* (Sacc.) Arx and D. Olivier var. *tritici* J. Walker], enhanced grain yield, male sterility, altered grain hardness, resistance to cyanamide, and altered carbohydrate metabolism. Whether these and additional traits will someday be available to wheat breeders remains more of a socioeconomic question rather than a scientific one. Ironically, many products made from other genetically engineered crops and organisms are nearly essential to the products made from wheat. For example, it would be virtually impossible to bake bread without transgenic soybean or cottonseed oil, high fructose corn syrup, and some of the added fortifying agents, which are produced by GM bacteria.

Similarly, the commonly used whey from milk might be produced from dairy cows protected by GM vaccines or given growth hormones to increase milk production.

7 Foundation Seed Production and Intellectual Property Issues

The last step in bringing a cultivar to market is increasing the amount of seed from that which is maintained as breeders' seed to commercial quantities. The intellectual property issues have been described in detail elsewhere (e.g., Baenziger et al., 2000), so, it is perhaps best to briefly mention the current practices and trends in intellectual property and seed increase. Historically, most cultivars that were protected were done so under national laws developed by nations to be in compliance with the International Union for the Protection of Varieties (http://www.upov.int/, verified May 19, 2008). For example, in the United States, the Plant Variety Protection Office (http://www.ams.usda.gov/Science/PVPO/PVPindex.htm, verified May 19, 2008) administers the Plant Variety Protection Act (PVPA). However, with more patented traits becoming available, many new cultivars are being protected by patents and by the PVPA. With continued interest in value-added traits and with transgenic wheat research continuing this trend of patent protection, in addition to PVPA, is likely to continue. Where there is dual ownership (e.g., one entity owns the patented trait and another entity owns the cultivar), additional intellectual property agreements, such as licenses to use traits, are involved. Furthermore, as cultivar development agencies become increasingly specialized, licenses for the increase of seed and sales are becoming more common. This trend is recognized in the case of publicly developed cultivars, which are frequently no longer fully marketed by the institutions that supported their development. Also, some countries with National Plant Breeders' Rights and who support collection of royalties for cultivars have reduced the incentive for the cultivar developer to have their own seed increase organization, at least one beyond the breeder or foundation seed classes.

For example, in the eastern soft winter wheat region, there has been a trend over the past 10 years to shift away from publicly released varieties to some sort of licensing (Whitt, 2005). Typically, agricultural experiment stations associated with public US institutions, will license marketing rights, either exclusively or nonexclusively depending on individual state seed laws. Most experiment stations have continued to release public cultivars, albeit at a much reduced frequency. Licensed cultivars generate royalties, and a percentage of these royalties are returned to the breeding programs to provide funding for continued research. Because experiment stations retain ownership and germplasm rights, this trend has seemingly not had an adverse impact on germplasm exchange in the region. However, from 2004 to 2006, some seed companies announced their intention to patent all of their varieties and to limit the extent to which these varieties can be used as parents in crosses. Because germplasm exchange among public wheat breeding programs is still vigorously maintained, this decision is expected to have a very limited impact in the eastern wheat production region of the United States.

The classes of seed during multiplication have various names, but most frequently they are known as breeder seed from which foundation seed is produced from which registered seed can be produced, from which certified seed is produced. Certified seed is sold to the producers. In the United States, the certification process can be found at the American Organization of Seed Certifying Agencies (AOSCA; http://www.aosca.org/, verified May 19, 2008). Each class has specific levels of purity and proximity to other cultivars (to avoid outcrossing) during the seed increase. Many certified seed producers will harvest the outer edge of the seed field as grain and the center for seed. Outcrossing is more likely at the edge of a field rather than in the center (where outcrossing will occur but the crosses will be sibmatings between plants of the cultivar rather than different cultivars). In addition, no matter how well a combine harvester is cleaned, there will always be some seed remaining in the combine that can mix with the next cultivar. By cleaning the combine and then harvesting the edge of the seed field as grain, the combine has effectively been "cleaned" by running the seed cultivar through the combine.

References

Acquaah, G. 2007. Principles of plant breeding and genetics. Blackwell, Malden, MA.

Allan, R.E., Peterson, C.J., Rubenthaler, G.L., Line, R.F., and Roberts, D.E. 1989. Registration of Madsen wheat. Crop Sci. 29(6): 1575.

Baenziger, P.S., Mitra, A., and Edwards, I.B. 2000. Protecting the value in value added crops: Intellectual property rights. In C.F. Murphy and D.M. Peterson (eds.) Designing crops for added value. American Society of Agronomy, Madison, WI, pp. 239–248.

Baenziger, P.S., Russell, W.K., Graef, G.L., and Campbell, B.T. 2006. Improving lives: 50 years of crop breeding, genetics and cytology (C-1). Crop Sci. 46: 2230–2244.

Baenziger, P.S., Shelton, D.R., Shipman, M.J., and Graybosch, R.A. 2001. Breeding for end-use quality: Reflection on the Nebraska experience. Euphytica 119: 95–100.

Bai, S., and Shaner, G. 1994. Scab of wheat: Prospects for control. Plant Dis. 78: 760–766.

Bernier, A.-M., and Howes, N.K. 1994. Quantification of variation in tyrosinase activity among durum and common wheat cultivars. J. Cereal Sci. 19: 157–159.

Brown-Guedira, G.L., Cox, T.S., Bockus, W.W., Gill, B.S., and Sears, R.G. 1999. Registration of KS96WGRC38 and KS96WGRC39 tan spot-resistant hard red winter wheat germplasms. Crop Sci. 39: 596.

Budak, N., Baenziger, P.S., Eskridge, K.M., Baltensperger, D., and Moreno-Sevilla, B. 1995. Plant height response of semidwarf and nonsemidwarf wheats to the environment. Crop Sci. 35: 447–451.

Chen, X.M. 2005. Epidemiology and control of stripe rust [Puccinia striiformis f. sp. tritici] on wheat. Can. J. Plant Pathol. 27: 314–337.

Cox, T.S., Sears, R.G., Gill, B.S., and Jellen, E.N. 1994. Registration of KS91WGRC11, KS92WGRC15, and KS92WGRC23 leaf rust resistant hard red winter wheat germplasms. Crop Sci. 34: 546–547.

Divis, L.A., Graybosch, R.A., Peterson, C.J., Baenziger, P.S., Hein, G.L., Beecher, B.B., and Martin, T.J. 2006. Agronomic and quality effects in winter wheat of a gene conditioning resistance to wheat streak mosaic virus. Euphytica 152: 41–49.

Friebe, B., Jiang, J., Raupp, W.J., McIntosh, R.A., and Gill, B.S. 1996. Characterization of wheat-alien translocations conferring resistance to diseases and pests. Current status. Euphytica 91: 59–87.

Gill, B.S., Friebe, B., Wilson, D.L., Martin, T.J., and Cox, T.S. 1995. Registration of KS93WGRC27 wheat streak mosaic virus-resistant hard red winter wheat germplasm. Crop Sci. 35: 1236–1237.

Graybosch, R.A. 1992. The high-molecular-weight glutenin composition of cultivars, germplasm and parents of U.S. red winter wheats. Crop Sci. 32: 1151–1155.

Graybosch, R.A. 1998. Waxy wheats: Origins, properties, and prospects. Trends Food Sci. Technol. 9: 135–142.

Graybosch, R.A. 2001. Uneasy Unions: Quality effects of rye chromatin transfers to wheat. J. Cereal Sci. 33: 3–16.

Hollenhorst, M.M., and Joppa, L.R. 1983. Chromosomal location of genes for resistance to greenbug in "Largo" and "Amigo" wheats. Crop Sci. 23: 91–93.

Hung, P.V., Maeda, T., and Morita, N. 2006. Waxy and high-amylose wheat starches and flours – Characteristics, functionality and application. Trends Food Sci. Technol. 17: 448–456.

Mahfoozi, S., Limin, A.E., and Fowler, D.B. 2001. Influence of vernalization and photoperiod responses on cold hardiness in winter cereals. Crop Sci. 41: 1006–1011.

Mares, D.J. 1992. Genetic studies of sprouting tolerance in red and white wheats. In M.K. Walker-Simmons and J.L. Reid (eds.) Pre-harvest sprouting in cereals 1992. American Association of Cereal Chemists, St. Paul, MN, pp. 21–29.

Mares, D.J., and Ellison, F.W. 1989. Dormancy and pre-harvest sprouting tolerance in white-grained and red-grained wheats. In K. Ringlund, E. Mosleth, and D.J. Mares (eds.) Fifth International Symposium on pre-harvest sprouting in cereals. Westview Press, Boulder, CO, pp. 75–84.

McIntosh, R.A., Dyck, P.L., and Green, G.J. 1976. Inheritance of leaf rust and stem rust resistances in wheat cultivars Agent and Agatha. Aust. J. Agric. Res. 28: 37–45.

Mesterházy, A., Bartók, T., Mirocha, C.G., and Komoróczy, R. 1999. Nature of wheat resistance to *Fusarium* head blight and the role of deoxynivalenol for breeding. Plant Breed. 118: 97–110.

Morris, C.F., and Konzak, K. 2001. Registration of hard and soft homozygous waxy wheat germplasm. Crop Sci. 41: 934–935.

Payne, P.I. 1987. The genetical basis of breadmaking quality in wheat. Aspects Appl. Biol. 15: 79–90.

Peterson, C.J., Graybosch, R.A., Baenziger, P.S., and Grombacher, A.W. 1992. Influence of genotype and environment on quality characteristics of hard red winter wheat. Crop Sci. 32: 98–103.

Peterson, C.J., Graybosch, R.A., Baenziger, P.S., Shelton, D.R., Worrall, W.D., Nelson, L.A., McVey, D.V., and Hatchett, J.H. 1993. Release of N86L177 wheat germplasm. Crop Sci. 33: 350.

Pike, P.R., and MacRitchie, F. 2004. Protein composition and quality of some new hard white winter wheats. Crop Sci. 44: 173–176.

Seah, S., Spielmeyer, W., Jahier, J., Sivasithamparam, K., and Lagudah, E.S. 2000. Resistance gene analogs within an introgressed chromosomal segment derived from Triticum ventricosum that confers resistance to nematode and rust pathogens in wheat. MPMI 13: 334–341.

Sears, E.R. 1953. Nullisomic analysis in common wheat. Am. Nat. 87: 245–252.

Singh, N.K., Donovan, R., and MacRitchie, F. 1990. Use of sonication and size-exclusion high-performance liquid chromatography in the study of wheat flour proteins. II. Relative quantity of gluten as a measure of breadmaking quality. Cereal Chem. 67: 161–170.

Souza, E. 1998. Host plant resistance to the Russian wheat aphid (Homoptera:Aphidae) in wheat and barley (pp. 122–147). In S.S. Quisenberry and F.B. Peairs (ed.) Response model for an introduced pest – The Russian wheat aphid. Thomas Say Publ. in Ento- mol.: Proc. Entomol. Soc. Am., Lanham, MD.

Souza, E., Smith, C.M., Schotzko, D., and Zemetra, R.S. 1991. Greenhouse evaluation of red winter wheats for resistance to the Russian wheat aphid. Euphytica 57: 221–225.

Whitt, D. 2005. Update on variety release procedures, branding and patents. Proc. of the Eastern Wheat Workers/Southern Small Grain Workers Conf. (C.S. Swanson, ed), May 9–12, Bowling Green, KY.

Wu, J., and Carver, B.F. 1999. Sprout damage and preharvest sprout resistance in hard white winter wheat. Crop Sci. 39: 441–447.

Van Sanford, D., Anderson, J., Campbell, K., Costa, J., Cregan, P., Griffey, C. Hayes, P. and Ward, R. 2001. Discovery and deployment of molecular markers linked to fusarium head blight resistance: An integrated system for wheat and barley. Crop Sci. 41: 638–644.

Yu, Y.J. 1982. Monosomic analysis for scab resistance and yield components in the wheat cultivar Soo-moo-3. Cereal Res. Commun. 10: 185–190.

Zhang, X.K., Liu, L., He, Z.H., Sun, D.J., He, X.Y., Xu, Z.H., Zhang, P.P., Chen, F., and Xia, X.C. 2008. Development of two multiplex PCR assays targeting improvement of bread-making and noodle qualities in common wheat. Plant Breed. 127: 109–115.

Triticale: A "New" Crop with Old Challenges

M. Mergoum, P.K. Singh, R.J. Peña, A.J. Lozano-del Río, K.V. Cooper, D.F. Salmon, and H. Gómez Macpherson

Abstract Triticale (X *Triticosecale*), a Man-made cereal grass crop obtained from hybridization of wheat (*Triticum spp*) with rye (*Secale cereale*). The hope was that triticale would combine the high yield potential and good grain quality of wheat, and the resistance/tolerance to the biotic and abiotic stresses of rye. Triticale grains can be used for human food and livestock feed. Since the last century, triticale has received significant attention as a potential energy crop. Today, research is currently being conducted includes the use of this crop biomass in bio-energy production. The aim of a triticale breeding programs mainly focuses on the improvement of economic traits such as grain yield, biomass, nutritional factors, plant height, as well as traits such as early maturity and high grain volume weight. Intense breeding and selection have made very rapid genetic improvements in triticale seed quality. The agronomic advantages and improved end-use properties of the triticale grains over wheat achieved by research and development efforts make triticale an attractive option for increasing global food production particularly, for marginal and stress-prone growing conditions. Details of the different breeding approaches utilized to enhance modern triticale cultivars for various uses are discussed in this chapter.

1 Introduction

Triticale (*X Triticosecale* Wittmack), a human-made crop, is a hybrid small grain produced between wheat and rye. The name "triticale" is an international crop name, with variations in pronunciation to suit the local language and dialect and is derived from the combination of the scientific classifications of the two genera involved, that is, wheat (*Triticum*) and rye (*Secale*). The triticale hybrids are all amphidiploid, which means the plant is diploid for two genomes derived from

M. Mergoum(✉)
North Dakota State University, Department of Plant Sciences, NDSU Dept. 7670, Po Box 6050, Fougo, ND 58108-6050

M.J. Carena (ed.), *Cereals*,
DOI: 10.1007/978-0-387-72297-9,

different species, in other words triticale is an allotetraploid. It is produced by doubling the chromosomes of the sterile hybrid that is produced using conventional plant breeding hybridizing techniques between wheat and rye. In earlier years most work was done on octoploid triticale; however, different ploidy levels have been created and evaluated over time. The octoploids showed little promise, but hexaploid triticale was successful enough to find commercial application.

Triticale cultivars, grown for forage as well as for grain, can be classified into three basic types: spring, winter, and intermediate (facultative). Spring types exhibit upright growth and produce much forage early in their growth. They are generally insensitive to photoperiod and have limited tillering. Winter types are generally planted in the fall, but also can be planted in the spring in some situations. Winter types have prostrate type of growth in the early stages of development. In general, winter types yield more forage than spring types mainly due to their long growth cycle. Intermediate (facultative) types, as the name implies, are intermediate to spring and winter types (Mergoum et al., 2004; Salmon et al., 2004).

Winter triticale differs from spring triticale because it requires vernalization to initiate heading. If winter types are spring seeded and there is no vernalization then the plants will remain vegetative and can be used for forage. Although the area under spring triticale acreage and the number of countries growing triticale has increased as a result of work conducted at the International Maize and Wheat Improvement Center (Centro Internacional de Mejoramiento de Maíz y Trigo, CIMMYT), Mexico, the majority of the world triticale acreage is still under winter types. While the CIMMYT program concentrated on spring and facultative triticales, an intensive effort in Poland rapidly progressed winter type triticale, and supplied breeding material around the world. This expanding area under winter triticale acreage includes mainly northern Europe and North America.

Although a relatively new crop, the history of triticale goes back to the late 1870s when the first crosses were attempted in Scotland. Detailed and fully referenced accounts of the fascinating history of triticale can be read in Ammar et al. (2004) and Oettler (2005). The first report describing the production of sterile hybrid plants between wheat and rye occurred in 1875 (Wilson, 1875). The first fertile, "true" triticale was produced by Rimpau in 1888, from crosses between *Triticum aestivum* and rye, followed by spontaneous chromosome doubling (Rimpau, 1891). Over the next 50 years, isolated experimentation and research occurred throughout Europe. It was not until the 1960s that the first commercial releases became available for producers. Commercially available triticale is almost always a second-generation hybrid, that is, a cross between two kinds of triticale (primary triticales). Generally, triticale combines the high yield potential and good grain quality of wheat with disease and environmental tolerance (including soil conditions) of rye. Triticale is, therefore, a crop which is particularly suited for marginal environments (acid- or drought-prone soils) or where disease pressure is high. Depending on the cultivar, triticale can more or less resemble either of its parents.

Current production is concentrated in Europe with nearly 90% of the world production and ~7 million acres harvested annually. US production is nearly 1 million acres, with the majority of the planted acres used for forage and pastures.

Production trends do show steady growth over the last 20 years with a 50% growth in production during the last decade. The leading producers of triticale worldwide are Germany, France, Poland, Australia, China, and Belarus. In 2005, according to the Food and Agriculture Organization (FAO), 13.5 million tons of triticale grain was harvested in 28 countries across the world. This is likely to be an underestimation as figures for Canada and the USA were not included.

2 Uses

Triticale, now a well-established crop internationally, is used for food, feed (monogastrics and ruminants), grazed or stored forage and fodder, silage, green-feed, and hay. In recent years, triticale has received attention as a potential energy crop and research is currently being conducted on the use of the crops biomass in bioethanol production. Interest in triticale has developed around two areas of potential use for the grain and its use as forage crop.

2.1 Feed Grain

The first area of interest is for use as a feed grain because it has proven to be a good source of protein, amino acids, and vitamin B. The protein content of triticale lines has ranged from 10% to 20% on a dry weight basis, which is higher than wheat. The amino acid composition of the protein is similar to wheat, but may be slightly higher in lysine. In addition, it is a better ruminant feed than other cereals due to its high starch digestibility. Results of feeding experiments indicate that pigs fed triticale-based diets had rates of gain and feed efficiencies similar to those of pigs fed corn-based diets. So, triticale has been found to be a palatable grain and can be used as either the partial or the sole grain source in diets for all classes of swine. Also, diets containing triticale grain are balanced to meet lysine rather than crude protein requirements (Myer, 2002; Boros, 2002). Modern triticale grain is an excellent feed grain for use in mixed poultry diets. Grain from modern triticale varieties has been reported to be comparable in energy value to other cereal grains for use in mixed diets of beef and dairy cattle, sheep, broilers and laying hens, and pigs and its protein is well utilized (Gursoy and Yilmaz, 2002; Myer and Lozano del Rio, 2004).

2.2 Food Grain

The second area of interest for triticale grain is in developing it as a food grain cereal that would exhibit unique baking traits. As a food grain, triticale has also been recognized as a hardy crop capable of helping combat world hunger. Triticale has potential in the production of bread and other food products such as pasta and breakfast cereals (Peña, 2004). The protein content is higher than that of wheat although the glutenin fraction is less. Assuming increased acceptance, the milling industry will have to adapt to triticale and develop milling techniques suited for triticales. While most of the varieties available are not suitable for leavened bread making on their own because of a weak and sticky gluten, they can be used in leavened products when blended with wheat flour. Triticale is suitable for producing a range of unleavened products such as cakes, cookies, biscuits, waffles, noodles, flour tortillas, and spaghetti (Skovmand et al., 1984). Triticale can be milled into flour using standard wheat or rye flour-milling procedures. Triticale cultivars possessing improved grain shape and plumpness produce flour yields equal or closer to those of wheat (Saxena et al., 1992). The low flour extraction rates commonly shown by soft triticale may be increased by milling wheat–triticale grain blends; mixing wheat–triticale at a 75:25 ratio prior to milling produces flour yields comparable to those of wheat milled alone (Peña and Amaya, 1992). Triticale has also been used alone or in blends with other cereal grains to manufacture high fiber snacks prepared by extrusion or by flaking triticale grains.

2.3 *Forage Crop*

Triticale has been and is increasingly grown for livestock grazing, cut forage (green chop), whole-plant silage, hay, and forage/grain dual purpose (Myer and Lozano del Rio, 2004). Triticale can be grown as a monocrop, winter/spring blend, and mixture with legumes, other cereal, and/or annual ryegrass. The advantage with blends is that the grazing season can be extended and/or forage nutritive value improved, in particular when blended with legumes.

In general, forage yield of triticale compares very favorably to other forage small-grain cereals in studies done all over the world (Varughese et al., 1996; Lozano et al., 1998). Research on the evaluation of triticale as a forage for ruminants has generally indicated comparative nutritive values to other forage cereal crops (Lozano et al., 1998). Spring triticale provides an excellent alternative to other spring cereals such as barley and oats. Spring triticale has been shown to be more drought tolerant than other spring cereals (Barary et al., 2002). Facultative and winter types are particularly well suited for grazing as they generally have a better distribution of forage over the growing season (Fig. 1).

The cutting and subsequent storage of triticale forage for silage is similar to that of any other small-grain forage. The best time to cut triticale for silage is in the boot to early-heading stage. Triticale cut earlier than the soft dough stage requires wilting in order to make high quality silage. Triticale like other small grains can provide a good source of hay when properly cut, cured, and baled (Fig. 2). For best results and quality, triticale should be cut between late boot and early heading stage.

Straw is an important by-product of triticale grain production and is often overlooked (Myer and Lozano del Rio, 2004). Triticale produces more straw than other small-grain cereals. Straw is frequently the only source of livestock feed

Fig. 1 Holstein heifers grazing AN-31, an intermediate-winter triticale, at Cuatrociénegas, Coahuila, México

Fig. 2 Haymaking on intermediate-winter triticale AN-31 at Ampuero Ranch, La Laguna region, México

in developing countries (Mergoum et al., 2004). Straw is also used in as construction material and even as a crop biomass source in alcohol or bioethanol production.

2.4 *Other Uses*

Triticale is important as a rotation crop for the reduction of soil pests, (e.g. nematodes), which may build-up on other crops. It can contribute to farmers' risk management, as triticale is likely to produce more bulk for grazing or haymaking in a drought year, and has relatively good grain retention in windy seasons. Triticale may be grown for environmental benefits because of its ability to capture soil nutrients and to reduce leaching into groundwater. Triticale can act as a soil improver, as its extensive root system binds erosion-prone soil and provides a good substrate for conversion into subsoil organic carbon by soil microbes (Salmon et al., 2004; Cooper, Per. Comm.). Although a new crop, the benefits of triticale production are enormous, and this is the reason for its adoption in more than 30 countries with an ever increasing acreage.

3 Genetics

Genetically triticale, an amphidiploid species resulting from a cross between wheat and rye, is self-pollinating (similar to wheat) and not cross-pollinating (like rye). This mode of reproduction results in a more homozygous genome. Cross-fertilization is also possible, but it is not the primary form of reproduction. The original or "primary" triticales are the fertile, true-breeding progenies of an intergeneric hybridization, followed by chromosome doubling between a seed parent from the genus *Triticum* and a pollen parent from the genus *Secale*. This makes it difficult to see the expression of rye genes in the background of wheat cytoplasm and the predominant wheat nuclear genome.

The great majority of today's triticales are descendants of primaries involving either common wheat (*Triticum aestivum* L., $2n = 42 = AABBDD$) or durum wheat (*Triticum durum* L., $2n = 28 = AABB$) as the seed parent and cultivated diploid rye (*Secale cereale* L., $2n = 14 = RR$) as the pollen parent. Hexaploid wheat-derived primaries, referred to as octoploid triticales ($2n = 56 = AABBDDRR$), were the first to be produced and extensively studied. However, in spite of very valuable breeding efforts during the first half of the twentieth century, they did not spread as cultivars to any substantial extent. Since the early 1950s, and to a greater extent during the last 40 years, the bulk of the breeding and research efforts has focused on developing and improving hexaploid triticales ($2n = 42 = AABBRR$), amphiploids originally made between tetraploid wheat and diploid rye. Hence, most of the currently available triticales are hexaploids due to their superior vigor and reproductive stability compared to the octoploid type.

What makes the history and evolution of triticale as a species so unique compared to that of wheat or other allopolyploids is that its evolution occurred during the last 130 years and its most dramatic evolutionary events (i.e. allopolyploidization as a result of intergeneric hybridization followed by chromosome doubling) were almost all directed by humans. Earlier "*in campo*" work, with wheat–rye crosses was difficult due to low survival of the resulting hybrid embryo and spontaneous chromosome doubling. To improve the viability of the embryo and thus avoid its abortion, in vitro culture techniques were developed. Colchicine was used as a chemical agent to double the chromosomes. These developments initiated the successful new era of triticale crop.

4 Early Triticale Breeding

The first triticale at the hexaploid level was reported in 1938, followed by others from various locations, for example, USA, Japan, Spain, Hungary, and the Russian Federation. Breeding efforts during the 1940s and early 1950s concentrated on the production and intercrossing of both octoploid and primary triticales, although these were of no commercial value. More intensive breeding programs with the explicit objective of developing triticale into a commercial crop were initiated during the 1950s in Spain (Sanchez-Monge, 1974), Canada (Shebeski, 1974), and Hungary (Kiss, 1974). These programs led to the release of the first cultivars, being Triticale numbers 57 and 64 in Hungary in 1968, followed by "Cachirulo" in Spain, and "Rosner" in Canada, both in 1969.

In 1953, the University of Manitoba, Winnipeg, Canada, began the first North American triticale breeding program working mostly with durum wheat–rye crosses. Early breeding efforts concentrated on developing a high yielding and drought-tolerant human food crop species suitable for marginal wheat-producing areas. Both winter and spring types were developed, with emphasis on spring types. Since Canada's program, other public and private programs have initiated both durum wheat–rye and common wheat–rye crosses. The major triticale development program in North America is now at CIMMYT in Mexico, with some private companies continuing triticale programs. The CIMMYT Triticale Improvement Program started in 1964 under the leadership of Dr. N.E. Borlaug, followed by Dr. F.J. Zillinsky in 1968 (Zillinsky and Borlaug, 1971). This program, in cooperation with the University of Manitoba, was funded initially by the Rockefeller Foundation. In 1971, the Government of Canada undertook complete funding of the CIMMYT Triticale Improvement Program. Breeding work diminished at the University of Manitoba and was later replaced by breeding and agronomic development programs of Alberta Agriculture, Food, and Rural Development (Field Crop Development Centre, Lacombe) and Agriculture and Agri-Food Canada (Swift Current and Lethbridge).

In the beginning, several major hurdles had to be overcome to tailor triticale to become a viable crop. Early triticales, though vigorous in growth habit, were extremely late, very tall, sterile, photoperiod sensitive, and possessed shriveled seeds. The first major breakthrough came by serendipity when a triticale plant resulting from a natural out-cross to unknown Mexican semi-dwarf bread wheat was selected in 1967. The selected line, designated "Armadillo," made a major contribution to triticale improvement worldwide since it was the first triticale identified to carry a chromosome substitution wherein the D-genome chromosome was substituted for the respective R-genome homeologue. Because of this drastic improvement in triticale germplasm, numerous cultivars were released, and the crop was promoted to farmers as a "miracle crop." By the late 1980s, data from international yield trials revealed that complete hexaploid triticale (AABBRR) was agronomically much superior to 2D(2R)-substituted hexaploid types, particularly under marginal growing conditions. Thereafter, triticale germplasm at CIMMYT was gradually shifted towards complete R-genome types to better serve these marginal environments. Today, CIMMYT is the principal supplier of improved spring triticale germplasm for many national and regional agricultural research systems around the world. The spring material also serves as an ancestral constituent of most winter triticales cultivars.

5 Achievements in Triticale Breeding

The last four decades of research on triticale initiated by CIMMYT in association with National Agricultural Research Systems (NARS) around the world have resulted in significant improvements of triticale crop. Triticale today is an international crop grown in more than 28 countries with the number of countries and the acreage under triticale production increasing. In 2003, triticale occupied nearly 3 million ha worldwide, compared to about 1 million ha in 1988 (Varughese et al., 1996; FAO, 2003). Results from many research and developments project demonstrate that triticale has potential as an alternative crop for different end-uses in a wide range of environments, particularly for marginal and stress-prone growing conditions.

5.1 Yield Increase

Major success in increasing the triticale yield has been attributed to research and development conducted at CIMMYT, Mexico. Under near optimal conditions at Ciudad Obregón, Mexico, a comparison of maximum-yield trials of triticale developed at CIMMYT revealed an average increase of 1.5% per year (Sayre et al., 1996) (Fig. 3). The genetic gain in yield potential was mainly due to a substantial increase in harvest index, grains/m^2, spikes/m^2, test weight, and a decrease in plant height.

Lodging resistance in triticale has been successfully introgressed using the dwarfing genes from both *Triticum* and *Secale* species. This has resulted in a

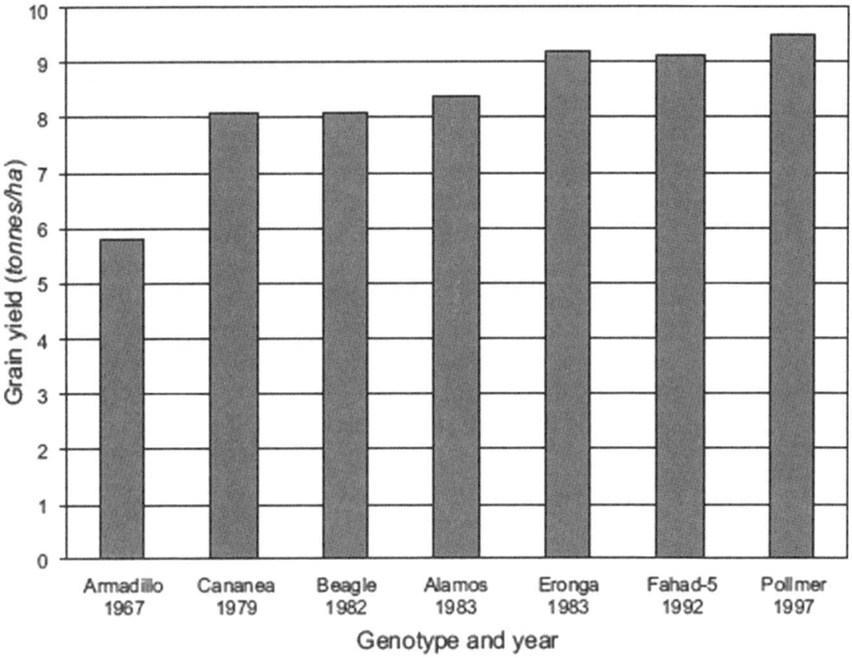

Fig. 3 Genetic enhancement in grain yield of spring triticale at CIMMYT. *Source:* Mergoum et al., 2004

decrease of up to 20 cm in plant height and increasing yield as the semi-dwarf cultivars are high yielding and more responsive to inputs. In 1968, at Ciudad Obregon, in northwest Mexico, the highest yielding triticale line produced 2.4 t/ha. Today, CIMMYT has released high yielding spring triticale lines (Pollmer-2) which have surpassed the 10 t/ha yield barrier under optimum production conditions (Hede, 2000).

5.2 Adaptation

Development of triticale cultivars expressing high and stable yields as a result of input efficiency and responsiveness, and resistance to a wide range of biotic and abiotic stresses, have resulted in increasing the acreage under triticale worldwide. Early maturity, a typical characteristic of modern triticale, allows escape from terminal developmental stresses, such as heat or frost, in highly productive environments, such as the irrigated subtropics and Mediterranean climates, which has contributed to triticale acceptance by farmers. Substituted or octoploid triticale attracted additional interest because of its R rye genome associated to high uptake of nutrients and D wheat genome associated to high efficiency of their metabolism

(Osborne and Rengel, 2002). Modern triticale cultivars have good tolerance of aluminum, which becomes increasingly available in acid soil conditions, and have good efficiency for accessing major nutrient (phosphorus) and trace elements (manganese, copper, zinc) in alkaline soils where these elements tend to become poorly available.

Many triticale cultivars have good water-logging tolerance, tolerance to periods of drought, and some Polish material has been demonstrated to have good tolerance to salinity (Koebner and Martin, 1996). Under marginal land conditions, where abiotic stresses related to environment (drought or temperature extremes) and soil conditions (extreme pH levels, salinity, toxicity, or deficiency of elements) are the limiting factors for grain production, modern triticale cultivars have consistently shown its advantages and has outperformed the existing cultivated cereal crops (Mergoum et al., 2004).

Since, triticale cultivation is similar to traditional cereal crops and it offers many more end-use alternatives for both humans and animals, triticale is more often grown in stressed environments with low input for grazing, grain, and straw production. In addition, triticale is grown as forage or for dual purposes involving grain and forage production. Research reveals an increase in the adaptation and successful production of triticale to stressed environments, particularly to water stress (Barary et al., 2002). Both successful breeding and management have resulted in acceptance of triticale as a major alternate crop to traditional cereal crops.

5.3 Enhanced Quality

Triticale breeding programs worldwide including CIMMYT have emphasized improving the product quality and developing triticale cultivars for specific end-uses. The most significant improvement was achieved for plumper grain. The test weight of the best Armadillo selections in 1970 at CIMMYT was 73.7 kg/hl compared to 65.8 kg/hl of the best line in 1968 (Zillinsky and Borlaug, 1971). Substantial progress has continued to improve test weight, and some modern triticales can reach 80 kg/hl under favorable environmental conditions (Mergoum et al., 2004). Since 1990, due to specific end-use and market requirements, more emphasis has been given to developing triticale for specific end-uses, such as milling and baking purposes, feed grain, dual purpose (forage and grain), and grazing types. Variability present in the triticale germplasm for preharvest sprouting and gluten quality has been exploited by breeders to develop cultivars with enhanced quality and sprouting resistance which has improved the bread-making qualities of triticale grain.

In general, winter triticale produces higher forage biomass than spring types. Therefore, their use for forage (grazing), cut forage, silage, and grain or hay has been improved through the release of several forage-specific cultivars. In addition, in many countries cereal straw is a major feed source for animals and in some years can have greater value than grain. Under arid and semiarid conditions, triticale has been shown consistently to produce higher straw yields than wheat and barley (Mergoum et al., 1992).

5.4 *Biotic Resistance*

Initially, biotic stresses did not appear to be a serious constraint to triticale production; however, as triticale acreage increased, most wheat and rye diseases started to infect triticale (Singh and Saari, 1991). In comparison with wheat, triticale appears to have good resistance to several common wheat diseases and pests including rusts (*Puccinia* sp.), Septoria complex, smuts (*Ustilago* and *Urocystis* sp.), bunts (*Tilletia* sp.), powdery mildew (*Blumeria graminis*), cereal cyst nematode (*Heterodera avenae*), and Hessian fly (*Mayetiola destructor*). It also resists virus diseases, such as barley yellow dwarf, wheat-streak mosaic, barley-stripe mosaic, and brome mosaic (Varughese et al., 1996; Skovmand et al., 1984). However, it gets disease in common with other cereals, but there is considerable varietal variation in terms of disease resistance. Triticale has relatively greater susceptibility than wheat to diseases such as spot blotch (*Bipolaris sorokiniana*), scab (*Fusarium* sp.), and ergot (*Claviceps purpurea*) and bacterial diseases caused by *Xanthomonas* sp. and *Pseudomonas* sp., which preclude the immediate commercial introduction of triticale in those areas where wheat is otherwise better adapted (e.g. Zambia and Brazil) (Skovmand et al., 1984). In the past, susceptibility to ergot was a major limitation to triticales expansion. However, its susceptibility in the past was linked to problems with floret sterility, and ergot is not seen as a major problem in current varieties.

The reaction of triticale to many diseases and pests meets the expectations of a combined resistance found in the two parental species. The disease and insect resistance reactions of one or the other of the parents is reflected in triticale progeny, or the reaction of triticale is intermediate between that of wheat and rye, as in the case of take-all (*Gaeumannomyces graminis*) and Russian wheat aphid (*Diuraphis noxia*). There is some evidence that triticale varieties vary in their resistance depending on the number of rye-genetic material (chromosomes) present with varieties that have a greater number of rye chromosomes having greater resistance to take-all (Wallwork, 1989). The considerable variability present in triticale germplasm for different diseases and pest is being exploited by breeders to develop durable resistant cultivar which has resulted in wide-scale production of triticale worldwide.

6 Breeding Strategies

For long-term success, a strategy for crop enhancement that emphasizes the maintenance and generation of genetic diversity, while carefully balancing diversity objectives required to ensure long-term progress with the relatively narrower frequency of favorable alleles necessary to achieve short-term breeding goals, is necessary. In triticale, the lack of genetic diversity may be overcome by the varied spectrum of human-introduced diversity. For example, spring and winter wheat and rye gene pools are accessed through direct interspecific (wheat × triticale) and intraspecific (winter triticale × spring triticale) crosses. Octoploid × hexaploid triticale crosses guarantee an influx of cytoplasmic variability. In addition, genetic variability can also be achieved by producing new triticales wherein some chromosomes from the R genome have been replaced by some from the D genome. Many modern triticale lines developed from such crosses carry D(A), D(B), and D(R) whole chromosome substitutions or chromosome translocations which add valuable traits to triticale. Furthermore, results from CIMMYT International Triticale Yield Nurseries suggest adaptive advantages of complete triticale carrying a 6D(6A) substitution (Mergoum et al., 2004).

Genetic traits from related wheat species are transferred into triticale via wheats carrying alien introgressions. Additional introgression involves the crossing of closely related plant relatives and results in the transfer of "blocks" of genes. Genes located in the proximal areas of chromosomes may be linked thus preventing or severely hampering genetic recombination which is necessary to incorporate the desirable genes. A weak colchicine chemical solution has been employed to increase the probability of recombination in the proximal chromosome regions and thus the introduction of the translocation to that region. The resultant translocation of smaller blocks that indeed carry the gene(s) of interest has decreased the probability of introducing unwanted genes. The optimal chromosomal constitution of triticale, the makeup of homeologous AA, BB, DD, and RR chromosomes or chromosome arms, has yet to be defined, and unique optimal chromosomal configurations for the diverse agroecological zones and end-uses are likely to emerge. In addition, unique combinations are being attempted in hybrid triticale.

The development of populations with specific traits facilitates the combination of desirable traits from different unadapted genotypes with adapted germplasm. Recently, more emphasis has been directed towards improving certain triticale agronomic traits, including grain-filling duration and rate, earliness and tillering capacity, but quality parameters have also to be addressed, such as test weight, protein content, and gluten strength enhancement (Boros, 2002). Although triticale has shown good resistance to most prevalent diseases and insects in most cereal-growing areas, with the spread of this crop and the race specialization of pests, triticale has become vulnerable to certain diseases or insects.

Breeding for the abiotic stresses of marginal lands (acid, sandy, or alkaline soils), trace element deficiencies (copper, manganese, and zinc), or trace element toxicity (high boron), and the different types of moisture stresses will still constitute a major effort in spring and winter/facultative triticale improvement worldwide. This can be achieved by exploiting key locations during selection, screening, and yield testing and through shuttle-breeding involving NARSs (e.g. Brazil for acid soils and sprouting and Morocco for terminal drought and sandy soils) (Mergoum et al., 2004). Further improvements, particularly in grain plumpness, grain color (white or amber), and gluten quantity and quality, are expected to make triticale more attractive as a food grain.

Triticale is a self-pollinated crop although some degree of cross-pollination is possible. Modern triticale-breeding programs follow different breeding strategies in order to develop superior triticale cultivars. Traditional breeding methodologies applicable for self pollinated crops like "Backcrossing selection," "Pedigree selection," "Bulk selection," and "Single Seed Descent" are followed in combination with breeding strategies like recurrent selection and hybrid triticale, which are predominantly followed in a cross-pollinated crop. In addition, modern breeding approaches involving shuttle breeding, double haploidy, marker-assisted selection (MAS), and genetic transformations are performed in most triticale-breeding programs. The following breeding strategies in combination with traditional breeding method are likely to further enhance the gains in triticale breeding.

6.1 Shuttle Breeding

Shuttle breeding, pioneered at CIMMYT, Mexico, was originally used to speed up the wheat breeding process by advancing and testing breeding material, at contrasting environments, has been successfully adopted in triticale breeding. Higher success in shuttle breeding is observed due to the exposure of the breeding germplasm to contrasting disease spectra, soil types, photoperiod length, and diverse environmental conditions. The success of shuttle breeding in triticale resulted in development in high yielding-adapted triticale cultivars worldwide and their large scale production in a relatively small time. Collaboration and sharing of germplasm among triticale breeders, due to participation of multiple triticale breeding centers in shuttle breeding, have further helped in development of superior triticale cultivars (Pfeiffer, 1995).

6.2 Hybrid Triticale

Based on the commercial success of other hybrid crops, the use of hybrid triticales as a strategy for enhancing yield in favorable as well as marginal environments has proven successful over time. Earlier research conducted by CIMMYT made use of a chemical hybridizing agent (CHA) in order to evaluate heterosis in hexaploid triticale hybrids (Mergoum et al., 2004). To select the most promising parents for hybrid production, testcrosses conducted in various environments are required. This is because the variance of their specific combining ability (SCA) under differing environmental conditions is the most important component in evaluating their potential as parents to produce promising hybrids.

Commercially exploitable yield advantages of hybrid triticale cultivars are dependent on improving parent heterosis and on advances in inbred-line development. Yield improvements of up to 20% have been achieved in hybrid triticale cultivars (Oettler et al., 2001, 2003) The identification of good combining ability or "heterotic groups" at an early stage in the breeding program can reduce the costs associated with "carrying" a large number of plants through the program and thus forms part of efficient selection.

6.3 Double Haploids

Double haploid (DH) plants have the potential to save much time in the development of inbred lines. This is achieved in a single generation in DH as opposed to many in case of traditional breeding methods which would otherwise occupy much physical space/facilities. Various techniques exist to create DHs. The androgenesis techniques involving in vitro culture of anthers and microspores is most often used in triticale (Bernard and Charmet, 1984; González and Jouve, 2000). Many cultivars within triticale are recalcitrant in that the success rate of achieving whole newly generated plants is very low. Genotype × culture–medium interaction is responsible for varying success rates, as is a high degree of microspore abortion during culturing (Gonzalez and Jouve, 2005). It is also known that the response of parental triticale lines to anther culture is correlated to the response of their progeny (González et al., 1997). Such information further help in optimizing the DH production process.

Chromosome elimination is another method of producing DHs and involves hybridization of triticale with maize (*Zea mays* L.) followed by auxin treatment and the artificial rescue of the resultant haploid embryos before they naturally abort. This technique is unfortunately less successful in triticale. However, *Imperata cylindrica* (a grass) was found to be just as effective as maize with respect to the production of DHs in both wheat and triticale (Pratap et al., 2005).

6.4 Marker-Assisted Selection

MAS is a form of indirect selection for a given trait which is becoming an important component of modern plant breeding. Triticale has not been well characterized with respect to molecular markers; although, an abundance of rye and wheat molecular markers are available and makes it possible to track segments/genes thereof within a triticale background. It is generally accepted that molecular markers are better predictors than morphological markers (agronomic traits) due to their insensitivity to variation in environmental conditions.

Comparative genome mapping has revealed a high degree of similarity in terms of sequence colinearity between closely related crop species. This allows the exchange of such markers within a group of related species such as wheat, rye, and triticale. One study established a 58% and 39% transferability rate to triticale from wheat and rye, respectively (Kuleung et al., 2004). "Transferability" refers to the phenomenon where the sequence of deoxyribonucleic acid (DNA) nucleotides flanking the simple sequence repeat (SSR) loci is sufficiently homologous or similar between genomes of closely related species. Thus DNA primers designed for one species can be used to detect SSRs in related species. Hence, there is great potential that SSR markers available in wheat and rye can be successfully used for triticale (Kuleung et al., 2004). Yong (2001) performed an amplified fragment length polymorphism (AFLP) analysis in order to identify 2RL-specific markers in a wheat–rye translocation line (2BS/2RL) developed for resistance to biotype L. of Hessian fly. He developed a sequence-tagged site (STS) primer for diagnostics of 2RL (SJ07), applicable in wheat, rye, and triticale.

6.5 Genetic Transformation

The genetic transformation of crops referred to as GMO (genetically modified organisms) involves the incorporation of "foreign" genes or, rather, very small DNA fragments compared to introgression discussed earlier. Amongst other uses, transformation is a useful tool to introduce novel traits/characteristics into the transformed crop. These novel traits/genes may be coming from any donor species which may not be transferred using conventional breeding due to lack of sexual compatibility between the two species. Two methods have been attempted to transform triticale, that is, *Agrobacterium*-mediated trasnformation and biolistics transformation. Triticale has been transformed via biolistics with a 3.3% success rate (Zimny et al., 1995), while a recent study by Nadolska-Orczyk et al. (2005) reported that *Agrobacterium*-mediated transformation was possible, however, the success rate was very low.

7 Future Challenges

Scientists and producers are interested in triticale because it is well adapted to harsh environmental conditions of high elevation, acid soil, salinity and aluminum toxicity, drought, and waterlogged soils (Mergoum et al., 2004). Triticale also has greater tolerance to common wheat diseases than wheat (Horlein and Valentine, 1995). Triticale grain also is high in essential amino acids, which makes it more nutritionally valuable than wheat, although the baking quality is inferior to that of bread wheat (Horlein and Valentine, 1995). Therefore, triticale is a promising crop and a valuable genetic resource for transferring ("bridge") desirable genes, particularly disease-resistance genes, from rye to wheat.

7.1 Adaptation

The relatively low adoption of triticale by farmers in several countries is in contrast with encouraging international nursery data, cultivar releases, and reports from NARS scientists and on-farm data, which indicate the high production potential of triticale, particularly for small farmers in marginal environments. However, there are several transitory social- and economic-related issues that limit triticale expansion in many countries. Improvements in several economic and biological traits are required in order to tailor this crop to fit farmers' needs and market requirements.

Triticale is the only man-made crop and just over 130 years of breeding, it is still in the process of evolving not only as a species but also in its utilization. Based on the genomes that contribute to triticale the crop has the potential to have all of the genetic variability that exists in the parental species, and it is relatively easy to introgress new genetic material from the parental species through the use of conventional pollination techniques.

In areas where opportunistic organisms such as ergot (*Claviceps purpurea*) are not a problem, hybrid triticale is feasible. Programs in CIMMYT (Ammar et al., 2006), Australia (Darvey et al., 2006), and Europe (Warzecha and Salak-Warzecha, 2006) as well as in other countries are working on the development of hybrid systems that are expected to provide yield advantages of up to 15% compared to conventional triticale. To date the complex nature of the triticale genome, which carries some minor partial restoration genes, has made it difficult to develop stable male sterile parents and maintainers. Once these problems are worked out, hybrid production system will not only provide increased yield for use in food, feed, fodder, and industrial applications but will also be of value to the developer for controlling seed distribution.

7.2 Uses

In the early stages of development during the last century a considerable amount of effort was placed on utilizing the adaptation of triticale to high stress environments to develop a crop for the direct production of food. However, triticale has become an important feed and fodder crop in most parts of the world where significant numbers of hectares are grown. Although triticale may not as yet have reached the goal of being a significant direct human food, animal feed grains and fodder are essential for the production of livestock which are a food source for humans. Consequently, triticale is in reality becoming a crop which can positively impact the world food supply. In addition to the more conventional uses, triticale is also being considered as a feedstock for ethanol production and in "Biorefining" and "Molecular Farming."

The current trend in the development of renewable fuels such as ethanol from grain or straw and plant biomass is in the most part based on sugarcane, maize, and sugarbeet. Triticale as a small-grained cereal is an excellent source of starch (Boros, 2006) or cellulose for ethanol production in areas were production of the most common sources of ethanol is not feasible, such as in the extreme northern and southern latitudes, higher altitudes, and other high stress environments. These are also areas where wheat and barley is grown, but the yield advantage and stability of triticale over wheat, under a range of environments, makes it a valuable source of feedstock for the fermentation process.

Another area in biorefining that triticale has a potential fit is the production of fiber from whole plant biomass or from straw as a harvest byproduct. "The Canadian Triticale Biorefinery Initiative" (Eudes, 2006), as well as similar programs elsewhere, plans to look not only at the potential of conventionally produced triticales but also at the genetically engineered triticale intended for purely industrial purposes. Given the concerns with development of GMOs this includes a detailed evaluation of biosafety issues related to potential contamination of food related and nonindustrial crops prior to any subsequent field production.

It is expected that the production of triticale both in the hectares grown and yield potential will continue to increase. The additional use of the crop for grain and biomass in renewable fuels and other industrial products as well as "Molecular Farming" has significant potential. Unfortunately, the land base available for crop production is finite, and changes in the environment such as "Global Warming" could have a serious impact on current crop production. Alternative uses for crops such as triticale will potentially compete with the production of human food and definitely will compete with the livestock feed and fodder supply.

7.3 Genetic Diversity

The large-scale triticale development program at CIMMYT, Mexico, has done much to increase the knowledge base on triticale and provide a range of triticale germplasm with wide adaptation to a range of environments. However, the genetic diversity in current programs is extremely narrow (Mergoum et al., 2004). This is not unexpected since the process of variety selection in a breeding program by its nature narrows the genetic base and not all regional/national breeding programs have the resources to conduct extensive germplasm development. New international initiatives in variety and germplasm development are expected to have a major impact on further increases in winter and spring triticale germplasm and variety development through collaborative work in conjunction with international germplasm programs such as CIMMYT.

7.4 Genomics

Genomics by definition involves the identification of all genes in a particular genome which in the case of hexaploid triticale involves the A, the B, the R, and portions of the D genomes. The production of genome maps in cereals allows for the development of systems such as genetic markers that can be used in MAS. Although a significant amount of work has been done in wheat and rye, more is required in triticale in order to effectively utilize MAS. Xue-Feng et al. (2002) found that genome modification or evolution is occurring in triticale since many of the RFLP bands normally found in rye are not expressed in triticale and that if the bands were found in both parents the expression could be up to three times greater than in the parents. Further research in this area may assist in understanding why traits from the parental species are either incompletely or not expressed in triticale.

There is no doubt that the use of genetic markers in triticale will have an important impact on continued development and selection of superior types. Although initially triticale tended to show higher levels of disease resistance than more commonly cultivated cereal species, the increase in area grown to triticale has resulted in triticale becoming susceptible to a range of diseases under increased disease pressure. This is exemplified by the susceptibility of triticale in Belgium (Haesaert et al., 2006) to powdery mildew (*Erysiphe graminis* f. sp. *tritici*), a pathogen, which triticale was highly resistant. The development of markers for disease resistance will have a significant impact combined with other selection criteria.

7.5 Health Issues

Triticale has not as yet become a staple food in the form of leavened bread, as originally intended when work was first initiated on the crop, but it does perform well in the production of unleavened products. The primary concern in these products has been color which does not result in the same appearance as traditional products (Bakhshi et al., 1998). Several programs have been developing hard white/ambers versions of triticale. Despite the fact that the uptake of triticale as a human food has been somewhat disappointing, triticale has excellent nutritional quality. Early studies have indicated that triticale has excellent vitamin balance and excellent amino acid content (Villegas et al., 1970). Other factors such as fiber and lipid quality have an important impact of human dietary health.

Salmon et al. (2002) discussed the importance of the high dietary fiber in triticale. High fiber content in food helps regulate blood glucose which is an advantage for individuals who are diabetic. Insoluble fiber assists in maintaining colon health and soluble fiber assists in reducing glucose release and absorption there by controlling blood cholesterol. In the same study the lipid content of triticale was found to be significantly higher than in wheat. This is very important since components of the lipid fraction can have antioxidant activity and result in metabolic regulation of cholesterol.

The utilization of triticale in a continually widening range of application bodes well for the expansion of the crop. Significantly, more work is required to take advantage of the nutritive health benefits as well as continuing the risk management aspects of food safety such as mycotoxins from organisms such as *Fusarium* species which appear to attacking small-grain cereals on a world-wide basis.

References

Ammar, K., Crossa, J. and Pfieffer, W.H. (2006). Developing a hybrid seed production system and evaluation of heterosis in hybrids from CIMMYT's spring triticale germplasm. In: Botes W. (ed.). Proc. 6th Int. Triticale Symp. September 3–7, 2006. Stellenbosch, SA.

Ammar, K., Mergoum, M. and Rajaram, S. (2004). The history and evolution of triticale. In: Mergoum, M. and Gómez-Macpherson H. (eds.). Triticale improvement and production. FAO Plant Production and Protection Paper No. 179, pp 1–9.

Bakhshi, A.K., Bakhshi, R., Saxena, A.K. and Dhindsa, G.S. (1998). Suitability of amber colored triticales in Indian ethnic food items. In: Juskiw P. (ed.). Proc. 4th Int. Triticale Symp. July 26–31, 1998. Red Deer, AB, Canada. pp 92–101.

Barary, M., Warwick, N.W.M., Jessop, R.S. and Taji, A.M. (2002). Osmotic adjustment and drought tolerance in Australian triticales. In: Arseniuk E. (ed.). Proc. 5th Int. Triticale Symp. June 30–July 5, 2002, Radzikow, Poland. pp. 135–141.

Bernard, S. and Charmet, G. (1984). Diallel analysis of androgenetic plant production in hexaploid triticale (X *Triticosecale*, Wittmack). Theor. Appl. Genet. 69, 55–61.

Boros, D. (2002). Physico-chemical quality indicators suitable in selection of triticale for high nutritive value. In: E. Arseniuk, ed. Proc. 5th Int. Triticale Symp., June 30–July 5, 2002, Radzikow, Poland, pp. 239–244.

Boros, D. (2006). Triticale of high end-use quality chance to increase its value on world cereal markets. In: Botes W. (ed.). Proc. 6th Int. Triticale Symp. September 3–7, 2006. Stellenbosch, SA.

Darvey, N.L., Zhao, X. and Trethowan, R. (2006). Isolated microspore culture in a cyclical breeding system for the production of inbred lines and hybrids. In: Botes W. (ed.). Proc. 6th Int. Triticale Symp. September 3–7, 2006. Stellenbosch, SA.

Eudes, F. (2006). Canadian triticale biorefinery initiative. In: Botes W. (ed.). Proc. 6th Int. Triticale Symp. September 3–7. Stellenbosch, SA.

FAO. (2003). FAOSTAT, FAO statistical datab agriculture. Available at http://apps.fao.org.

González, J.M. and Jouve, N. (2000). Improvement of anther culture media for haploid production in triticale. Cereal Res. Commun. 28, 65–72.

Gonzalez, J.M. and Jouve, N. (2005). Microspore development during *in vitro* androgenesis in triticale. Biologia Plantarum 49, 23–28.

González, J.M., Hernández, I. and Jouve, N. (1997). Analysis of anther culture response in hexaploid triticale. Plant Breeding 116, 302–304.

Gursoy, U. and Yilmaz, A. (2002). Determination of energy values and digestibility characteristics of triticale varieties. J. Anim. Sci. 80, 396.

Haesaert, G., Derycke, V., Latre, J. and Hermans, B. (2006). Evolution of disease pressure on triticale under Belgian growing conditions: overview of the last 20 years. In: Botes W. (ed.). Proc. 6th Int. Triticale Symp. September 3–7, 2006. Stellenbosch, SA.

Hede, A.R. (2000). A new approach to Triticale improvement. Available at http://www.cimmyt.org.

Horlein, A. and Valentine, J. (1995). Triticale (*X Triticosecale*). In: Williams J.T. (ed). Cereals and pseudocereals. Chapman and Hall, New York, NY. pp 187–221.

Lozano, A.J., Zamora, V.M., Diaz-Solis, H., Mergoum, M. and Pfeiffer, W.H. (1998). Triticale forage production and nutritional value in the northern region of Mexico. In: Proc. 4th Int. Triticale Symp.Vol. 2. Alberta Agric. Rural Dev. Ctr., Lacombe, AB, Canada. p. 259.

Kiss, A. (1974). Triticale breeding experiments in Eastern Europe. In: MacIntyre R. and Campbell M. (eds.). Triticale: Proc. Int. Symp. 1973. International Development Research Centre, El Batan, Mexico; Ottawa, Ont., Canada. pp. 41–50.

Koebner, R.M.D. and Martin, P.K. (1996). High levels of salt tolerance revealed in triticale. In: H. Guides-Pinto et al., eds. Proc. 3rd Int. Triticale Symp., June 1994, Lisbon, Portugal, pp. 429–436.

Kuleung, C., Baenziger, P.S. and Dweikat, I. (2004). Transferability of SSR markers among wheat, rye, and triticale. Theor. Appl. Genet. 108, 1147–1150.

Mergoum, M., Pfeiffer, W.H., Peña, R.J., Ammar, K. and Rajaram, S. (2004). Triticale crop improvement: the CIMMYT programme. In: Mergoum, M. and Gómez-Macpherson H. (eds.). Triticale improvement and production. FAO Plant Production and Protection Paper No. 179. Food and Agriculture Organization of United Nations, Rome. pp. 11–26.

Mergoum, M., Ryan, J. and Shroyer, J.P. (1992). Triticale in Morocco: potential for adoption in the semi-arid cereal zone. J. Nat. Res. Life Sci. Edu. 21, 137–141.

Myer, R.O. (2002). Triticale grain in young pig diets. In: Proc. Fifth Internat'l Triticale Symposium. Vol. I. Plant Breeding and Acclimatization Inst., Radzikow, Poland. p 272.

Myer, R. and Lozano del Rio, A.J. (2004) Triticale as animal feed. In: Mergoum, M. and Gómez-Macpherson H. (eds.). Triticale improvement and production. FAO Plant Production and Protection Paper No. 179. Food and Agriculture Organization of United Nations, Rome. pp. 49–58.

Nadolska-Orczyk, A., Przetakiewicz, A., Kopera, K., Binka, A. and Orczyk, W. (2005). Efficient method of *Agrobacterium*-mediated transformation for triticale (*X Tritosecale* Wittmack). J. Plant Growth Regul. 24, 2–10.

Oettler, G. (2005). The fortune of a botanical curiosity-Triticale: past, present and future. J. Agric. Sci. 143, 329–346.

Oettler, G., Becker, H.C. and Hoppe, G. (2001) Heterosis for yield and other agronomic traits of winter triticale F_1 and F_2 hybrids. Plant Breed. 120, 351–353.

Oettler, G., Burger, H. and Melchiner, A.E. (2003). Heterosis and combining ability for grain yield and other agronomic traits in winter triticale. Plant Breed. 122, 318–321.

Osborne, L.D. and Rengel, Z. (2002) Screening cereals for genotypic variation in efficiency of phosphorus uptake and utilization. Aust. J. Agric. Res. 53, 295–303.

Peña, R.J. (2004). Food uses of triticale. In: Mergoum, M. and Gómez-Macpherson H. (eds.). Triticale improvement and production. FAO Plant Production and Protection Paper No. 179. Food and Agriculture Organization of United Nations, Rome. pp. 37–48.

Peña, R.J. and Amaya, A. (1992). Milling and bread making properties of wheat-triticale grain blends. J. Sci. Food Agric. 60, 483–487.

Pfeiffer, W.H. (1995). Triticale breeding at CIMMYT. In: Rajaram S. and Hettel G.P. (eds.). Wheat breeding at CIMMYT: commemorating 50 years of research in Mexico for global wheat improvement. Wheat Special Report No. 29, Mexico, DF, CIMMYT. pp. 87–95.

Pratap, A., Sethi, G.S. and Chaudhary, H.K. (2005). Relative efficiency of different gramineae genera for haploid induction in triticale and triticale x wheat hybrids through the chromosome elimination technique. Plant Breeding 124, 147–153.

Rimpau, W. (1891). Kreuzungsprodukte landwirthschaftlicher Kulturpflanzen. Landwirthschaftlicher Jahrbűcher 20, 335–371.

Salmon, D.F., Mergoum, M. and Gómez Macpherson, H. (2004). In: Mergoum M. and Gómez-Macpherson H. (eds.). Triticale production and management. FAO Plant Production and Protection Paper No. 179. pp. 27–36.

Salmon, D.F., Temelli, F. and Spence, S. (2002). Chemical composition of western Canadian triticale varieties. In: Arseniuk E. (ed.). Proc. 5th Int. Triticale Symp. June 30–July 5, 2002. Radzikow, Poland. pp. 445–450.

Sanchez-Monge, E. (1974). Development of triticales in western Europe. In: MacIntyre R. and Campbell M. (eds.). Triticale: Proc. Int. Symp. 1973. International Development Research Centre, El Batan, Mexico; Ottawa, Ont., Canada. pp. 31–39.

Saxena, A.K., Bakhshi, A.K., Sehgal, K.L. and Sandha, G.S. (1992). Effect of grain texture on various milling and end use properties of newly bred advanced triticales (wheat x rye) lines. J. Food Sci. Tech. 29, 14–16.

Sayre, K., Pfeiffer, W.H. and Mergoum, M. (1996). Triticale: grain potential and response to input management levels. In: Triticale Topics. Int. Ed. No. 14, pp. 9–16.

Shebeski, L.H. (1974). Future Role of triticales in Agriculture. In: MacIntyre R. and Campbell M. (eds.). Triticale: Proc. Int. Symp. 1973. International Development Research Centre, El Batan, Mexico; Ottawa, Ont., Canada. pp. 247–250.

Singh, R.P. and Saari, E.E. (1991). Biotic stresses in triticale. In: Proc. 2nd Int. Triticale Symp., Passo Fundo, Rio Grande do Sul, Brazil, October 1–5, 1990. Mexico, DF, CIMMYT. pp. 171–177.

Skovmand, B., Fox, P.N. and Villareal, R.L. (1984). Triticale in commercial agriculture: progress and promise. Adv. Agron. 37, 1–45.

Varughese, G., Pfeiffer, W.H. and Peña, R.J. (1996). Triticale (Part 1): a successful alternative crop. Cer. Foods World 41, 474–482.

Villegas, E., McDonald, C.E. and Gilles, K.A. 1970. Variability in Lysine content of wheat, rye and triticale. Cer. Chem. 47, 746.

Wallwork, H. (1989). Screening for resistance to take-all in wheat, triticale and wheat-triticale hybrid lines. Euphytica 40, 103–110.

Warzecha, R. and Salak-Warzecha, K. (2006). Progress in CMS development for hybrid triticale. In: Botes W. (ed.). Proc. 6th Int. Triticale Symp., September 3–7, 2006. Stellenbosch, SA.

Wilson, A.S. (1875). Wheat and rye hybrids. In: Transactions and Proceedings of the Botanical Society, Edinburgh 12, 286–288.

Xue-Feng, M., Rodrigues, M., Milla, A. and Gustafson, J.P. (2002). AFLP-based genome studies of triticale following polyploidization. In: Arseniuk E. (ed.). Proc. 5th Int. Triticale Symp. June 30–July 5, 2002. Radzikow, Poland. pp. 89–94.

Yong, W.S. (2001). Development of AFLP and STS markers for identifying wheat-rye translocations possessing 2RL. Euphytica 121, 279–287.

Zillinsky, F.J. and Borlaug, N. (1971). Progress in developing triticale as an economic crop. CIMMYT Res. Bull. 17. Mexico, DF, CIMMYT. 27 pp.

Zimny, J., Becker, D., Brettschneider, R. and Lörz, H. (1995). Fertile, transgenic Triticale (*X Triticosecale* Wittmack). Molecular Breeding 1, 155–164.

Section II
Adding Value to Breeding

Statistical Analyses of Genotype by Environment Data

Ignacio Romagosa, Fred A. van Eeuwijk, and William T.B. Thomas

Abstract We introduce in this chapter a series of linear and bilinear models for the study of genotype by environment interaction (GE) and adaptation. These models increasingly incorporate available genetic, physiological, and environmental information for modelling genotype by environment interaction (GE). They are based on analyses of variance and regression and can be formulated in most standard statistical packages. We use the data of a series of trials for 65 barley genotypes (G) grown in 12 environments (E) for illustration and interpretation of the output of such analyses. We aim at identifying key environmental covariables to explain differential phenotypic responses as well as to estimate genotypic sensitivities to these covariables. Using genetic covariables in the form of molecular markers, we partition genotypic main effect terms and GE terms into main effects for quantitative trait loci (QTL) and QTL by environment interaction (QTL.E). The QTL.E estimates can be further regressed on environmental covariables to target differential QTL expression potentially related to environmental factors. We believe that the statistical models that describe GE in direct association to genetic, physiological, and environmental information provide insight in GE and facilitate the development and deployment of new breeding strategies

1 Introduction

Despite recent advancements in molecular marker-assisted selection, applied cereal breeding still relies largely on direct phenotypic selection of advanced genotypes. Breeders focus in the first segregating generations on highly heritable traits, such as height, spike morphology, phenology, to concentrate later on complex traits like grain yield and end-use quality. A major objective in plant breeding programs is to assess the suitability of advanced lines or potential cultivars for agricultural purposes across a range of agro-ecological conditions. To this purpose breeders

I. Romagosa(✉)
Centre UdL-IRTA, University of Lleida, Lleida, Spain. e-mail: iromagosa@pvcf.udl.es

M.J. Carena (ed.), *Cereals*,
DOI: 10.1007/978-0-387-72297-9, © Springer Science + Business Media, LLC 2009

perform so-called Multi Environment Trials (METs). In METs, a set of genotypes is evaluated in a series of trials that sample the target environmental range in order to identify those that are partially or wholly adapted (specific and wide adaptation, respectively). Data from METs are typically summarized in the form of genotype by environment tables of means. Simple inspection of such tables of means will often reveal the presence of genotype by environment interaction (GE), that is, differences between genotypes that are trial dependent. GE weakens association between phenotype and genotype, reducing genetic progress in breeding programs. Most studies in which the magnitude of GE for cereal grain yield has been measured have detected a large and statistically significant interaction; only studies with limited genotypic and/or environmental diversity found negligible or non-significant interaction. Therefore, identification of superior genotypes largely depends on extensive METs, conducted over years at different locations. This identification is hindered in the presence of GE Statistical models for MET data need to contain facilities for describing GE.

Statistical analyses that detect and describe GE have been one of the most extensively reviewed areas of applied statistics in breeding [for example in Annicchiarico (2002); Cooper and Hammer (1996); Fox et al. (1997); Gauch (1992); Kang (1990, 1998); Kang and Gauch (1996); Kempton and Fox (1997); Romagosa and Fox (1993); van Eeuwijk (1996, 2006); van Eeuwijk et al. (1996, 2005, 2007); Voltas et al. (2002)]. Past studies were largely empirical, describing postdictively genotypic performances across a sample of environments in the form of two-way genotype-environment tables of means. Such statistical characterizations of genotypic responses across environments, while frequently deployed by breeders, do not provide any physiological insight into the basis of the response.

Recent efforts have searched for the genetic factors underlying GE in the form of genetic covariables defined on the genotypes to describe GE patterns. Quantitative trait loci (QTL) responsible for adaptation have been reported in several populations for most crop species (see, e.g. Paterson, 1998). QTL related to adaptation show different effects in different environments. The magnitude of individual QTL effects (expressed as the amount of GE variation explained by a particular QTL) varied among populations. Some QTL underlying GE were coincident with QTL for the genotypic main effect, that is, QTL with constant expression across environments, within given populations, but the agreement between QTL locations across populations was low. Because QTL locations vary across environments and populations, implementation of selection methodologies for such QTL in applied breeding programs generally remains a challenge.

A landmark publication offering new perspectives on the integration of genetical, statistical, and physiological approaches to plant breeding is the book by Cooper and Hammer (1996). Modern GE studies have introduced external environmental, physiological, and/or genetic information to develop statistical models whose parameters relate better to physiological knowledge (see Spiertz et al., 2007), and therefore offer better possibilities for implementation of QTL selection methodologies in breeding programs.

Our approach in this chapter is practical; we introduce a series of statistical models for the analysis of GE and adaptation that increasingly use additional genetic, physiological, and environmental information for modelling GE. We believe that the more advanced statistical models that describe GE in direct relation to genetic, physiological, and environmental information provide more insight in GE and facilitate the development of appropriate breeding strategies. We illustrate the models via analyses of a barley data set. For didactical reasons, we take a somewhat simplified approach by avoiding more elaborate, and in the end more powerful mixed model methodology for understanding GE as described in Verbyla et al. (2003), Malosetti et al. (2004), Piepho (1997, 2000), Piepho and Pillen (2004), Smith et al. (1999, 2005), and Boer et al. (2007). We will focus on analysis of variance and regression types of models, because the basics of our approach are best illustrated in this slightly simplified context. After mastering the material described in this chapter, interested readers should consult the mixed model papers noted above.

We use a real worked example based on a subset of a large barley trial which formed part of a European Union international cooperation project called 'Mapping Adaptation of Barley to Droughted Environments' (EU FP5 INCO-MED ICA3-CT2002-10026). In this project, a population of 192 genotypes, composed of landraces, older, and contemporary cultivars sampling key regions around the Mediterranean basin and the rest of Europe, was grown in 28 environments with varying degrees of stress from which a series of papers are now being prepared (Comadran et al., 2008; Pswarayi et al., 2008). For the purposes of this chapter, we will restrict our analysis to grain yield to just the 65 modern cultivars grown in the less stress prevalent sites. Again, the purpose is to facilitate comprehension of the statistical methods described and the interpretation of key parameters, rather than produce a comprehensive and integrated analysis of the physiological and genetic bases of adaptation in this species.

All data files as well as the GenStat 9.1 (Payne et al., 2006) codes for generating the results discussed below are available from the authors.

2 An Example Data Set: Grain Yield of 65 Modern Barley Cultivars Grown in 12 Mediterranean Environments

Barley (*Hordeum vulgare* L.) (see Chap. 7 in this book and Slafer et al., 2002) is the fourth most widely grown cereal crop after wheat, rice, and maize. More than any of these three crops, it is well adapted over a large range of growing environments being sown from the most fertile areas to the poorest marginal environments. Barley is a highly self-pollinated diploid species ($2n = 2x = 14$) that has been proposed as a model species for other cereal crops. It shows a high degree of natural and easily inducible variation. The chromosomes are large (6–8 μm) as is the genome (5×10^9 bp DNA, @1,200 cM) with a physical/genetic distance ratio of between 0.1 and 300 Mb/cM (Kleinhofs and Han, 2002). In this chapter, we will carry out analyses on a subset of a large cultivar trial carried out in the Mediterranean basin. In particular, yields for 65 modern cultivars (released after 1990) were used for this study. Thirty-one of these cultivars were releases from North Mediterranean countries, 14 from other European countries, 9 from South Mediterranean, 9 from Turkey, and the last 2 from Jordan. The genotypes were multiplied at the ICARDA field site in Tel Hadya, Syria, for harvest year 2003 to produce sufficient seed for trialling in the subsequent two years. Grain yields from 12 of the original 28 sites different sites (Table 1) selected for moderate levels of abiotic stress are used in this study.

2.1 Genotyping

Each of the 65 cultivars was genotyped with a stratified set of 50 polymorphic genomic and EST-derived Simple Sequence Repeat (SSR) molecular markers that gave good coverage, 6–8 per chromosome, of the barley genome (Russell et al., 2004). These markers provided a coarse genome-wide survey of the genetic diversity represented in the 65 modern cultivars. To increase genomic coverage further, Diverse Arrays Technology (DArT®, www.diversityarrays.com) was used as a high throughput assay that had already been successfully used to map over 1,000 markers in barley (Wenzl et al., 2006). DArT® analysis produced 1,130 biallellic markers with corresponding PIC values ranging from 0.061 to 0.500 with an average diversity value of 0.387. Fifty-six alleles were present in less than 10% of the 65 modern cultivars and 10 were particularly rare as they only appeared in less than 5% of the cultivars. Out of the 1,130 markers, 811 markers were located on the DArT consensus map. Overall map coverage was good (Fig. 1), but there was a notable lack of markers on chromosome 4H with gaps greater than 20 cM. Nevertheless, the DArT® platform provided a rapid and cost-effective means of generating sufficient markers for a reasonably dense genome-wide scan of marker trait associations.

As the SSR markers were sufficient in number to cover the barley genome and they were not closely linked, they were used to identify underlying population substructure among the cultivars. The program *Structure* (Pritchard et al., 2000; Falush et al., 2003, 2007) uses genotype data consisting of unlinked markers to implement a model-based clustering method for inferring and identifying distinct genetic populations and for assigning individuals to populations. *Structure* assumes a model in which there are K populations (where K may be unknown), each of which is characterized by a number of loci with population-specific distributions of allele frequencies. Individuals in the sample are characterized with respect to the identified genetic populations in terms of membership probabilities; the higher the membership probability with regard to a particular population, the more likely it is that the individual belongs to that population. The genetic populations are constructed such as to achieve Hardy-Weinberg and linkage equilibrium within the identified populations. The program is freely available and well documented and can be downloaded from http://pritch.bsd.uchicago.edu/software.

Table 1 Description of the 12 environments used

Site	Code	Latitude	Longitude	Altitude (m)	Sowing date	Yield (t/ha)	Intrablock error	H^2[1]
El Khroub, Algeria	A4	36°32′N	06°42′E	596	05/12/2003	5.29	0.76	0.38
El Khroub, Algeria	A5	36°32′N	06°42′E	596	19/02/2005	3.26	0.33	0.49
Gimenells, Spain	E4	41°35′N	00°32′E	260	17/12/2003	6.79	0.13	0.54
Gimenells, Spain	E5	41°35′N	00°32′E	260	24/11/2004	3.05	0.05	0.53
Foggia, Italy	I4	41°27′N	15°34′E	57	13/01/2004	4.77	0.10	0.83
Foggia, Italy	I5	41°27′N	15°34′E	57	16/12/2004	5.26	0.19	0.53
Settat, Morocco	M4	33°07′N	07°37′W	240	16/12/2003	4.23	0.66	0.63
Settat, Morocco	M5	33°07′N	07°37′W	240	11/12/2004	1.61	0.20	0.50
Tel Hadya, Syria	S4	36°01′N	36°56′E	362	11/12/2003	3.97	0.24	0.09
Tel Hadya, Syria	S5	36°01′N	36°56′E	362	01/12/2004	5.08	0.22	0.54
Haymana, Turkey	T4	39°36′N	32°40′E	1214	01/03/2004	6.53	0.44	0.54
Esenboga, Turkey	T5	40°08′N	33°01′E	953	21/03/2005	4.88	0.43	0.53
					Median	4.82	0.23	0.53

[1] Calculated from variance components analysis; strictly, these are repeatabilities

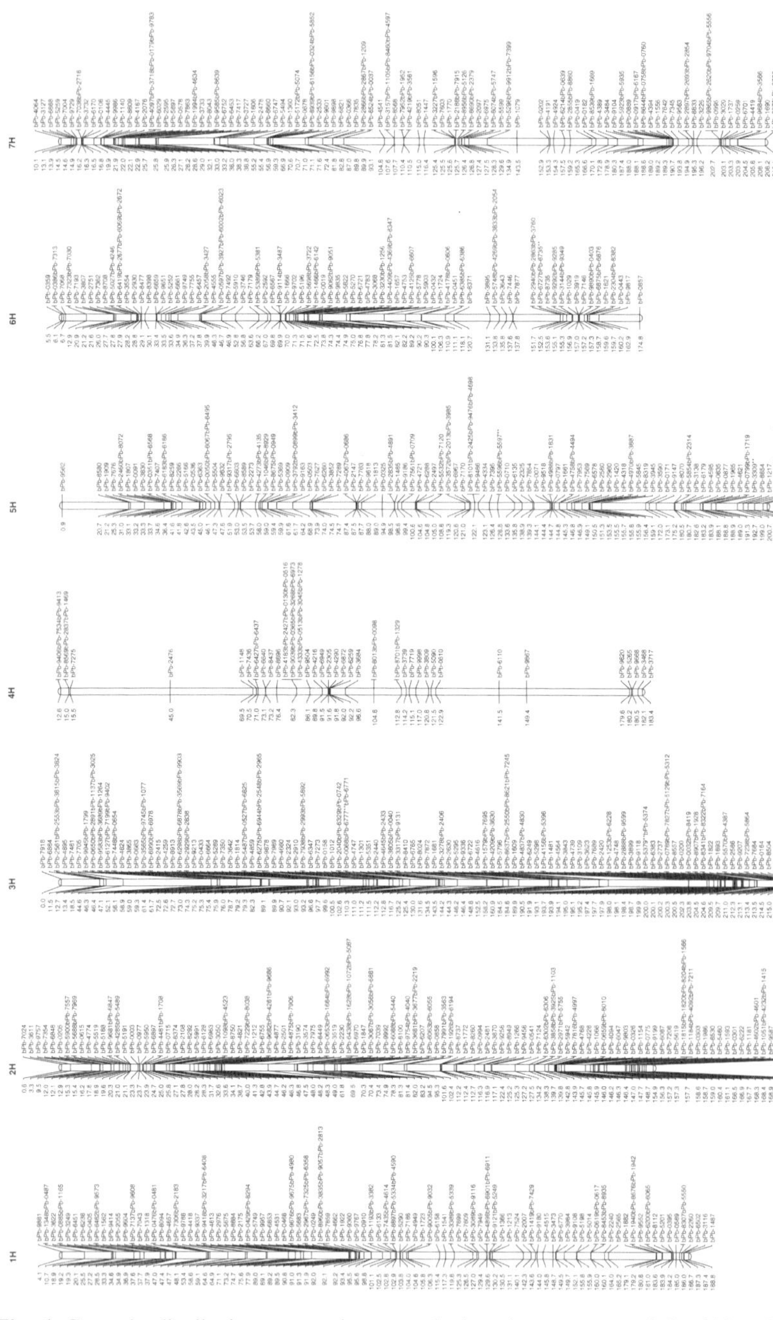

Fig. 1 Genomic distribution across the seven barley chromosomes of the 811 polymorphic DArT® markers used for linkage disequilibrium mapping

On the basis of *Structure*, four groups were identified (Fig. 2) which made geographical and physiological sense. Key factors distinguishing groups were their winter versus spring habit, phenotypically determined by late spring sowing in Northern Europe, and whether they were two- versus six-row ear type. The four groups consisted of (a) eight of the nine Turkish cultivars (Tk); (b) twenty-four North Mediterranean two- and other six-row types, which included three genotypes from Spain and North Africa (NMW) and were mainly winter habit; (c) twelve spring cultivars, mainly six-row types from the South and West Mediterranean, including the only two genotypes from Jordan (SW); and (d) twenty-one North Mediterranean two-row spring types (NMS).

2.2 Phenotyping

The original 192 genotypes from which our 65 form a subset were multiplied at the ICARDA field site in Tel Hadya, Syria, for harvest year 2003 to produce sufficient seed for trials in subsequent years. This common seed source was used for sowing trials across the Mediterranean region to estimate yield in a wide range of environments (Table 1).

The experimental designs for individual trials consisted of, first, an unreplicated trial for the 192 entries, augmented by four repeated checks that were included in a diagonal fashion. Second, as a special feature, a partial replicate, built up in a similar way as the initial full replicate, was added, containing a quarter of the entries to make 300 trial plots. The whole trial was sown in a rectangular grid of 15 rows and 20 columns at each site but with a different randomization and composition of the partial replication. Three of the four repeated checks varied across sites, with one being a specific landrace, and two others being an old and a modern cultivar that were specific to the region in which the trial was being grown. The fourth check was cv 'Rihane', which was grown at every site. The partial replication together with the repeated checks served to estimate the experimental error and correct for any spatial patterns for each trial. Trials were sown in plots of 6 m^2 at each site and were grown according to local practise for sowing rate and other inputs.

We conducted a two-step statistical analysis, where we first obtained by mixed model analysis spatially corrected Best Linear Unbiased Predictors (BLUPs) for grain yield of the genotypes in each trial. Second, we organized the genotypic BLUPs in a two-way genotype by environment table of means, the analysis of which will be the topic of this chapter. The mixed model analysis of the individual trials is relevant and interesting and will be communicated elsewhere.

Repeatabilities for grain yield at each site varied between 0.09 and 0.83 in Syria 2004 and Italy 2004, respectively, with a median value of 0.54, considered relatively high for this trait. To assess the relative importance of the different terms of the model, a rough estimate of the experimental error across trials was required. We used the median of the average plot error divided by the average number of replicates. Given that in the complete trials designs at each site there were 300 plots and 192 entries, we approximated the number of replicates by 1.5. Thus, an intra

Entryname	Origin	Spike type	Phenology
M19 Orza96	TUR	2	S
M14 Anadolu98	TUR	2	S
M17 Karatay	TUR	2	S
M18 Tarm92	TUR	2	S
M13 Efes98	TUR	2	S
M21 Yesevi93	TUR	2	S
M16 Sahin91	TUR	2	S
M20 Bulbul89	TUR	2	S
M33 Aliseo	ITA	6	W
M48 Solen	ITA	6	W
M38 Federal	ITA	6	W
M07 Manitou	FRA	6	W
M12 Intro	NLO	2	W
M40 Grecale	ITA	2	W
M42 Kelibia	ITA	2	W
M49 Sonora	ITA	6	W
M45 Nure	ITA	2	W
M34 Amillis	ITA	2	W
M06 Igri	DEU	2	W
M53 Ultra	ITA	2	W
M39 Gotic	ITA	6	W
M44 Mattina	ITA	6	W
M30 Reinette	FRA	2	W
M10 Siberia	FRA	6	W
M65 Manel	DZA	6	S
M54 Vertige	ITA	2	W
M05 Fanfare	GBR	2	W
M25 Hispanic	FRA	2	W
M46 Naturel	ITA	2	W
M23 Dobla	ESP	6	S
M22 Candela	ESP	6	S
M15 Aydanhanim	**TUR**	**2**	**W**
M60 Oussama	MAR	6	S
M58 Massine	MAR	2	S
M59 Rabat01	MAR	6	S
M01 Arig8	MAR	6	S
M57 Merzaga07	MAR	6	S
M61 ASCAD 176	JOR	6	S
M31 Steptoe	USA	6	S
M29 Orria	ESP	6	S
M64 Alanda01	DZA	6	S
M62 Rum	JOR	6	S
M56 Amalou	**MAR**	**6**	**S**
M55 Aglou	**MAR**	**2**	**S**
M35 Apex	ITA	2	S
M50 Tea	ITA	2	S
M36 Barke	ITA	2	S
M41 Grosso	ITA	2	S
M11 Triumph	DEU	2	S
M43 Magda	ITA	2	S
M28 Nevada	GBR	2	S
M04 Chariot	GBR	2	S
M02 Atem	NLO	2	S
M47 Otis	ITA	2	S
M03 Alexis	DEU	2	S
M63 Aramir	NLO	2	S
M09 Scarlett	DEU	2	S
M52 Tremois	ITA	2	S
M32 Zaida	ESP	2	S
M24 Graphic	GBR	2	S
M27 Kym	GBR	2	S
M08 Optic	GBR	2	S
M26 Kika	ESP	2	S
M51 Tidone	**ITA**	**2**	**S**
M37 Dasio	**ITA**	**2**	**S**

Fig. 2 Inferred population Structure based on 50 markers using *Structure* (Pritchard et al., 2000). Each individual is represented by a *line partitioned in four coloured segments* that represent the individual's estimated membership probability in relation to four genotypic clusters. Entries in **bold** type were misclassified according to standard hierarchical cluster analysis

block error across trials value, derived from Table 1, equal to 0.23 is available for a number of specific purposes below. The experimental error across environments did not seem particularly homogeneous. However, for the sake of simplicity we will ignore that complication.

2.3 Explicit Environmental Characterization

We recorded a series of meteorological variables at each site. They were maximum, minimum, and average daily mean temperature, rainfall, reference evapotranspiration (ET0), and radiation for three consecutive growth periods: tillering, jointing, and grain filling. To characterize physically the environment, we then derived ten different variables at each of the three phenological stages. They were number of days with minimum temperature below 0° (dTb0); number of days with average daily temperature below 4° (base temperature) (dTbb); number of days with maximum temperature over 30° (dTo30); average maximum temperature (TMx); average minimum temperature (TMn); difference between the maximum and the minimum temperature (Tdif); total growing degree days (GDD); total rainfall plus irrigation (mm) (WT); a measurement of the ratio between available water and evapotranspirative demand, $100 \times$ total water/ET0, (WDT); average photo thermal quotient (PQ) defined as solar radiation divided by average daily temperature. The above environmental characterizations were calculated separately for each of three developmental stages: tillering (1), jointing (2), and grain filling (3).

As data for jointing and physiological maturity were not available for each trial, we arbitrarily defined jointing as heading date minus three weeks and physiological maturity as heading date plus two weeks. Thus, to reveal the meteorological conditions during tillering we considered the period from sowing to heading minus three weeks. We averaged daily variables for the three weeks before heading to determine the meteorological variables during jointing. Finally, we averaged the daily meteorological variables from heading to two weeks after heading to characterize the grain-filling period.

Figure 3 shows a biplot that allows graphical exploration of relationships between environments (squares), between meteorological variables (circles), and of environments together with meteorological variables. The environments (objects) and meteorological variables are positioned on the biplot according to their scores from a principal components analysis, where the variables were standardized. The distances between environments correspond to the differences in meteorological conditions. Similar environments are plotted together and different ones are plotted further apart. The direction of a meteorological variable shows how the value for that variable changes across the plot. Projecting environments orthogonally on the variable representations allows a direct assessment of the relative values for that variable for each of the environments. The origin represents for each variable its average. Environments projecting above the origin (in the direction of the arrow) were above average for that variable, environments projecting below the origin were below average. Such an interpretation requires that the

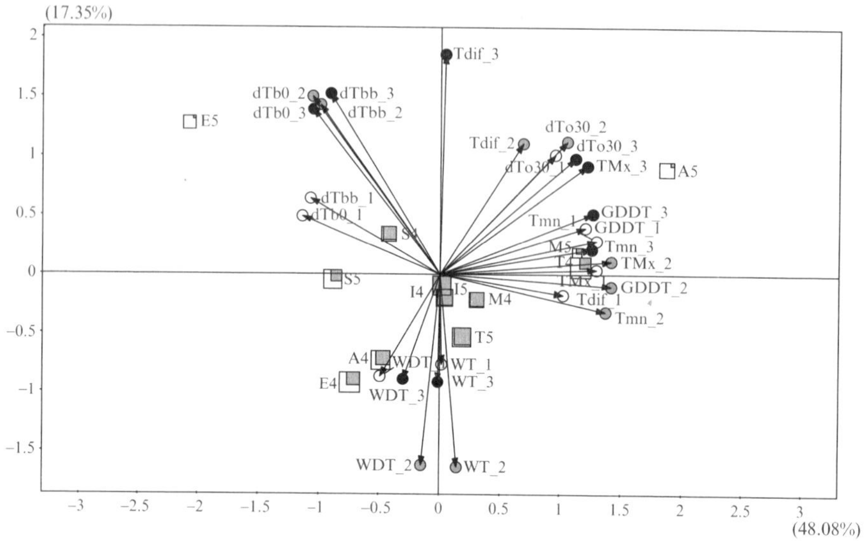

Fig. 3 Biplot of the Principal Component Analysis run on standardized meteorological variables. *White, grey, and black circles* represent variables taken during tillering, jointing, and grain-filling phases. *Squares* represent the environments. The size of each square is proportional to its average yield. The unexplained variance for each site is shown as a *grey cut-out* (see text for the acronyms of meteorological variables)

first two principal component axes account for most of the variability shown among environments for the set of meteorological variables.

A number of inferences can be drawn from the biplot in Fig. 3 that explained approximately two thirds of the variability. The first axis seems related to temperatures and the second to water status variables. Low temperatures characterized the Spanish 2005 site (E5) while the Algerian 2005 site (A5) had the highest temperatures in the second part of the growth cycle. We detected high correlations between meteorological variables across growth periods, particularly for temperature-derived variables. As the size of the solid internal square represents the unexplained variance associated with each environment, we see that the first two principal components did not account for much of the variation in the four sites which were located close to the origin (I4, I5, M4, and T5).

3 Phenotype-Based Statistical Analyses of Two-Way GE Tables: Assessment and Partitioning of the Variability

3.1 *The Additive Model*

A very simple model for the description of phenotypic responses across environments is the additive model. In this model, the expected phenotypic response for genotype i ($i = 1 \ldots I$) in environment j ($j = 1 \ldots J$), μ_{ij}, is defined as

$$\mu_{ij} = \mu + G_i + E_j \tag{1}$$

with μ the general mean, G_i the genotypic main effect, and E_j the environmental main effect (both expressed as deviations from the general mean).

For balanced data, the estimate for the main effect of genotype i is the average across environments of the phenotypic observations indexed by i minus the general mean. Likewise, the estimate for the main effect of environment j follows from the average across genotypes of observations indexed by j. So, genotypic main effects depend on the particular set of environments that were included in the experiments, while environmental main effects depend on the genotypes that were included.

The main purpose of the additive model is to interpret phenotypic differences in terms of mean differences between the genotypes on the one hand and mean differences between the environments on the other hand. The additive model is a benchmark model for all other models and is really applicable only in the unlikely event of the absence of GE or, maybe, in the presence of moderate GE. The additive model describes the phenotypic responses for a set of genotypes to a set of environments as a set of parallel lines. The analysis of variance for this model applied to our data is shown in Table 2 (i). Both the environments and genotypes were highly significant explaining 85.2% and 2.5% of the total sum of squares.

3.2 *The Full Interaction Model*

Means across environments are relevant indicators of genotypic performance when there is no GE. If, however, GE is present, the use of means across environments ignores the differential reaction of genotypes to environmental changes. Hence, a common way to extend the additive model is to add a term for each combination of genotype and environment:

$$\mu_{ij} = \mu + G_i + E_j + (G.E)_{ij} \tag{2}$$

Model (2) has as many independent parameters as genotype by environment combinations. With this model, predictions of phenotypic responses for environments that were not in the set of trial environments is impossible, because there will be no estimates for the particular $(G.E)_{ij}$ terms. In our example, the analysis of variance table for the full interaction model is shown in Table 2 (ii). In terms of sums of squares, G.E was approximately six times greater than G. If we use the rough estimate of the experimental error across trials identified above, 0.23 t/ha, every term in the model was highly significant.

To assess the relative importance of each term in the model, an alternative to the magnitude of sum of squares in the ANOVA table, or equivalently, their R^2, which depends on the number of degrees of freedom associated to each term, is the use of variance components (and their associated standard errors). Useful variance com-

Table 2 Analyses of variance for the two way GE tables according to 7 alternative linear models. Significances of the calculated F- values are shown as $-\log_{10}(p\text{-value})$. Thus a p-value of 0.0001 would translate into 4. Residuals from each model were used as denominators for the F tests

(i) Additive model

Souce of variation Equation (1)	d.f.	Sum of squares	R^2	Mean squares	Variance ratio	$-\log_{10}(p\text{-value})$
Total	779	1785.04		2.29		
Environment [E]	11	1522.17	85.3	138.38	445.85	*>100*
Genotype [G]	64	44.36	2.5	0.69	2.23	*6.33*
Residual	704	218.50	12.2	0.31		

(ii) Full interaction model

Source of variation Equation (2)	d.f	Sum of squares	R^2	Mean squares	Variance ratio	$-\log_{10}(p\text{-value})$
Environment [E]	11	1522.17	85.3	138.38	445.85	*>100*
Genotype [G]	64	44.36	2.5	0.69	2.23	*6.33*
G.E	704	218.50	12.2	0.31		

(iii) Reduced interaction method: Clusters indentified according to Corsten & Denis (1990)

Source of variation Equation (3)	d.f	Sum of squares	R^2	Mean squares	Variance ratio	$-\log_{10}(p\text{-value})$
Environment [E]	11	1522.17	85.3	138.38	445.85	*>100*
Ecluster [EC]	*2*	*24.57*	*1.6*	*12.29*	*39.58*	*16.29*
E'	*9*	*1497.60*	*98.4*	*166.40*	*536.13*	*>100*
Genotype [G]	64	44.36	2.5	0.69	2.23	*6.35*
Gcluster [GC]	*2*	*22.71*	*51.2*	*11.36*	*36.59*	*>100*
G'	*62*	*21.64*	*48.8*	*0.35*	*1.12*	*0.61*
G.E	704	218.50	12.2	0.31	1.45	*6.24*
EC.GC	*4*	*68.21*	*31.2*	*17.05*	*79.42*	*>100*
Residual	*700*	*150.29*	*68.8*	*0.21*		

(iv) Reduced Interaction method: Clusters indentified with Structure (Pritchard et al. 2000)

Source of variation Equation (3)	d.f	Sum of squares	R^2	Mean squares	Variance ratio	$-\log_{10}(p\text{-value})$
Environment [E]	11	1522.17	85.3	138.38	445.85	*>100*
Ecluster [EC]	*2*	*24.57*	*1.6*	*12.29*	*39.58*	*16.29*
E'	*9*	*1497.60*	*98.4*	*166.40*	*536.13*	*>100*
Genotype [G]	64	44.36	2.5	0.69	2.23	*6.35*
$Structure_4$	*3*	*17.76*	*40.0*	*5.92*	*19.08*	*>100*
G'	*61*	*26.59*	*60.0*	*0.44*	*1.40*	*1.58*
G.E	704	218.50	12.2	0.31	1.29	*3.34*
EC.Structure	*6*	*49.95*	*22.9*	*8.33*	*34.48*	*>100*
Residual	*698*	*168.55*	*77.1*	*0.24*		

(*continued*)

Table 2 (Continued)

(v) Regression on the the mean model

Source of variation Equation (4)	d.f	Sum of squares	R^2	Mean squares	Variance ratio	$-\log_{10}(p\text{-value})$
Environment [E]	11	1522.17	85.3	138.38	442.07	*>100*
Genotype [G]	64	44.36	2.5	0.69	2.21	*6.15*
G.E	704	218.50	12.2	0.31	0.99	*0.26*
JRA	*64*	*18.17*	*8.3*	*0.28*	*0.91*	*0.17*
Residual	*640*	*200.34*	*91.7*	*0.31*		

(vi) Additive main effect multiplicative (AMMI) model

Source of variation Equation (7)	d.f	Sum of squares	R^2	Mean squares	Variance ratio	$-\log_{10}(p\text{-value})$
Environment [E]	11	1522.17	85.3	138.38	445.85	*>100*
Genotype [G]	64	44.36	2.5	0.69	2.23	*6.35*
G.E	704	218.50	12.2	0.31	2.89	*29.80*
IPCA1	*74*	*95.40*	*43.7*	*1.29*	*6.60*	*44.94*
IPCA2	*72*	*38.60*	*17.7*	*0.54*	*3.54*	*17.75*
IPCA3	*70*	*24.20*	*11.1*	*0.35*	*2.80*	*11.23*
IPCA4	*68*	*15.20*	*7.0*	*0.22*	*2.08*	*5.67*
Residual	*420*	*45.10*	*20.6*	*0.11*		

(vii) GGE

Source of variation Equation (8)	d.f	Sum of squares	R^2	Mean squares	Variance ratio	$-\log_{10}(p\text{-value})$
Environment [E]	11	1522.17	85.3	138.38	404.30	*>100*
G.E	768	262.86	14.7	0.34	3.19	*35.46*
GGE1	*75*	*118.23*	*45.0*	*1.58*	*7.55*	*45.21*
GGE2	*73*	*38.82*	*14.8*	*0.53*	*3.12*	*12.95*
GGE3	*71*	*33.62*	*12.8*	*0.47*	*3.60*	*16.28*
GGE4	*69*	*16.93*	*6.4*	*0.25*	*2.13*	*5.67*
Residual	*480*	*55.25*	*21.0*	*0.12*		

ponents may be easily determined using a random model for the two-way GE table of means in which E, G and G.E are considered random effects. Variance components have the additional advantage of being directly comparables as they have the same scale. For the current data set the estimates for E, G, and G.E are 2.124 $\pm$ 0.908, 0.032 $\pm$ 0.010, and 0.310 $\pm$ 0.016, respectively, which clearly shows the greater importance of G.E over G, but both are much less than E, as expected.

3.3 *Reduced Interaction Model: Clustering of Genotypes and Environments*

An improvement on the full interaction model can be attained by grouping genotypes and environments in such a way that the majority of the interactions are represented by the interactions between the groups of genotypes and environments. If meaningful clusters of genotypes (GC) and environments (EC) are identified, such a reduced interaction model would help in understanding the nature of GE. In model terms, the expected phenotypic response for a genotype i ($i = 1 \ldots I$) that belongs to a given cluster of genotypes identified as GC_k ($k = 1 \ldots K$, where K is hopefully much smaller than I) in the environment j ($j = 1 \ldots J$) that belongs to a cluster of environments EC_l ($l = 1 \ldots L$, where $L < J$), $\mu_{i(j)k(l)}$, would be defined as

$$\mu_{i(k)j(l)} = \mu + [\mathrm{GC}_k + \mathrm{G}'_{i(k)}] + [\mathrm{EC}_l + \mathrm{E}'_{j(l)}] + [(\mathrm{GC.EC})_{kl} + (\mathrm{G.E})'_{i(k)j(l)}] \quad (3)$$

where μ stands for the general mean, GC_k is the genotype-grouping main effect expressed as a deviation from the general mean, $G_{i(k)}$ stands for a residual genotypic main effect or deviation of the mean of genotype i from the mean of its GC_k group, and should be noticeably smaller than the original G_i in Eq. (2) if the genotype groups are to be useful. Likewise EC_l represents the environmental-grouping main effect expressed as a deviation from the general mean. $E'_{j(l)}$ symbolizes the deviation of environment j from the mean of the environmental group EC_l. Each of the square bracket pairs in Eq. (3) reflects an orthogonal partitioning of G, E, and GE. The most important term for our purposes is $(GC.EC)_{kl}$. This interaction term gives a deviation from the simple additive model for the combination of genotype group k and environment group l. When successful, the portion of the interaction explained by $(GC.EC)_{kl}$ should be substantial in relation to the whole of the initial GE.

Corsten and Denis (1990) developed a useful algorithm to simultaneously cluster genotypes and environments in an orthogonal balanced two-way table in order to identify groups of genotypes and environments that maximally explain GE. Starting with a significant interaction, the procedure goes through a sequence of steps in each of which the mean square for interaction is calculated for all possible sub-tables consisting of a pair of rows (genotypes) or a pair of columns (environments) from the full table. The pair of rows or columns with the minimal mean square for interaction is merged, giving an updated table, and the process is repeated. Thus, a sequence of amalgamations of rows and columns is produced, eventually leading to a 2×2 table. In this way, the total sum of squares for the interaction is made up of orthogonal increments. The pattern of amalgamations can be visualized in the form of two dendrograms. When an estimate for error is available, a cut-off value for group identification can be calculated. The resulting genotype and environment groups hopefully provide more insight in the driving forces behind GE. The clustering procedure is conceptually very simple, but laborious to implement in standard statistical programs. The Biometris library of GenStat, www.biometris.wur.nl, includes an easily implemented procedure, named CINTERACTION and developed by J. Thissen and de J. de Bree, that runs the Corsten and Denis (1990) algorithm.

The top of Fig. 4 illustrates the CINTERACTION-derived dendograms for genotypes and sites. The horizontal axis shows the cumulative interaction sum of squares built up in the agglomerative hierarchical clustering steps. The first genotypic cluster (GC_1) was made up of a mixture of predominantly winter and a few spring genotypes; the second (GC_2) of mainly spring types. The third cluster contained all Turkish lines and a few winter types (GC_3). A graphical representation of the reduction of complexity of the GE is shown at the bottom of Fig. 4. Genotypes in GC_3 interacted positively with environments in EC_1 and negatively with Morocco 2004 (M4), the latter defining EC_2. As mentioned, genotypes in GC_3 are late winter and Turkish entries that did well in a series of temperate sites, and very poorly under warm conditions.

Table 2 (iii) shows the partitioning of the variation and corresponding tests when the genotype and environment groups from the clustering above are introduced as *a priori* defined groups in model (3), that is, when the groups would have been defined independent of the data. As the groups were actually obtained from analysis of the data, that is, *a posteriori*, the tests will inflate the importance of the groups in describing the interaction. However, for general exploratory purposes, Table 2 (iii) is reliable enough. We see that three genotypic and environment clusters explained, with just four out of the total of 704 degrees of freedom, more than 30% of the G.E sum of squares. To test the grouping effect on the interaction, we used a model with fixed genotypic and environmental clusters in combination with random genotypes and sites within the respective clusters. We then tested the significance of the portion of G.E explained by the groups against their residuals, which is a more appropriate test than testing it against the intra-trial experimental error. This revealed that a highly significant portion of the interaction was associated with the groups.

Throughout this chapter, we will use the Corsten and Denis (1990) derived groups. Similar group interaction models could be defined using alternative genotypic and/or environmental groupings, but due to the nature of the Corsten and Denis algorithm, alternative groupings will explain less of the G.E. Table 2 (iv) shows the partitioning of GE using the four *Structure*-defined genotype groups and the three above-defined environment groups. These four genotypic groups combined with the three environmental clusters detected above, explained with six degrees of freedom, just over 20% of the interaction sum of squares. The reason for the lesser importance of the *Structure* solution is clear. Contrary to the Corsten and Denis (1990) clusters, the *Structure* clusters are based on all genetic information, some of which is unlikely to be related to grain yield, which is the response process under current study.

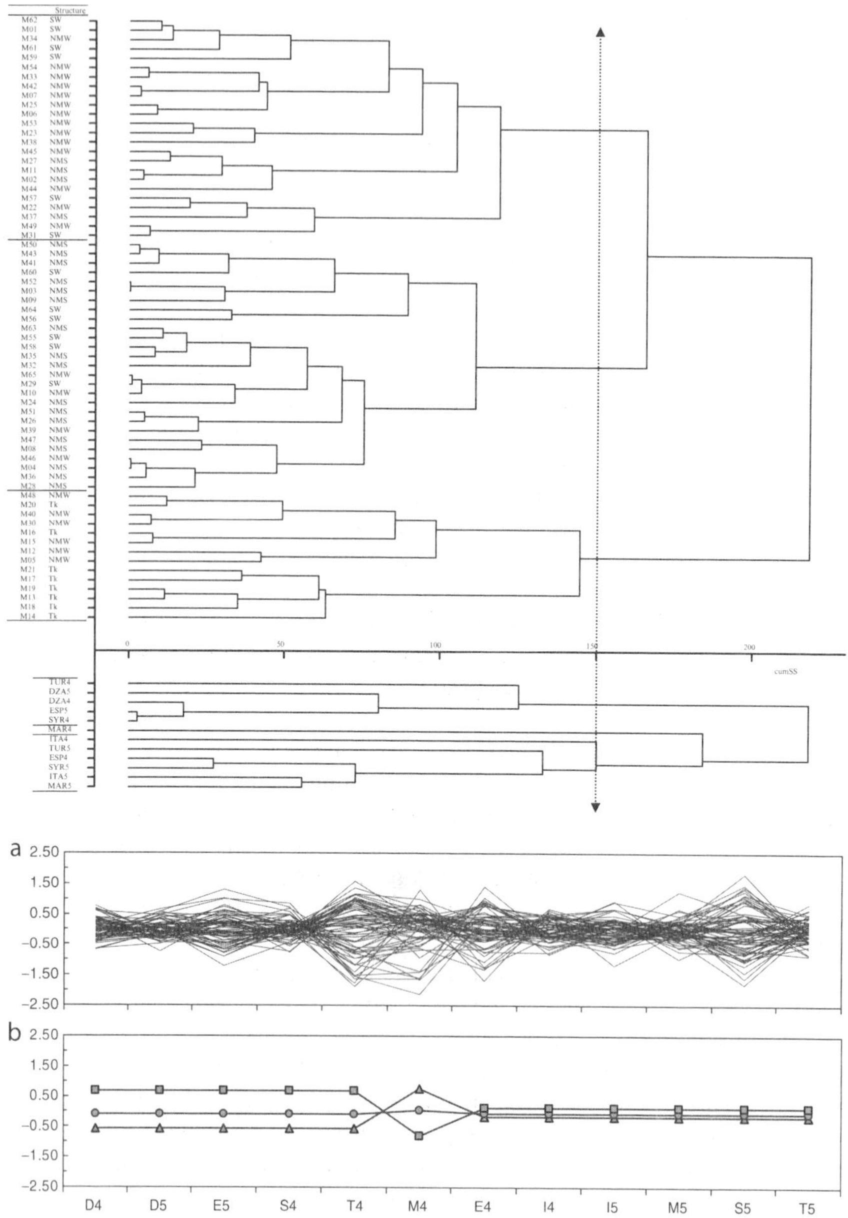

Fig. 4 *TOP*: Parallel genotypic and environmental dendograms and identification of clusters according to the Corsten and Denis' (1990) procedure. The second column for each genotype shows the *Structure* grouping. *BOTTOM*: **(a)** Original G.E deviations for the 65 genotypes in the 12 environments; **(b)** estimated G.E of the three clusters in the 12 sites

3.4 Modelling the Interaction Using Phenotypic Characterizations of the Environment

A popular modelling approach to GE in plant breeding is the regression on the mean analysis, or joint regression analysis, made popular by Finlay and Wilkinson (1963). The model describes phenotypic responses as straight lines differing in intercept (genotype main effect) and slope (environmental sensitivity). The principle behind the model is that, in the absence of explicit environmental information (physical or meteorological), a good approximation of the agronomical quality of an environment may be given by the average phenotypic performance of all genotypes in that environment. The phenotypic responses of individual genotypes are then regressed on the average genotypic performance across the full set of environments. GE is revealed by differences in the slopes of individual genotypes. We can define this model in two equivalent ways:

$$\mu_{ij} = \mu + \mathrm{G}_i + \mathrm{E}_j + \beta_i \mathrm{E}_j \tag{4}$$

$$\mu_{ij} = \mu + \mathrm{G}_i + \mathrm{E}_j + \beta_i \mathrm{E}_j = (\mu + \mathrm{G}_i) + (1 + \beta_i)\mathrm{E}_j = \mathrm{G}_i^* + \beta_i^* \mathrm{E}_j \tag{5}$$

In (4), GE is modelled by the differential genotypic sensitivities, represented by the parameter β_i (with average zero), to the environmental characterization E_j. Eq. (5) emphasizes the non-parallelism of the genotypic responses in the regression on the mean model. The average sensitivity in (5) will be unity. The additive model can be obtained from (4), by taking all β_i as zero, or from (5) by taking all β_i^* as one.

Regression-on-the-mean models are conceptually simple: the differential genotypic responses are summarized by their slopes. Models (4) and (5) partition the GE of the full interaction model, $(\mathrm{G.E})_{ij}$, into a part due to regression on the environmental main effect (environmental index), $\beta_i \mathrm{E}_j$, and a new orthogonal residual, $(\mathrm{G.E})'_{ij}$, which is considered to be random with mean zero. The statistical success of the regression on the mean model directly depends on the proportion of GE that can be described by the differential environmental sensitivities of the genotypes. In practical terms, the use of the model should, however, be restricted to those rare cases in which environmental differences are driven by just a single major biotic or abiotic factor. In this case, the linear regression on the mean model may reflect linear differences in relation to a stress factor of interest.

In our example data set, differences in the slopes [shown in Table 2 (v)] as Joint Regression Analysis (JRA) only explained 8.3% of the G.E sum of squares, while the residual was still significant. Figure 5 shows a box plot of slopes for the 65 genotypes organized by the three genetic clusters, GC_1, GC_2, and GC_3.

In the regression on the mean model we can group genotypes with similar responses to produce:

$$\mu_{i(k)j} = \mu + [\mathrm{GC}_k + \mathrm{G}'_{i(k)}] + [\beta_k^* \mathrm{E}_j + \beta_{i(k)}^* \mathrm{E}_j] \tag{6}$$

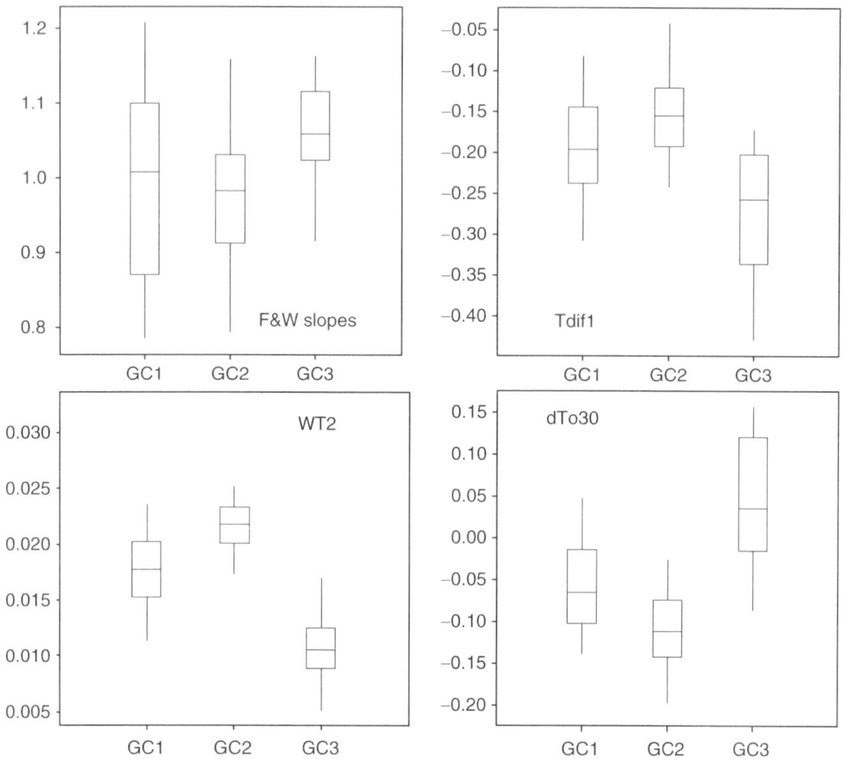

Fig. 5 Box plots for the slopes of the regression on the means model (F&W Slopes) and the factorial regression derived genetic sensitivities for the Tdif1, WT2, and dTo30 variables, classified according to the genotypic clusters GC_1 to GC_3 (for the acronyms of meteorological variables see text)

GC_k is the main effect for the genotype group k expressed as a deviation from the general mean, $G'_{i(k)}$ stands for a residual main effect for genotype i within group k, β^*_k represents the sensitivity of the genotypic cluster k to the environmental characterization, E_j, and $\beta^*_{i(k)}$ represents a deviation in sensitivity for genotype i with respect to the sensitivity of the group to which it belongs, k.

3.5 Other Linear–Bilinear Models

The regression on the mean model is rather limited in its possibilities. GE is included in the model by differential sensitivity to a one-dimensional linear representation of the environmental factors affecting the phenotypic responses. The regression on the mean model is a member of the family of linear–bilinear models (Gabriel, 1978, 1998; van Eeuwijk, 1995a; van Eeuwijk et al., 1995; Denis and

Gower, 1996; Crossa and Cornelius, 2002). Other models of this class allow more flexible and powerful characterizations of the environment. All linear–bilinear models describe GE by differential genotypic sensitivities to one or more environmental characterizations that are simple linear functions of the phenotypic data themselves.

Linear–bilinear models contain simple additive (linear) and multiplicative (bilinear) terms. In the regression on the mean model, using the regression formulation (5), $G^*_i + \beta^*_i E_j$, the linear part of the model is given by G^*_i, while the bilinear part is given by $\beta^*_i E_j$. The latter term is not an ordinary regression term, because both genotypic and environmental parameters have to be estimated simultaneously. A bilinear model becomes a standard linear model in the genotypic parameters upon fixation of the environmental parameters, and it becomes linear in the environmental parameters upon fixation of the genotypic parameters. This property forms the basis for a general estimating procedure for the parameters (Gabriel and Zamir, 1979; Gabriel, 1998; van Eeuwijk, 1995b).

Additional bilinear terms provide higher flexibility for the modelling of GE. A popular example of a linear–bilinear model with a facility for multiple bilinear terms is the Additive Main effects and Multiplicative Interaction effects model, or AMMI model (Gollob, 1968; Mandel, 1969; Gabriel, 1978; Gauch, 1988). The model is defined as follows:

$$\mu_{ij} = \mu + G_i + E_j + \sum_{k=1}^{K} a_{ki} b_{kj} \tag{7}$$

with a_{ki} and b_{kj} genotypic and environmental parameters (scores) for the bilinear term k, which in this case represents the number of multiplicative terms necessary for an adequate description of GE. Similar to the multiplicative term in the regression on the mean model, the genotypic scores, a_{ki}, can be interpreted as sensitivities, and the environmental scores, b_{kj}, are environmental characterizations. From a statistical point of view, the environmental scores for the first bilinear term represent the best environmental characterization possible for the description of GE. It is the environmental variable with maximally different genotypic sensitivities. The second bilinear term represents the second best environmental characterization, etc.

The parameter estimates for an AMMI model with two bilinear terms, $K = 2$, can conveniently be visualized by means of a biplot (Kempton, 1984; Fox et al., 1997). The first and second bilinear terms are often called IPCA1 and IPCA2, where IPCA stems from interaction principal component analysis. The position of the point of genotype i in the biplot is given by the estimates for the genotypic scores, a_{1i} and a_{2i}; similarly, the point coordinates for environment j originate from the estimates for the environmental scores (b_{1j}, b_{2j}). Distances from the origin (0, 0) are proportional to the amount of interaction due to genotypes over environments or to environments in relation to genotypes. Genotypes that are located close to each other in the biplot behave similarly with regard to adaptation patterns. Environments that are located close to each other reflect similar interaction patterns. Assuming a vector representation for the environments, the interaction effect of genotype i in environment j is approximated by projecting the genotype point (a_{1i}, a_{2i}) onto the line determined by the environmental vector, which has a slope b_{2j}/b_{1j}. The distance of the projected point to the origin provides information about the magnitude of the interaction of genotype i in environment j, with positive interaction when the genotype projects above the origin (in the direction of the arrow) and negative interaction when below. The angle between the vectors of genotype i and environment j provides information about the interaction; they interact positively for acute angles, negatively for obtuse angles, with a negligible interaction for right angles, provided that much of the G.E is accounted for by IPCA1 and IPCA2.

Table 2 (vi) shows the partitioning of the variation in our data set according to the AMMI model. Each bilinear term was tested as a mean square with degrees of freedom equal to $I+J-1-2k$ against a residual term that was constructed from the remaining G.E sum of squares divided by the remaining degrees of freedom for GE (Mandel, 1969). An alternative and equally simple way of testing for the number of bilinear terms is by retaining only those terms that explain more than the expected average percentage of GE sums of squares. This figure can be found by dividing 100% by the minimum of the number of genotypes minus one ($I-1$) or the number of environments minus one ($J-1$). For our data, the expected average is 100/11% or about 9%. According to both of our criteria, the first two bilinear terms, or axes, were clearly significant, explaining together over 60% of the G.E sum of squares. The third axis explained an additional 11% and was also significant. The fourth axis, with 7.0% of the GE sum of squares, had an associated mean square term equal to 0.22, very close to the pooled across environments intra block error, and therefore was not used in further analyses. The AMMI biplot for IPCA1 and IPCA2 is displayed in Fig. 6. Genotypes are represented by circles (open, grey, and black representing the three clusters identified previously). The triangles represent the means of the three genotypic groups. Information on the mean yield performance of genotypes (generally with small differences) and environments (much greater differences) can be added to the biplot by making the area for each symbol proportional to this mean. Furthermore, the fraction of the interaction sum of squares for each environment that was not explained by the first two bilinear terms can be shown by cut-outs in the upper right corner of the symbols. For our data set, the AMMI K = 2 model was driven by the four environments that were furthest away from the origin. The first IPCA was clearly associated with differences between the three genotypic clusters. GC_2 was strongly different from GC_3. GC_3 was particularly well adapted to Turkey 2004 as well as to the other sites with negative IPCA1 scores. Genotypes from GC_2, spring types, were specifically adapted to one of the Moroccan environments. Genotypes from GC_1 did not show as much interaction with the environments as the others as they are closer to the origin.

The environmental characterizations in bilinear terms are estimated by a purely statistical criterion (least squares minimization) and, therefore, they may not have a direct agroecological meaning. Despite this, regressing the environmental scores on

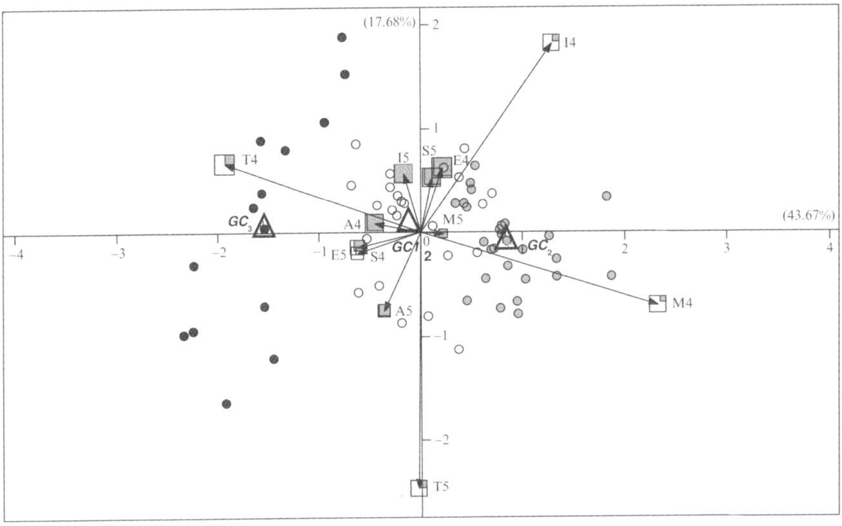

Fig. 6 AMMI biplot. Genotypes are represented in *circle* [*open*, *grey*, and *dark* representing the three Corsten and Denis (1990) clusters identified in the text]. The *triangles* represent the mean for the three clusters. Environments are shown in *squares* with areas proportional to their average yield. Within each square the *cut-out portion* is a representation of the amount of the sum of squares for each environment which is not explained by the axes under consideration

explicit environmental measurements may allow the genotype by environment interaction to be related to physiological processes (Vargas et al., 1999; Voltas et al., 1999a, 1999b, 2002). Table 3 shows the correlation coefficients between the first three AMMI environmental scores with every one of the meteorological variables recorded. There is, however, no easy interpretation of the results. IPCA1 is not particularly related to any specific variable; the largest correlation is with days of maximum temperature above 30 °C during grain filling, but still the magnitude is low. IPCA2 is more closely associated with the temperature range during jointing and water availability during grain filling. IPCA3 is associated with high temperatures in the first growth phase, tillering, and water status during the second, jointing.

Closely related to the AMMI model is the so-called GGE model. The GGE model has become popular through the extensive use of the biplot associated with it (Yan et al., 2000, 2001; Yan and Kang, 2003). The GGE model applies a principal components analysis to a two-way genotype by environment table with the genotypes being the objects and the environments being the variables. The variables are not standardized. The model is given in Eq. (8):

$$\mu_{ij} = \mu + E_j + \sum_{k=1}^{K} a_{ki} b_{kj} \tag{8}$$

Table 3 Correlation coefficients, r_{ik} between the kth AMMI environmental scores, IPCAe[k] and the ith explicit meteorological variable (see text for the acronyms of meteorological variables)

Meteo Variable [i]	R_k^2:	IPCAe[1] *43.66*	IPCAe[2] *17.67*	IPCAe[3] *11.08*	$\sum_{k=1}^{3} r_{ik}^2 \times R_k^2$
Tillering					
dTb0		−0.07	0.23	0.09	*1.27*
dTbb		−0.19	0.13	0.06	*1.88*
dTo30		−0.04	−0.19	0.63	*5.13*
TMx		0.17	−0.32	−0.12	*3.26*
Tmn		0.07	−0.08	−0.04	*0.33*
Tdif		0.22	−0.45	−0.17	***6.11***
GDDT		0.09	0.20	0.26	*1.82*
WT		−0.12	0.14	0.35	*2.32*
WDT		−0.22	0.04	0.20	*2.58*
PQ		0.00	−0.07	0.05	*0.11*
Jointing					
dTb0		−0.23	−0.05	0.07	*2.37*
dTbb		−0.19	−0.03	0.03	*1.65*
dTo30		−0.15	−0.19	0.54	*4.96*
TMx		−0.06	−0.23	0.09	*1.14*
Tmn		−0.04	−0.01	0.02	*0.07*
Tdif		−0.07	−0.64	0.23	***7.98***
GDDT		−0.08	−0.12	0.04	*0.53*
WT		0.29	−0.03	−0.50	***6.31***
WDT		0.15	0.05	−0.35	*2.36*
PQ		0.14	0.22	0.14	*1.96*
Grain Filling					
dTb0		−0.15	0.01	0.14	*1.19*
dTbb		−0.19	−0.06	0.01	*1.56*
dTo30		−0.46	−0.12	0.18	***9.69***
TMx		−0.36	−0.15	0.08	*6.27*
Tmn		−0.36	−0.04	−0.03	*5.84*
Tdif		−0.04	−0.29	0.27	*2.30*
GDDT		−0.38	−0.06	−0.01	*6.40*
WT		0.07	−0.66	−0.30	*9.00*
WDT		−0.04	−0.63	−0.26	*7.93*
PQ		0.28	0.31	0.17	*5.31*

The last column shows a weighted average of the r_{ik}'S using as weights the proportion of G.E sum of squares explained by the different AMMI scores, R_k^2.

One may debate about the relative superiority of GGE over AMMI for prediction purposes (Gauch, 2006; Yan et al., 2007), but a more fruitful approach is to use both as exploratory tools for visualization of adaptation patterns. GGE biplots are easier to interpret than AMMI biplots, as all relevant information on the genotypes, G and G.E, can be shown simultaneously.

4 Models for Interaction Using Explicit Environmental Characterizations

4.1 Factorial Regression Models

Bilinear models for interaction can be used for exploratory analysis of GE. We attempted to construct hypothetical environmental characterizations to which genotypes have different sensitivities. The results of analysis with bilinear models do not necessarily have an interpretable physiological basis. To ascertain whether we can invoke a physiological explanation, we can check the relationship of the environmental main effects and scores with explicit characterizations of the environment (physical and meteorological).

Suppose that for a particular data set the regression on the mean model gives an adequate description of GE and that the environmental main effect is a direct reflection of the average daily temperature, T_j. Eq. (4) can then be modified with the parameter E_j in the GE part of the regression on the mean model becoming a function of T_j: $\mu_{ij} = \mu + G_i + E_j + \beta_i f(T_j)$. Describing the interaction as driven linearly by temperature then leads to $\mu_{ij} = \mu + G_i + E_j + \beta_i T_j$, with β_i the sensitivity of genotype *i* to changes in temperature. These statistical models for GE that include differential genotypic sensitivity to explicit environmental variables belong to the class of factorial regression models (Denis, 1988; van Eeuwijk et al., 1996).

Extension to more environmental variables and more complex response curves are conceptually simple. For our example data this is, however, somewhat complicated by the reduced number of environments. If we consider the GE a resultant of three variables: 'average difference between daily maximum and minimum temperature during jointing', z_{1j}, 'total water during jointing', z_{2j}, and 'number of days with temperature above 30 °C during grain filling', z_{3j}, then the following model may be appropriate:

$$\mu_{ij} = \mu + G_i + E_j + \beta_{1i} z_{1j} + \beta_{2i} z_{2j} + \beta_{3_i} z_{3j} \tag{9}$$

In model (9), β_{1i}, β_{2i}, and β_{3i} are the genotypic sensitivities to these three variables, respectively. The model resembles closely a linear–bilinear model with three bilinear terms for the interaction, $\mu_{ij} = \mu + G_i + E_j + a_{1i}b_{1j} + a_{2i}b_{2j} + a_{3i}b_{3j}$. In this bilinear model, the environmental scores b_{1j}, b_{2j}, and b_{3j} are, theoretically, the best environmental covariables for explaining GE. Physiological understanding of GE requires us to interpret these scores in terms of explicit environmental characterizations.

4.2 Variable Selection

A central question when using factorial regression models is the choice of covariables for description of GE. Continuous monitoring of the environment is becoming increasingly common, so the question arises how to summarize the most relevant features of the environment from a GE point of view (Cooper and Hammer, 1996). Purely statistical approaches as variable subset-selection procedures will not solve this problem, because they often lead to incomprehensible models. Ecophysiological understanding of the genotypes and environments under study should therefore complement statistical considerations and we should use such knowledge to delimit the collection of potentially useful sets of environmental covariables. Even then, different combinations of variables may give similar goodness of fit (Voltas et al., 1999a,b).

A further point of consideration is that yield arises from an integration of growth processes over the entire crop life cycle, from which it may be concluded that the order of inclusion of variables should reflect the sequence of growth stages. We would therefore begin by testing for inclusion of variables observed during tillering, followed by those variables recorded during jointing, after having corrected for the tillering variables, and finally including those variables related to the grain-filling phase. For barley, Voltas et al. (1999a,b) presented examples of the use of factorial regression using physiological knowledge in analysis of adaptation and GE.

Our goal is not to identify the 'best' factorial regression model, but to illustrate the use of such models in practical analyses. In this context, Table 4 exemplifies analysis of variance tables for factorial regression models that linearly incorporate one by one the ten explicit meteorological variables recorded at each of the three growing phases. We used the genotypic groups obtained from the Corsten and Denis (1990) clustering procedure to partition the variation explained by factorial regression on the environmental covariables into a part due to regression for the genotypic clusters (one response for each of the three groups) and a part due to residual genotypic variation within clusters (residual genotypic deviations from the group response). The best individual variable was 'days of temperature over 30°C during grain filling'. However, the amount of the G.E sum of squares explained, although highly significant, was limited (13.54%; 8.27% for differential responses between groups and 5.27% for residual genotypic differences within groups). The three best variables were relatively uncorrelated and highly significant when incorporated into a multiple factorial regression model. Altogether they explained 33% of the G.E sum of squares (Table 5). Figure 5 shows the box plots for the fitted sensitivities of the 65 entries for these three variables separately, taking into account the genetic groups. From Fig. 5 the different adaptive behaviour of GC_3 group is again evident. Similar observations on the differences between the groups can be made using other techniques, see for example Fig. 4. This difference could be casually or causally attributed to one or more of the meteorological variables identified here.

For identifying subsets of environmental covariables, multiple (factorial) regression variable subset procedures may be computationally and conceptually complex. Furthermore, estimated genotypic parameters may be difficult to interpret within elaborate regression models. An alternative approach to variable selection would be to correlate the estimates for the environmental scores from linear–bilinear models (e.g. the first AMMI IPCA scores) with a candidate set of environmental covari-

Table 4 Partitioning of the G.E. interaction in Table 1 using factorial regression for a collection of ten meteorological variables at three sequential growth phases (Tillering, Jointing and Grain Filling)

Explicit variable: Meteo	GC. Meteo			Genotype(GC).Meteo		
	Sum of Squares	R^2	$-\log_{10}(p)$	Sum of Squares	R^2	$-\log_{10}(p)$
Tillering						
dTb0	0.00	0.29	0.43	0.00	4.26	0.00
dTbb	0.00	1.36	1.98	0.00	3.43	0.00
dTo30	0.00	0.61	0.91	0.00	7.08	0.06
TMx	3.34	1.53	2.27	12.57	5.75	0.01
Tmn	0.29	0.13	0.20	12.34	5.65	0.00
Tdif	6.37	2.92	4.36	12.04	5.51	0.00
GDDT	0.75	0.34	0.51	12.83	5.87	0.01
WT	2.46	1.13	1.62	5.87	2.69	0.00
WDT	5.14	2.35	3.43	7.29	3.34	0.00
PQ	0.01	0.00	0.01	8.00	3.66	0.00
Jointing						
dTb0	4.76	2.18	3.13	4.42	2.02	0.00
dTbb	3.26	1.49	2.12	3.25	1.49	0.00
dTo30	3.74	1.71	2.57	14.33	6.56	0.03
TMx	0.47	0.21	0.31	10.29	4.71	0.00
Tmn	0.50	0.23	0.33	7.69	3.52	0.00
Tdif	0.19	0.08	0.13	22.00	10.07	0.70
GDDT	0.71	0.33	0.47	8.65	3.96	0.00
WT	11.30	5.17	7.63	6.99	3.20	0.00
WDT	3.69	1.69	2.46	7.88	3.60	0.00
PQ	0.94	0.43	0.62	6.59	3.01	0.00
Grain Filling						
dTb0	2.54	1.16	1.66	5.39	2.47	0.00
dTbb	2.90	1.33	1.89	3.12	1.43	0.00
dTo30	18.07	8.27	12.69	11.51	5.27	0.01
TMx	10.99	5.03	7.63	12.61	5.77	0.01
Tmn	10.39	4.76	7.14	11.05	5.06	0.00
Tdif	0.35	0.16	0.23	9.70	4.44	0.00
GDDT	11.58	5.30	8.01	11.76	5.38	0.00
WT	2.53	1.16	1.80	22.05	10.09	0.75
WDT	0.47	0.21	0.33	22.79	10.43	0.85
PQ	4.18	1.91	2.89	15.08	6.90	0.05

The GC. Meteo term assesses if the three different groups of genotypes, GC, identified by the Corsten & Denis (1990) procedure, have the same sensitivity to each explicit meteorological variable. The Genotype (GC).Meteo term evaluates if all genotypes within a group have equal sensitives (see text for the acronyms of meteorological variables).

ables (Table 3). If highly correlated, the resulting coefficients may help determining the candidate environmental variables for factorial regression models. To develop an integrated criterion, across AMMI IPCA's, for the identification of suitable environmental covariables to be included in factorial regression models, we

Table 5 Partitioning of the G.E interaction in Table 1 according to a multiple factorial regression based on the average difference between daily maximum and minimum temperature during Tillering (Tdif), Total Water (WT) at Jointing and days with temperature over 30ºC (dTo30) at grain filling.

Source of variation	d.f.	Sum of squares	Semipartial R^2	Mean squares	variance ratio	$-\log_{10}(p)$
Environment [E]	*11*	*1522.2*	*85.3*	*138.4*	*445.8*	
Tdif (Tillering)	1	138.5	9.1	138.5	446.3	*>100*
WT (Jointing)	1	426.8	28.0	426.8	1375.2	*>100*
dTo30 (Grain Filling)	1	121.0	7.9	121.0	389.8	*>100*
E'	8	835.9	54.9	104.5	336.6	*>100*
Genotype [G]	*64*	*44.4*	*2.5*	*0.69*	*2.23*	
G.E	*704*	*218.5*	*12.2*	*0.31*	*0.99*	
Tdif.GC	2	6.4	2.9	3.19	11.18	*4.75*
WT.GC	2	6.7	3.1	3.37	11.81	*5.02*
dTo30.GC	2	28.1	12.9	14.05	49.33	*19.59*
Tdif.Genotype (GC)	62	12.0	5.5	0.19	0.68	*0.01*
WT.Genotype (GC)	62	9.8	4.5	0.16	0.56	*0.00*
dTo30.Genotype (GC)	62	9.6	4.4	0.15	0.54	*0.00*
Residual	512	145.8	66.7	0.28		

calculated a weighted average of the squared correlation coefficients of each meteorological variable with the AMMI scores, using as weights the proportion of G.E sum of squares explained by the kth score. Variables identified by this criterion (Table 3) indeed were earlier found to play a role in factorial regression (see Table 4).

5 Models for Interaction Incorporating Explicit Genotypic Information

The identification of genetic covariables whose variation contributes substantially to mean differences between genotypes, G, and to environment dependent differences between genotypes, GE, is critical for a genetic and physiological interpretation of G and GE effects. In Sect. 4, we estimated the sensitivity of genotypes to changes in environmental covariables. In this section, we will investigate how to partition G and GE effects by the use of genotypic covariables. These covariables can have various forms. They can be genotypic means of other recorded phenotypic variables, where these means can refer to all or just a subset of environments. One example is days to heading as assessed in a specific trial under suitable conditions and another is a physiological measurement recorded under controlled conditions. Alternatively, genotypic covariables can represent molecular marker information, where markers can either be DNA polymorphisms in anonymous DNA sequences or be functional genes. Whatever the covariable, factorial regression can then be used to detect and locate gene/QTL main effects and gene/QTL by environment interactions.

Consider a co-dominant marker in a diploid crop with genotypes MM, Mm, and mm. To represent this information in a factorial regression model, we can create a covariable with values x_i for genotype i of 2, 1, and 0 to represent genotypes MM, Mm, and mm, respectively. The genotypic covariable takes a value equal to the number of M alleles. In a factorial regression model that includes this covariable x_i, under the assumption that the marker coincides with a QTL, the interpretation of the accompanying slope, say ρ, will be that of the effect of a QTL allele substitution of m by M. Effectively, ρ stands for the additive genetic effect of the QTL. However, typically markers do not coincide with QTL. To identify QTL, we need to fit factorial regression models for a grid of markers and virtual markers along the genome. The virtual markers are constructed from flanking marker information following well-known procedures for standard biparental crosses between inbred lines (see Lynch and Walsh, 1998).

Genotypic covariables derived from molecular marker information can be introduced in an additive model for the phenotype, model (1), to describe mean differences between genotypes across environments as follows:

$$\mu_{ij} = \mu + [x_i\rho + \mathrm{G}'_i] + \mathrm{E}_j \qquad (10)$$

where $x_i\rho + \mathrm{G}'_i$ represents a partitioning of the genotypic main effects in (1), G_i. Model (10) is fitted at a grid of points across the genome. The genotypic covariable x_i changes in relation to the genomic position that we are testing for a possible QTL expression. When (10) is fitted at positions close to QTL, the regression on x_i will explain a significant part of the genotypic differences. The slope ρ is the potential QTL main effect at the genomic position corresponding to x_i and G'_i is a residual genotypic main effect that should be smaller than the original G_i.

For a full genome scan, model (10) should be fitted a large number of times and this requires a multiple test correction of the significance level for assessing significance of x_i. A very conservative Bonferroni correction would simply take the significance level for individual markers (including virtual markers) as the genome-wide level divided by the total number of markers. In this approach, markers are assumed to be independent. Less conservative corrections attempt to consider dependence between markers, as will be the case due to linkage even with relatively few markers in a linkage group. One such method is estimating an effective number of independent tests across the genome and then dividing the genome-wide significance level by the effective number of independent tests. The latter number can be estimated in various ways. An interesting approach, based on eigenvalue decomposition of the correlation matrix of the full set of marker-derived covariables, is presented by Li and Ji (2005). Alternative approaches use simulation or permutation. An approximate rule could be to consider tests independent when they are more than a certain genetic distance apart, for example, 20 or 30 cM in the case of populations derived from biparental crosses or as little as 1-10 cM when considering a diverse pannel of genotypes. Such a rule would lead to a significance level for individual tests that is equal to the genome-wide significance level divided by a number between 50 and 200, according to the nature of the genetic population used, for a genome of size 1,000 cM, which is roughly equivalent to the reported map length of barley in many mapping studies.

For the sake of simplicity, we will only apply simple marker regression to our data, that is, we will take model (10) using observed markers as the basis for our QTL models. Model (10) is a single QTL model. To construct multiple QTL models, a composite interval mapping strategy can be developed using a straightforward extension of model (10), where we add so-called co-factors, markers that correct for QTL elsewhere on the genome. Co-factors can be selected from the genetic predictors corresponding to the QTL identified in a preliminary genome scan by marker regression or simple interval mapping.

Besides terms for QTL main effects, factorial regression models can also include terms for QTL by environment interaction:

$$\mu_{ij} = \mu + [x_i\rho + \mathrm{G}'_i] + \mathrm{E}_j + [x_i\rho_j + (\mathrm{G.E})'_{ij}] \tag{11}$$

The $(\mathrm{G.E})_{ij}$ term from model (2) is partitioned in a term due to differential QTL expression across environments, $x_i\rho_j$, and a random residual, $(\mathrm{G.E})'_{ij}$. In the presence of significant QTL by environment interaction, the new parameter ρ_j adjusts the average QTL expression across environments, ρ, to a more appropriate level for the individual environment j.

Models (10) and (11) can be used for any type of segregating population provided that appropriate genetic predictors are constructed. The same models are useful for QTL mapping in the situation of a collection of genotypes without a clearly defined family structure: linkage disequilibrium mapping, or association mapping. The problem with collections of genotypes with arbitrary structure is that linkage disequilibrium between markers and traits does not necessarily result from genetic linkage between a marker and a QTL. When the collection of genotypes consists of various sub populations, such as winter and spring types in the case of barley, false marker trait associations can occur due to differences in marker allele frequencies between the sub populations. For correct inference on marker trait associations, we therefore need to correct for any potential population structure prior to QTL detection.

A popular way to identify population structure is described by Pritchard et al. (2000), whose approach is available within the package *Structure*. However, alternative methods of defining population structure usually perform as well as the *Structure* approach. For example, as we have seen in Sect. 2, one may define population structure on the basis of the origin of the material.

Assuming that the genotypes are grouped according to a population structure definition, then an additive model which incorporates QTL main effects, see model (10), corrected for population structure is

$$\mu_{i(k)j} = \mu + [\mathrm{GC}_k + x_{i(k)}\rho + \mathrm{G}'_{i(k)}] + \mathrm{E}_j \tag{12}$$

where μ stands for the general mean, GC_k is the sub population to which genotype i belongs expressed as a deviation from the general mean, $x_{i(k)}$ represent the marker information for genotype i within sub population k, ρ is the QTL main effect and $\mathrm{G}'_{i(k)}$ corresponds to a residual genotypic effect.

In an analogous way, the $(\mathrm{G.E})_{ij}$ term from the full interaction model can be partitioned into a term for the interaction of sub population with environment, a term for differential QTL expression across environments, $x_{i(k)}\rho_j$, and a residual, $(\mathrm{G.E})'_{i(k)j}$. The complete model for marker trait association analysis incorporating QTL main effects and QTL by environment interactions is

$$\mu_{i(k)j} = \mu + [\mathrm{GC}_k + x_{i(k)}\rho + \mathrm{G}'_{i(k)}] + \mathrm{E}_j + [(\mathrm{GC.E})_{kj} + x_{i(k)}\rho_j + (\mathrm{G.E})'_{i(k)j}] \tag{13}$$

When the environments also have a structure, we can generalize model (13) to become

$$\begin{aligned}\mu_{i(k)j(l)} = \mu &+ [\mathrm{GC}_k + x_{i(k)}\rho + \mathrm{G}'_{i(k)}] + [\mathrm{EC}_l + \mathrm{E}'_{j(l)}] \\ &+ [(\mathrm{GC.EC})_{kl} + (\mathrm{GC.E})'_{kj(l)} + x_{i(k)}\rho_l + x_{i(k)}\rho_{j(l)} + (\mathrm{G.E})'_{i(k)j(l)}]\end{aligned} \tag{14}$$

In (14), ρ stands for consistent QTL effects across all environments, ρ_l is a deviation of the main effect QTL for environment group l, and $\rho_{j(l)}$ stands for a residual QTL effect in environment j. Interaction between genotype and environment groups is represented by $(\mathrm{GC.EC})_{kl}$. $(\mathrm{GC.E})'_{kj(l)}$ gives an environment-specific correction to the genotype by environment group interaction. $(\mathrm{G.E})'_{i(k)j(l)}$ gives a final residual GE term.

To demonstrate our approach, we applied models (10) to (14) to our data. To account for multiple testing, we used a significance criterion for QTL detection of $-\log_{10}(p\text{-value}) > 3$, that is, p-value < 0.001; this criterion corresponded empirically to a Bonferroni correction based on 50 independent tests across the full genome, taking about 30 cM as the distance at which marker trait association tests become independent. We used 811 genetic covariables, DArT® markers of known genomic position, for a genome-wide scan. The genetic covariables took the values of 1 or 0, depending on the (homozygous) presence or absence of each anonymous DArT sequence.

Figure 7 shows the number of DArT® markers which, when utilized in Eq. 10 (uncorrected for structure) and Eq. 12 (corrected), produced a significance level [$\log_{10}(p$-value)] and accounted for a fraction of the phenotypic variation (R^2) greater than a particular value shown on the X axes. For example, 144 of the markers had a $\log_{10}(p$-value) greater than 4 in the uncorrected data, while 40% of the markers, 316 out of 811, explained more than 5% of the original uncorrected differences in yield. Approximately 5% of the markers, with at least one marker located on each of barley's seven chromosomes, explained individually more than 20% of the phenotypic differences for yield (data not shown). In fact, significant

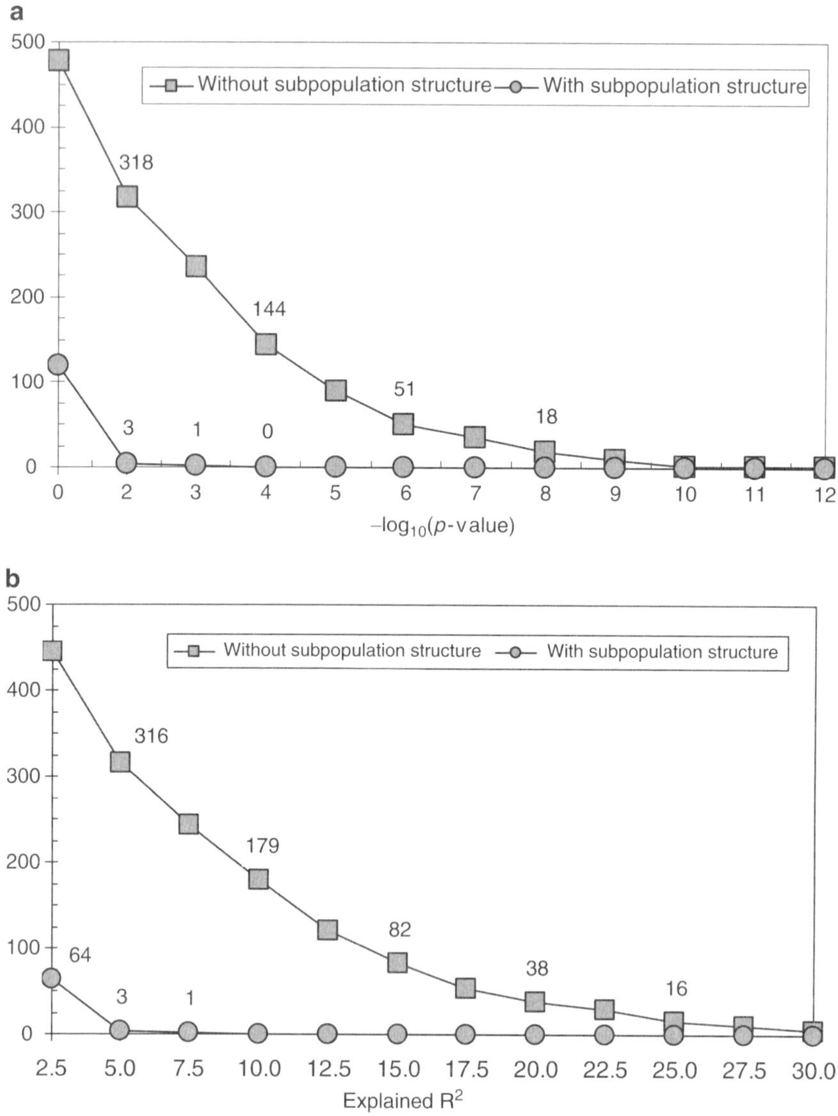

Fig. 7 Number of DArT® markers with $-\log_{10}(p\text{-value})$ (a) and proportion of the genotypic R^2 explained greater than any given value (b) in the association mapping of grain yield for 65 barley varieties grown in 12 sites according to an additive main effect QTL model (Eqs. 10 and 12) and simple marker regressions. *Squares* represent data not corrected for population substructure and *circles* data corrected for substructure

DArT® markers could be found in the proximity of most major developmental genes of known map position, which were fixed within a given genetic sub population (data not shown). With such large numbers, most genomic regions, represented by the barley bin map of Kleinhofs et al. (1998), contained at least one significant QTL. Many of these QTL would not have any direct use in breeding as their putative associations with yield are unlikely to be causal.

Few main effect QTL were detected when the sub population structure was included (model 12). Only three markers explained more than 5% of the phenotypic variation observed for grain yield. The lack of significant QTL for the main effects on population corrected data is not that surprising given the high variability of environments across the entire Mediterranean basin. QTL that interacted with the environments were more frequent than QTL main effects both in the uncorrected (applying Eq. 11) and, particularly, for the corrected data (Eq. 13) (Fig. 8). In the latter case, 99 DArT markers had a p-value smaller than 0.0001, and 55 below 10^{-6}. Twenty-nine markers explained more than 5% of the G.E interaction, and 12 each explained more than 7.5% of G.E.

The outputs for the different models are listed in Table 6 for the specific DArT® marker bPb0429, located in bin 6 of chromosome 1H. It may be worth mentioning that no major developmental gene has been detected so far in this region, which makes this marker particularly interesting. Sequential orthogonal partitioning of G, E, and G.E according to Eqs. 11–14 in population structure uncorrected and corrected data are shown in this table. The number of degrees of freedom and the total variation did not coincide with those in Table 2, as a number of entries could not be genotyped with this marker and, thus, these entries were not included in the analyses. bPb0429 explained more than 20% of the main effect genotypic differences in the uncorrected data (Table 6, Eqs. 10 and 12), but only 1.1% of the genotypic main effects in the corrected (Table 6, Eqs. 11, 13, and 14). Genetic effects were more tightly associated with differential QTL expression across individual environments than to the groups of environments and to QTL main effects. The interaction bPb0429.E represented 20% of G.E on uncorrected data (Table 6, Eq. 12) and 10% of G.E on corrected (Table 6, Eq. 13). The effect of this marker varied significantly among groups of environments and, particularly, between environments within each environmental group. This latter term of the model, bPb0429.E in Eq. 14 (Table 6), had a $-\log_{10}(p\text{-value})$ equal to 16.54 and explained almost 9% of G.E.

Table 7 shows estimates for QTL main effects and QTL.E associated with bPb0429 according to Eqs. 10–14. Some of these values should be interpreted with care as they represent deviations from specific levels. QTL effects were larger in uncorrected data. Presence of marker bPb0429 translated into a yield decrease of 0.27 t/ha across environments once corrected for population substructure, versus 0.34 t/ha on uncorrected data. The QTL effects varied between environment groups, with the estimated effect in environment group 3 being 1.09 t/ha larger than in group 1, which was 0.22 t/ha larger than in group 2 (Table 7, Eq. 14). They also varied significantly across specific environments within and across environmental groups, being particularly large for Italy 2004, in which the difference associated to this marker was 1.75 t/ha and of different sign than that for Turkey 2005, with an associated effect of +1.00 t/ha. These genetic effects based on population structure corrected data are often not readily observable from the raw data. This fact can be observed in Fig. 9, in which separate box

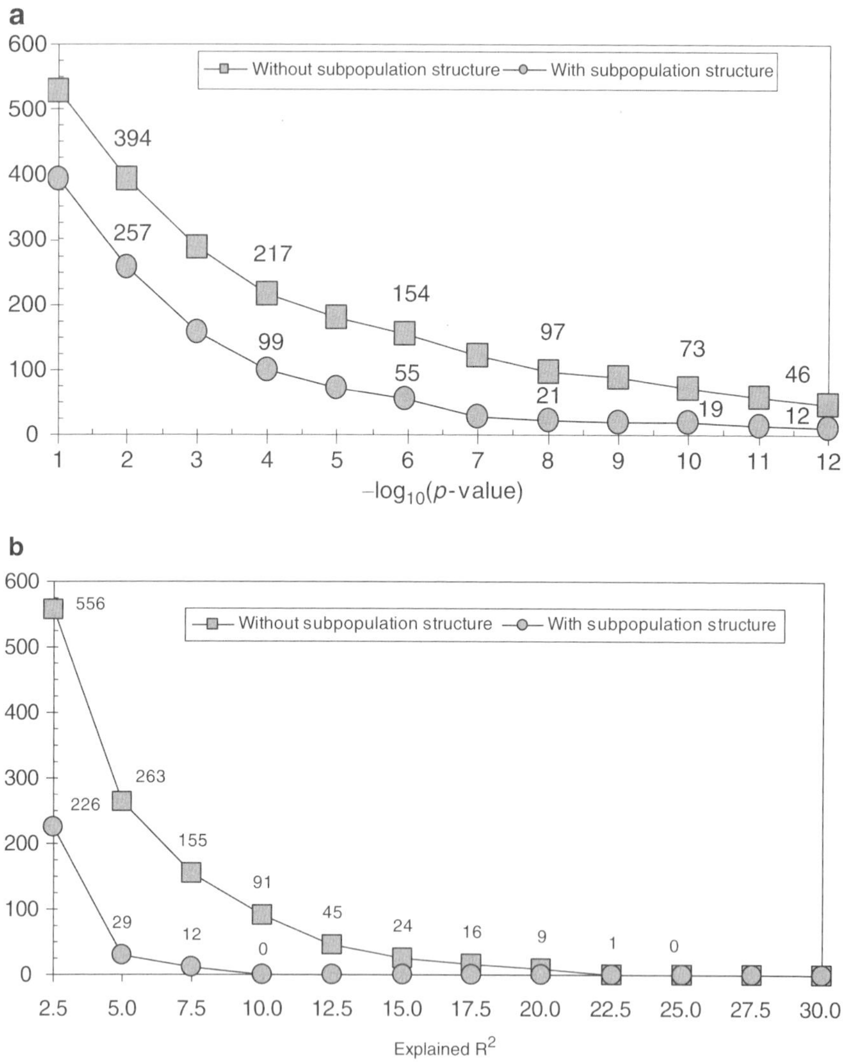

Fig. 8 Number of DArT® markers with $-\log_{10}(p\text{-value})$ (**a**) and proportion of the genotypic R^2 explained greater than any given value (**b**) in the association mapping of grain yield for 65 barley varieties grown in 12 sites for the QTL.E term (Eqs. 11 and 13) and simple marker regressions. *Squares* represent data not corrected for population substructure and *circles* data corrected for substructure

plots are shown for the original yield values for varieties carrying and not carrying marker bPb0429 in each of 12 environments. Considering uncorrected data, presence of bPb0429 was particularly negative in Italy and Morocco, with yield reductions of up to 2 t/ha, and increases in Turkey 2004 and 2005 of 0.41 and 0.58 t/ha, respectively. These effects were drastically changed at some sites when using population corrected data (Table 7).

Table 6 Partitioning of the G.E interaction in Table 1 according to a genetic covariable (the DArT marker bPb0429) using alternative linear models described in the text. Residuals from each model were used as denominators for the F tests

Source of variation Equation (10)	d.f.	Sum of squares	Mean squares	Variance ratio	$-\log_{10}$ (p -value)
Environment [E]	11	1475.8	134.16	423.2	*>100*
bPb0429	1	9.9	9.90	31.2	*7.48*
G'	61	34.1	0.56	1.8	*3.31*
Residual	682	216.2	0.32	1.0	
Source of variation Equation (11)	**d.f.**	**Sum of squares**	**Mean squares**	**Variance ratio**	**$-\log_{10}$ (p -value)**
Environment [E]	11	1475.8	134.16	518.4	*>100*
bPb0429	1	9.9	9.90	38.2	*8.96*
G'	61	34.1	0.56	2.2	*5.64*
bPb0429.E	11	42.6	3.87	15.0	*25.44*
Residual	671	173.7	0.26		
Source of variation Equation (12)	**d.f.**	**Sum of squares**	**Mean squares**	**Variance ratio**	**$-\log_{10}$ (p-value)**
Environment [E]	11	1475.8	134.16	423.2	*>100*
GC	2	22.9	11.44	36.1	*14.89*
bPb0429	1	0.5	0.46	1.4	*0.64*
G'	59	20.7	0.35	1.1	*0.55*
Residual	682	216.2	0.32		
Source of variation Equation (13)	**d.f.**	**Sum of squares**	**Mean squares**	**Variance ratio**	**$-\log_{10}$ (p-value)**
Environment [E]	11	1475.8	134.16	776.4	*>100*
GC	2	22.9	11.44	66.2	*26.16*
bPb0429	1	0.5	0.46	2.6	*0.98*
G'	59	20.7	0.35	2.0	*4.69*
GC.E	22	83.0	3.77	21.8	*63.17*
bPb0429.E	11	21.0	1.91	11.1	*18.20*
Residual	649	112.2	0.17		
EC	2	24.0	12.00	67.0	*26.46*
E'	9	1451.8	161.31	901.2	*>100*
GC	2	22.9	11.44	63.9	*25.33*
bPb0429	1	0.5	0.46	2.5	*0.95*
G'	59	20.7	0.35	2.0	*4.29*
GC.EC	4	68.3	17.08	95.4	*63.11*
GC.E'	18	14.7	0.76	4.2	*7.87*
bPb0429.EC	2	2.1	1.03	5.7	*2.47*
bPb0429.E'	9	19.0	2.11	11.8	*16.54*
Residual	649	112.2	0.18		

Table 6 (Continued)

Source of variation Equation (15)	d.f.	Sum of squares	Mean squares	Variance ratio	$-\log_{10}$ (p-value)
EC	2	24.0	12.00	67.0	*26.46*
mTdif2	1	775.8	775.77	4333.9	
E'	8	676.0	84.50	472.1	*>100*
GC	2	22.9	11.44	63.9	*25.33*
bPb0429	1	0.5	0.46	2.5	*0.95*
G'	59	20.7	0.35	2.0	*4.29*
GC.EC	4	68.3	17.08	95.4	*63.11*
mTdif2.bPb0429	1	7.6	7.59	42.4	*9.83*
bPb0429.EC	2	0.7	0.37	2.1	*0.89*
bPb0429.E'	8	14.0	0.76	4.2	*4.26*
GC.mTdif2	2	2.1	1.03	5.7	*2.47*
G'.mTdif2	59	11.8	0.20	1.1	*0.57*
Residual	606	111.8	0.18		

6 Models for Interaction Simultaneously Incorporating Explicit Environmental and Genotypic Information

In the previous section, we have discussed how the G and the G.E terms of the analysis of variance model can be partitioned by means of genetic covariables, x_i, into QTL main effects (ρ) and in QTL.E (ρ_j). In the presence of QTL by environment interaction, the parameter ρ_j adjusts the average QTL expression across environments, ρ, to a more appropriate level for the individual environment j as shown in Table 7. The QTL.E parameters, ρ_j, can be regressed on an environmental covariable, z, to link differential QTL expression directly to key environmental factors causing GE. This can be done by replacing the QTL by environment interaction term, $x_i\rho_j$ by a regression term $x_i(\lambda z_j)$ and a residual term, $x_i\rho'_j$,

$$\mu_{ij} = \mu + [x_i\rho + \mathrm{G}'_i] + \mathrm{E}_j + [x_i(\lambda z_j) + x_i\rho'_j + (\mathrm{G.E})'_{ij}] \tag{15}$$

The residual term $x_i\rho'_j$ will disappear from the expectation when ρ'_j is assumed to be random. The parameter λ is a proportionality constant that determines the extent to which a unit change in the environmental covariable z influences the effect of a QTL allele substitution.

Model (15) predicts differential genotypic responses to environmental changes from marker information characterizing the genotypes and environmental covariables characterizing the environments. Van Eeuwijk et al. (2001, 2002) provide an example of differential QTL expression in relation to the minimum temperature during flowering for yield in maize data from the CIMMYT program on drought

Table 7 Significance of alternative terms and estimates of QTL main effects and QTL.E effects associated to a genetic covariable (the DArT® marker bPb0429) according to Equation 10 and 11 (without population substructure adjustment), and 12 to 14 (upon population substructure adjustment). Estimates for each each environment are given both as deviates and as absolute terms. See Table 6 for detailed analyses of variances

Model	Term	$-\log_{10}$ (p-value)	QTL effect	QTL-E effects															Average Effect
				EC1	*M4*	*EC2*	D4	D5	E5	S4	T4	*EC3*	E4	I4	I5	M5	S5	T5	
			ρ	ρ_1	$\rho_{1(1)}$	ρ_2	$\rho_{2(1)}$	$\rho_{2(2)}$	$\rho_{2(3)}$	$\rho_{2(4)}$	$\rho_{2(5)}$	ρ_3	$\rho_{3(1)}$	$\rho_{3(2)}$	$\rho_{3(3)}$	$\rho_{3(4)}$	$\rho_{3(5)}$	$\rho_{3(6)}$	
Equation 10	bPb0429	*7.48*	−0.34																−0.34
Equation 11	bPb0429	*8.96*	−0.06																
	bPb0429*E	*25.44*			−1.41		0.00	0.44	0.20	0.34	0.47		−0.59	−1.96	−0.44	−0.38	−0.63	0.57	
	Absolute effects				−1.47		−0.06	0.38	0.14	0.28	0.41		−0.65	−2.02	−0.50	−0.44	−0.69	0.51	−0.34
Equation 12	bPb0429	*0.64*	−0.27																−0.27
Equation 13	bPb0429	*0.98*	−0.31																
	bPb0429*E	*18.20*			0.22		0.00	0.67	0.22	0.39	−0.07		−0.19	−1.44	−0.27	0.11	−0.46	1.31	
	Absolute effects				−0.08		−0.31	0.36	0.09	0.08	0.38		−0.50	−1.75	−0.58	−0.20	−0.76	1.00	−0.27
Equation 14	bPb0429	*0.95*	−0.08																
	bPb0429*EC	*2.47*		0.00		−0.22						1.09							
	bPb0429*E′	*16.54*			0.00		0.00	0.67	0.22	0.39	−0.07		−1.50	−2.75	−1.58	−1.20	−1.77	0.00	
	Absolute effects				−0.08		−0.31	0.36	−0.09	0.08	−0.38		−0.50	−1.75	−0.58	−0.20	−0.76	1.00	−0.27

Estimates for ach environment are given both as deriates and as absolute terms. See Table 6 for detailed analyses of variances

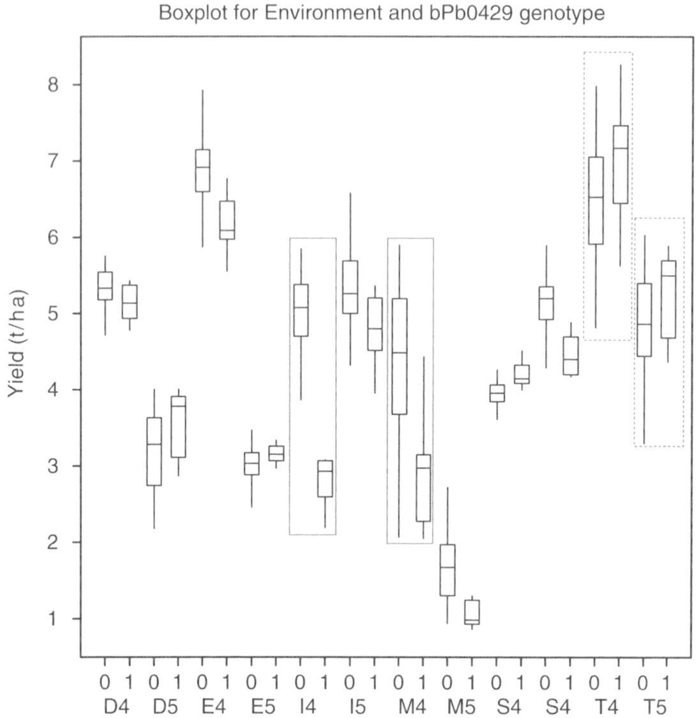

Fig. 9 Box plot for grain yield of lines carrying alternate alleles of DArT® marker bPb0429 in bin 6 of chromosome 1H for each of the 12 environments described in Table 1. On the *X* axes the *numbers* in the first row identify whether the bPb0429 marker was absent, 0, or present, 1; the two characters in the second row identify the 12 environments

stress. Similarly, Malosetti et al. (2004) analysed yield data from the North American Barley Genome Project with added environmental information. In the latter case using yield data from the Steptoe × Morex double haploid population grown at 10 sites, a significant QTL.E at chromosome 2H was found to depend on the temperature range during heading. A QTL allele substitution increased/decreased yield by 0.112 t/ha for every degree Celsius that the temperature range increased.

We applied Eq. 15 for every combination of the 30 environmental and 811 genetic covariables used before. A number of markers significantly interacted with environmental variables. DArT® marker bPb0429 located on chromosome 1H significantly interacted with the temperature range during jointing (Table 6, last section). This term was highly significant; with just one out of the 682 degrees of freedom the $-\log_{10}(p\text{-value})$ was larger than 9. However, although extremely significant, its associated R^2 was not that high, explaining around 4% of the G.E sum of squares. Alternate QTL alleles increased/decreased yield by 0.25 t/ha for every degree Celsius that the temperature range increased (Fig. 10).

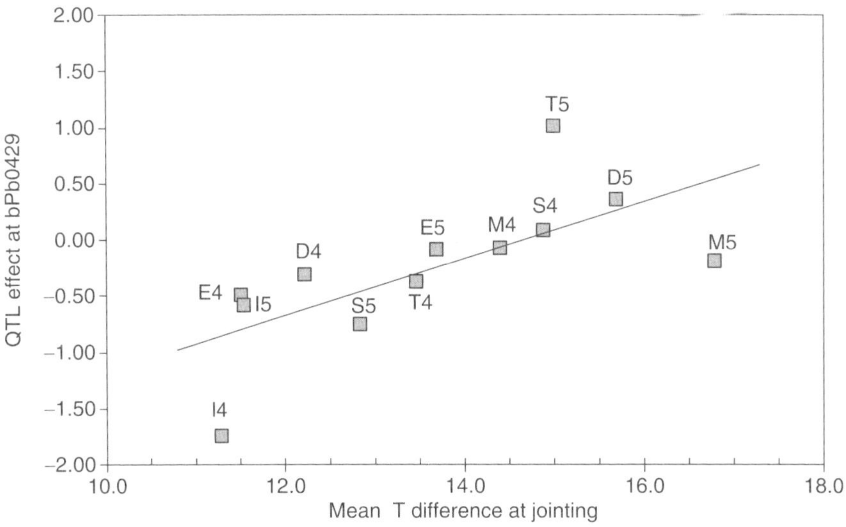

Fig. 10 Regression of QTL.E effects as determined for DArT® marker bPb0429 located on chromosome 1H for grain yield in barley on environmental covariable average temperature range, difference between maximum and minimum daily temperature, during jointing

7 Conclusions

We have shown in this chapter how we can integrate environmental and genetic information into a series of general linear models of increasing complexity based on analyses of variance and regression, which can be easily formulated using standard statistical packages. These models identify key environmental variables to explain differential phenotypic responses and estimate the genotypic sensitivities to them. They can also partition, by means of genetic covariables, the G and the G.E terms of the analysis of variance model into QTL main effects and QTL.E interaction terms which are then readily estimable. The QTL and QTL.E estimates can be further regressed on any environmental covariable to identify differential QTL expression potentially related to environmental factors. Critical analysis of these models may result in new applied breeding strategies for adaptation. However, we have just focused on modelling the expected responses in terms of their dependence on genotypic and environmental covariables. No attention has been given to the variance–covariance section of the data. The mixed model framework, combining modelling of mean and variance, provides a more powerful tool to study G.E and QTL.E. It offers greater flexibility with regard to *a priori* basic assumptions on homoscedasticity of residual variances and lack of correlations across environments and improves precision of genotypic estimates.

Nevertheless, we have demonstrated the value of including meteorological parameters in our models that can lead to greater insight into genomic regions underlying interactions with the environment. This has been achieved despite a common assignment of the onset of jointing and senescence for all genotypes under study. As there is genetic variation in the duration of the jointing and grain-filling growth phases, we expect that inclusion of genotypic specific jointing and grain filling times as genetic covariables will improve the fit of our models.

Acknowledgements Contribution of the other partners of the EU FP5 INCO-MED project 'Mapping Adaptation of Barley to Droughted Environments' in assembling this data set is highly appreciated, namely, Salvatore Ceccarelli and Stefania Grando from ICARDA, Syria; Michele Stanca from the Istituto Sperimentale per la Cerealicoltura, Italy; José Luis Molina-Cano and Alexander Pswarayi from the Centre UdL-IRTA, Spain; Taner Akar from the Central Research for Field Crops, Turkey; Adnan Al-Yassin from NCARTT, Jordan; Abdelkader Benbelkacem from ITGC, Algeria; Mohammed Karrou and Hassan Ouabbou from INRA, Morocco; Nicola Pecchioni and Enrico Francia from the Università di Modena e Reggio Emilia, Italy; Wafaa Choumane from Tishreen University, Latakia, Syria; and Jordi Bort and José Luis Araus from the University of Barcelona. We also want to express our gratitude to Jordi Comadran, Joanne Russell from SCRI for providing the marker data used for association mapping and to Christine Hackett from BioSS for fruitful statistical discussions. The above work was funded by the European Union-INCO-MED program (ICA3-CT2002-10026). The Centre UdL-IRTA forms part of the Centre CONSOLIDER on Agrigenomics and acknowledges partial funding from grant AGL2005-07195-C02-02 from the Spanish Ministry of Science and Education. Fred van Eeuwijk wants to acknowledge funding of the Generation Challenge Program (project G4007.09: Methodology and software development for marker-trait association analyses).

References

Annicchiarico, P. (2002) Genotype x Environment Interactions – Challenges and Opportunities for Plant Breeding and Cultivar Recommendations. Food and Agriculture Organization of the United Nations, Rome.

Boer, M., Wright, D. Feng, L., Podlich, D., Luo, L., Cooper, M. and van Eeuwijk, F.A. (2007) A mixed model QTL analysis for multiple environment trial data using environmental covariables for QTLxE, with an example in maize. Genetics 177, 1801–1813.

Comadran, J., Russell, J.R., van Eeuwijk, F., Ceccarelli, S., Grando, S., Stanca, A.M., Francia, E., Pecchioni., N., Akar, T., Al-Yassin, A., Benbelkacem, A., Choumane, W.,Ouabbou, H., Bort, J., Araus, J.L., Pswarayi, A., Romagosa, I., Hackett, C.A. and Thomas, W.T.B. (2008) Mapping adaptation of barley to droughted environments. Euphytica 161, 35–45.

Cooper, M. and Hammer, G.L. (Eds.) (1996) 'Plant Adaptation and Crop Improvement'. CAB International, Wallingford, UK.

Corsten, L.C.A. and Denis, J.B. (1990) Structuring interaction in two-way tables by clustering. Biometrics 46, 207–215.

Crossa, J. and Cornelius, P. (2002) Linear–bilinear models for the analysis of genotype–environment interaction. In: Kang, M.S. (Ed.) Quantitative genetics, genomics and plant breeding. pp. 305–322. CAB International, Wallingford, UK.

Denis, J.B. (1988) Two-way analysis using covariates. Statistics 19, 123–132.

Denis, J.B. and Gower, J.C. (1996) Asymptotic confidence regions for biadditive models: interpreting genotype-environment interactions. Applied Statistics 45, 479–492.

Falush, D., Stephens, M., and Pritchard, J.K. (2003) Inference of population structure: Extensions to linked loci and correlated allele frequencies. Genetics, 164, 1567–1587.
Falush, D., Stephens, M., and Pritchard, J.K. (2007) Inference of population structure using multilocus genotype data: dominant markers and null alleles. Molecular Ecology Notes 7, 574–578.
Finlay, K.W. and Wilkinson, G.N. (1963) The analysis of adaptation in a plant breeding programme. Australian Journal of Agricultural Research 14, 742–754.
Fox, P.N., Crossa, J. and Romagosa, I. (1997) Multi-environment testing and genotype environment interaction. In: R.A. Kempton and P.N. Fox (Eds.) Statistical Methods for Plant Variety Evaluation. pp. 117–137. Chapman and Hall, London.
Gabriel, K.R. (1978) Least squares approximation of matrices by additive and multiplicative models. Journal of the Royal Statistical Society, Series B 40, 186–196.
Gabriel, K.R. (1998) Generalised bilinear regression. Biometrika 85, 689–700.
Gabriel, K.R. and Zamir, S. (1979) Lower rank approximations of matrices by least squares with any choice of weights. Technometrics, 21, 489–498.
Gauch, H.G. (1988) Model selection and validation for yield trials with interaction. Biometrics, 44, 705–715.
Gauch, H.G. (1992) Statistical Analysis of Regional Yield Trials. Elsevier, Amsterdam.
Gauch, H.G. (2006) Statistical Analysis of Yield Trials by AMMI and GGE. Crop Science, 46, 1488–1500.
Gollob, H.F. (1968) A statistical model which combines features of factor analysis and analysis of variance techniques. Psychometrika, 33, 73–115.
Kang, M.S. (Ed.). (1990) Genotype-By-Environment Interaction and Plant Breeding. Louisiana State University, Baton Rouge, Louisiana.
Kang, M.S. (1998) Using genotype-by-environment interaction for crop cultivar development. Advances in Agronomy, 62, 199–252.
Kang, M.S. and Gauch, H.G. (1996) Genotype by Environment Interaction: New Perspectives. CRC Press, Boca Raton, FL.
Kempton, R.A. (1984) The use of biplots in interpreting variety by environment interactions Journal of Agricultural, 103, 123–135.
Kempton, R.A. and Fox, P.N. (1997) Statistical Methods for Plant Variety Evaluation. Chapman and Hall, London.
Kleinhofs, A. and Han, F. (2002) Molecular Mapping of the Barley Genome. In: Slafer G.A., Molina-Cano, J.L., Savin, R., Araus, J.L. and Romagosa, I. (Eds.) Barley Science: Recent Advances from Molecular Biology to Agronomy of Yield and Quality. pp. 31–63. Haworth Pres, Binghamton, NY.
Kleinhofs, A., Kudrna, D.A. and Matthews, D. (1998) Co-ordinators report: Integrating barley molecular and morphological/physiological marker maps. Barley Genetics Newsletter, 28, 89–91.
Li, J. and Ji, L. (2005) Adjusting multiple testing in multilocus analyses using the eigenvalues of a correlation matrix. Heredity, 95, 221–227.
Lynch, M. and Walsh, J.B. (1998) 'Genetics and analysis of quantitative traits'. Sinauer Associates, Sunderland, Massachusetts.
Malosetti, M., Voltas, J., Romagosa, I., Ullrich, S.E. and van Eeuwijk, F.A. (2004) Mixed models including environmental variables for studying QTL by environment interaction. Euphytica, 137, 139–145.
Mandel, J. (1969) The partitioning of interaction in analysis of variance. J Res NBS 73B, 309–328.
Paterson, A.H. (Ed.) (1998) Molecular Dissection of Complex Traits. CRC Press., Boca Raton, FL.
Payne, R.W., Harding, S.A., Murray, D.A., Soutar, D.M., Baird, D.B., Welham, S.J., Kane, A.F., Gilmour, A.R., Thompson, R., Webster, R., Tunnicliffe, E., Wilson, G. (2006) GenStat release 9 reference manual, part 2 directives. VSN International, Hemel Hempstead, UK.
Piepho, H.P. (1997) Analyzing genotype-environment data by mixed models with multiplicative effects. Biometrics 53, 761–766.

Piepho, H.P. (2000) A mixed model approach to mapping quantitative trait loci in barley on the basis of multiple environment data. Genetics 156, 2043–2050.

Piepho, H.P. and Pillen, K. (2004) Mixed modeling for QTL x environment interaction analysis. Euphytica 137, 147–153.

Pritchard, J. K., Stephens, M. and Donnelly, P. (2000) Inference of population structure using multilocus genotype data. Genetics 155, 945–959.

Pswarayi, A., van Eeuwijk, F., Ceccarelli, S., Grando, S., Comadran, J., Russell, J.R., Stanca, A.M., Francia, E., Pecchioni, N., Akar, T., Al-Yassin, A., Benbelkacem, A., Choumane, W., Karrou, M., Ouabbou, H., Bort, J., Araus, J.L., Molina-Cano, J.L., Thomas, W.T.B., and Romagosa, I. (2008) Barley adaptation and improvement in the Mediterranean basin. Plant Breeding 127, 554–556.

Romagosa, I. and Fox, P.N. (1993) Genotype-environment interaction and adaptation. In: Hayward, M.D., Bosemark, N.O., and Romagosa, I. (Eds.) Plant Breeding, Principles and Prospects. Pp. 373–390. Chapman and Hall, London.

Russell, J., Booth. A., Fuller, J., Harrower, B., Hedley, P., Machray, G., and Powell, W. (2004) A comparison of sequence-based polymorphism and haplotype content in transcribed and anonymous regions of the barley genome. Genome 47, 389–398.

Slafer, G.A., Molina-Cano, J.L., Savin, R., Araus, J.L. and Romagosa, I. (Eds.) (2002) Barley Science: Recent Advances from Molecular Biology to Agronomy of Yield and Quality. Haworth Press, Binghamton, NY.

Smith, A.B. (1999) Multiplicative mixed models for the analysis of multi-environment trial data. PhD thesis, Dpt of Statistics, University of Adelaide, South Australia.

Smith, A.B., Cullis, B.R. and Thompson, R. (2005) The analysis of crop cultivar breeding and evaluation trials: an overview of current mixed model approaches; Journal of Agricultural Science Cambridge 143, 1–14.

Spiertz, J.H.J., Struik, P.C. and van Laar, H.H. (Eds.) (2007) Scale and Complexity in Plant Systems Research. Gene-Plant-Crop relations. Wageningen UR Frontier Series. Vol 21. Springer, Dordrecht, The Netherlands.

van Eeuwijk, F.A. (1995a) Linear and bilinear models for the analysis of multi-environment trials: I An inventory of models. Euphytica 84, 1–7.

van Eeuwijk, F.A. (1995b) Multiplicative interaction in generalized linear models. Biometrics 51, 1017–1032.

van Eeuwijk, F.A. (1996) Between and Beyond Additivity and Non-Additivity: the Statistical Modelling of Genotype by Environment Interaction in Plant Breeding. PhD Thesis. Wageningen, The Netherlands.

van Eeuwijk, F.A. (2006) Genotype by environment interaction: basics and beyond. In: Lamkey, K. and Lee, M. (Ed.) Plant Breeding: The Arnell Hallauer International Symposium, pp. 155–170. Blackwell Publishing, Oxford.

van Eeuwijk, F.A. Crossa, J., Vargas, M. and Ribaut, J.M. (2001) Variants of factorial regression for analysing QTL by environment interaction. In: Gallais, A., Dillmann, C. and Goldringer, I. (Eds.) 'Eucarpia, Quantitative Genetics and Breeding Methods: the way Ahead'. pp. 107–116. INRA Editions Versailles Les Colloques series 96.

van Eeuwijk, F.A., Crossa, J., Vargas, M. and Ribaut, J.M. (2002) Analysing QTL by environment interaction by factorial regression, with an application to the CIMMYT drought and low nitrogen stress programme in maize. In: Kang, M.S. (Ed.) 'Quantitative Genetics, Genomics and Plant Breeding'. pp. 245–256. CAB International, Wallingford, UK.

van Eeuwijk, F.A., Denis, J.B. and Kang, M.S. (1996) Incorporating additional information on genotypes and environments in models for two-way genotype by environment tables. In Kang, M.S. and Gauch H.G. (Eds.) 'Genotype-by-Environment Interaction'. pp. 15–50. CRC Press, Boca Raton, FL.

van Eeuwijk, F.A., Keizer, L.C.P. and Bakker, J.J. (1995) Linear and bilinear models for the analysis of multi-environment trials. II. An application to data from the Dutch Maize Variety Trials. Euphytica 84, 9–22.

van Eeuwijk, F.A., Malosetti, M., Yin, X., Struik, P.C. and Stam, P. (2005) Statistical models for genotype by environment data; From conventional ANOVA models to eco-physiological QTL models. Australian Journal of Agricultural Research 56, 883–894.

van Eeuwijk, F.A., Malosetti, M. and Boer, M.P. (2007) Modelling The Genetic Basis Of Response Curves Underlying Genotype By Environment Interaction. In: Spiertz, J.H.J., Struik, P.C. and van Laar, H.H. (Eds.) Scale and Complexity in Plant Systems Research. Gene-Plant-Crop relations. Wageningen UR Frontier Series. Vol 21. Springer, Dordrecht, The Netherlands.

Vargas, M., Crossa, J., van Eeuwijk, F.A., Ramírez, M.E. and Sayre, K. (1999) Using AMMI, factorial regression, and partial least squares regression models for interpreting genotype x environment interaction. Crop Science 39, 955–967.

Verbyla, A., Eckermann, P.J., Thompson, R. and Cullis, B. (2003) The analysis of quantitative trait loci in multienvironment trials using a multiplicative mixed model. Australian Journal of Agricultural Research 54, 1395–1408.

Voltas, J., van Eeuwijk, F., Igartua, E., Garcia del Moral, L.F., Molina-Cano, J.L. and Romagosa, I. (2002) Genotype by Environment Interaction and Adaptation in Barley Breeding: Basic Concepts and Methods of Analysis. In: Slafer G.A., Molina-Cano, J.L., Savin, R., Araus, J.L. and Romagosa, I. (Eds.) Barley Science: Recent Advances from Molecular Biology to Agronomy of Yield and Quality. pp. 205–241. Haworth Pres. Binghamton, NY.

Voltas, J., van Eeuwijk, F.A., Sombrero, A., Lafarga, A., Igartua, E. and Romagosa I (1999a) Integrating statistical and ecophysiological analysis of genotype by environment interaction for grain filling of barley in Mediterranean areas I. Individual grain weight. Field Crops Research 62, 63–74.

Voltas, J., van Eeuwijk, F.A., Araus, J.L. and Romagosa, I. (1999b) Integrating statistical and ecophysiological analysis of genotype by environment interaction for grain filling of barley in Mediterranean areas II. Grain growth. Field Crops Research 62, 75–84.

Wenzl, P., Li, H., Carling, J., Zhou, M., Raman, H., Paul, E., Hearnden, P., Maier, C., Xia, L., Caig, V., Ovesná, J., Cakir, M., Poulsen, D., Wang, J., Raman, R., Smith, K.P, Muehlbauer, G.P, Chalmers, K.J., Kleinhofs, A., Huttner, E. and Kilian, A. (2006) A high-density consensus map of barley linking DArT markers to SSR, RFLP and STS loci and agricultural traits. BMC Genomics 7, 206.

Yan, W. and Kang, M.S. (2003) GGE Biplot Analysis: A Graphical Tool for Breeders, Geneticists, and Agronomists. CRC Press. Boca Raton, FL.

Yan, W., Cornelius, P.L., Crossa, J. and Hunt, L.A. (2001) Two Types of GGE Biplots for Analyzing Multi-Environment Trial Data. Crop Science 41:656–663.

Yan, W., Hunt. L.A., Sheng, Q. and Szlavnics, Z. (2000) Cultivar evaluation and mega-environment investigation based on the GGE biplot. Crop Science 40, 597–605.

Yan, W., Kang, M.S., Baoluo Ma B, Woods, S. and Cornelius, P.L. (2007) GGE Biplot vs. AMMI Analysis of Genotype-by-Environment Data. Crop Science 47, 643–653.

Breeding for Quality Traits in Cereals: A Revised Outlook on Old and New Tools for Integrated Breeding

Lars Munck

Abstract Breeding for complex multigenic phenotypic quality characters in cereals by chemical analyses and functional pilot tests is traditionally a slow and expensive process. The development of new instrumental screening methods for complex quality traits evaluated by multivariate data analysis has during the last decades revolutionised the economy and scale in breeding for quality. The traditional explorative plant breeding view is pragmatically oriented to manipulate *the whole plant* and its environment by "top down" observation and selection to improve complex traits, such as yield and baking quality. The new molecular and biochemical techniques are promising in increasing the genetic variation by breaking the barriers of species and in explaining the chemical and genetic basis of quality. In molecular biology traits are seen "bottom up" from the genome perspective, for example, to find genetic markers by quantitative trait loci (QTL). To improve efficiency the plant breeder can now complement his classical tools of observation by overviewing *the whole* physical–chemical composition of the seed by near infrared spectroscopy (NIRS) from a Principal Component Analysis (PCA) score plot to connect to genetic, (bio)chemical, and technological data through pattern recognition data analysis (chemometrics). Genes and genotypes can also be directly evaluated as imprints in NIR spectra. Recent applications in NIR technology by "data breeding" demonstrate manual selection for complex high-quality traits and seed genotypes directly from a PCA score plot. New equipment makes automatic analysis and sorting for complex quality traits possible both in bulk and on single seed basis. Seed sorting can be used directly in seed production and to speed up selection for quality traits in early generations of plant breeding and to document genetic diversity in gene banks.

L. Munck
University of Copenhagen, Faculty of Life Sciences, Department of Food Science, Quality and Technology, Spectroscopy and Chemometrics Group, Frederiksberg, Denmark, e-mail: lmu@life.ku.dk

M.J. Carena (ed.), *Cereals*,
DOI: 10.1007/978-0-387-72297-9, @ Springer Science + Business Media, LLC 2009

1 Introduction: The Need for an Upgrading of the Classical Holistic Tools of the Plant Breeder to Breed for Complex Quality Traits

Until recently, the plant breeder was an integrated member of the agricultural food producing society where the whole production chain could be overviewed "top down," including plant husbandry, cleaning, milling, cooking, baking, and brewing. Information from all these operations in food production and utilization were self-evidently related. They were integrated to the immediate sensory benefits of taste, smell, and mouth feel and to the longer perspective of satiety, health, and growth of children. The total phenomenological experience of society could be fixed into language, equipment and habits and expressed in selected cereal cultivars with specific advantages to be exploited and perfected in future generations.

The invention of man-made crossbreeding and artificial fertilizers to fix nitrogen from the air, laid the foundation to the *first generation of modern plant breeders* that greatly increase yield and quality in cereal production throughout the twentieth century (Olsson, 1986). However, it takes at least 10–15 years to produce a new variety by conventional breeding. A new *second generation* of "plant breeders" (Anderson, 1996) are, therefore, suggesting that one should design the biological diversity needed "bottom up" from the deoxyribonucleic acid (DNA) by biochemistry, genetic engineering, and gene transfer (Shewry and Casey, 1999; Horvath et al., 2002), rather than searching for solutions pragmatically in gene banks and in the field by expensive selection. Along these lines, molecular genetic screening methods by DNA markers through quantitative trait loci (QTL) (Arus and Gonzales, 1993) have been developed to breed for the more and less complex quality traits in practice. However, these methods have hereto not been cost effective enough (Thomas, 2002) to be used directly in the breeding work for selection of complex quality traits in great plant numbers.

The plant breeder's task is to produce a whole functional plant which combines high and reliable yield with quality. *One should therefore ask how holistic plant breeding at all can absorb, combine, and prioritize the great many fragments of knowledge regarding, for example, gene sequences and proteins, characterized by a precise but limited scope, that are produced by so many skilled scientists?*

2 Analyses and Data Models in Screening for Simple and Complex Quality Traits and the Genes Behind

2.1 Screening and Validation Methods for Technological and Physical–Chemical Quality

The classical maize breeding work in Illinois selecting for high and low protein and oil since the 1890s (Dudley and Lambert, 1969), demonstrates the great genetic flexibility of cross-fertilized populations and the interdependence in seed synthesis between the different chemical components locked in within the limits of the seed coat. Because protein and oil has 2.1 and 2.5 times the energy costs of starch for plant dry-matter (DM) production, selection for yield indirectly results in high starch, low protein, and oil composition of seeds (MacKey, 1981). Breeding for yield is thus not quality neutral. Behind the traditional seemingly univariate coarse, chemical criteria protein, oil, starch, crude fiber, and ash lie a complex abundance of chemical components that are analyzed by chemical separation methods, such as electrophoresis, extraction, and (gas) chromatography for a deeper definition of quality. Most of these analyses are because of costs not directly included in the selection work but are rather used by breeders, scientists, and the industry for an explanatory evaluation of the final products – the marketed varieties.

From a physiological and sampling point of view the individual seed should be considered as the ultimate unit behind grain composition. It is now possible to analyze 300 g of seeds in 3 min on an individual basis by image analysis gaining six form factors and three colors (Graincheck, Foss A/S, Hillerød, Denmark). A destructive seed hardness measurement instrument also analyzes seed width and water content of individual seeds (SKCS 4100, Perten Instruments, Inc., Reno, NV, USA). These two instruments are of great use in defining the physical seed quality (Sect. 7.3) including risk assessment for fungal infection.

A range of complex functional food analyses aims at visualizing the technological trait important for industrial use of the final product. This can be done by miniature versions of dehullers, rice polishers, roller mills, sifters, pasta extruders, cookers, amylographs, wheat dough quality instruments (farinographs, mixographs, extensiographs), baking machines, micro maltings, and pilot brewing facilities (Wrigley and Morris, 1996; Bergman et al., 2003).

In earlier generations, small sample screening physical–chemical screening methods are used, such as the falling number test (for field germination), the Zeleny sedimentation test (for wheat protein quality), and the alkali test (for rice cooking quality).

2.2 Nondestructive Screening for Quality Traits and Improved Genotypes by NIR Spectroscopy Evaluated by Pattern Recognition Data Analysis (Chemometrics)

The introduction of near infrared spectroscopy (NIRS) since 1973 (Williams, 2002; Møller Jespersen and Munck, 2008), for prediction of simple and complex seed quality characteristics, has fundamentally changed the economy of quality assessment for both plant breeders and the grain and food industries. The NIRS instruments come in two principal versions: near infrared transmission (NIT) 850–1,050 nm applicable to whole seeds (in batches or single) and near infrared reflection (NIR) 400–2,500 nm working on milled seeds also including the visual spectral area. Both NIT and NIR spectra give a reproducible and informative log 1/R fingerprint of "the whole" seed phenotype reflecting physical texture and patterns of chemical bonds that can be interpreted by indicative wavelength lists for chemical components published in the spectroscopic literature.

Classical statistics based on variance cannot handle the highly covariate information that is behind complex quality traits, for example, as represented by thousands of spectral wavelengths. The necessary data analytical methods, Principal Component Analysis (PCA) for classification and Partial Least Squares Regression (PLSR) for prediction, are included in chemometrics originating from the social sciences (Martens and Næs, 1989, 2001).

The NIRS instruments are working as multimeters and can be calibrated to predict several parameters using PLSR or neural nets (Table 1). Most plant breeders buy calibration software together with the instrument for prediction of water, protein, starch, fiber, and hardness optimized for different cereals.

The NIRS technology as demonstrated by specialized cereal laboratories can also be used to predict amylose/amylopectin in starch, β-glucan and lysine, etc. for which there are no commercial NIR calibrations in the market.

Other calibrations for complex traits, such as gluten strength, falling number, hot water extract, digestibility, frost-damaged kernels, and Fusarium blight (Table 1) are even more resource sparing but are difficult to obtain commercially because such calibrations are often only valid for local material and seasons.

In order to develop NIRS prediction methods on site, the plant breeding facility needs a technician trained in spectroscopy and in chemometric data analysis. Development and maintenance of NIRS prediction models needs precautions in handling outliers and access to a cereal laboratory for calibrations and checks. Such an investment will, however, pay off after a few years when the prediction models have been stabilized and the laboratory controls can be reduced to a few percent of the total analyses.

Table 1 Near Infrared (NIR) spectral predictions of technological quality parameters[a]

	r^2	SEP	High	Low
Kernel color wheat	0.96	0.30	20.5	11.7
Kernel texture wheat	0.94	2.38	75.2	36.6
Farinograph wheat:				
Water absorption (%)	0.91	1.93	71.8	53.4
Development time	0.71	1.2	13.0	1.0
Mixing tolerance	0.92	17.7	200	15
Extensiograph: Max height	0.72	85	905	180
Malt fine grind HWE extract (%)	0.52	1.00	78.5	69.5
True metabolic energy (barley) (cal)	0.61	0.15	15.4	13.4
Groat (%) (oats)	0.82	0.95	79.5	70.0
Falling number (s)	0.85	42.5	500	110
Fusarium head blight	0.76	1.10	6.8	0.1
Frost damaged kernels (%)	0.82	4.62	65.1	0.1

[a]Data from Williams (2002)

There is, however, another application of existing NIR instruments (Møller Jespersen and Munck, 2008) for sample classification without the need for commercial calibrations that requires a minimum of training and that has the potential of being even more important than prediction of specific quality traits.

Very few, if any plant breeders are aware of that they without expensive calibration software can use their NIR instruments as an explorative extension of their sight to evaluate the total physical–chemical fingerprint of seeds from their breeding lines as compared to high- and low-quality standard varieties grown in the same field.

Such a classification of simple and complex quality traits by "data breeding," as demonstrated in Sects. 7.1–7.3, is possible by a PCA score plot seen on a computer display (Martens and Næs, 1989; Munck and Møller, 2005; Munck, 2005, 2006). A complete automation of the NIRS measurement is possible. The breeder can now use the NIR technology in the early stages of breeding and wait with the most expensive conventional analyses for quality until a year before the official yield analyses.

2.3 QTL Analyses for Complex Traits Revitalized by Chemometrics

Phenotypic data, such as NIR spectra from biological individuals, are highly informative because they are compressible into multivariate representations ranging from chemical components to complex technological traits as shown in Table 1. The fundamental compressibility of phenotypic nature to indicate heritability explains why plant breeding based on inspection of the whole plant phenotype has been so successful through the pass of thousands of years as judged from the present wealth of cultivars adapted to a great variety of uses.

Let it be no doubt. Classical analysis of variance in genetics assuming normal distribution and free variability of variables, analyzing genes, and traits pair-wise has been extremely successful in mapping the statistical gene by linkage maps as confirmed at the physical low level of resolution mapping with Giemsa stain and at the high level by DNA sequencing (Kleinhofs and Han, 2002). When DNA technology enabled applications of multivariate restriction fragment length polymorphism (RFLP) and PCR markers the idea of QTL emerged based on classic statistics of variance (Arus and Gonzales, 1993), where genes near to the markers are correlated to more and less complex quantitative phenotypic traits.

QTL seem to function well when both the genotypic and phenotypic traits are simple. However, the localization of a QTL is hereto only considered as approximate, as discussed by Kleinhofs and Han (2002), and a more accurate localization should be performed by additional backcrosses. Several other researchers, such as Thomas (2002) in barley and Darrah et al. (2003) in maize, mention several arguments to explain why QTL are not yet widely used as a tool in plant breeding, such as lack of laboratory resources and the too limited population size applied in crosses.

However, it now stands clear from the NIR experience (Martens and Næs, 2001) that classical statistics cannot handle complex covariate data, because of the distributional assumptions, which are an inseparable part of the statistical model of variance used in the classical QTL analysis. An explorative data program that can define gene expression unsupervised as patterns in the phenotype is thus essential (Munck, 2007).

As discussed by Darrah et al. (2003), the QTL approach can now be revitalized as "*association genetics*," a term originating from medical genetics. It includes pattern recognition data analysis now in plant genetics using data reduction by least-squares solutions (Knott and Haley, 2000) in PCA (Wilson et al., 2004) and in PLSR (Bjørnstad et al., 2004). The chemometric data models based on pattern recognition are much better suited for evaluation of complex QTLs than classical statistics based on variance.

> Thus different mathematical models should be used for the gamete and zygote levels of biological organization where classical probability statistics effective on the gene recombination level in populations should be complimented by pattern recognition analysis (chemometrics) for analyzing interactive gene expression at the phenotype level with the biological individual as the ultimate unit. (Sects. 7.1–7.4, Munck and Møller, 2005)

Wilson et al. (2004) recently demonstrated a combination of NIR and DNA mapping evaluated by PCA in maize.

The widening of the genetic basis by gene transfer across the barriers of species by molecular techniques (Horvath et al., 2002; Anderson, 1996) could constitute another turning point in plant breeding, if the sceptic arguments against genetically modified organism (GMO) from the public can be refuted. However, transferred genes as well as mutants often have pleiotropic side effects, for example, on yield that cannot be foreseen. New compensating gene backgrounds have therefore to be bred by classical pragmatic plant breeding to optimize the expression of each gene with regard to quality without compromising yield. As will be exemplified in the following the NIR–PCA model makes such an adaptation feasible in practice in plant breeding on a mass scale.

2.4 Characterizing and Connecting Complex Genetic, Biochemical, and Technological Traits in Cereal Variety Testing

Genetic engineering is focused on *specific traits* such as transformation of a gene for heat-tolerant β-glucanase from a *Bacillius sp.* to barley (Horvath et al., 2002) of potential importance for both the feed and the brewing industries. In manipulating *complex technological traits*, such as baking quality in wheat, it is difficult to be able to predict the final result of a gene manipulation.

Thus Rook et al. (1999) over expressed the high molecular weight (HMW) subunit 1D×5 in transgenic wheat which resulted in a fourfold increase in this protein fraction but also a corresponding increase in the proportions of the total HMW proteins and glutenins.

Such a new radically changed genotype may inspire technologists to find new uses. However, in traditional baking the over expression of the 1D×5 gene made the dough strength of the resulting wheat too strong to be practically useful. There is thus need for a data model that can integrate and optimize the biochemical, genetic, and technological information. Chemometrics that is used in NIRS (Sect. 2.2) in evaluating complex covariate spectral traits can also be applied here (Sects. 7.1–7.3) (Munck, 2007; Munck and Møller Jespersen, 2009).

The data challenge is enormous. As overviewed in Fig. 1, different scientific disciplines now produce a gigantic network of data considering the cereal crop. It starts by gene sequences (A) and gene expression (B) at the different "omic-levels" of biological organization as affected by environment and further moves toward the chain of technolological, sensory, and nutritional utilization and acceptance. The ultimate strive for the "bottom up" path modeling approach to gene expression can be exemplified for the transcriptomics part of area B in Fig. 1 by the atlas of gene

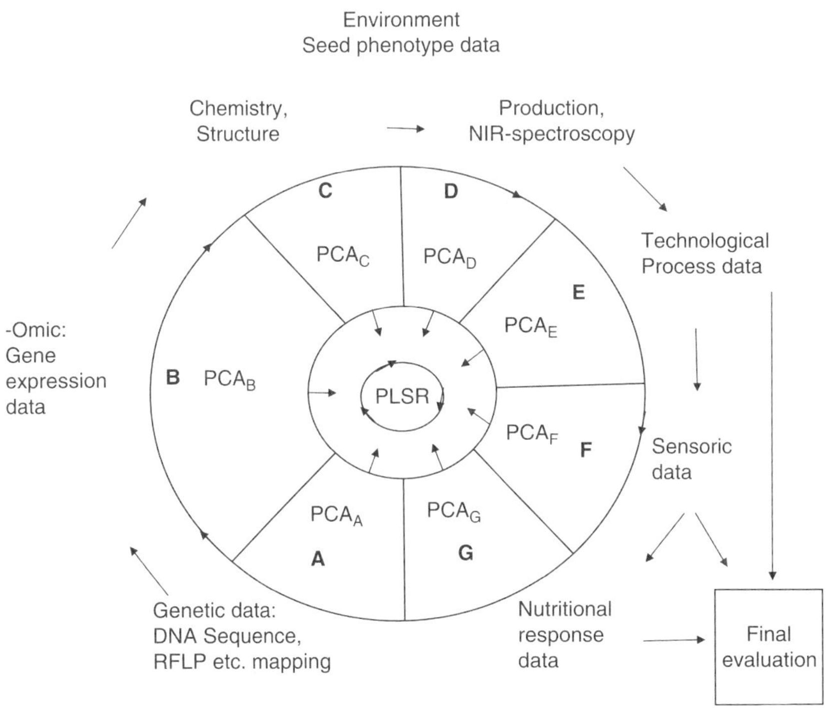

Fig. 1 Different data sets in plant breeding integrated as *association genetics* as evaluated by chemometric data analysis

expression of the barley variety Morex (Druka et al., 2006). This consortium of scientists used the Affymetrix Barley 1 GeneChip to express at least 21,439 genes in 15 tissues at 8 developmental stages. The barley data set is now available at the Internet. The question is now how such data may benefit the plant breeder *directly* in his selection work and how it should be evaluated?

Chemometric data analysis (Fig. 1) provides a method through principal components (PCs) to sum up in *functional traits* the interactions of the many chemical and genetic entities, which science has so skillfully defined.

The use of chemometrics in wheat proteomics [two-dimensional electrophoresis (2DE), mass spectrometry, and NIR data from gluten], to classify cultivars according to baking quality, is reviewed by Gottlieb et al. (2004). There are in the rich literature of cereal proteomics only a few applications (Sect. 7.3) on using chemometrics to connect to other data sets according to the model in Fig. 1. The exploratory pattern recognition data approach starts unsupervised with *data reduction* to let the data set speak for itself. The task is to explore relationships that cannot easily be anticipated by the usual strategy of *problem reduction*. The patterns of manifest variables from the samples in the dataset are explored *as a whole*, for example, by a PCA *score plot* with a minimum of specific assumptions. A latent, hidden world behind data is assumed and is indirectly observed by PCs. They are composed automatically by finding the major directions in the data set and are characterized by combining different amounts of variables. The complex manifest data in biology can almost always be compressed in a sequence of *a few* (Sects. 2–5) latent, orthogonal PCs numbered 1,2.3, . . . according to their falling share of the total variance of the data set (Martens and Næs, 1989). The PCs are plotted in *x–y score plots* (e.g., PC1 to PC2 and PC2 to PC3) where the samples are classified according to their score values at the PC axes. The nature of the PCs is revealed as *functional factors* by introducing prior knowledge to interpret the combination of manifest variables that are composing the PCs (Sects. 7.1–7.3). A PCA biplot demonstrates the relationships between samples and variables.

PCAs from seven different data sets of cereal quality trait evaluation are suggested in Fig. 1 where the compressibility of each data set into more or less functional PCs is investigated. Multiple PCs from the data sets A–G in Fig. 1 may communicate with each other by PLSR analysis that also is built on principal components (Martens and Næs, 1989). It is thus possible by PLSR to validate to what degree the same data structure is present in two or more data sets together and to use one to predict the parameters in the other as demonstrated in Sect. 7.3. In fact NIRS introduces a complimentary approach to System Biology (Munck, 2007; Munck and Møller Jespersen, 2009).

In exploiting the widely different data sets outlined in Fig. 1 to back up plant breeders, data *quality* is the limiting factor beside the costs of analyzes. While protein electrophoresis (2DE) spots are quite tricky to reproduce and digitize (Gottlieb et al., 2004), the strength in NIRS is its very high reproducibility and physical–chemical relevance as revealed by chemometrics.

3 Quality Traits in Cereal Technology and Plant Breeding

3.1 *Wheat*

Today's cultivated hexaploid bread wheat (*Triticum aestivum*) originates from the fertile crescent in the Middle East since about 8000 BC. Wheat is together with rice and maize the three major cereals in the world with a production of about 570 million tons/year (Williams, 2002). For pasta the tetraploid wheat *T. turgidum* (*durum*) is widely used. The good taste of bread from ancient wheats *T. monococcum* (diploid) and tetraploid *T. dicoccum* (emmer) are now increasingly enjoyed as specialty foods in Europe and the USA (Abdel-Aal and Wood, 2005). The major characteristics of bread wheat are: winter and spring habit, color, red or white, of bran (testa), and degree of hardness. A special roller milling process for wheat flour developed 150 years ago is permitting a clear-cut separation between the endosperm (white, low ash flour) and the outer seed coat (bran) facilitating high volume bread with a white bread crumb (Schofield, 1994). Sprouted kernels in the field are detrimental to bread and pasta quality leading to increased levels of α-amylase and other enzymes. Hard seed and high protein (gluten) content is characteristic for special high-quality bread varieties and further supported by a dry climate. The glassy, transparent hard wheat seed makes high starch damage in the roller milling process and results in high water absorption of the dough and in a large bread volume (Schofield, 1994). The hard texture of the wheat seed is due to a single gene (*Ha*) in Chromosome 5D. Schofield (1994) has identified a protein – friabiline – that is lining the surface of the starch granule. It seems to protect the granule to be cut through when the seed is divided with a knife and is indicative for soft seeds. The friabiline trait is firmly associated to the *Ha* gene.

The classical early work in the protein biochemistry of wheat by Payne et al. (1983), clarified that the strength and elasticity of gluten is under control of endosperm protein loci for HMV and low molecular weight (LMV), glutenines (*Glu*), and gliadines (*Gli*). Five loci were identified located on the first and sixth chromosomes. Since then biochemical and molecular research have described most of the proteins in wheat and barley endosperm and identified several of the genes and gene sequences controlling them (Shewry and Casey, 1999; Shewry, 1992; Schofield, 1994). The genetic variation of wheat, barley, and maize has been instrumental in this pioneering work on seed proteins also including a wide range of barley and maize mutants. The rapidly increasing biochemical and molecular knowledge with regard to *specific* proteins genes has been of great explanatory importance for plant breeders and food technologists to choose the right varieties. As discussed in Sects. 7.3–7.4, it is now possible to exploit this knowledge directly in the wheat breeding work.

3.2 Barley

Barley (*Hordeum vulgare*) is among the oldest cereals exploited by mankind with a great adaptability to both cold temperate and hot-arid climate zones. It has traditionally been used as a food after dehulling, polishing, and milling (Bhatty, 1993). In Japan and Korea, the whole polished barley seed is still used as a substitute for rice. About 160 million tons of barley is produced annually worldwide about 80% of which is used for feed and ~15% for the brewing and distilling industries (Williams, 2002). Because of the earlier focus in the brewing industry on research, there is now an abundance of detailed information regarding, starches, proteins, plant hormones, enzymes, and β-glucans (Shewry, 1992; Munck, 1992a, b; Swanston and Ellis, 2002), as well as about the estimated hereditability of traits (Kleinhofs and Han, 2002; Ullrich, 2002), for the plant breeder and the brew masters to exploit. As discussed in Sect. 3.1 such information is of great explanatory importance to define high quality lines to be used in crosses. However, to be exploited directly in quality breeding and in improving technology the *functionality* of all these variables have to be revealed as *complexes of relations*. Thus a major collection of barley seed enzymes of importance in starch synthesis, germination, and malting is induced by the same plant hormone – gibberellic acid (Swanston and Ellis, 2002). This implies a data reduction because all these enzymes are dependent on the same hormone mechanism for their expression.

From the maltsters point of view, barley malt is evaluated according to 11 parameters obtained from the EBC (European Brewing Convention) wort extraction method in the laboratory. Malt quality is traditionally evaluated according to specification limits given for each of the 11 single values. Still problems might arise in full-scale beer production even if the malt is fulfilling all individual specifications. A PCA study on the 11 malt quality variables from a set of 186 malts (Munck, 1991), all following the specifications, revealed that malt quality should not been evaluated as 11 independent variables but instead as *three functional factors* or PCs explaining 66% of the variation namely

PC1 "Chemistry" – extract, wort color, soluble N in wort, Kolbach index – all characteristic for starch content and enzyme activity,

PC2 "Physics" – Friability, wort β-glucan and viscosity, malt modification and homogeneity, extract difference – all variables dependent on malt hardness, cell wall thickness (β-glucan), and resistance toward malt modification, and

PC3 protein in malt.

The PCA score plot sorted out all malt samples from the barley variety Minerva, which gave problems in full-scale production in spite of fulfilling the malt specifications. It had a bad "chemistry" (PC1) value that could be detected as a specific pattern in a PCA (PC1–PC2) score plot when all the 11 variables were evaluated together (Munck, 1991).

The resistance toward cell wall breakdown, which is a significant part of the "physics" quality trait of malt modification, can be visualized as seen in a light box by staining the endosperm cell walls of half malt seeds by calcofluor (Aastrup et al., 1981). The β-glucan-containing cell walls are broken down by enzymes excreted both from the germ and from the aleuron layer. A β-glucan-reduced barley mutant with slender endosperm cell walls (Aastrup and Munck, 1984) had a faster modification in spite of unchanged β-glucanase activity. In barleys with live embryo the physical–chemical composition of the barley endosperm as revealed by NIT spectroscopy is the limiting factor (Munck and Møller, 2004) for both malt modification rate and germination velocity (vigor). Thus a simple 1 day (vigor) to 3 days (viability) *x–y* germination plot is able to predict and classify pilot malting performance (extract and wort β-glucan) of barley varieties grown in different years. This exemplifies the importance of a separate representation of the main functional criteria as simple *x–y* relations in evaluating quality traits by classification of varieties from plots rather than by a single score value for malt quality as was further discussed by Munck and Møller (2004).

3.3 Rye

Rye (*Secale cereale*), originating from the wild form *S. montanum*, has been much later introduced in agriculture than barley and wheat (Scoles et al., 2001). The worldwide annual production for feed and food in the late-1990s is slightly below 30 million tons mainly in North Eastern Europe and in North America (Williams, 2002). Rye cannot form gluten after mixing with water. Instead, the water absorbing pentosans (arabinoxylans) are important for the baking performance of rye bread. Rye is relatively well defined with classical genetic chromosome maps as well as by RFLP- and PCR-based markers (Scoles et al., 2001). The long rye bread process needs hard work if made manually. The long shelf life of the rye bread gives a basis for an industrial process that facilitates a wide distribution of the products.

Rye bread production is sensitive for weather conditions because of its tendency for precocious sprouting giving high α-amylase, protease, and pentosanase activities, as reflected in a low falling number (viscometric test of flour). There are three loci for α-amylase in rye (Scoles et al., 2001). The sourdough fermentation and addition of lactic acid to the dough counteract the effects of sprouted seeds in dark rye bread, but may not be able to suppress effects of taste and odor from damaged seeds of the final dark rye bread (Seibel and Weipert, 2001). Rye crisp bread is especially sensitive to sprouted seeds. New hybrid rye varieties are more resistant to sprouting and perform well in milling and baking processes (Seibel and Weipert, 2001).

Rye flour of different granulations and whiteness are produced with a shortened wheat roller milling process. Low extraction rye flours are mixed with wheat flour to produce lighter form of loaf with an attractive rye flavor (Seibel and Weipert, 2001).

3.4 Oats

Like many small grains mainly used for feed, oats (*Avena sativa,*) production has decreased due to competition with the more high-yielding cereals, wheat and maize (Burrows, 1986). Around 30 million tons are produced worldwide as an average (Williams, 2002). In later years oats food products have been in focus as a dietary food because of their attractive taste combined with a high-quality dietary fiber (β-glucan; McCleary and Prosky, 2001), a great wealth of antioxidants (tocopheroles, phenols, and flavones derivatives (Collins, 1986), and high-quality seed oil and protein (Burrows, 1986). The oat hull is a large part of the seed (23–28%), which covers the dehulled seed (groat) including the endosperm. There are naked seed varieties that are more attractive as food and feed raw materials, if field scattering can be avoided by breeding (Burrows, 1986). Rolled oat flakes, oat flour for cakes, as well as oat bran and oat milk-like drinks enriched in β-glucan are the main food products. The oats seeds have to be steamed before dehulling in order to eliminate the lipase activity of the groats to secure the shelf life of the final products. Clinical trials have demonstrated the positive health effects of oats (McCleary and Prosky, 2001), which led to a decision in 1997 of the Food and Drug Administration in the USA that oat food labels may carry a claim that oat products may reduce heart disease when combined with a low-fat diet.

3.5 Rice

Rice (*Orytza sativa)* is a semiaquatic plant originating from South-East Asia, now grown widely on all continents. It is one of the leading food crops in the world and the staple food for more than half of the world's population (Childs, 2003). Of a world harvest of ~600 million metric tons, only about 23 million tons are traded on the world market compared to about 105 million tons for wheat and 72 million tons for maize (Williams, 2002). There are 420,500 rice samples of rice and related species kept in germplasm collections which guarantee a rich source of genetic variation for quality traits (Bergman et al., 2003). The rice genome has recently been completely sequenced in 430,000 base pairs in the 12 $\times$ 2 chromosomes, an international effort coordinated by the International Rice Research Institute from the Philippines. This breakthrough is fundamental, as a tool in comparative studies to understand quality traits in other cereals.

Rice is mainly used for cooking after dehulling the seed (paddy) into brown rice, which is polished to white rice. The yield of white rice is only about 67%. The rice industry thus produces a large tonnage of by-products, including rice oil, rice flour (from broken rice) and fiber. Starch that constitutes about 90% of the milled rice is a key component, where the relations between the starch components amylopectin (glutinous sticky) and amylose (firm nonsticky) determine cooking quality and gelatinization temperature (Fitzgerald, 2003). In order to fully evaluate the subtleness of rice cooking quality, sensory panels are necessary for estimating mouth-feel/texture, smell, and taste. The special aromatic cultivars of rice of Basmati and Jasmine type preferred in India, Pakistan, and in Thailand contain 2-acetyl-1-pyrrolidine.

The physical form of the rice kernels is not genetically associated with their cooking and processing qualities. However, traditionally seed form is associated to cooking quality on the rice markets. Seed form, for example short kernels of Italian rice, is associated to low amylose–high amylopectin glutinous products suitable for porridge. Consumers in North and South America, Southern China, and Europe prefer a long rice kernel with an intermediate amylose content, which after cooking is firm and fluffy. In Japan, Northern China, and Korea, a soft, moist, and sticky-cooked product is estimated that is easily eaten by chopsticks (Bergman et al., 2003). These rice seeds are of medium length. Another style of rice preferred by Japanese and Koreans has short grains and low amylose and gelatinization temperature. The cooked grain has a high degree of glossiness, lack of off-flavor, distinct aroma, and sticky but smooth texture that are maintained after cooling involving a minimum of retrogradation of starch. Kernel discoloration is an important quality attribute that can be corrected by photoelectric seed sorters (Bergman et al., 2003). During rice polishing of brown rice, important vitamins and minerals are removed from the product. Milled rice has a longer shelf life than brown rice. Wetting and preheating rice in the parboiling process allows vitamins and minerals to migrate into the endosperm creating a more nutritious product. Parboiling also decrease seed breakage by healing cracks and improving milling yield. Endosperm texture and crack formation are important for the milling yield in rice.

3.6 Maize

Maize or corn (*Zea mays*) is the second most important grain of commerce cereal with a yearly world production of ~600 million tons (Williams, 2002), 40% of which is produced in the USA. It originates from Mexico (Eckhoff and Paulsen, 1999) and is today mainly used as a feed in industrial countries (in USA including export 80%). The most important cultivar classes in yellow maize are dent (semi-hard most US maize), floury (soft), flint, (very hard), and popcorn (small hard seeds). Maize is used extensively in many countries as a food cereal. White maize is the staple food in South Africa where up to 14 million tons are produced in a season (Williams, 2002). The large maize seeds are best stored on intact cobs as in traditional maize farming. The combiner introduced the danger of mechanical damage and cracks, which makes the kernel more sensitive for molding (Paulsen et al., 2003). Mold toxins in maize, such as aflatoxine, are a major problem for livestock and humans in humid climates. Cracks in seeds result in difficulties to isolate the whole intact germ in the dry milling process and in a high fat content and shorter shelf life of the grits and flour products. When the combiner was introduced grain driers were needed to reduce water content. Gentle slow drying at moderate temperature is needed to keep the structure of the large maize kernel intact. Hard seeds are more resistant to mold and insect damage than soft and keep germination vigor for a longer time during storage (Paulsen et al., 2003).

Maize, like most cereals, is not a complete food for most livestock and humans because it is deficient in essential amino acids especially lysine and tryptophane (see Sect. 5). It is the raw material basis for large-scale industrial wet and dry milling operations for production of starch, food grade oil, maize gluten (for feed), flour, grits, and alcohol (Sect. 4). Industrial food products made from maize also include sweeteners, syrups, cornflakes, tortillas, salty snacks (the market value \$20.6 billion in USA in 2000), puddings, and a wide range of convenience foods (Rooney and Serna-Salvidar, 2003). These authors describe several versions of traditional maize food products (including their local names) locally prepared worldwide, such as whole grain products ($n = 4$), thin unfermented porridges (6), ditto fermented (3), thick porridges (7), snack foods (3), steamed food-couscous (2), unfermented breads (8), fermented bread (1), fermented dough (2), and beer (8).

Maize with all its endosperm mutants (Darrah et al., 2003) is a favorable genetic model for studying carbohydrate synthesis in cereals. A wide range of specific mutants with multiple pleiotrophic effects is expressed in the endosperm. Analogues are to some extent present in rice (Fitzgerald, 2003) and barley (Munck et al., 2004). The genetically buffered allopolyploid wheat does not show the same variation in carbohydrate genes although waxy wheat is now available (Graybosch, 1998). The starch granule (plastid) population in maize is rather uniform in maize (A-starch) which makes a high yield in wet starch milling process compared to wheat and barley that, additionally, have a population of smaller starch granules (B-starch) that are more difficult to isolate. The starch granule consists of branched amylopectin starch with a high swelling and gelatinization capacity (favorable as a component in frozen foods) compared to the linear amylose component which is a preferred component in industrial applications, for example, for membranes as a substitute for plastics. The waxy (*wx*) recessive gene may produce 100% amylopectin. The endosperm mutant gene amylose extender (*ae*) plus modifiers increase the amylose in maize up to 80%.

An endosperm mutant that results in a decrease in starch is an indication to look for other components that the plant may synthesize as compensation. In maize many such mutants [e.g., sugary *su,* sugary extender *se*, shrunken (*sh*)-2, and brittle (*bt*)-1 and 2] tend to produce sugars (up to 35%) together with phytoglycogen that is an amylopectin with reduced molecular size (Eckhoff and Paulsen, 1996). Instead, in barley starch mutants a compensative production of the polymer β-glucan is common (Munck et al., 2004). A wide range of special purpose cereals based on endosperm mutants (in maize involving about 12 genes) is now commercially available (Eckhoff and Paulsen, 1996). It is notable that most of these mutants that have lesions in the DNA coding for specific starch synthesizing enzymes are of commercial interest because of their unforeseeable pleiotropic multivariate effects on the entire endosperm synthesis (see Sect. 7).

3.7 Sorghum and Millets

Grain sorghum (*Sorghum bicolor*) production is about 60 million tons annually (Williams, 2002). It is a grass originating from Africa, which is able to give yield under dry and hot conditions where maize will not thrive. High-yielding sorghums are grown abundantly in hot climates as a raw material to the feed industry. Sorghum is, however, equally important as a reliable food source for subsistence farmers in the hot and dry tropics of Africa and Asia where an abundance of different cultivars and food-processing habits have been developed throughout historical time.

Millets (world production about 30 million tons/year (Williams, 2002) is a collective name for nine small-seeded grass species (House, 1995) including Pearl millet (*Pennisetum glaucum*) and Foxtail millet (*Setaria italica*) that are not directly related. They have a low yield, but are even more drought and heat resistant than sorghum and are of fundamental importance for the survival of the poorest farmers in the most difficult locations.

Polyphenolic compounds including tannins produced in the testa layer are a characteristic element in the genetic diversity of sorghum (Serna-Saldivar and Rooney, 1995). Sorghum due to specific genes may be almost tannin free or contain different amounts of tannins, which may diffuse from the testa into the endosperm. Only inspection of the testa layer by a knife cut in the seeds gives a safe indication of the state of tannins in sorghum. The color of the outer pericarp of sorghum grain is not related to the high polyphenol/tannine trait. Very high tannin sorghums are toxic and lethal for birds and rodents but resistant to insects and fungal infection. They are carcinogenic to humans. High tannin sorghum is grown for bird control to protect the harvest. In spite of these harsh conditions, humans have succeeded in creating food processes based on soaking, lime treatment, malting, and fermentation to make these seeds edible (Murthy and Kumar, 1995).

Opaque, low alcohol-fermented beers containing yeast and bran are produced in great amounts by the population in south of Sahara in Africa based on sorghum and millets. Several hundreds of liters of such beers are consumed per capita in Western Africa. Because of the germination and fermentation process, the negative effects of the harsh polyphenols and the often-contaminated water are neutralized. The liquid is further supplemented by the essential amino acids and vitamins from the yeast to make a nutritional product approaching the value of cow's milk.

Sorghum endosperm has the lowest content of the essential amino acid lysine of all cereals [down to 1.8% (Munck, 1995)] but is rich in starch. The kafirin proteins of sorghum are highly cross-linked low lysine storage proteins that retards digestion of the other components of a meal, for example starch. Such a diet takes time to get used to but has the advantage of keeping satiety for a long time. Adequate protein supplementation by, for example, pulses produces nutritious foods when combined by dehulled and milled products from low and medium tannin sorghums (Munck, 1995).

Cooking sorghum into porridges and to fermented *kisra* as in East Africa increase insoluble dietary fiber due to bound protein (mainly kaferins) and enzyme-resistant starch (Bach Knudsen and Munck, 1985). The sorghum proteins associate with dietary fiber and are transported as "dietary" components to the lower intestines for microbiological digestion, strongly increasing the volume of the fecal stools (Munck, 1995; Bach Knudsen and Munck, 1985). If sorghum and millet foods could compete with maize and wheat products in the cities by establishing local milling industries (Munck, 1995), many countries in Africa would have had a chance of being self sufficient in food grains.

4 Quality Aspects in Breeding Cereals for Whole Crop Utilization in the Nonfood and Food Industries

The technological development in agriculture and industry has a decisive influence on plant breeding. The invention of artificial nitrogen fertilizer based on fossil energy 150 years ago is the prerequisite for feeding today's expanding world population. It is *the* most effective way to invest fossil resources to trap carbon dioxide and to boast the utilization of energy from the sun. In subsistence plant husbandry, all parts of the crops were needed for survival and were, therefore, carefully utilized for food and nonfood purposes (Munck, 1995). Still tenths of million of tons of starch and plant oils are used for nonfood purposes world wide for, for example, paper and the detergents. In the order of 8% of the current world production of paper pulp is based on straw (~30 million tons Munck, 1992a).

However, cheep energy-downgraded straw, to be burnt in the field, when combiners where introduced. In the industrial revolution, local biological and agricultural production chains, which to a great extent were self-sufficient yet low producing, were broken up due to new technology, transport, and trade. During the 1950–1960s the agricultural raw materials for nonfood purpose were to a large extent exchanged by substitutes based on coal, mineral oil, and gas. Now in 2006 when oil prices exceed US $50 per barrel, the whole plant utilization concept (Munck, 2004, 1993) is starting to be economically feasible as outlined in Fig. 2 with maize as an example.

Very large-scale maize and wheat industrial units (above 500,000 tons/year and unit) for starch, oil, and ethanol manufacturing with feed as a byproduct are now in operation in USA, Europe, and South America. There is a wealth of possibilities for utilizing the starch polymer after modification by means of organic chemistry to cation and anion starches for the paper industry or by microbiological transformations to, for example plastic molds, ethanol, acetone, and butanol (Fig. 2). Unfractionated straw has a mediocre value for feed as well as for paper. However, the leaf fraction has improved protein and energy value for feed, and the internode part has a content of α-cellulose as high as in wood (Petersen and Munck, 1994). The internode fraction is excellent for paper and for fiberboards. A simple disc mill plus a sifter is able to separate straw into fine leaf meal and coarse

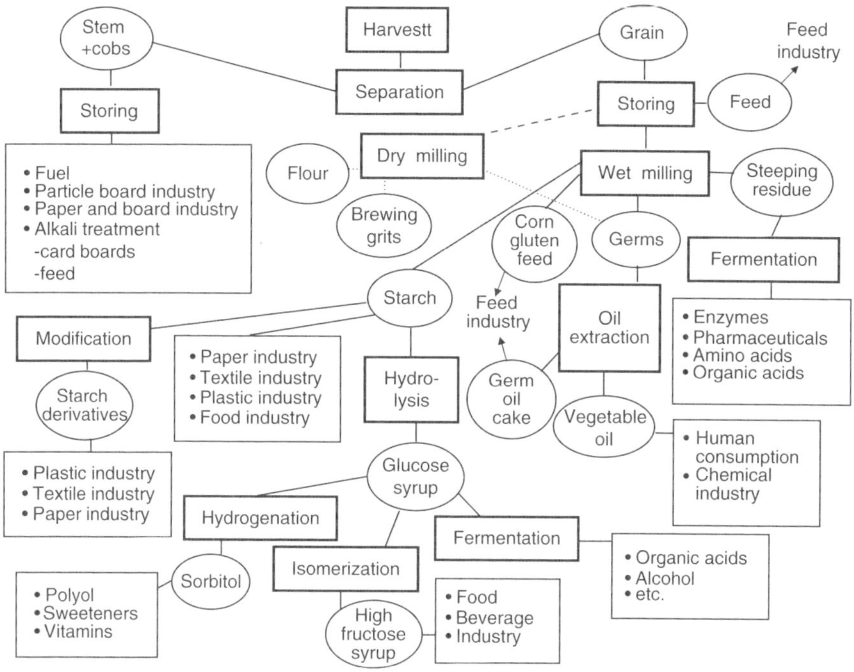

Fig. 2 Maize as an example of a raw material for whole plant industrial utilization

internode chips. There should, therefore, be an economic incentive for a value-added fractionated of straw, tailored to different uses.

In order to solve the logistic problems of whole crop utilization, the whole plant should be harvested by a self-propelling harvesting chopper and transported in containers to a local *Biorefinery* (Munck, 2004; Munck and Rexen, 1990) that could separate seeds and straw and distribute the raw material fractions to larger industrial units for production of starch, oil, feed concentrates, and ethanol. Such a local unit should be energy self-sufficient because the leaf meal should be enough to dry the whole harvest. This would make plant production less sensitive for weather conditions, for example, making it possible to grow maize for grain production in Northern Europe to exploit its high yield potential. The number of crops as well as the harvesting and processing season should thus be able to be expanded. The efficiency of the biorefinery should not be judged on the basis of individual products, but on the integrated total output from a flexible diversified production to a great number of alternative receivers. From a plant breeding perspective a range of new breeding concepts for future can be visualized, including low silicon, long cellulose fiber internodes for paper pulp and, specially, bred varieties for specific fatty acids, and starches for plastics. In 2008 at the time of climate change, global warming fossil energy shortage there is a focus on renewing the global infrastructure including a total use of the renewable plant resources for food, feed, energy and manufacture products as discussed by Munck and Møller Jespersen 2009.

5 Breeding for Nutritional Quality

In defining nutritional needs of humans and animals, cereals have played an important role (Munck, 1972). The importance of B-vitamins was elucidated due to deficiency symptoms that were introduced when brown rice was polished and the germ and aleuron was removed. The essential amino acids were defined when deficient maize protein was fed to young rats and adding lysine and tryptophane could restore the growth. During the 1930s, practically, all nutritionally important elements were defined and could be purified and put together in synthetic diets to maintain growth in rats.

In the 1960s, there was a major focus in science on alleviating the amino acid deficiencies of cereal protein by plant breeding. Remarkably a range of "high lysine" mutants were isolated in maize, barley, and sorghum (Axtell, 1981; Sect. 6), which drastically improved the nutritional value of the proteins by changing the balance between the proteins high and low in essential amino acids. The maize mutants, such as opaque-2 and floury-2, were previously known as morphological mutants with defects in grain filling and with decreasing starch and increasing sugar content indicating a complex pleiotropic effect of the mutant gene.

The majority of the cereal production is probably used for feed that has not been adequately reflected in plant breeding for cereal quality (Ullrich, 2002). However, as discussed above, starch content is related to yield because starch is the most economical way for the plant to produce dry substance. Carbohydrates and their availability is, therefore, the main target of quality for the purpose of feed (Rudi et al., 2006). In poultry and pig feeding, cereals are mainly used as an energy (starch) source with some protein, vitamins, and minerals which has to be supplemented by, for example, soybean press cake and phosphorous and calcium. Availability of energy in feeding cereals is highly dependent on the thickness and composition of the endosperm cell walls containing β-glucan, arabinoxylans, and cellulose surrounding the starch granules as well as on the hardness of the particles from the seed milled to flour. NIRS can estimate these physical–chemical factors and is thus able to predict digestibility (Table 1). A soft endosperm with slender cell walls preferred for malting should thus be preferred also for feed.

Intense animal production gives local pollution problems when the limited land has difficulties in absorbing the manure. It has until, recently, been overseen that the large content of nonessential amino acids such as glutamine and aspartic acid in maize and barley is not adequately utilized by soybean protein supplementation targeted for an optimization of the limited amino acid lysine, that also carries with it large amounts of unessential amino acids. There is thus an overfeeding of protein. The very high lysine barley mutant Risø 1508 (gene *lys3a*, 5.2% lysine g/16 g N) allows for optimal growth in pigs at a much lower protein content after some supplementation. In fact, it is a gene for 15–20% less nitrogen pollution in feeding slaughter pigs (Munck, 1992b). The same effect can be obtained by supplementing with microbiologically produced pure lysine.

Phosphorus is mainly bound to phytic acid in cereals and cannot be utilized by the animal. Mutants have been obtained in barley and maize (Pilu et al., 2003, Rasmussen and Hatzac, 1998) that reduced phytic acid and led to an increase in inorganic phosphorus that can be utilized. Thus, addition of phosphorus to the diet can be diminished as well as the total amount excreted in the feces and in the urine (Poulsen et al., 2001). Another approach to the same problem is demonstrated in wheat with a transgenic expression of an Aspergillus phytase (Brinch-Pedersen et al., 2000) that is able to degrade phytate during digestion.

The high β-glucan barley mutants (up to 20% DM; Munck et al., 2004) are detrimental for poultry and swine that can not utilize this kind of dietary fiber but may be of interest in feeding live stock if the β-glucan and can be utilized as a slow-release carbohydrate by the microbiological digestion in the rumen.

Plant breeding and nutrition as sciences are based on complex interactions between many elements that need a multivariate approach to be understood. An optimal diet for growth of young children is certainly not optimal for the maintenance of health in adults. During the last 20 years, attention has been given to foods with a slow release of glucose during digestion (low glycemic index, GI) to avoid stress in the insulin production that may lead to diabetes. The constituents of bran and the endosperm cell walls (McCleary and Prosky, 2001) that are indigestible in upper part of the digestive system function in several ways in the diet. One function is trapping starch for a more slow digestion. The other function is as filler stimulating the gut to increase the flow through the digestive system washing out cholesterol and carcinogenic substances. A third function is to serve as a source of energy for the microbes and as a water absorber in the colon. One could conclude that the biological variation in the composition of the cereal seed is a great source of inspiration also in nutritional research.

6 Mutation Breeding for Endosperm Quality Traits

Natural endosperm mutations with attractive sensory traits, such as sugary-2 in maize and high lysine sorghum (Axtell, 1981; Darrah et al., 2003), has always attracted the consumers and were propagated and bred. There were great expectations when artificial mutants were produced in the 1930s. Now, one should be able to induce new genes in high-yielding genotypes to obtain a shortcut in breeding (van Harten, 1998). This way of thinking also prevails in today's genetic engineering concept. There are unexpected pleiotropic side effects of mutants and transferred genes, for example, on yield and seed quality. However, the flexibility of nature is great. It is in most cases possible to find "a happy home" for the new gene (Ramage, 1987) by recurrent crossing and selection. Very few mutation breeders and genetic engineers believe that such a tedious, less glamorous work would at all be successful. They, therefore, tend to concentrate their work in inducing new mutants and transferring new genes.

The breeding cases of high lysine (opaque-2) quality protein maize (QPM) from Centro Internacional de Mejoramiento de Maíz y Trigo (CIMMYT) (Vasal, 1999) and of high lysine barley (Risø mutant 1508, gene *lys3a*) from Carlsberg Research Laboratory (Munck, 1992b) demonstrate that the negative pleiotropic effects on yield and seed quality of these mutants can be compensated by optimizing the gene background to these mutants by intensive classical cross breeding and selection. It is not the plant-breeding prospects but the limited notion of quality control of the market and competition with soybean meal and industrially produced amino acids that have prevented the use of these high lysine varieties in the feed industry.

From a theoretical point of view mutations (Munck, 2005, 2006; Sects. 7.1–7.2) with their near isogenic backgrounds give a much more clear cut insight into the multivariate aspect of pleiotropy than QTL analysis combined with backcrosses (Kleinhofs and Han, 2002). Pleiotropy has tended to be underestimated by geneticists and molecular biologists because of lack of tools and data programs to overview the phenotype. For the first time, pleiotropy of specific genes can now be studied as physical–chemical imprints in the endosperm tissue by NIRS and chemometrics as demonstrated in Sect. 7 (Munck et al., 2004; Munck, 2005, 2006).

7 Four Examples on How NIR Technology Supports Advances in Plant Breeding, Seed Sorting, and Plant Science

7.1 "Data Breeding": NIR Spectra of Barley Endosperm Mutants Evaluated by PCA Support a Selection for Complex Traits and Genotypes Based on a Physical–Chemical Interpretation of Spectral Data

In 1999, the barley "high lysine" mutation collection selected 1965–1989 by the dye-binding method (Munck et al., 1970) at Svalöf, Risø, and Carlsberg was used as a test case for NIRS and chemometrics (Munck et al., 2001; Munck, 2003; Munck et al., 2004).

The spectral analysis of the 28 barley samples grown in greenhouse in Fig. 3 is performed unsupervised. The NIR spectra 1,100–2,500 nm (Foss-NIR Systems 6500, USA) from milled samples are outlined in Fig. 3a. A PCA classification of the spectral patterns (every second wavelength was omitted) is presented in Fig. 3b. The samples are now identified by consulting the field book. There are three main clusters: N for normal barley, P for high lysine mutants, such as Risø mutants 8 (*lys4d*) and 1508 (*lys3a*) in Bomi, and, finally, the cluster C for starch-reduced mutants with only a slight lysine improvement including Risø mutants 13 (*lys5f*) and 16 in Bomi and mutant 29 (*lys5g*) in Carlsberg II.

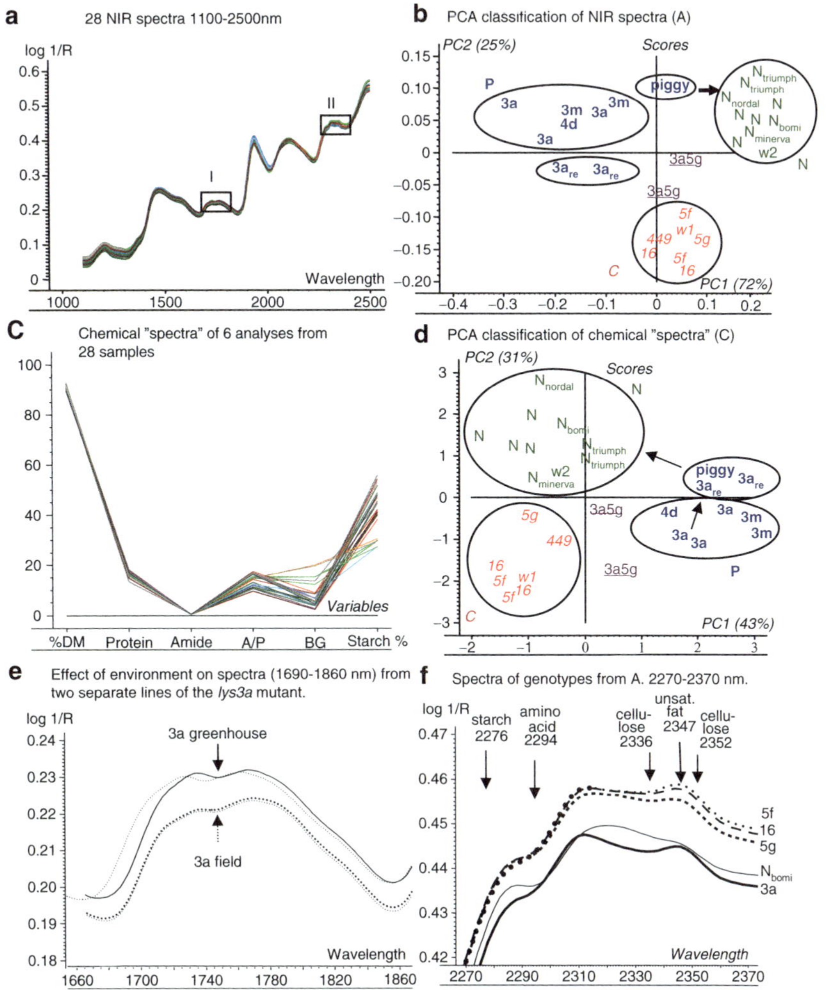

Fig. 3 Principal Component Analysis (*PCA*) classification of spectral and chemical data from 28 genotypes grown in green house from Table 2, for sample identification

Table 2 Chemical analyses of N, C, and P clusters and selected genotypes from Fig. 3 (greenhouse)

	n	Percentage DM	Protein	Amid	A/P	BG	Starch
N	8	91.12 ± 1.27	15.63 ± 1.22	0.42 ± 0.03	16.9 ± 0.4	5.6 ± 1.4	50.78 ± 3.34
C	7	91.70 ± 0.59	16.40 ± 0.90	0.40 ± 0.04	15.2 ± 0.3	15.9 ± 2.9	31.61 ± 5.86
P	6	90.40 ± 0.52	17.57 ± 0.64	0.31 ± 0.05	11.1 ± 1.5	3.8 ± 1.4	40.15 ± 1.12
Piggy	1	89.51	16.12	0.32	12.4	3.9	44.60

DM dry matter

The C mutants have been extensively used in studies of starch synthesis in the developing barley endosperm (see review by Rudi et al., 2006). It was found that Risø 16 lacks one of the adenosine diphosphate (ADP)-glucose pyrophosphorylase (AGPase) genes that are necessary for starch synthesis while *lys5f* and *lys5g* that are allelic in the *lys5* locus are low in starch because they have an inactive ADP–glucose membrane transporter. It was therefore surprising to find that all genotypes located in cluster C (Fig. 3b) more or less compensated the loss in starch by an overproduction of β-glucan as shown in Table 2. A new regulative pathway for the biochemists to evaluate was thus anticipated (Munck et al., 2004) by revealing six β-glucan compensating low starch mutants of the C type.

The spectral outliers outside the N, P, and C clusters in Fig. 3b are the double recessive *lys3a5g* recombinants and three recombinants from the Carlsberg high lysine barley-breeding program (1974–1989) for improved seed quality and starch content. The very high lysine "Piggy" *lys3a* recombinant (45% increase in lysine) is moving in the PCA spectral score plot (Fig. 3b) from the P toward the N cluster indicating an improvement in starch (Table 2) due to 15 years selection work for improved seed quality.

However, now when the NIR technology has been introduced, selection of improved varieties can be made directly by interpreting their position in a spectral PCA score plot in relation to high quality controls by "data breeding" (Munck and Møller, 2005; Munck et al., 2004). The validation of the NIR PCA score plot (Fig. 3b) is made by a parallel data set of six analyses (Table 2) that is represented as 28 "chemical spectra" in Fig. 3c. A PCA on this data set (Fig. 3d) makes a classification equal to that of NIR (Fig. 3b).

It may be surprising to plant geneticists and breeders that individual samples and genotypes can be evaluated by direct visual evaluation of the patterns of NIR spectra as in Figs. 3e–f and Fig. 4. However, in the spectroscopic literature, wavelengths are tentatively assigned to represent specific spectral bonds and components as indicated in Fig. 3f. The spectral reproducibility of two separate lines of the *lys3a* mutant is demonstrated for two environments in Fig. 3e for the small area I in Fig. 3a enlarged in Fig. 3e. The environmental effect is mainly seen as an offset (Munck et al., 2001). The spectral signatures 2,270–2,370 nm (area II in Fig. 3a) of four mutants and the Bomi control are visualized in Fig. 3f. The spectral patterns of the high β-glucan C mutants 16 (16.6%) and *lys5f* (19.8%) are almost identical (see discussion in Sect. 7.2) while the *lys5g* mutant with a lower β-glucan content (13.3%) is different. The *lys3a* P spectrum has a distinct pattern deviating from the C and the N (Bomi) spectra. The bulb at 2,347 nm characteristic for all the four mutant spectra in Fig. 3f indicates an increase in unsaturated fat that was verified as a pleiotropic effect (Munck et al., 2004).

The heterogeneity in physical–chemical representation in the 2,190–2,400 nm NIR area is demonstrated in Table 3 for six chemical components for an enlarged barley material. The spectral prediction coefficients of the six chemical analyses are listed for each of seven 30 nm intervals in a PLSR (iPLS) evaluation (Nørgaard et al., 2000) explaining the physical–chemical basis of the spectral patterns of the genes and genotypes visualized in Figs. 3e–f and 4.

Table 3 Confirming the representation of NIR spectra as chemical patterns by iPLS correlation coefficients (r, less significant coefficients marked in bold) in seven 30-nm intervals (2,190–2,400 nm) for the barley material classified by iECVA in Fig. 4[a]

PLSR	DM	BG	Amide	A/P	Protein	Starch
Analytical range	87.7–92.8	2.5–20.0	0.2–0.5	10.5–17.7	9.7–19.7	27.2–60.4
N	69	73	66	66	68	37
Spectral range (*n* = 92)						
1,100–2,498	0.95 (3)	0.97 (8)	0.97 (8)	0.94 (5)	0.99 (10)	0.97 (4)
2,190–2,218	**0.65** (2)	**0.58** (4)	**0.89** (4)	**0.83** (4)	0.95 (6)	**0.83** (3)
2,220–2,248	0.92 (2)	0.94 (4)	0.91 (3)	**0.86** (4)	0.94 (3)	0.93 (3)
2,250–2,278	0.93 (5)	0.94 (3)	**0.89** (5)	0.90 (5)	**0.89** (5)	0.96 (5)
2,280–2,308	0.93 (5)	0.92 (5)	0.93 (5)	0.93 (4)	0.92 (5)	0.95 (5)
2,310–2,338	0.95 (2)	0.95 (4)	**0.47** (2)	**0.89** (5)	**0.77** (5)	0.97 (5)
2,340–2,368	0.94 (3)	0.92 (4)	0.77 (5)	**0.85** (4)	**0.77** (5)	0.97 (5)
2,370–2,396	0.94 (3)	**0.85** (3)	**0.21** (3)	**0.57** (4)	**0.49** (4)	0.95 (3)

[a]Correlation coefficients: *r* (*n* = PCs)

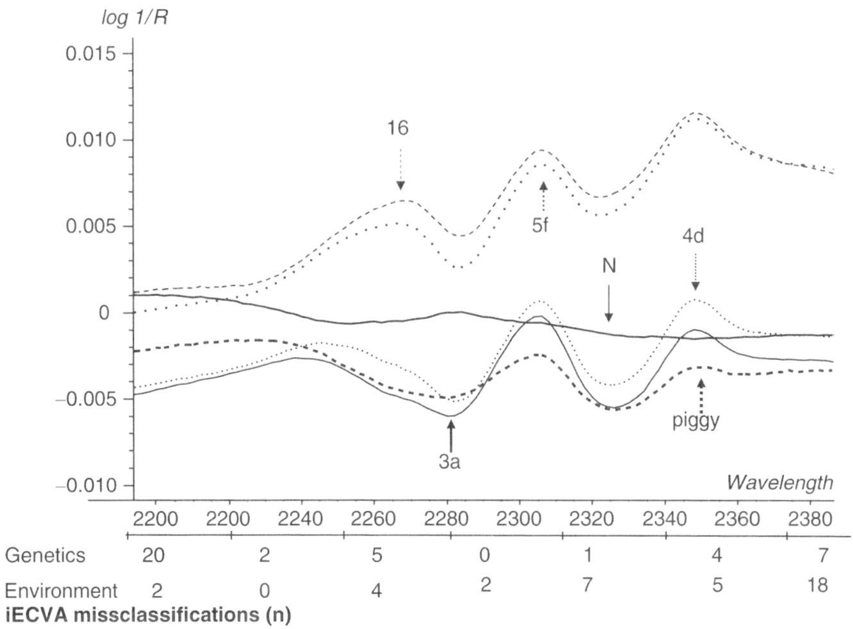

Fig. 4 Demonstrating pleiotropy for the Bomi endosperm mutants *lys3a*, *lys4d*, *lys5f*, and Risø mutant 16 by differential spectra to Bomi 2,200–2,380 nm subtracting the spectrum of Bomi. The spectral effect of changed gene background for the *lys3a* mutant bred into the recombinant Piggy is shown. Below a mutant/normal barley near infrared (*NIR*)-spectral material (*n* = 92, greenhouse *n* = 69; field = 23) is classified by *iECVA* (interval Extended Canonical Variates Analysis) in seven spectral intervals. Number of misclassifications (*n*) by iECVA are indicated for genetics (N, C, and P classes; Fig. 3b) and environment (greenhouse/field). See Table 3

7.2 The Chemical Composition of the Endosperm Is a Response Interface for Mutants and Genotypes that Facilitates Spectral NIR Definitions of Pleiotropy, the Phenome, and of Complex Quality Traits (Munck, 2007)

The physical–chemical relevance of gene-specific spectral patterns from the endosperm now makes it possible to evaluate the complete pleiotropic expression of four mutant genes in Fig. 4 by subtracting the spectrum for the near isogenic Bomi background. The spectra cover the wavelength area 2,200–2,380 nm. The chemical composition is indicated in Table 2. The Bomi background is a straight line at zero in Fig. 4. The mean spectrum of eight normal varieties (N) is slightly deviating from the zero line. There are two main patters of spectra: P (*lys3a*, *lys4d*) and C (*lys5f*, mutant 16) spectra that are classified in the PCA in Fig. 3b and discussed in Sect. 7.1. The difference within these pairs looks small. However, the reproducibility of NIRS is very high and it is likely that a larger material will be able to verify the small differences observed and interpret them in chemical terms. NIR spectra from endosperm genotypes are suggested to represent "the digitized phenome" (Munck et al., 2004; Munck, 2005, 2006) and constitutes a new exploratory approach to the phenome in Systems Biology (Munck, 2007).

The effect of "data breeding" in visualizing selection for improved yield, seed quality, and starch on the high lysine *lys3a* recombinant Piggy (Table 2) is clearly seen as a normalization and flattening out of the spectrum (Fig. 4). The differential spectrum between the spectra of *lys3a* and Piggy in Fig. 4 is a holistic representation of the changed gene background that can be interpreted in physical and chemical terms. The highly reproducible NIR spectra contains repetitive confounded information on the level of chemical bonds which to some degree can be interpreted by consulting spectral literature (Williams, 2002; Martens and Næs, 2001) and by using PLSR correlations to all kinds of measurements as indicated in Fig. 1 and Table 3. Below in Fig. 4, the number of misclassifications in seven spectral intervals for genotype and environment for N, C, and P barleys ($n = 92$) are indicated using the newly developed interval Extended Canonical Variates Analysis (iECVA) model by Nørgaard et al. (in press). There are large differences in classification ability throughout the relative small 2,190–2,396 areas for the seven intervals.

The classification in each area is chemically interpreted in Table 3 by iPLS correlation coefficients to six chemical analyses. The areas 2.220–2248 nm and 2,280–2,308 nm that have the lowest number of misclassifications for genetics and environment are also the most versatile with regard to chemical representation as seen by the high correlation coefficients approaching those of the whole spectrum 1,100–2,498 nm given above in Table 3.

Statistical models such as analysis of variance and PCA are destructive and are not able to represent the finely tuned, reproducible spectra in the barley endosperm model (Munck, 2005, 2006). A careful visual evaluation of each spectrum with controls is therefore essential in a dialogue with chemical analyses and prior genetic knowledge. The genotype should be evaluated as a whole "*genetic milieu*" as suggested by Chetverikov already in 1926. All genes may in principle more or less interact with the expression of all other genes by the principle of pleiotropy. This concept is operationally adopted by the classical plant breeders such as Ramage (1987); however, it is far from the theories in current plant science. Gene interaction on the level of chemical composition can now be quantified as a whole by NIR technology and chemometrics as demonstrated in the barley endosperm model (Munck et al., 2004; Jacobsen et al., 2005). The screening and interpretation of technological traits by NIRS is further discussed for wheat in Sects. 3.3 and 3.4.

7.3 Classification of Wheat Genotypes from a Gene Bank by Their Spectral and Physical-Chemical Fingerprints Correlated to Quality Traits

There are today millions of accessions in cereal gene banks that are waiting for a classification of their physical–chemical composition by NIRS and automatic single-seed imaging and hardness instruments. A collection of diploid, tetraploid, and hexaploid wheat cultivars from the Nordic Gene bank in Lund, Sweden, grown in the field 1999 was analyzed (Fig. 5), involving the Foss-NIR-6500 reflection, spectrograph, and the single-seed instruments, Grain check (Foss A/S, Hillerød, Denmark) and SKCS 4100 (Perten North America, Reno, USA) single seed hardness device. The chemometric strategy of Fig. 1 was followed with PCA classification of separate data sets involving spectral (n = 750; Fig. 5a) and physical–chemical variables (n = 18; Fig. 5b) data connected with a PLSR correlation plot of hardness (Fig. 5c) and other variables.

It is clear from Fig. 5a that a PCA on NIR spectra is able to almost perfectly classify the wheat collection according to their chromosome number with the diploid species to the right, the tetraploid to the left, and the hexaploid in the middle. Note that the *T. carthlicum* sample (ca; n = 28) encircled to the right in Fig. 5a in the spectral PCA is classified as an outlier of the tetraploid family to the left.

The PCA on the 18 physical–chemical variables in Fig. 5b is a biplot where the variables are marked. If a variable appears near to a cluster of samples they are all high in this analysis. Thus the hardness, protein (P), amide (A), A/P-index, and DM variables is placed in a tetraploid cluster below to the right together with emmer (em), and dicoccoides (di), and timopheevii (ti) wheat's indicating that these cultivars tends to have a hard seed texture and a high protein level. On the opposite side down to the left in the biplot, most of the seed form parameters like width, volume, length, and diagonal are located marking that wheat's located in this direction are more large seeded, such as polish wheat (n = 28; po) and some spelts (n = 42; sp). Above to the left (Fig. 5b), some common wheat's (n = 42; wh) are placed together with the intensity and color parameters indicating a red seed coat. The roundness seed variable is situated above in the middle of the biplot near to a collection of diploid and hexaploide wheats with round seeds. There is a reasonable

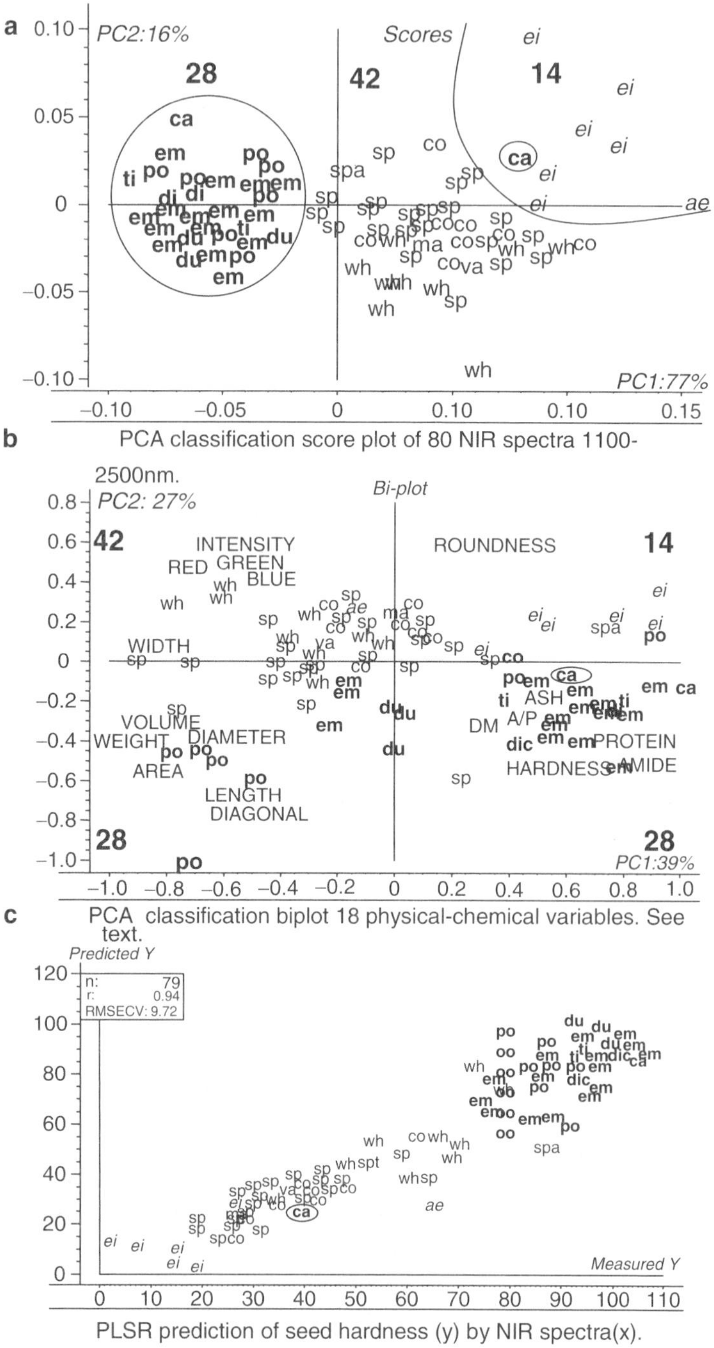

Fig. 5 The aspect of "association genetic" tested on 80 gene-bank samples of ancient and modern wheat's by chemometric data analysis (Principal Component Analysis, *PCA*, and Partial

"ploidy" classification in the PCA (Fig. 5b) of the physical–chemical data set with hexaploid species in the first diploid in the second quadrant and tetraploid in the third and fourth quadrants.

The resemblance in discrimination between the two parallel PCAs suggests of that NIR spectra could represent the physical–chemical data set as previously discussed (Fig. 4; Table 3). This is confirmed in the PLSR prediction plot in Fig. 5c where hardness (y) is correlated to NIR spectra 1,100–2,500 nm (x) with a regression coefficient of $r = 0.94$. The differentiation in hardness between the soft diploid and the hard tetraploid cultivars with the hexaploid in between is confirmed. The soft character of the tetraploid *T. carthlicum* (ca) outlier (encircled) discussed above (Fig. 5a) is verified (Fig. 5c). Significant PLS spectral predictions are obtained for protein $r = 0.98$, amide $r = 0.98$, A/P-index $r = 0.75$, ash $r = 0.69$, weight $r = 0.67$, roundness $r = 0.50$, red reflection $r = 0.74$, and for total reflection intensity $r = 0.67$.

The results from the wheat material in Fig. 5 should be interpreted by a wheat geneticist and completed with NIR studies on genetically defined wheat lines. The profitability of such an approach with NIT and PLSR has been demonstrated in the identification of different chromosomal wheat–rye translocations by Delwiche et al. (1999). The NIRS approach interpreted by chemometrics is an economic and promising tool for gene banks to define the genetic variation of physical–chemical traits in seeds.

Chemometrics is also used for correlating proteins from 2DE separations with technological and genetic data (Gottlieb et al., 2004), according to Fig. 1. A limited number of the many publications on the biochemistry of cereal seed proteins have utilized the multivariate option to explain quality.

One of the first was Mosleth and Uhlen in 1990 who used PLSR to predict Zeleny sedimentation and extensiograph values from 2DE analyses of protein bands in wheat. A more recent example is by Mosleth Færgestad et al. (2004) where the effect of storage protein composition was related to wheat dough rheology by PLSR. Specific protein bands within the glutenine and gliadine subunits were found to positively and negatively influence mixograph peak time that could explain the quality differences between wheat varieties. In the future, the role of friabiline (Schofield, 1994) and the many glutenine and glutelin proteins (Mosleth Færgestad et al., 2004; Shewry and Casey, 1999) should be evaluated in *relation to the functional technological traits* and the genes behind (Fig. 1) using NIR-spectra as a data merger in analyzing a gene bank material (Fig. 5).

Fig. 5 (Continued) Least Squares Regression, *PLSR*). Separate PCA classification through 1,400 spectral (NIR 1,100–2,500 nm) and 18 physical chemical variables are compared (see text). Sample identification; diploid 2 (n = 14) eincorn *T. monococcum* (ei); subsp. *aegilopoides* (ae), tetraploid 4 (n = 28), emmer *T. turgidum*; subsp. *dicoccum* (em); wild emmer subsp. *dicoccoides* (di); *T. polonium* (po, p); *T. durum* (du); *T. timopheevii* (ti), hexaploid 6 (n = 42) *T. vavilovii* (va); *T. aestivum* (wh); subsp. *spharococcum* (spa), *compactum* (co), *spelta* (sp)

Table 4 Individual seed sorting of a winter wheat sample (protein 11.7) with a Bomill AB NIR pilot seed sorter calibrated to bread volume (see discussion in text)

	Yield (%)	Farinograph		Extensiograph	
		Dough stability time	Water uptake (%)	Dough elasticity height	Gluten content (%)
Fraction 1	35	1.7	53.1	100	17.4
Fraction 2	45	5.5	56.7	129	22.7
Fraction 3	20	8.4	59.7	146	27.6

NIR near infrared

7.4 Seed Sorting for Complex Quality Traits by NIR Technology

Near infrared sensors and satellite Global Positioning System (GPS) control are now used in harvesting in *precision agriculture* (Stafford, 1999). With value-added sorting, it is now also possible to separate the local wheat harvest in bulks suitable for baking quality (Table 1) or feed by a NIR sensor mounted on a combiner with two separate bins. Chemical composition can also be studied on the basis of single seeds by NIT spectroscopy. In a collection of wheats grown at two locations in Denmark, NIT calibrations (Pram Nielsen et al., 2003) demonstrated great variation in single seed hardness (−28.8/+101.5 units), protein (6.8–17.0%), and density (0.99–1.25 g/cm^3).

It is now possible to program a NIR/NIT spectrograph/computer/single-seed sorter to select for a complex trait such as baking quality on single seed basis. A pilot machine (BOMILL AB, Lund, Sweden; Löfqvist and Pram Nielsen, 2003) was calibrated by a set of samples of wheat varieties with a wide range of bread volume.

The result (Table 4) demonstrate the effect of single-seed sorting for baking and feeding purpose of a genetically homogeneous winter wheat variety where three fractions were analyzed with regard to dough parameters. There are pronounced environmental effects on single-seed quality, which can be exploited by value-added sorting to improve dough stability, water uptake and elasticity, and gluten content for baking (Table 4) in fractions 2 + 3 (65%). Fraction one (35%) with lower baking value could be sold for biscuits or for feed. New seed sorters based on NIR/NIT and chemometric data evaluation with the capacity of several tons an hour are underway (Pram Nielsen and Löfqvist, 2006). The genetic versus the environmental effect on single-seed sorting for different quality traits should be studied to define the possibilities and limits of the new technology. As an example, the new technology could be used analytically to support data from yield trials by sharpening the selection in breeding for high yield with seed density distribution and high starch content as indicators (Sect. 2.1).

8 The Economy in Breeding and Sorting for Complex Quality Traits in Cereals in the Future

NIRS evaluated by chemometrics is an extension of breeder's vision to the micro world by "high tech" tools. It includes chemometric pattern recognition data analysis that will also considerably sharpen QTL analysis and connect to genomic data. NIR screening and "*association genetics*" will link the practical breeding work on the physical–chemical phenotype level to molecular and biochemical data evaluated as whole traits from the computer interface. The biological diversity of gene banks will be able to be defined and documented by NIR spectra. Selection by "data breeding" of high-quality genotypes as whole spectroscopic patterns in a PCA is extremely cost-effective. The instruments already available now are, however, with few exceptions only used for specific analytes.

In order to fully introduce the advantages of the new technology, the conservative market on cereal handling and processing have to be convinced of the advantages for seeds tailored and marketed for specific uses. Sorting individual seeds in full production scale for complex quality traits by NIRS launches new opportunities for added value in cereal production if the process can be made economical. Already, the introduction of the now available pilot-scale NIR seed sorters in early generation selection will drastically change theory in genetics and the logistics of quality improvement in plant breeding. In fact, NIRS introduces a new exploratory view on the phenome in systems biology (Munck, 2007). *The new challenge for the universities and the industry is to create a renaissance in classical plant breeding by the new high-tech direct tools for observation and selection. A second generation of plant breeders should be educated who can combine the traditional phenomenological "top down" and the molecular "bottom up" perspectives bound together by the advanced data and screening technology that now is available.*

Acknowledgments The contribution to figures, tables, and language correction, from my colleagues Birthe Møller, Lars Nørgaard and Gilda Kischinovsky, is gratefully acknowledged. Bo Løfqvist, A.B. Bomill, and Lund Sweden has kindly supplied the data in Table 4. I am indebted to the great number of friends, coworkers, and employers in Sweden, Denmark, and internationally, who have inspired me when writing this chapter.

References

Aastrup, S. and Munck, L. (1984) A β-glucan mutant in barley with thin cellwalls. In: R.D. Hill and L. Munck (Eds.), *New Approaches to Research on Cereal Carbohydrates*. Elsevier, Amsterdam, pp. 291–296.

Aastrup, S., Gibbons, G.C. and Munck, L. (1981) A rapid method for estimating the degree of modification in barley by measurement of cell wall breakdown. *Carlsberg Research Communications* 46, 77–86.

Abdel-Aal, E. and Wood, P.J. (Eds.) (2005) *Specialty Grains for Food and Feed*. American Association of Cereal Chemists, St. Paul, MN.

Anderson, O. (1996) Molecular approaches to cereal quality improvement. In: R.J. Henry and P.S. Kettlewell (Eds.), *Cereal Grain Quality*. Chapman and Hall, London, pp. 371–404.

Arus, P. and Gonzales, J. (1993) Marker-assisted selection. In: M.D. Hayward, N.O. Bosemark and I. Romagosa (Eds.), *Plant Breeding: Principles and Prospects*. Chapman and Hall, London, pp. 314–331.

Axtell, J.D. (1981). Breeding for improved nutritional quality. In: K.J. Frey (Ed.), *Plant Breeding II*. Iowa State University Press, Ames, IA, pp. 365–432.

Bach Knudsen, K.E. and Munck, L. (1985) Dietary fibre contents and compositions of sorghum and sorghum-based foods. *Journal Cereal Science* 3, 153–164.

Bergman, C.J., Bhattacharya, K.R. and Ohtsubo, K. (2003) Rice and end use quality analysis. In: E.T. Champagne (Ed.), *Rice: Chemistry and Technology*. American Association of Cereal Chemists, St. Paul, MN, pp. 415–472.

Bhatty, R.S. (1993) Nonmalting uses of barley. In: A.W. MacGregor and R.S. Bhatty. *Barley: Chemistry and Technology*. American Association of Cereal Chemists, St. Paul, MN, pp. 355–418.

Bjørnstad, Å., Westad, F. and Martens, H. (2004) Analysis of genetic marker-phenotype relationships by jack-knifed partial least squares regression (PLSR). *Hereditas* 141, 149–165.

Brinch-Pedersen, H., Olesen, A., Rasmussen, S.K. and Holm, P.B. (2000) Generation of transgenic wheat for constitutive accumulation of an Aspergillus phytatase. *Molecular Breeding* 6, 195–206.

Burrows, V.D. (1986) Breeding oats for food and feed: conventional and new techniques and materials. In: F.H. Webster (Ed.), *Oats: Chemistry and Technology*. American Association of Cereal Chemists, St. Paul, MN.

Chetverikov, S.S. (1926) On certain aspects on the evolutionary process. Translated from the original article in Zurnal Eksperimentalńoi Biologii, A2, -54 by M. Barker. In: Proc. American Phil. Soc. 105(2), 167–195.

Childs, N.W. (2003) Production and utilization of rice. In: E.T. Champagne (Ed.), *Rice: Chemistry and Technology*. American Association of Cereal Chemists, St. Paul, MN, pp. 1–23.

Collins, F.W. (1986) Oat phenolics: structure, occurrence and function. In: F.H. Webster (Ed.), *Oats: Chemistry and Technology*. American Association of Cereal Chemists, St. Paul, MN, pp. 227–291.

Darrah, L.L., MacMullen, M.D. and Zuber, M.S. (2003) Breeding genetics and seed corn production. In: P.J. White and L.A. Johnson (Eds.), *Corn: Chemistry and Technology*. American Association of Cereal Chemists, St. Paul, MN, pp. 35–67.

Delwiche, S.R., Graybosch, R.A. and Peterson, J. (1999) Identification of wheat lines possessing the 1AL.1RS or 1BL.1RS wheat-rye translocation by near-infrared spectroscopy. *Cereal Chemistry* 76(2), 255–260.

Druka, A., Muehlbauer, G., Druka, I., Caldo, R., Baumann, U., Rostoks, N., Schreiber, A., Wise, R., Close, T., Kleinhofs, A., Graner, A., Schulman, A., Langridge, P., Sato, K., Hayes, P., McNicol, J., Marshall, D. and Waugh, R. (2006) An atlas of gene expression from seed to seed through barley development. *Functional and Integrative Genomics* 6(3), 202–211. Available at http://www.barleybase.org

Dudley, J.W. and Lambert, R.J. (1969) Genetic variability after 65 generations of selection in Illinois high oil–low oil, high protein and low protein strains of *Zea mays*. *Journal of Crop Science* 9(2), 179–181.

Eckhoff, S.R. and Paulsen, M.R. (1996) Maize. In: R.J. Henry and P.S. Kettlewell (Eds.), *Cereal Grain Quality*. Chapman and Hall, London, pp. 77–112.

Fitzgerald, M. (2003) Starch. In: E.T. Champagne (Ed.), *Rice: Chemistry and Technology*. American Association of Cereal Chemists, St. Paul, MN, pp. 109–141.

Gottlieb, D.M., Schultz, J. Bruun, S.W., Jacobsen, S. and Søndergaard, I. (2004). Multivariate approaches in plant science. *Phytochemistry* 65, 1531–1548.

Graybosch, R.A. (1998) Waxy wheatÕs: origin, properties, prospects. *Trends in Food Science and Technology* 9, 135–142.

Horvath, H., Huang, J., Wong, O.T. and von Wettstein, D. (2002) Experiences with genetic transformation of barley and characteristics of transgenic plants. In: G.A. Slafer, J-L. Molina-Cano, R. Savin, J.L. Araus and I. Romagosa (Eds.), *Barley Science: Recent Advances from Molecular Biology to Agronomy of Yield and Quality*. Food Products Press, New York, NY, pp. 143–204.

House, L.R. (1995) Sorghum and millets: history, taxonomy and distribution. In: D.A.V. Dendy (Ed.), *Sorghum and Millets: Chemistry and Technology*. The American Association of Cereal Chemistry, St. Paul, MN, pp. 1–10.

Jacobsen, S., Søndergaard, I., Møller, B., Desler, T. and Munck, L. (2005) A chemometric evaluation of the underlying physical and chemical patterns that support near Infrared spectroscopy of barley seeds as a tool for explorative classification of endosperm genes and gene combinations. *Journal of Cereal Science* 42(3), 281–299.

Kleinhofs, A. and Han, F. (2002) Molecular mapping of the barley genome. In: G.A. Slafer, J.-L. Molina-Cano, R, Savin, J.L. Araus and I. Romagosa (Eds.), *Barley Science: Recent Advances from Molecular Biology to Agronomy of Yield and Quality*. Food Products Press, New York, NY, pp. 31–64.

Knott, S.A. and Haley, C.S. (2000) Multitrait least squares for quantitative trait loci detection. *Genetics* 156, 899–911.

Löfqvist, B. and Pram Nielsen, J. (2003). A method of sorting objects comprising organic material. *European Patent* EC B07C5/34; G01N21/35G.

MacKey, J. (1981) Cereal production. In: Y. Pomeranz and L. Munck (Eds.), *Cereals: A Renewable Resource*. American Association of Cereal Chemists, St. Paul, MN, pp. 5–20.

Martens, H. and Ns, T. (1989) *Multivariate Calibration*. Wiley, Chichester.

Martens, H. and Ns, T. (2001) Multivariate calibration by data compression. In: P. Williams and K. Norris (Eds.), *Near Infrared Technology in the Agricultural and Food Industries*. American Association of Cereal Chemists, St. Paul, MN, pp. 59–100.

McCleary, B.V. and Prosky, L (2001) *Advanced Dietary Fiber Technology*. MPG Books Bodmin, Cornwall.

Mosleth, E. and Uhlen, A.-K. (1990) Identification of quality-related gliadins and prediction of bread-making quality from the electrophoretic patterns of gliadins and high molecular weight subunits of glutenin. *Norwegian Journal of Agricultural Science* 4, 27–45.

Mosleth Færgestad, E., Solheim Flæte, N.E., Magnus, E.M., Hollung, K., Martens, H. and Uhlen, A.-K. (2004) Relationships between storage protein composition, protein content, growing season and flour quality of bread wheat. *Journal of the Science of Food and Agriculture* 84, 887–886.

Munck, L. (1972) Improvement of nutritional value in cereals. *Hereditas* 72, 1–128.

Munck, L. (1991) Quality criteria in the production chain from malting barley to beer. *Ferment* 4, 235–241.

Munck, L. (1992a) The contribution of barley to agriculture today and in the future. In: L. Munck (Ed.), *Barley Genetics VI*. Muncksgaard Int. Publish. Ltd., Copenhagen, Denmark, pp. 1099–1109.

Munck, L. (1992b) The case of high-lysine barley breeding. In: P.R. Shewry (Ed.), *Barley: Genetics, Biochemistry, Molecular Biology and Biotechnology*. C. A. B. Int., Great Britain, pp. 573–601.

Munck, L. (1993) On the utilization of renewable plant resources. In: M.D. Hayward, N.O. Bosemark and I. Romagosa (Eds.), *Plant Breeding Principles and Prospects*. Chapman and Hall, London, pp. 500–522.

Munck, L. (1995) New milling technologies and products whole plant utilization by milling and separation of the botanical and chemical components. In: D.A.V. Dendy (Ed.), *Sorghum and Millets: Chemisty and Technology*. The American Association of Cereal Chemistry St. Paul, MN, pp. 223–281.

Munck, L. (2003) Detecting diversity – a new holistic exploratory approach. In: R. von Bothmer, T. van Hintum, H. Knupffer and K. Sato (Eds.), *Diversity in Barley*. Elsevier Science B.V., Amsterdam, pp. 227–245.

Munck, L. (2004) Whole plant utilisation. *Encyclopaedia of Grain Science*, Elsevier, Amsterdam, pp. 459–466.

Munck, L. (2005) *The revolutionary aspect of exploratory chemometric technology*. The Royal and Veterinary University of Denmark, Narayana Press, Gylling, Denmark, pp. 352.

Munck, L. (2006) Conceptual validation of self-organisation studied by spectroscopy in an endosperm gene model as a data driven logistic strategy in chemometrics. *Chemometrics and Intelligent Laboratory Systems* 84, 26–32.

Munck, L. (2007) A new holistic exploratory approach to Systems Biology by near infrared spectroscopy evaluated by chemometrics and data inspection. *Journal of Chemometrics* 21, 406–426.

Munck, L. and Møller, B. (2004) A new germinative classification model of barley for prediction of malt quality amplified by a near infrared transmission spectroscopy calibration for vigour "on-line" both implemented by multivariate data analysis. *Journal of the Institute of Brewing* 110(1), 3–17.

Munck, L. and Møller, B. (2005) Principal component analysis of near infrared spectra as a tool of endosperm mutant characterisation and in barley breeding for quality. *Czech Journal of Genetics and Plant Breeding* 41(3), 89–95.

Munck, L. and Rexen, F. (Eds.). (1990) *Agricultural Refineries: A Bridge from Farm to Industry*. The Commission of the European Communities, EUR 11583 EN.

Munck, L., Karlsson, K.E., Hagberg, A. and Eggum, B.O. (1970) Gene for improved nutritional value in barley seed protein. *Science* 168, 985–987.

Munck, L., Pram Nielsen, J., Møller, B., Jacobsen, S., Søndergaard, I., Engelsen, S.B., Nørgaard, L. and Bro, R. (2001) Exploring the phenotypic expression of a regulatory proteome-altering gene by spectroscopy and chemometrics. *Analytica Chimica Acta* 446, 171–186.

Munck, L., M¿ller, B., Jacobsen, S. and S¿ndergaard, I. (2004) Near infrared spectra indicate specific mutant endosperm genes and reveal a new mechanism for substituting starch with (1→3, 1→4)-β-glucan barley. *Journal of Cereal Science* 40, 213–222.

Murthy, D.S. and Kumar, K.A. (1995) Traditional uses of sorghum and millets. In: D.A.V. Dendy (Ed.), *Sorghum and Millets: Chemistry and Technology*. The American Association of Cereal Chemistry, St. Paul, MN, pp. 185–222.

Nørgaard, L., Saudland, A., Wagner, J., Nielsen, J.P., Munck, L. and Engelsen, S.B. (2000) Interval partial least squares regression (iPLS): a comparative chemometric study with an example from near-infrared spectroscopy. *Applied Spectroscopy* 54(3), 413–419.

Nørgaard, L., Bro, R., Westad, F. and Balling Engelsen, S. (2006) A modification of canonical variates to handle highly collinear multivariate data. *Journal of Chemometrics* 20, 425–435.

Olsson, G. (Ed.). (1986) Svalöf *1886–1986*. LTs Publishers, Stockholm.

Paulsen, M.R., Watson, S.A. and Singh, M. (2003) Measurement and maintenance of Corn Quality. In: P.J. White and L.A. Johnson (Eds.), *Corn: Chemistry and Technology*. American Association of Cereal Chemists, St. Paul. MN, pp. 159–212.

Payne, P.I., Thompson, R., Bartels, R., Harberd, N., Harris, P. and Law, C. (1983) The high molecular weight subunits of glutenin: classical genetics, molecular genetics and the relationship with bread making quality. *Proceedings of the 6th International Wheat Genetics Symposium, Japan*, pp. 827–834.

Petersen, P.B. and Munck, L. (1994) Whole crop utilization of barley including new potential uses. In: A.W. MacGregor and R.S. Bhatty (Eds.), *Barley Chemistry and Technology*. The American Association of Cereal Chemistry, St. Paul, MN, pp. 437–474.

Pilu, R., Panzeri, D., Gavazzi, G., Rasmussen, S.K., Consonni, G. and Nielsen, E. (2003) Phenotypic, genetic and molecular characterization of a maize low phytic acid mutant (lpa241). *Theoretical and Applied Genetics* 107, 980–987.

Poulsen, H.D., Johansen, K.S., Hatzack, F. Boisen, S. and Rasmusssen, S.K. (2001) The nutritional value of low phytate barley evaluated in rats. *Acta Agriculturae Scandinavica – Section A, Animal Science* 51, 53–58.

Pram Nielsen, J. and Lšfqvist, B. (2006) *Method and device for sorting objects*. European Patent EC B07C5/36C1; BO75/34.

Pram Nielsen, J.P., Pedersen, D.K. and Munck, L. (2003) Development of non-destructive screening methods for single kernel characterisation of wheat. *Cereal Chemistry* 80, 274–280.

Ramage, R.T. (1987) A history of barley breeding methods. *Plant Breeding Reviews* 5, 95–138.

Rasmussen, S.K. and Hatzak, F. (1998) Identification of two low phytate barley grain mutants by TLC and genetic analysis. *Hereditas* 129, 107–112.

Rooke, M.S., Bekes, F., Fido, R., Barro, F., Gras, P., Tatham, A.S., Barcelo, P., Lazzeri, P. and Shewry, P.R. (1999) Over expression of a gluten protein in transgenic wheat results in greatly increased dough strength. *Journal of Cereal Science* 30, 115–120.

Rooney, L.W. and Serna-Salvidar, S.O. (2003) Food use of whole corn and dry-milled fractions. In: P.J. White and L.A. Johnson (Eds.), *Corn: Chemistry and Technology*. American Association of Cereal Chemists, St. Paul, MN, pp. 496–535.

Rudi, H., Uhlen, A.K., Harstad, O.M. and Munck, L (2006) Genetic variability in cereal carbohydrate compositions and potentials for improving nutritional value. *Animal Feed Science and Technology* 130(1–2), 55–65.

Schofield, J.D. (1994) Wheat proteins: structure and functionality in milling and bread making. In: W. Bushuk and V.F. Rasper (Eds.), *Wheat – Production, Properties and Quality*. Chapman and Hall, Glasgow, pp. 73–105.

Scoles, G.J., Gustafson, J.P. and McLeod, J.G. (2001) Genetics and breeding. In: W. Bushuk (Ed.), *Rye: Production, Chemistry and Technology*. American Association of Cereal Chemists, St. Paul, MN, pp. 9–35.

Seibel, W. and Weipert, D. (2001) Bread baking and other food uses around the world. In: W. Bushuk (Ed.), *Rye: Production, Chemistry and Technology*. American Association of Cereal Chemists, St. Paul, MN, pp. 147–211.

Serna-Saldivar, S. and Rooney, L.W. (1995) Structure and chemistry of sorghum and millets. In: D.A.V. Dendy (Ed.), *Sorghum and Millets: Chemistry and Technology*. The American Association of Cereal Chemistry, St. Paul, MN, pp. 69–124.

Shewry, P.R. (1992) Barley seed proteins. In: P.R. Shewry (Ed.), *Barley: Genetics, Biochemistry, Molecular Biology and Biotechnology*. C.A.B. International, Wallingford. U.K., pp. 319–335.

Shewry, P.R. and Casey, R. (Eds.) (1999) *Seed Proteins*. Kluwer, ISBN 0-4128-1570-2.

Stafford, J.V. (Ed.). (1999) *Precision Agriculture 99*. Part II and I. Sheffield Academic Press, Sheffield.

Swanston, J.S. and Ellis, R.P. (2002) Genetics and breeding of malt quality attributes. In: G.A. Slafer, J.-L. Molina-Cano, R. Savin, J.L. Araus and I. Romagosa (Eds.), *Barley Science: Recent Advances from Molecular Biology to Agronomy of Yield and Quality*. Food Products Press, New York, NY, pp. 85–114.

Thomas, W.T.B. (2002) Molecular marker-assisted versus conventional selection. In: G.A. Slafer, J.-L. Molina-Cano, R. Savin, J.L. Araus and I. Romagosa (Eds.), *Barley Science: Recent Advances from Molecular Biology to Agronomy of Yield and Quality*. Food Products Press, New York, NY, pp. 177–204.

Ullrich, S. (2002) Genetics and breeding of barley feed quality attributes. In: G.A. Slafer, J.-L. Molina-Cano, R. Savin, J.L. Araus and I. Romagosa (Eds.), *Barley Science: Recent Advances from Molecular Biology to Agronomy of Yield and Quality*. Food Products Press, New York, NY, pp. 115–142.

van Harten, A.M. (1998) *Mutation Breeding – Theory and Applications*, Cambridge University Press, Cambridge.

Vasal, S.K. (1999) Qualiy maize story. In: *Improving Human Nutrition Through Agriculture*. Intern. Rice Res. Inst., Los Banos, Philippines, pp. 1–19.

Williams, P.C. (2002) Near infrared spectroscopy of cereals. In: J.M. Chalmers and V.R. Griffiths (Eds.), *Handbook of Vibrational Spectroscopy*. Wiley, Chichester. Vol. 5, pp. 3693–3719.

Wilson, L.M., Whitt, R.S., Ibanez, A.M., Rocheford, T.R., Goodman, M.M. and Buckler IV, E.S. (2004) Dissection of maize kernel composition and starch production by candidate gene association. *The Plant Cell* 16, 2719–2733.

Wrigley, C.W. and Morris, C.F. (1996) Breeding for quality improvement. In: R.J. Henry and P. Kettlewell (Eds.), *Cereal Grain Quality*. Chapman and Hall, London, pp. 326–369.

Breeding for Silage Quality Traits in Cereals

Y. Barrière, S. Guillaumie, M. Pichon, and J.C. Emile

Abstract Forage plants are the basis of ruminant nutrition. Among cereal forages, maize cropped for silage making is the most widely used. Much research in genetics, physiology, and molecular biology of cereal forages is thus devoted to maize, even if silage of sorghum or immature small-grain cereals and straws of small-grain cereals are also given to cattle. Cell wall digestibility is the limiting factor of forage feeding value and is, therefore, the first target for improving their feeding value. Large genetic variation for cell wall digestibility was proven from both in vivo and in vitro experiments in numerous species. Among the regular maize hybrids [excluding brown-midrib (bm) ones], the cell wall digestibility nearly doubled from 32.9% to 60.1%. Genetic variation has also been proven in cell wall digestibility of sorghum and wheat, barley or rice forage, or straw, with lower average values than in maize. Despite lignin content is well known as an important factor making cell wall indigestible, breeding for a higher digestibility of plant needs the use of specific traits estimating the plant cell wall digestibility. Quantitative trait loci (QTL) analysis, studies of single-nucleotide polymorphism (SNP) × feeding value traits relationships, studies of mutants and deregulated plants, and expression studies will contribute to the comprehensive knowledge of the lignin pathway and cell wall biogenesis. Plant breeders will then be able to choose the best genetic and genomic targets for the improvement of plant digestibility. Favorable alleles or favorable QTL for cereal cell wall digestibility will thus be introgressed in elite lines through marker-assisted introgression. Efficient breeding of maize and others annual forage plants demands a renewing of genetic resources because only a limited number of lines are actually known with a high cell wall digestibility. Among bm genes, the bm3 mutant in maize and the bmr12 (and possibly bmr18) mutant in sorghum, which are both altered in the caffeic acid O-methyltransferase (COMT) activity, appeared as the most efficient in cell wall digestibility improvement. Genetic engineering is both an inescapable tool in mechanism understanding and an efficient way in cereal breeding for improved feeding value. Moreover, gene mining and genetic engineering in model plant

Y. Barriére(✉)
Unité de Génétique et d'Amélioration des Plantes Fourragéres, INRA, Route de Saintes, BP6, F-86600 Lusignan, France, e-mail: yves.barriere@lusignan.inra.fr

M.J. Carena (ed.), *Cereals*,
DOI: 10.1007/978-0-387-72297-9, © Springer Science + Business Media, LLC 2009

and systems (Arabidopsis, Zinnia, Brachypodium, ...) are also essential complementary approaches for improvement of cell wall digestibility in grass and cereal forage crops.

1 Introduction

Forage plants and cereals are the basis of energy nutrition of ruminant. However, although forages contain almost the same amount of gross energy as do grains per unit of dry matter (DM), the energy value of forages is lower and much more variable, ranging approximately from 80% (leafy ray grass) to 33% (wheat straw) of maize grain value. Silage maize energy value, which is among the highest forage values, reached an average value of 6.3 MJ/kg DM, but is nearly equal only to 75% of grain maize value. This difference results from the high content of cell walls in forage plants and to the limited digestion of this fiber part by the microorganisms of rumen and, to a lesser degree, of large intestine of animals. The quantitative importance of lignins in the cell wall, their variable structure, and a variety of cross-linkages between cell wall components all have variable depressive effects on cell wall carbohydrate degradation by microorganisms in the rumen and/or large intestines of herbivores (Barrière et al., 2003a, 2004a, b; Grabber et al., 2004; Ralph et al., 2004). The energy supplied by a forage plant in animal diets is thus related to the forage or silage intake and digestibility. For a given animal, intake and digestibility are plant characteristics resulting of plant growth and cell wall development. Both traits are subject to plant genetic variation and are, therefore, of major interest in breeding for silage quality in cereals.

Protein content is also a trait of major interest in animal nutrition. Observed variation between grass genotypes are mostly related to the nitrogen dilution law [nitrogen = $3.40 \times (\text{yield}^{-0.37})$], with lower nitrogen content in plants when the DM yield is higher (Plénet and Cruz, 1997). True variation for protein content is low, especially in maize, and programs devoted to the improvement of protein content in whole plant of cereals did seemingly not really succeeded. However, the low protein content of ensiled cereal diets is easily corrected by cattle cakes (soya, sunflower, and rapeseed). Moreover, the availability of sunflower and rapeseed cakes is expected to increase with oleaginous plants cropping for biofuel production. An alternative to the use of cattle cakes for the improvement of silage protein content is the mixed cropping and ensiling of small grain cereals with legumes such as vetch or pea.

Among cereals cropped for silage making, maize is the most widely used. Sorghum and immature small-grain cereals (wheat, barley, triticale, ...) are also given to cattle after ensiling. Straws, including rice straws in tropical areas, are also used for cattle feeding after grain harvest. Because of the economic importance of the "corn" crop worldwide, and of the economic importance of forage maize in Europe, much research in physiology, genetics, and molecular biology of cereals and grasses silage quality traits is devoted to maize. However, due to the close phylogenic positions of grasses, breeding targets of interest in maize should easily be extrapolated to other C_4 and C_3 grasses. The focus of this chapter will be on maize, as there are more little data available on cell wall digestibility improvement in other cereals, but information on other cereals cropped for silage will also be reported when available.

2 Genetic Variations for Cell Wall Digestibility in Cereals

2.1 Devising an Estimate of Cell Wall Digestibility

Cell wall digestibility, which is the limiting factor of the energy availability in cattle, is the key target for improving the energy value of ensiled cereal crops. This trait is also free of digestible starch and soluble carbohydrate contents that are subject to extensive environmental variation. Moreover, due to rumen microorganism ecology and correlative acidosis risks, the optimal grain content in cereal silages has to be adjusted according to the extra starch content of the diet, and according to the proportion of by-pass starch. Higher grain content in the cereal silages is favorable if the diet included grass silage, whereas the optimum starch content in maize is lower and was thus proved to be close to 30% when no other raw food is given to dairy cattle (Barrière and Emile, 1990; Barrière et al., 1997). This result, which was shown in maize, is very likely true in immature small-grain cereals which have a lower content of by-pass starch.

The more relevant assessments of plant digestibility are done with animals, and these measurements were mostly often based on sheep in digestibility crates. For practical and financial reasons, digestibility assessments done during breeding cycles have to be performed using in vitro tests of digestibility and must be easily and accurately predicted through near infrared reflectance spectroscopy (NIRS). The first in vitro digestibility trait (IVDMD) was proposed by Tilley and Terry (1963) and was based on plant sample degradation by rumen fluid taken from fistulated cows. Different whole plant enzymatic IVDMD were also developed in Europe, including the one of Aufrère and Michalet-Doreau (1983) used in France for hybrid registration, which are of easier management and lower costs as they do not required anaerobic conditions or the maintenance of animals giving rumen fluid. NIRS calibrations for both Tilley and Terry and enzymatic IVDMD were developed in different European and US labs. Correlations between these different enzymatic IVDMD are high and most often greater than 0.90 (INRA Lusignan, unpublished data). For plant breeding purpose, cell wall digestibility can be easily computed, based on a Tilley–Terry or an enzymatic IVDMD and on content in cell wall or noncell wall constituents of the plants (all traits predicted through NIRS calibrations). As proposed by Struik (1983) and Dolstra and Medema (1990), the in vitro neutral detergent fiber digestibility (IVNDFD) can be computed assuming that the non-NDF part (NDF; Goering and van Soest, 1971) of plant material is completely digestible

$$[\text{IVNDFD} = 100 \times (\text{IVDMD} - (100 - \text{NDF}))/\text{NDF}].$$

Complementarily, according to Argillier et al. (1995) and Barrière et al. (2003a), the in vitro digestibility of the "non starch, non soluble carbohydrates, and non crude protein" part (DINAGZ) is computed assuming these three constituents are completely digestible.

$$[\text{DINAGZ} = 100 \times (\text{IVDMD} - \text{ST} - \text{SC} - \text{CP})/(100 - \text{ST} - \text{SC} - \text{CP})]$$

where ST, SC, and CP are starch, soluble carbohydrates, and crude protein contents, respectively.

Either for evaluation of genetic resources or during successive generation of elite hybrid breeding, lignin content and cell wall digestibility estimates are easier and cheaper to obtain from lines rather than after topcrossing. Moreover, variance of traits is greater in lines than in hybrids. Correlations between hybrid values and per se values ranged between 0.62 and 0.94 for cell wall digestibility traits and between 0.63 and 0.87 for lignin content in maize, while similar correlations were low for starch content and did not exceed 0.30 (Barrière et al., 2003a). These results strengthened the relevance of choice of lines from their per se value in breeding cycle for the improvement of forage cell wall digestibility in maize. This result is also very likely true in sorghum.

Reported correlations between Tilley–Terry and enzymatic IVDMD ranged in maize from 0.50 and 0.84, while correlations between enzymatic IVDMD and in vivo organic matter digestibility ranged from 0.57 to 0.82 (Barrière et al., 2003a). An important concern is therefore that in vivo and in vitro methods does rank, or not, genotypes in a similar order. Comparisons of hybrids ranking based either on in vivo data (Barrière et al., 2004a) or on in vitro correlative values (INRA Lusignan, unpublished data) showed that both NDF digestibility (NDFD) and IVNDFD or DINAGZ traits allowed similarly to the elimination of hybrids with poor cell wall digestibility, or to the choice of hybrids with high cell wall digestibility, including bm3 hybrids. Breeding for higher cell wall digestibility is thus efficient when it is based on an in vitro trait, such as IVNDFD, DINAGZ, or a Tilley–Terry-based estimate. However, in restricted ranges of variation such as within subsamples of hybrids of low, intermediate, or high cell wall digestibility, respectively, genotype ranking often partly differed whether an in vivo or in vitro trait was used. The plant cell wall is not completely similarly degraded when subjected to in vivo and in vitro conditions. This fact, which does not impede breeding efficiency, could be more limiting during registration processes if new hybrids are compared to a threshold value, inducing the possibility of rejecting hybrids not significantly different from the accepted ones when they would be fed to cattle, or the reverse.

2.2 Genetic Variation for Cell Wall Digestibility in Maize

Data giving variation for maize in vivo organic matter digestibility (OMD) are available from several investigations. Conversely, in vivo cell wall digestibility variation was rarely investigated in maize or other cereals. From a study based on 478 hybrids (Barrière et al., 2004a), the in vivo cell wall digestibility in maize (estimated as NDFD) nearly doubled from 32.1% to 60.4% with an average value equal to 48.8%. Whereas the genotype effect for NDFD was highly significant, the NDFD genotype × year interaction was not significant, strengthening the interest of cell wall traits during breeding programs. Studies of genotypic correlations showed that OMD was related to NDFD (r = 0.76) but not to grain content (r = −0.16). Similarly, the correlation between NDF content and NDFD was also low (r = 0.10), highlighting that no significant relationship existed between the cell wall digestibility and the cell wall content for maize plants harvested at a similar maturity stage. Based on the results obtained in ruminants, the genetic progress in plant energy value appears thus directly related to NDFD improvement. Besides these in vivo investigations, much research has shown large genetic variations in the in vitro cell wall digestibility of maize (Argillier et al., 2000; Barrière et al., 1997), with similarly, small genotype × environment interaction effects compared to main effects. Heritability of in vivo and in vitro cell wall digestibility traits was high, ranging between 0.65 and 0.80, and it was at least equal to that of yield (Roussel et al., 2002). Breeding for higher in vitro cell wall digestibility values should therefore be very efficient, and the expected progress for the first selection cycle of breeding for cell wall digestibility could thus reach 3.0% points.

The genetic variations in cell wall digestibility of maize silage have consequences on young bull or dairy cow performances, even if maize was not the only constituent of the diet (Barrière et al., 1995a, b; Emile et al., 1996; Hunt et al., 1993; Istasse et al., 1990), strengthening the interest of breeding silage maize for higher cell wall digestibility. All other factors being equal, when comparing hybrids with low or high cell wall digestibility in dairy cows, fat-corrected milk (FCM) yields could differ from 1 to 3 kg among hybrids. The protein contents in milk were also equal or higher in hybrids with higher cell wall digestibility. In a similar way, differences in average daily gains of young bulls reached 100 g/day among hybrids.

2.3 Genetic Variation for Cell Wall Digestibility in Sorghum and Small-Grain Cereals

Cell wall digestibility was shown lower in sorghum silages than in maize silages, with values ranging between 40% and 45% when maize values ranged between 39% and 59% (Barriere et al., 2003a). Sorghum silage had similarly lower OMD values than maize, despite the fact that some grain sorghum silages had higher grain content than maize (Barriere et al., 2003a). This could be hypothetically related to the different morphology of the two plants. Maize bears one ear at the lower third of the plant when sorghum bears grainy panicle at its upper part with higher mechanical constraints inducing likely a greater need of lignification and rigidity of the stalk. Consequently, in most studies that compared sorghum with maize silage

(Aydin et al., 1999), milk production was consistently higher for cows fed maize silage than for those fed sorghum silage. However, results of Mahanta and Pachauri (2005) showed that some varieties of sorghum had a significantly higher cell-wall digestibility than that of current varieties, leading to higher silage digestibility and intake in sheep. As it was the case for maize few years, a higher silage energy value is rarely a trait considered in sorghum-breeding programs.

Genetic variation in cell wall digestibility of small-grain cereals was rarely investigated, either in silage, even most often in straws. From Barrière et al. (2003a), average NDFD in triticale and wheat silage were close to 49%, and close to 46% in rye, but these values were considered as significantly overestimated because the low or very low forage intakes of awned plants by animals. Genetic variation in cell wall digestibility of rice straw has been reported by Abou-el-Enin et al. (1999) from 53 varieties with *in sacco* NDFD ranging from 21.2% to 31.1%. Differences in IVDMD between varieties of barley and between varieties of oats harvested at the soft-dough stage have been reported by Tingle and Dawley (1974), likely related to difference in cell wall digestibility as plants were harvested at a similar stage of maturity. Large differences in IVDMD of barley straw were also reported by Capper et al. (1988). These differences were due to variations in cell wall digestibility because the NDF content of straw is higher than 80%. Similarly, varietal differences in IVDMD of rice straw have been reported and ranged from 23.6% to 36.9% (Vadiveloo, 1992) or from 23.6% to 35.6% (Agbagla-Dohnani et al., 2001; *in sacco* OMD). When it was investigated, the variation in feeding values of straws of different varieties of cereal crops affected the performance of cattle (Capper et al., 1988; Orskov et al., 1988; Reid et al., 1988 quoted in Capper et al., 1992; Schiere et al., 2004).

Cell wall digestibility of straw could not be used directly as a breeding criterion in small-grain cereal improvement programs. This would induce extra costs that could not be paid off by seed sales. However, identification of varieties with more digestible straws is of interest for cattle breeders using their farm-produced straws. Especially, in lands with limited availability of water during summer where ensiled small-grain cereals could be an alternative to maize, varietal information on stem cell wall digestibility can be obtained at low costs by cereals breeders or merchants with important economical benefit in cattle feeding (Schiere et al., 2004). In addition, small-grain cereals seems significantly used in complex mixture often including wheat or triticale, oat, forage pea, and vetch, giving silages of higher yield than pure legumes and of higher nitrogen content than pure cereals. However, conversely to maize or sorghum, of which energy value varied little according to the date of ensiling in a 27–35% interval of DM content, great decreases in cell wall digestibility and energy content are observed in small-grain cereal silages, due to the rapid decrease of stem digestibility during plant maturation. Cropping of mixture of cereals and legumes can contribute partly to reduce the negative susceptibility of plants to a small delayed harvest and improved the digestibility of the mixed diet (Droushiotis, 1989).

3 Intake as a Primary Nutritional Factor of Cattle Fed Cereal Silages or Straws

3.1 Genetic Variation for Intake in Cereal Silages

Ruminants consuming forage diets, high in cell wall content, often are unable to eat sufficient quantities of food to meet their energy demands. Voluntary intake is thus a primary nutritional factor controlling animal production. DM content is the first factor of intake variation in any silage. Moreover, DM contents are also involved in optimal silage fermentation and conservation, in silage palatability. Maize DM content between 32% and 37% allowed satisfactory compromises for these different traits. For a given DM content, genetic variation in intake was first observed in interspecific comparisons. Most studies that compared sorghum with maize silage have shown that DM intake was consistently higher for cows fed maize silage than for those fed sorghum silage, with lower cell wall digestibility. The average DM intakes of sorghum silage were 81 and 85% that of maize, when fed to heifers or dairy cows in the Cummings and McCullough (1969) and Aydin et al. (1999) experiments, respectively. However, unpublished recent results at INRA Lusignan have shown that intake of grain sorghum silage could be as high as intake of maize silage, even if the milk production was lower or only equal to that of maize with sorghum silage. Within species, Blaxter et al. (1961) and Hawkins et al. (1964), respectively, first reported that voluntary intake was positively correlated with plant digestibility and negatively correlated with its lignin content. Later, intake of maize hybrids of low cell wall digestibility was shown lower than the intake of hybrids of higher cell digestibility (Barrière et al., 1995a, b, 2003b, 2004c; Emile et al., 1996). However, for a given and rather high cell wall digestibility, some rare hybrids were shown to have indeed a higher intake in dairy cows than most of other ones. Ciba-semences (1990, 1995) have shown a higher intake for the kindred hybrids, Briard and Bahia, close to 0.5 and 1.0 kg, respectively, compared to a commonly used hybrid. More demonstratively, the voluntary intake of hybrid DK265 in cattle was proved to be greater than that of all other hybrids (Barrière et al., 1995a, 2004c). When maize silage was given as about 80% of the diet, dairy cows fed DK265 silage had an average intake reaching nearly 1.5 kg/day more than hybrids with the same DM and grain contents, and, in two comparisons, with the same cell wall digestibility.

3.2 Devising a Breeding Criterion for Genetic Improvement of Intake

Intake can be truly measured only with cattle. Mostly, due to the great impossibility for plant breeders to work with cattle, there was then "a failure of most scientists to recognize the importance of voluntary intake" (Minson and Wilson, 1994). The regulation of intake in cattle is above all a physical regulation. The intake of a forage is thus controlled by the time it needs to be broken in the mouth so to be swallowed and the time this forage is retained in the rumen and ruminated until particles reach a size close to 1 mm and escape out of the rumen through the digestive tract (Fernandez et al., 2004; Jung and Allen, 1995; Minson and Wilson, 1994). All traits that make fiber particles physically strong and difficult to reduce in size can be considered to be involved in variation of intake. Variations in cell wall digestibility (NDFD) thus explained nearly one-half of intake variations in cows (Barrière et al., 2003b). Scattered but convergent results allow hypothesizing that the second half of genetic variations for intake are explained by plant tissue friability and susceptibility to crushing, specific characteristics likely present at a high level in hybrids, such as DK265, and explaining its extra intake. Intensity of cross-linking within arabinoxylan chains and between arabinoxylans and lignins through ferulic and diferulic acid bridges are probably linked to the stiffness and mechanical properties of tissues (MacAdam and Grabber, 2002). Improvement of cell wall digestibility in maize (and very likely in other cereal forage plants) will bring about an improvement in intake. Complementarily, lowering cross-linkages between cell compounds would also allow specifically an improvement of intake. Breeding for lower ferulate cross-links is possible (Casler and Jung, 1999), even if it is difficult to correlate, directly, content of ferulate release by solvolytic methods and intensity of linkage in plant tissues (Grabber et al., 2004).

4 Genetic Resources for Cell Wall Digestibility Improvement

4.1 Necessity of Specific Genetic Resources for the Improvement of Feeding Value Traits

Maize is likely the plant species in which the genetic improvement for agronomic traits was the most remarkable during the last five decades in Europe (Barrière et al., 1987, 2005, 2006; Derieux et al., 1987), and in the last century in the USA (Russell, 1984; Troyer, 1999, 2002). In forage maize (Barrière et al., 1987, 2004a, 2005), the genetic progress was close to 0.17 t/ha/year for hybrids registered in France between 1986 (the first year with registration after forage maize official trials) and 2000. In the period before 1986, forage yield improvement was correlative to the genetic progress in grain and was nearly equal to 0.10 t/ha/year (Barrière et al., 1987). However, feeding value was not considered for forage maize registration until 1998 in France, even if little earlier in more northern countries, and a significant drift of hybrids toward lower cell wall digestibility values was observed (Table 1) in the last two or three decades (Barrière and Argillier, 1997; Barrière et al., 2004a). In the USA, Lauer et al. (2001) highlighted an annual rate of forage yield increase of 0.13–0.16 t/ha since 1930. But they did not find any change in the cell wall digestibility of plants, despite major improvement in stalk standability, and in stalk-rot resistance, were achieved during the same period. The discrepancy

Table 1 Average values for agronomic and quality traits in early and medium-early maize registered in France in five successive eras from 1958 to 2002[a]

Registration era	nbr	OMD %	NDFD %	Grain %	C protein %	Yield t/ha
1958–1980	22	70.9	51.1	43.8	8.2	12.5
1981–1988	43	70.7	49.9	42.9	8.1	14.4
1989–1993	60	69.8	48.4	44.9	8.0	16.1
1994–1999	77	69.7	47.6	44.5	7.9	16.4
1999–2002	44	69.0	45.7	45.1	7.7	18.1
1958–2002	246	69.9	48.2	44.4	8.0	15.9

[a]Adapted from Barrière et al. (2005), *nbr* = number of investigated hybrids, *OMD* = in vivo organic matter digestibility, *NDFD* = in vivo NDF digestibility with *NDF* = neutral detergent fiber, and *C protein* = crude protein

between European and US results is likely due to different evolutions of hybrid germplasm in Europe and in the USA. The maize improvement for agronomic traits in the USA was carried without major germplasm changes, and continuously based on the Reid and Lancaster groups, even if the Iodent subgroup have got a greater place. Conversely, dent lines in modern European hybrids are now more related to Iodent and Reid origins than were old early dent lines used in Europe, with higher cell wall digestibility. Old European flint lines of high cell wall digestibility, such as F7, are not involved in the modern flint germplasm, due to their lower combining ability values for yield, stalk rot or lodging resistance. Moreover, early flint European lines are now often introgressed by dent germplasm (Barrière et al., 2005, 2006). Improvement of maize cell wall digestibility in the USA or in Europe requested the targeted (re)introduction of original germplasm in currently used elite germplasm. No data are available showing such a drift in sorghum or small-grain cereals. However, similar results could be considered because similar progresses in stalk standability were obtained for all these species.

4.2 Availability of Genetic Resources for Cell Wall Digestibility Improvement

Whereas most parental lines currently used in commercial hybrids are of medium or weak cell wall digestibility, a great range of cell wall digestibility is available when including lines of lower agronomic values. Cell wall digestibility (DINAG trait) values ranged between 53.0% and 64.5%, and 68.7% including bm3 lines in a set of 125 early and medium-early maize lines (INRA Lusignan, unpublished data). Among flint early or medium-early lines, F7, F286, and F324 were shown to have a high cell wall digestibility, whereas F4 had a exceptionally high cell wall digestibility equal or higher to the one of bm3 lines (Fontaine et al., 2003; Méchin et al., 1998, 2000). Conversely, only few dent-related lines of high cell wall digestibility were shown today, and public medium-early resources of interest with a significantly higher cell wall digestibility are likely F7019, F7058, and F7074 (INRA Lusignan, unpublished data). In later germplasm, lines are available from the Wisconsin Quality Synthetic (Frey et al., 2004). W94129 and W95115 lines also appeared of high cell wall digestibility in European (Lusignan) conditions, with lignin contents significantly lower than lines of similar earliness. Progress in cell wall digestibility in both flint and dent lines is thus possible, because the germplasm used in maize breeding only represents a small part of the available genetic resources in maize. Most of this germplasm corresponds to resources used in grain maize breeding, even different breeding companies have also programs specifically devoted to silage use. However, older accessions, and older lines bred from the early cycles of breeding, had to be investigated for cell wall digestibility traits. The objective is to discover, in accessions or lines that were considered not suitable for grain breeding, new alleles of interest for cell wall digestibility and silage intake. The use of genetic distance based on molecular markers will help to classify the genetic resources and thus to highlight those that were not related to lines of low cell wall digestibility. Because there is obviously a great gap in agronomic value between lines of interest for feeding value traits and elite modern lines, specific strategies of introgressing feeding value traits in elite germplasm have to be considered. Even if such investigations can be considered in maize and, possibly, in sorghum, it is weakly probable that it could be done in small-grain cereals for economical reasons.

4.3 Feeding Value Improvement Based on Brown-Midrib Mutations

The brown-midrib (bm) plants exhibit a reddish brown pigmentation of the leaf midrib and stalk pith, associated with lignified tissues. Four bm genes were described in maize between 1924 and 1947 (bm1, Jorgenson, 1931; bm2, Burnham and Brink, 1932; bm3, Emerson, 1935; and bm4, Burnham, 1947), while no new bm mutants were seemingly found (or published) since this period, despite the intensive use of transposon tagging in maize reverse genetics. The four bm genes segregate as monogenic Mendelian recessive traits. The effect of maize bm mutations on lignin content and feeding value was first evidenced by Kuc and Nelson (1964) and Barnes et al. (1971), respectively. In Sorghum, 19 independently occurring bm mutants were obtained from chemically treated seeds of two lines (Porter et al., 1978). Some of the mutant lines had significantly reduced lignin contents, and/or a significantly higher cell wall digestibility. Bm mutants in pearl millet also originated from chemically induced mutations Cherney et al., 1988). Many studies were then made on bm plants, which proved very early to be powerful models in cell wall digestibility and lignification studies. In-depth descriptions of their specific lignification patterns were thus made (review in Barrière et al., 2004b).

The improvement of cell wall digestibility in bm3 maize ranged from 0.9% to 17.9% points, with an average improvement equal to 8.7% points (Table 2) and a tendency to a lower efficiency of the mutant gene when normal hybrids were of higher cell digestibility (Barrière et al., 2004a). The improvement in performances of cattle fed bm maize plants was mostly established with the maize bm3 mutant,

Table 2 Comparison of normal and bm3 hybrids for digestibility and agronomic traits[a]

	OMD (%)		NDFD (%)		Yield (t/ha)		Grain (%)	
	N	bm3	N	bm3	N	bm3	N	bm3
31 hybrid mean	70.0	73.5	49.4	58.1	14.3	12.4	43.8	41.8
Mini	66.0	67.2	43.1	50.9	7.8	4.7	28.2	25.5
Maxi	73.5	76.3	58.6	64.2	19.8	16.6	55.1	53.5
Inra258 (1958)	72.2	74.5	53.8	60.1	11.7	11.2	44.0	46.4
LG11 (1970)	71.5	74.3	50.8	60.4	12.7	11.6	45.5	45.3
Adonis (1984)	70.4	73.9	48.7	56.2	16.2	13.5	45.5	42.2
Dk265 (1987)	71.4	75.4	50.0	61.5	13.7	12.1	45.9	42.5
Rh162 (1990)	67.4	72.0	43.1	54.1	17.1	14.8	44.8	43.0
Helix (1993)	68.6	74.9	46.0	58.2	15.9	13.2	44.8	46.5

[a]Adapted from Barrière et al. (2004a), *N* = normal hybrid, registration year in brackets, *OMD* = in vivo organic matter digestibility, *NDFD* = in vivo NDF digestibility with *NDF* = neutral detergent fiber

probably because, compared to other maize bm mutants; the maize bm3 mutant appeared to be especially improved in cell wall digestibility (Table 3). The intake of bm3 silage by dairy cows was always higher than the intake of normal silage, even if the difference was not always significant (Table 3). Higher milk yield of cows fed bm3 hybrids were reported in 11 out of 15 experiments, ranging from 0.5 to 3.3 kg/day. Milk yields were always at least equal with the bm3 diet. Moreover, every time this trait was recorded, increase of body weight was observed in cattle fed bm3 silage. The primary apparent benefit of the bm3 mutation in cattle feeding efficiency is from an increased silage intake. Consequently, bm3 hybrids indeed appear of a greater efficiency than normal hybrids in dairy cows, when maize silage is a significant ingredient in the diet, and when the supply of concentrates is correlatively reduced, because the extra intake of silage, and taking into account the higher digestibility and energy value of bm3 hybrids. Comparisons involving the other different maize bm genes with meat or dairy cattle are very rare. From one experiment with fattening bulls, a bm1 hybrid was slightly more efficient than its normal counterpart, but much lower efficient than its bm3 counterpart (Barrière et al., 1994). The interests in cattle feeding of bm2 and bm4 hybrids have seemingly not been investigated.

A higher digestibility of bm plants was also observed in sorghum and pearl millet (Akin et al., 1991; Fritz et al., 1981; Oliver et al., 2005a; Watanabe and Kasuga, 2000). Correlatively, from different experiments with bm sorghum or pearl millet in the cattle diets, DM intakes were higher with bm diets than with standard diets (Aydin et al., 1999; Cherney et al., 1990; Grant et al., 1995; Lusk et al., 1984). Conversely, no effect in the diet intake was observed in the recent experiment of Oliver et al. (2004) comparing maize, and normal, bmr6, and bmr18 sorghum silages. However, milk yields were higher in bm sorghum and maize silages than in normal sorghum silages.

Whereas the higher efficiency of bm3 maize for cattle feeding was clearly established, breeders were for a long time disappointed by the lower yield, somehow irregular earliness, susceptibility to bending, and susceptibility to dry conditions of

Table 3 Feeding efficiency of bm3 maize silage in dairy cattle, from experiments published since 1976[a]

	Silage % diet		IV NDFD bm3-N	Maize intake bm3-N	FCM bm3-N	ADG bm3-N
	N	bm3				
Frenchick et al. (1976)	49	49	–	0.2	-0.1°	88
Rook et al. (1977)	60	60	–	1.1	-0.1°	14
Rook et al. (1977)	85	85	–	2.7	0.7°	42
Keith et al. (1979)	75	75[b]	10.5	0.6	$0.9^{\circ\circ}$	–
Sommerfeldt et al. (1979)	55	57	10.0	0.7	-0.5°	106
Block et al. (1981)	65	65	–	3.5	$1.2^{\circ\circ}$	755
Stallings et al. (1982)	49	47	15.0	0.6	$-0.6^{\circ\circ}$	80
Hoden et al. (1985)	80	80	8.9	1.0	$0.7^{\circ\circ}$	165
Hoden et al. (1985)	78	86	8.9	1.7	$0.5^{\circ\circ}$	0
Weller and Phipps (1986)	69	70	14.6	0.6	$3.3^{\circ\circ}$	90
Oba and Allen (1999)	45	45[b]	9.7	2.1	2.6°	100
Bal et al. (2000)	32	40[b]	–	0.0	0.5°	40
Oba and Allen (2000)	51	56	9.4	1.4	3.2°	20
Tine et al. (2000)	60	60	7.0[c]	2.4	1.7°	170
Ballard et al. (2001)	31	31	10.9	0.5	2.5°	–
Barrière et al. (2003b)	75	75	8.3	2.6	–	–
Moreira et al. (2003)	40	40	–	1.9	2.0°	–
Barrière et al. (2004c)	76	76	8.5[d]	1.3	–	–
Taylor and Allen (2005)	38	38	12.6	0.5	0.9	95

[a]Comparisons were done between isogenic hybrids, except in Bal et al. (2000) and Ballard et al. (2001). [*Conc* = concentrates, *IVNDFD* = in vitro NDF digestibility with *NDF* = neutral detergent fiber, *FCM* = fat-corrected milk at 3.5° or $4.5^{\circ\circ}$ %, *ADG* = average daily gain (g/day)]

[b]Concentrate giving were similar in normal and bm3 diets except (1) in Keith et al. (1979) where cows fed bm3 silage were given 0.4 kg/day soybean meal less and 0.4 kg/day ground maize more than cows fed isogenic normal hybrid, (2) in Oba and Allen (1999) where cows fed bm3 hybrids were given 0.1 kg/day soybean meal less and 0.1 kg/day high moisture maize more than cows fed isogenic normal hybrid, and (3) in Bal et al. (2000) where cows fed bm3 hybrids were given 1.3 kg/day alfalfa silage more and 3.6 kg/day concentrate less

[c]Apparent digestibility measured in lactating cows

[d]In vivo digestibility measured in sheep

bm3 hybrids. A recent and renewed interest in bm3 hybrids for dairy cattle feeding is illustrated by the new experiments done since 1998 especially in the USA, while no were published between 1987 and 1998 (Table 3). The great improvement in agronomic value of maize germplasm in the last 25 years, with the simultaneous lower feeding value of the parental lines used in modern medium-late and late hybrids, strengthened the possibility and the interest of breeding bm hybrids. With normal hybrids of good standability, whose potential farm yields are higher or equal to 15 t/ha, it is conceivable to breed related bm3 hybrids whose yield will be reduced by about 2 or 3 t/ha, but whose cell wall digestibility will be increased by about 8% points. Ballard et al. (2001) and Cox and Cherney (2001) thus reported a yield reduced by 2–3 t/ha with a cell wall digestibility improved by at least 10%, allowing an increase of the FCM yield, in bm3 hybrids. The availability of bm3 hybrids on the seed market in the USA has proved the feasibility of the use of this particular genetic resource for cell wall digestibility improvement of commercial hybrids, at least for late or medium-late hybrids. But the higher seed costs of bm3 commercial hybrids in the USA have obscured their economic interest. In Europe, the reputation of bm3 genotypes is still poor, and they are always suspected of a greater susceptibility to lodging, on top of their lower yields. An experimental medium-early bm3 hybrid (F7026bm3 × F2bm3) bred at INRA Lusignan (Barrière et al., 2003b) with a yield close to 13 t/ha, had thus a NDFD close to 59% and an intake in dairy cows equal to 17.9 kg DM/cow/day, with an acceptable standability, when normal hybrids of similar earliness yielded about 17 t/ha, with an NDFD equal or lower than 47%, and an intake nearly equal to 15 kg DM/cow/day. Improvement in yield, but also in standability, can be expected since the two parental lines of this bm3 hybrid are representative of nearly 15-year old germplasm. From comparison of bmr6 and bmr12 sorghum in different genetic background, Oliver et al. (2005a) and Oliver et al. (2005b) observed a reduced lignin content and an improvement of cell wall digestibility in both bmr6 and bmr12 plants. Moreover, the bmr12 gene had less negative impact on agronomic traits and greater positive impact on quality traits. The genes bmr12 in sorghum and bm3 in maize both correspond both to an alteration of the caffeic acid O-methyltransferase (COMT) gene (Vignols et al., 1995; Bout and Vermerris, 2003). Breeding bm sorghum with improved feeding value is likely of greater short-term impact than breeding bm maize, because of the lower feeding value of sorghum compared with maize. Recent registration of bmr6 and bmr12 sorghum in the USA, simultaneously with an increasing interest for bmr12 sorghum in France and southern Europe, thus illustrated the interest of having more drought-tolerant forage cereals, such as sorghum (Pedersen et al., 2006a, b, c), especially before further improvements of maize in drought tolerance.

Nevertheless, the choice of using lower-yielding hybrids of higher feeding value, is a matter of strategy which has yet to be agreed on, especially so in more friendly environmental conditions of plant cropping and cattle rearing. The water need of plants is linked to its yield. In C4 grasses, each millimeter of transpired water allows the biosynthesis of 40 kg DM/ha. Plant yield has to be adjusted to present and future water availability. A decrease in maize or sorghum yield by 5 t/ha corresponds to a reduced water use equal to 125 mm/ha, that could be economically compensated by a significantly higher cell wall digestibility and silage intake in such hybrids.

5 Investigating Quantitative Trait Loci for Cell Wall Digestibility Improvement

Once lines of different feeding values and/or different genetic background are identified, different recombinant inbred line (RIL) progenies can be developed in order to determine the genomic location involved in feeding value traits.

Quantitative trait loci (QTL) for cell wall digestibility and/or lignification traits in maize are available at least from data in RIL progenies by Lübberstedt et al. (1997), Méchin et al. (2001), Roussel et al. (2002), and unpublished results from the INRA – ProMaïs and Génoplante networks. Six major clusters of IVNDFD QTL were thus found of decreasing importance according to both their limit of detection (LOD) values in bins 6.06, 4.08/09, 1.02/04, 8.07, 9.02, and 7.03, explaining from 6% to 40% of the phenotypic variation for this trait (Table 4).

Additional less-important locations were also involved in cell wall digestibility for these four RIL progenies, located in eight other bins. The number of locations involved in IVNDFD variations is not known, but a meta-analysis, based on data from eight RIL progenies in per se value experiments, has shown that at least 43 locations were involved in lignin content of maize plants (Barrière et al., 2007). From published and unpublished data, QTL for lignin content and cell wall digestibility might colocalize in half to two-third of occurrences. Cross-linkages between arabinoxylan chains and arabinoxylan chains and guaiacyl monomeric units of lignins, likely explain the second half of IVNDFD variations which is not explain by lignin content variations.

QTL for lignin content were also given from progenies developed for corn borer tolerance studies (Cardinal et al., 2003; Krakowsky et al., 2004, 2005). Conflicting situations in maize breeding for cell wall digestibility will probably result from different colocalizations between QTL involved in wall lignification and digestibility, and QTL for European corn borer tolerance. Nearly 50% of locations involved in wall digestibility and/or lignin content were also described as involved in *Ostrinia nubilalis* tolerance (tunneling length or stalk damage rating). Today, it cannot be dismissed that some genotypes with high cell wall digestibility will be

Table 4 Putative major QTL for IVNDFD observed in four recombinant inbred lines progenies experimented in per se value[a]

IVNDFD QTL	chr-pos	bin	Closest marker	Dist clo-m	LOD	R^2	Line (+)
F288 × F271	1–92	1.02	bnlg1627	−7	3.1	10.3	F288
F838 × F286	1–84	1.02	bnlg1178	10	3.3	6.1	F286
F7025 × F4	1–78	1.04	bnlg2238	−2	5.7	10.8	F4
Io × F2	4–174	4.08	sc82	−1	2.0	6.5	Io
F7025 × F4	4–136	4.08	bnlg2162	9	7.0	12.9	F7025
F288 × F271	6–184	6.06	bnlg345	7	14.6	40.2	F288
Io × F2	7–36	7.03	umc116	10	3.3	11.3	F2
F7025 × F4	7–28	7.03	bnlg1305	1	2.5	4.9	F7025
F838 × F286	8–142	8.07	bnlg1065	31	8.6	15.0	F838
F288 × F271	9–100	9.02	bnlg1401	−1	4.1	13.4	F271

[a]*IVNDFD* = in vitro NDF digestibility with *NDF* = neutral detergent fiber, distance is given as cM to the closest marker with positive/negative value from left/right flanking marker, line (+) increased the value of the trait. Data from Méchin et al. (Io × F2), Roussel et al. (F288 × F271), and unpublished data of INRA Lusignan

QTL quantitative trait loci, *LOD* limit of detection

more susceptible to pest damages, especially if corn borers susceptibility will not be estimated simultaneously during cell wall digestibility improvement programs.

The genes underlying QTL for cell wall digestibility are not yet really known. Several known genes of the maize lignin pathway have been found colocalizing with QTL, but the biological significance is limited by the fact that most of the genes of this pathway belong to large multigenic families. Except works with bm1 and bm3 mutants, and transgenic COMT antisense constructs (Piquemal et al., 2002; He et al., 2003; Pichon et al., 2006), no functional analysis with lignin pathway genes were seemingly published in maize. However, even genes underlying QTL are still unidentified, their marker-assisted introgression based on the two flanking markers into an elite genetic background is possible as soon as a QTL has been detected. The efficiency of a breeding scheme based on anonymous markers depends on the linkage phase between markers and target locus alleles.

6 Targeted Investigations of Genetic Resources for Cell Wall Digestibility Improvement

Deregulation of gene expression through genetic engineering is an essential way toward the understanding of lignification and cell wall biosynthesis in plants and, therefore, of future improvements of cell wall digestibility in plants. Boudet (2000), Chen et al. (2001), Dixon et al. (2001), and Halpin (2004) have recently published extensive reviews of genetic engineering of the lignin pathway, with the resulting consequences on lignin content and structure of altered transgenic plants. Even most studies have been performed on dicotyledonous plants, including model plants such as tobacco or *Arabidopsis*, the efficiency of antisense or silencing strategies in increasing the cell wall digestibility of plants has been clearly established. Most of recent significant understanding of the monolignol biosynthesis has been obtained from both disrupted (transgenic) mutants and down- or upregulated plants (Chen et al., 2006; Hoffmann et al., 2004; Reddy et al., 2005; Schoch et al., 2001). Correlatively, the validation of a gene involvement in variation of cell wall digestibility through genetic engineering or transposon tagging strengthens the interest of investigating its natural allelic variation in available germplasm. Association studies between single-nucleotide polymorphism (SNP) or insertion–deletion polymorphism (INDEL) in cell wall-related genes, and cell wall digestibility, give functional markers more efficiently used in marker-assisted selection than anonymous markers (Andersen and Lübberstedt, 2003).

Lignin pathway in plants and grasses begins after the shikimate pathway with the deamination of L-phenylalanine into cinnamic acid. Successive steps including hydroxylation and methylation on the aromatic ring lead to the production of three monolignols (*p*-hydroxyphenyl, coniferyl, and syringyl alcohols), which are polymerized into lignins. Moreover, grass lignins are typified by both the acylation of the syringyl units by *p*-coumaric acid, and by numerous cross-linkages between arabinoxylans and guaiacyl units by ferulic and diferulic acids. Deregulation of genes involved at each step of the pathway is thus a way to select candidates of interest in cell wall digestibility improvements.

According to opinions of Halpin et al. (1995) and Casler and Kaeppler (2001), the alteration of early steps in lignin and phenylpropanoid metabolism (PAL, phenylalanine ammonia-lyase; C4H, cinnamate 4-hydroxylase), which are clearly involved in other important processes in plants, could lead to too many adverse pleiotropic effects to be useful for cell wall digestibility improvement of plants. However, at least four map positions are available for PAL genes in the maizeGDB database (http://www.maizeGDB.prg), in bin 5.05 (PAL1, bl17.23a), 2.03 (PAL2, bnl17.23b), 4.05 (PAL3, bnl17.23c), and 4.05 (PAL, csu358b), likely corresponding to different orthologs, which were differentially expressed in different tissues and times of growth (Guillaumie et al., 2007a). Silking bm3 plants, which have a nearly null COMT expression, were shown simultaneously to have a significant decrease in expression of two PAL genes out of four investigated, likely as a consequence of the disrupted pathway toward the syringyl alcohol formation (Table 5).

In *Arabidopsis*, the disruption of two PAL genes induced a decrease of lignin content, with a complex transcriptomic adaptation of phenylpropanoid, carbohydrate, and amino acid gene expression (Rohde et al., 2004) The PAL gene orthologs, which manage a key step of lignin biosynthesis and regulate the carbon flux channeled in the pathway, could therefore be of significant interest to reduce the flux of lignin precursors. Complementarily, Andersen et al. (2007) have shown a significant association with a SNP in the PAL (MZEPAL) gene and maize digestibility.

The hydroxylation/methylation reactions along the lignin pathway are not really elucidated in maize, despite the strategic interest of these steps in both identifying key genes controlling the S/G ratio and the formation of ferulic acid and subsequent cross-links in the cell wall. Caffeoyl-CoA, the key compound of the pathway, is synthesized from coumaroyl-CoA through the formation of quinate or shikimate esters by a reverse-active hydroxycinnamoyl transferase (HCT). Hydroxylation of

Table 5 COMT and PAL genes expressed in ear internode of silking maize plants, and their expression in the F2 bm3 mutant as compared to normal INRA F2 line

	mRNA	Expression	
		F2	F2bm3/F2
COMT	M73235	142203	0.05
Phenylalanine ammonia lyase (MZEPAL)	L77912	187353	0.22
Phenylalanine ammonia lyase	AC185453	207907	0.44
Phenylalanine ammonia lyase	CF631905	102659	0.90
Phenylalanine ammonia lyase	AY104679	10421	0.67

Normalized expression values are given for the F2 line and bm3 mutant values are expressed as ratios of signal intensity compared to normal plants. Genes were considered as significantly differentially expressed when expression ratio values were lower than 0.5 or higher than 2.0
COMT caffeic acid O-methyltransferase, *PAL* phenylalanine ammonia lyase, *mRNA* messenger ribonucleic acid

these esters to caffeoyl analogues is catalyzed by a *p*-coumaroyl-shikimate/quinate 3′-hydroxylase (C3′H) (Schoch et al., 2001; Hoffmann et al., 2003; Mahesh et al., 2007). Disruption of HCT or C3′H genes led to stunted plants with H lignins (Hoffmann et al., 2004; Shadle et al., 2007). HCT or C3′H weak alleles are, therefore, of higher interest in breeding than null alleles. Methylation of caffeoyl-CoA is driven by caffeoyl-CoA O-methyltransferase (CCoAOMT) enzymes, which are encoded in maize by at least five genes differentially expressed throughout the time and plant organs (Guillaumie et al., 2007a). Moreover, the previously described CCoAOMT1 and CCoAOMT2 genes (Civardi et al., 1999) were not the most-expressed genes in numerous cases (Guillaumie et al., 2007a, b), and the respective roles of each orthologous genes are not known. Downregulations of each CCoAOMT orthologs, and studies of knocked-out mutants, are thus of interest for both theoretical and breeding topics. COMT has been extensively studied based on the bm3 mutant and different downregulations. Among conclusions, COMT is very likely not involved is the biosynthesis of ferulic acid in maize. Conversely, COMT appears as a target of interest in breeding for a higher cell wall digestibility, based on weak alleles or regulation rather than on null expression, in order to avoid or diminish negative agronomic consequences. Piquemal et al. (2002) thus reported COMT downregulated maize plants with 30% COMT residual activity and a 9% point increase in maize cell wall digestibility, a value similar to the one observed in bm3 isogenic lines. The drawback of COMT downregulation or silencing is the correlative S/G decrease, because a higher S/G ratio could impact positively the cell wall digestibility in maize (Méchin et al., 2000), possibly through different linkage types and stereochemical arrangements of S units compared to G units. CCoAOMT could be considered a priori as an even better target than COMT, because CCoAOMT downregulation in plants would logically result in lower lignin contents without a decrease in S/G ratio, as observed in alfalfa (Guo et al., 2001). However, while the respective involvement of CCoAOMT and (C)OMT genes in S-unit biosynthesis is not currently understood (Chen et al., 2006; Do et al., 2007), the most important improvements in cell wall digestibility of cereals have been obtained today with COMT mutations or downregulations.

CCR (cinnamoyl-CoA reductase) and CAD (cinnamyl-alcohol dehydrogenase), the last two enzymes involved in monolignol biosynthesis, have been considered as potentially suitable targets for cell wall digestibility improvement (Halpin et al., 1995). In maize, the bm1 mutant, which exhibited lower CAD activity (Halpin et al., 1998), was recently proved to alter in fact the expression of numerous CAD genes (Guillaumie et al., 2007b). Bm1 lignins thus substantially incorporate coniferaldehyde and, to a lower extent, sinapaldehyde and have substantially more carbon–carbon interunit linkages (Barrière et al., 2004b; Halpin et al., 1998; Kim et al., 2002). The feeding value of the bm1 mutant was always significantly lower than the one of bm3 plants (Barrière et al., 1994). In tall fescue, IVDMD was increased by 7.2–9.5% in CAD downregulated lines (Chen et al., 2003). In maize, after the description of the CCR1 and CCR2 genes, this later being little involved in constitutive lignification (Pichon et al., 1998), several CCR or putative CCR were found differentially expressed in different tissue or stage of development (Table 6).

Table 6 CCR and CAD/SAD genes normalized expression values in ear internodes of silking plants of the maize INRA line F2[a]

	mRNA	Expression
CCR1, ZmCINNRED	X98083	37894
CCR	AY108351	13755
CCR	AY103770	11730
CCR	AI881365	9973
CCR	DV490994	8886
CCR	BT018028	8736
CCR	AI737052	8414
CCR2	Y15069	8776
ZmCAD2 type	Y13733	30285
Putative CAD	AY107977	13998
Putative CAD	AY110917	9826
Putative CAD	CX129557	8210
ZmCAD1 type	AY106077	16082
SAD	AY104431	17398
SAD	CD995201	9165

[a]Based on data of Guillaumie et al. (2007a)

CCR cinnamoyl-CoA reductase, *CAD* cinnamyl-alcohol dehydrogenase, *SAD* sinapyl-alcohol dehydrogenase

Similarly, CAD genes, which encode enzymes involved in the last step of monolignol biosynthesis, also belong to a multigene family (Table 6). However, while the role of ZmCAD2 genes is established in lignin biosynthesis, the role of ZmCAD1- or SAD-type genes is less understood (Li et al., 2001; Damiani et al., 2005).

CAD gene mutation and deregulation, as observed in maize bm1 mutant and fescue deregulated plants, had variable effects on cell wall digestibility, but it is not known if this difference is related to deregulation of several members of the family in maize, whereas it is probably one gene in fescue. The efficiency of CCR deregulation in cell wall digestibility improvement of grasses is currently not known. In any way, it is necessary to further elucidate the respective specificity of different CCR and CAD/SAD enzymes, and the independence (or not) of pathways leading to guaiacyl and syringyl units of lignins, in order to target the choice of members in each multigene family for CCR and CAD gene engineering or the search of weak alleles.

The polymerization reactions may also be considered as good targets, even though laccases and peroxidases are also encoded by multigene families. The disruption of the ZmPox3 peroxidase, located in bin 6.06, due to a miniature inverted repeat transposable element (MITE) insertion in the first exon, was shown to be related to a higher cell wall digestibility of flint early lines (Guillet-Claude et al., 2004a). This result was recently corroborated by analyses of RNAi ZmPox3 downregulated plants (Génoplante, unpublished data). The downregulation of one laccase in poplar led to plants with highly altered xylem fiber cell walls and modified mechanical properties of the wood. Such a laccase was supposed to be involved in the formation of certain types of phenoxy radicals leading to cross-linking in xylem fibers (Ranocha et al., 2002). Laccase downregulated plants could, therefore, be considered as resources of reduced cross-linked fibers, and should be considered as potential targets in forage digestibility and intake improvements.

Regulatory genes of lignification are also potential targets for cell wall digestibility improvement in plants. Myb transcription factors are involved in regulating phenylpropanoid metabolism. Lignification was thus heavily reduced in tobacco plants overexpressing the Antirrhinum Myb 308 transcription factor (Tamagone et al., 1998), while the overexpression of EgMYB2 in tobacco plants induced a great increase in secondary wall thickness (Goicoechea et al., 2005). Moreover, Guillaumie et al. (2007b) have shown that other regulatory genes (Lim factor, Argonaute, Shatterproof, . . .) have modified expressions in bm mutants and could thus be new targets in cereal breeding for quality traits. Similarly, genes involved in regulation of tissue patterning or those involved in the transport of constituents to the cell wall should be considered as candidate in feeding value improvement of forage cereals.

While the importance of ferulate cross-linkages in cell wall digestibility and in forage intake of grasses is now established, the pathway leading from *p*-coumaric acid to ferulic acid is still largely unknown. In *Arabidopsis*, the ref1 mutant, which has a reduced content in soluble sinapate esters, was shown to be affected in an aldehyde dehydrogenase (ALDH) gene, and that the REF1 protein exhibited both sinapaldehyde and coniferaldehyde dehydrogenase activities (Nair et al., 2004). Sinapic and ferulic acids in *Arabidopsis* thus derived from oxidation of the corresponding aldehydes. Whether this sinapate and ferulate ALDH pathway also exists in grasses is currently not established, even if at least eight ALDH genes have been described in maize (Skibbe et al., 2002). Correlatively, the bm3/COMT mutation does not affect ferulate content of maize plants. In alfalfa, ferulic acid content, which is nearly 100 times lower than in maize, was significantly decreased in C3′H downregulated plants, but not in CCoAOMT downregulated plants (Chen et al., 2006). However, no information allowed, excluding that one CCoAOMT specifically devoted to ferulic acid biosynthesis, has escaped to the deregulation. Complementarily to phenylpropanoid components, reduced cross-linkages in grass cell walls could be considered based on reduced arabinoxylan availability. However, no gene has been proven to be involved in arabinoxylans feruloylation, and only candidate genes specifically expressed in grasses have been identified for this step by Mitchell and Shewry (2007), based on a bioinformatics approach on rice, wheat, and barley ESTs, comparatively to dicotyledons. In any way, the breeding targets toward a reduced content of ferulic acid in grasses remain currently unknown.

Complementarily, engineering the expression of fungal ferulic acid esterase in transgenic ryegrass has been investigated as an alternative strategy, with an increase digestibility of transformed plants compared to normal ones (Buanafina et al., 2006).

Allelic variations resulting from SNP, or INDEL, have been related to variations in lignin content and/or cell wall digestibility. Allelic variation studies in the COMT gene have shown that this gene was greatly variable not only with many SNP and INDEL in its unique intron but also with several variations in exons leading to several amino acid changes. Association studies between these allelic modifications and the cell wall digestibility have shown that one INDEL, located in the intron, explained 32% ($P = 0.0017$) of the observed cell wall digestibility variation (Guillet-Claude et al., 2004b). Similarly, one INDEL polymorphism within the COMT intron has revealed significant association with stover digestibility in another set of maize lines (Lübberstedt et al., 2005). A 1-bp deletion in the second exon of PAL, introducing a premature stop codon, has been also associated with higher plant digestibility (Andersen et al., 2007). Whether these associations are related to a causal modification in the candidate gene sequence, or to linkage disequilibrium with a causal factor closely linked to the favorable SNP, they illustrated the possibility of breeding for weak alleles in the lignin pathway toward the improvement of maize and cereal cell wall digestibility.

7 Conclusion

In the search for a forage ideotype in cereals, the breeding effort to be placed, respectively, on either biomass yield or biomass digestibility is open to debate. However, a high biomass yield can lead to significant disillusion if dairy cows yield not much milk because of low intake and digestibility of the silage. A high intake and digestibility should also allow farmers to provide lower amounts of expensive concentrates to cattle. Cell wall digestibility is thus, undoubtedly, one of the major targets for the improvement of feeding value in silage of cereal plants. Because lignin content is not the only trait involved in cell wall digestibility, breeders should use a trait directly related to cell wall digestibility, such as IVNDFD or DINAGZ.

Breeding for quality traits in forage cereals should be considered at two different levels, according forage is, or not, one of the main purpose of the cereal use. Even if several lines with high feeding traits are available in maize, new investigations of genetic resources, including lines or germplasm forgotten after decades of breeding for agronomic value and/or grain yield, are required for a successful breeding of maize and sorghum for silage quality traits. Available genetic backgrounds are rich in gene clusters giving good yield and standability, even whole plant yield has been counter-selected in semidwarf or dwarf grain sorghum varieties. Conversely, original alleles giving high feeding value have probably greatly disappeared from available genetic backgrounds in modern maize, sorghum. In small-grain cereals, breeding varieties for a specific whole plant or straw uses as forage is likely economically not possible. However, it should be of interest to have studies of the genetic variation for cell wall digestibility in best-adapted genotypes, and a preferential use of them in cropping for forage. These lines should be used first in further crossing toward breeding new varieties for both grain and forage utilization. For a given quantity of inputs (nitrogen fertilization, water availability, ...), a forage ideotype resembling bm3 maize or bmr12 sorghum would maximize the production of cattle efficient energy with high intake and digestibility, increasing the profit of productions. Such varieties could be obtained with the use of specific normal germplasm. Breeding directly bmr12 sorghum with improved feeding value is likely more easy than breeding bm3 maize because of sorghum lower feeding value compared with maize, and likely lower adverse effect in sorghum than in maize. Recent registration of bmr6 and bmr12 sorghum in the USA thus illustrated the interest of breeding more digestible sorghum.

QTL analysis, studies of SNP × feeding value traits relationships, studies of mutants and deregulated plants will contribute to the comprehensive knowledge of the lignin pathway and cell wall biogenesis. Plant breeders will then be able to choose the best genetic and genomic targets for the improvement of plant digestibility. Favorable alleles or favorable QTL for cereal cell wall digestibility will thus be introgressed in elite lines through marker-assisted introgression. Genetic engineering is both an inescapable tool in mechanism understanding and an efficient way in cereal breeding, but the social acceptability of genetically modified plants is greatly different according to the country.

Up to now, most of the researches in plant lignification have been done in dicotyledonous and woody plants. However, grass breeders must consider the specificity of the grass cell wall, with the importance of cross-linkages by ferulic acid bridges. Because a great advance in genomic, maize may thus be considered as a model plant for lignification and digestibility studies in all cereals. At present, similar research efforts are not being made in cell wall biosynthesis on other annual or perennial grass forage plants, neither in rice. Because of the synteny between rice and maize (Wilson et al., 1999), the availability of the rice genome will bring very valuable complementary information, until the maize genome will be completely available. Moreover, gene mining and genetic engineering in model plant and systems (Arabidopsis, Zinnia, Brachypodium, ...) are also complementary approaches for improvement of cell wall digestibility in grass and cereal forage crops.

References

Abou-el-Enin, O.H., Fadel, J.G. and Mackill, D.J. (1999) Differences in chemical composition and fibre digestion of rice straw with, and without, anhydrous ammonia from 53 rice varieties. Anim. Feed Sci. Technol. 79:129–136.

Agbagla-Dohnan, A., Nozière, P., Clément, G. and Doreau, M. (2001) *In sacco* degradability, chemical and morphological composition of 15 varieties of European rice straw. Anim. Feed Sci. Technol. 94:15–27.

Akin, D.E., Rigsby, L.L., Hanna, W.W. and Gates, R.N. (1991) Structure and digestibility of tissues in normal and brown midrib pearl millet (*Pennisetum glaucum*). J. Sci. Food Agric. 56:523–538.

Andersen, J. and Lübberstedt, T. (2003) Functional markers in plants. Trends Plant Sci. 8:554–560.

Andersen, J.R., Zein, I., Wenzel, G., Krutzfeldt, B., Eder, J., Ouzunova, M. and Lübberstedt, T. (2007) High levels of linkage disequilibrium and associations with forage quality at a phenylalanine ammonia-lyase locus in European maize (Zea mays L.) inbreds. Theor. Appl. Genet. 114:307–319.

Argillier, O., Barrière, Y. and Hébert, Y. (1995) Genetic variation and selection criteria for digestibility traits of forage maize. Euphytica 82:175–184.

Argillier, O., Méchin, V. and Barrière, Y. (2000) Genetic variation, selection criteria and utility of inbred line per se evaluation in hybrid breeding for digestibility related traits in forage maize. Crop Sci. 40:1596–1600.

Aufrère, J. and Michalet-Doreau, B. (1983) In vivo digestibility and prediction of digestibility of some by products. EEC seminar, September 26–29, Mlle Gontrode, Belgium.

Aydin, G., Grant, R.J. and O'Rear J. (1999) Brown midrib sorghum in diets of lactating dairy cows. J. Dairy Sci. 82:2127–2135.

Bal, M.A., Shaver, R.D., Al-Jobeile, H., Coors, J.G. and Lauer, J.G. (2000) Corn silage hybrid effects on intake, digestion, and milk production by dairy cows. J. Dairy Sci. 83:2849–2858.

Ballard, C.S., Thomas, E.D., Tsang, D.S., Mandevu, P., Sniffen, C.J., Endres, M.I. and Carter, M.P. (2001) Effect of corn silage hybrid on dry matter yield, nutrient composition, *in vitro* digestion, intake by dairy heifers, and milk production by dairy cows. J. Dairy Sci. 84:442–452.

Barnes, R.F., Muller, L.D., Bauman, L.F. and Colenbrander, V.F. (1971) *In vitro* dry-matter disappearance of brown midrib mutants. J. Anim. Sci. 33:881–884.

Barrière, Y. and Argillier, O. (1997) In vivo silage feeding value of early maize hybrids released in France between 1958 and 1994. Euphytica 99:175–182.

Barrière, Y. and Emile, J.C. (1990) Effet des teneurs en grain et de la variabilité génétique sur la valeur énergétique du maïs ensilage mesuré par des vaches laitières. Agronomie 10:201–212.

Barrière, Y., Gallais, A., Derieux M. and Panouillé A. (1987) Etude de la valeur agronomique en plante entière au stade de récolte ensilage de différentes variétés de maïs grain sélectionnées entre 1950 et 1980. Agronomie 7:73–79.

Barrière, Y., Argillier, O., Chabbert, B., Tollier, M.T. and Monties, B. (1994) Breeding silage maize with brown-midrib genes. Feeding value and biochemical characteristics 3. Agronomie 14:15–25.

Barrière, Y., Emile, J.C., Traineau, R. and Hébert, Y. (1995a) Genetic variation in the feeding efficiency of maize genotypes evaluated from experiments with dairy cows. Plant Breed. 114:144–148.

Barrière, Y., Emile, J.C. and Hébert Y. (1995b) Genetic variation in the feeding efficiency of maize genotypes evaluated from experiments with fattening bulls. Agronomie 15:539–546.

Barrière, Y., Argillier, O., Michalet-Doreau, B., Hébert, Y., Guingo, E., Giauffret, C. and Emile, J.C. (1997) Relevant traits, genetic variation and breeding strategies in early silage maize. Agronomie 17:395–411.

Barrière, Y., Guillet, C., Goffner, D. and Pichon, M. (2003a.) Genetic variation and breeding strategies for improved cell wall digestibility in annual forage crop. A review. Anim. Res. 52:193–186.

Barrière, Y., Emile, J.C. and Surault, F. (2003b) Genetic variation of silage maize ingestibility in dairy cattle. Anim. Res. 52:489–500.

Barrière, Y., Emile, J.C., Traineau, R., Surault, F., Briand, M. and Gallais, A. (2004a) Genetic variation for organic matter and cell wall digestibility in silage maize. Lessons from a 34-year long experiment with sheep in digestibility crates. Maydica 49:115–126.

Barrière, Y., Ralph, J., Méchin, V., Guillaumie, S., Grabber, J.H., Argillier, O., Chabbert, B. and Lapierre, C. (2004b) Genetic and molecular basis of grass cell wall biosynthesis and degradability. II. Lessons from brown-midrib mutants. C. R. Biol. 327:847–860.

Barrière, Y., Dias-Goncalvès, G., Emile, J.C. and Lefèvre, B. (2004c) Higher ingestibility of the DK265 corn silage in dairy cattle. J. Dairy Sci. 87:1439–1445.

Barrière, Y., Alber, D., Dolstra, O., Lapierre, C., Motto, M., Ordas, A., Van Waes, J., Vlasminkel, L., Welcker, C. and Monod, J.P. (2005) Past and prospects of forage maize breeding in Europe. I. The grass cell wall as a basis of genetic variation and future improvements in feeding value. Maydica 50:259–274.

Barrière Y., Alber, D., Dolstra, O., Lapierre, C., Motto, M., Ordas, A., Van Waes, J., Vlasminkel, L., Welcker, C. and Monod, J.P. (2006) Past and prospects of forage maize breeding in Europe. II. History, germplasm evolution and correlative agronomic changes. Maydica 51:435–449.

Barrière Y., Riboulet C., Méchin V., Maltese S., Pichon M., Cardinal A., Martinant J.P., Lübberstedt T. and Lapierre C. (2007) Genetics and genomics of lignification in grass cell walls based on maize as a model system. Genes, Genomes and Genomics 1:133–156.

Blaxter, K.L., Wainman, F.W. and Wilson, R.S. (1961) The regulation of food intake by sheep. Anim. Prod. 3:51–61.

Block, E., Muller, L.D., Griel, L.C., Garwood, J.R. and Garwood, D.L. (1981) Brown-midrib3 corn silage and heat-extruded soybeans for early lactating dairy cows. J. Dairy Sci. 64:1813–1825.

Boudet, A.M. (2000) Lignin and lignification: selected issues. Plant Physiol. Biochem. 38:81–96.

Bout, S. and Vermerris, W. (2003) A candidate-gene approach to clone the sorghum brown midrib gene encoding caffeic acid O-methyltransferase. Mole. Genet. Genomics 269:205–214.

Buanafina, M.M., Langdon, T., Hauck, B., Dalton, S.J. and Morris, P. (2006) Manipulating the phenolic acid content and digestibility of Italian ryegrass (Lolium multiflorum) by vacuolar-targeted expression of a fungal ferulic acid esterase. Appl. Biochem. Biotechnol. 10.1007/978-1-59745-268-7_34:129–132,416–426.

Burnham, C.R. (1947) Maize genetics. Cooperation Newsletter 21:36.

Burnham, C.R. and Brinks, R.A. (1932) Linkage relations of a second brown-midrib gene (*bm2*) in maize. J. Am. Soc. Agric. 24:960–963.

Capper, B.S., Thomson, E.F. and Herbert, F. (1988) Genetic variation in the feeding value of barley and wheat straw. In: Reed J.D., Capper, B.S. and Neate, J.H. (Eds.), Plant breeding and the nutritive value of crop residues, Addis Abeba, Ethiopia, International livestock for Africa, pp. 177–193.

Capper, B.S., Sage, G., Hanson, P.R. and Adamson A.H. (1992) Influence of variety, row type and time of sowing on the morphology, chemical composition and *in vitro* digestibility of barley straw. J. Agric. Sci. 188:165–173.

Cardinal, A.J., Lee, M. and Moore, K.J. (2003) Genetic mapping and analysis of quantitative trait loci affecting fiber and lignin content in maize. Theor. Appl. Genet. 106:866–874.

Casler, M. and Jung H. (1999) Selection and evaluation of smooth bromegrass clones with divergent lignin or etherified ferulic acid concentration. Crop Sci. 39:1866–1873.

Casler, M.D. and Kaeppler, H.F. (2001) Molecular breeding for herbage quality in forage crops, In: Spangenberg G. (Ed.), Molecular breeding of forage crops, pp. 175–188.

Chen, C., Baucher, M., Christensen, J.H. and Boerjan, W. (2001) Biochetchnology in trees: towards improved paper pulping by lignin engineering. Euphytica 118:185–195.

Chen, F., Srinivasa Reddy, M.S., Temple, S., Jackson, L., Shadle, G. and Dixon, R.A. (2006) Multi-site genetic modulation of monolignol biosynthesis suggests new routes for formation of syringyl lignin and wall-bound ferulic acid in alfalfa (*Medicago sativa* L.). Plant J. 48:113–124.

Chen, L., Auh, C.K., Dowling, P., Bell, J., Chen, F., Hopkins, A., Dixon, R.A. and Wang, Z.Y. (2003) Improved forage digestibility of tall fescue (*Festuca arundinacea*) by transgenic down-regulation of cinnamyl alcohol dehydrogenase. Plant Biotechnol. J. 1:437–449.

Cherney, J.H., Axtell, J.D., Hassen, M.M. and Anliker K.S. (1988) Forage quality characterization of chemically induced brown-midrib mutant in pearl millet. Crop Sci. 28:783–787.

Cherney, D.J.R., Patterson, J.A. and Johnson K.D. (1990) Digestibility and feeding value of pearl millet as influenced by the brown-midrib, low lignin trait. J. Anim. Sci. 68:4345–1351.

Ciba-semences. (1990) Valorisation laitière d'une variété de maïs en ensilage. Synthesis of an experimentation conducted by the EDE of Vendée during 1988–89–90, 13 p.

Ciba-semences. (1995) Comparaison de la valorisation par des vaches laitières de deux hybrides de maïs, Miscellaneous paper, 7 p.

Civardi, L., Rigau, J. and Puigdomenech, P. (1999) Nucleotide sequence of two cDNAs coding for caffeoyl coenzyme A O-methyltransferase (CCoAOMT) and study of their expression in *Zea mays*. Plant Physiol. 120:1206.

Cox, W.J. and Cherney, D.J.R. (2001) Influence of brown midrib, leafy and transgenic hybrids on corn forage production. Agron. J. 93:790–796.

Cummings, D.G. and McCullough, M.E. (1969) A comparison of the yield and quality of corn and sorghum silage, University of Georgia, College of agriculture experimental station. Res. Bull. 67:5–19.

Damiani, I., Morreel, K., Danoun, S., Goeminne, G., Yahiaoui, N., Marque, C., Kopka, J., Messens, E., Goffner, D., Boerjan, W., Boudet, A.M. and Rochange, S. (2005) Metabolite profiling reveals a role for atypical cinnamyl alcohol dehydrogenase CAD1 in the synthesis of coniferyl alcohol in tobacco xylem. Plant Mol. Biol. 59:753–769.

Derieux, M., Darrigand, M., Gallais A., Barrière, Y., Bloc, Y. and Montalant, Y. (1987) Estimation du progrès génétique réalisé chez le maïs grain en France entre 1950 et 1985. Agronomie 7:1–11.

Dixon, R.A., Chen, F., Guo, D. and Parvathi, K. (2001) The biosynthesis of monolignols, a "metabolic grid", or independent pathways to guaiacyl and syringyl units. Phytochem. 57:1069–1084.

Do C.T., Pollet, B., Thévenin, J., Sibout, R., Denoue, D., Barrière, Y., Lapierre, C. and Jouanin, L. 2007 Both caffeoyl coenzyme A 3-O-methyltransferase 1 and caffeic acid O-methyltransferase 1 are involved in redundant functions for lignin, flavonoids and sinapoyl malate biosynthesis in Arabidopsis. Planta. 226:1117–1129.

Dolstra, O. and Medema, J.H. (1990) An effective screening method for genetic improvement of cell-wall digestibility in forage maize. In: Proceedings of the 15th congress maize and sorghum section of Eucarpia, Baden, Austria, June 4–8, pp. 258–270.

Droushiotis, D.N. (1989) Mixtures of annual legumes and small-grained cereals for forage production under low rainfall. J. Agric. Sci. Camb. 113:249–253.

Emerson, R.A. (1935) Cornell University. Agric. Exp. Stn. Memoir No. 180.

Emile, J.C., Barrière, Y. and Mauries, M. (1996) Effects of maize and alfalfa genotypes on dairy cow performances. Ann. Zootechn. 45:17–27.

Fernandez, I., Martin, C., Champion, M. and Michalet-Doreau, B. (2004) Effect of corn hybrid and chop length of whole-plant corn silage on digestion and intake by dairy cows. J. Dairy Sci. 87:1298–1309.

Fontaine, A.S., Bout, S., Barrière, Y. and Vermerris, W. (2003) Variation in cell wall composition among forage maize (*Zea mays* L.) inbred lines and its impact on digestibility: analysis of neutral detergent fiber composition by pyrolysis-gas chromatography-mass spectrometry. J. Agric. Food Chem. 51:8080–8087.

Frenchick, G.E., Johnson, D.G, Murphy, J.M. and Otterby, D.E. (1976) Brown midrib corn silage in dairy cattle ration. J. Dairy Sci. 59:2126–2129.

Frey, T.J., Coors, J.G., Shaver, R.D., Lauer J.G., Eilert, D.T. and Flannery, P.J (2004) Selection for silage quality in the Wisconsin Quality Synthetic and related maize populations. Crop Sci. 44:1200–1208.

Fritz, J.O., Cantrell, R.P., Lechtenberg, V.L., Axtell, J.D. and Hertel, J.M. (1981) Brown midrib mutants in sudangrass and grain sorghum. Crop Sci. 21:706–709.

Goering, H.K. and van Soest, P.J. (1971) Forage fiber analysis (apparatus, reagents, procedures and some applications). Agric. Handb. No. 379. US Government Print Office, Washington, DC.

Grabber, J., Ralph J., Lapierre, C. and Barrière, Y. (2004) Genetic and molecular basis of grass cell-wall degradability. I. Lignin-cell wall matrix interactions. C. R. Biol. 327:455–465.

Grant, R.J., Haddad, S.G., Moore, K.J. and Pederson, J.F. (1995) Brown midrib sorghum silage for midlactation dairy cows. J. Dairy Sci. 78:1970–1980.

Guillaumie, S., San Clemente, H., Deswarte, C., Martinez, Y., Lapierre, C., Murigneux, A., Barrière, Y., Pichon, M. and Goffner, D. (2007a) MAIZEWALL, a database and developmental gene expression profiling of cell wall biosynthesis and assembly maize genes. Plant Physiol. 143:339–363.

Guillaumie, S., Pichon, M., Martinant, J.P., Bosio, M., Goffner, D. and Barrière, Y. (2007b) Differential expression of phenylpropanoid and related genes in brown-midrib bm1, bm2, bm3, and bm4 young isogenic mutant maize plants. Planta 266:235–250.

Guillet-Claude, C., Birolleau-Touchard, C., Manicacci, D., Rogowsky, P.M., Rigau, J., Murigneux, A., Martinant, J.P. and Barriere, Y. (2004a) Nucleotide diversity of the ZmPox3 maize peroxidase gene: relationships between a MITE insertion in exon 2 and variation in forage maize digestibility. BMC Genet. 5:19.

Guillet-Claude, C., Birolleau-Touchard, C., Manicacci, D., Fourmann, M., Barraud, S., L'Home-det, J., Carret, V., Martinant, J.P. and Barrière, Y. (2004b) Nucleotide diversity associated in silage corn digestibility for three O-methyltransferase genes involved in lignin biosynthesis. Theor. Appl. Genet. 110:126–135.

Goicoechea, M., Lacombe, E., Legay, S., Mihaljevic, S., Rech, P., Jauneau, A., Lapierre, C., Pollet, B., Verhaegen, D., Chaubet-Gigot, N. and Grima-Pettenati, J. (2005) EgMYB2, a new transcriptional activator from *Eucalyptus* xylem, regulates secondary cell wall formation and lignin biosynthesis. Plant J. 43:553–597.

Guo, D., Chen, F., Inoue, K., Blount, J.W. and Dixon, R.A. (2001) Downregulation of caffeic acid 3-O-methyltransferase and caffeoyl CoA 3-O-methyltransferase in trangenic alfalfa. Impacts on lignin structure and implications for the synthesis of G and S lignin. Plant Cell 13:73–88.

Halpin, C. (2004) Re-designing lignin for industry and agriculture. Biotechnol. Genet. Eng. Rev. 21:229–245.

Halpin, C., Foxon, G.A. and Fentem P.A. (1995) Transgenic plants with improved energy characteristics. In: Chesson A., Wallace R.J. (Eds.), Biotechnology in animal feeds and animal feeding, VCH Publishers, Weinheim, pp. 279–293.

Halpin, C., Holt, K., Chojecki, J., Olivier, D., Chabbert, B., Monties, B., Edwards, K., Barakate, A. and Foxon, G.A. (1998) Brown-midrib maize (bm1), a mutation affecting the cinnamyl alcohol dehydrogenase gene. Plant J. 14:545–553.

Hawkins, G.E., Parr, G.E. and Little, J.A. (1964) Composition, intake, digestibility and prediction of digestibility of coastal Bermudgrass hay. J. Dairy Sci. 47:865–870.

He, X., Hall, M.B., Gallo-Meagher, M. and Smith, R.L. (2003) Improvement of forage quality by downregulation of maize O-methylteransferase. Crop Sci. 43:2240–2251.

Hoden, A., Barrière, Y., Gallais, A., Huguet, L., Journet, M. and Mourguet, M. (1985) Le maïs brown-midrib plante entière. III Utilisation sous forme d'ensilage par des vaches laitières. Bull. Tech. CRZV Theix, INRA 60:43–58.

Hoffmann, L., Maury, S., Martz, F., Geoffroy, P. and Legrand, M. (2003) Purification, cloning, and properties of an acyltransferase controlling shikimate and quinate ester intermediates in phenylpropanoid metabolism. J. Biol. Chem. 278:95–103.

Hoffmann, L., Besseau, S., Geoffroy, P., Ritzenthaler, C., Meyer, D., Lapierre, C., Pollet, B. and Legrand, M. (2004) Silencing of hydroxycinnamoyl coenzyme A shikimate/quinate hydro-xycinnamolyltransferase affects phenylpropanoid biosynthesis. Plant Cell 16:1446–1465.

Hunt, C.W., Kezar, W., Hinnam, D.D., Combs, J.J., Loesche, J.A. and Moen, T. (1993) Effects of hybrids and ensiling with and without a microbial inoculant on the nutritional characteristics of whole-plant corn. J. Anim. Sci. 71:39–43.

Istasse, L., Gielen, M., Dufrasne, L., Clinquart, A., Van Eenaeme, C. and Bienfait, J.M. (1990) Ensilage de maïs plante entière, comparaison de 4 variétés. 2. Performances zootechniques. Landbouwtijdschrift – Revue de l'Agriculture 43:996–1005.

Jorgenson, L.R. (1931) Brown midrib in maize and its lignage relations. J. Am. Soc. Agron. 23:549–557.

Jung, H.G. and Allen M.S. (1995) Characteristics of plant cell wall affecting intake and digestibil-ity of forages by ruminants. J. Anim. Sci. 73:2774–2790.

Keith, E.A., Colenbrander, V.F., Lechtenberg, V.L. and Bauman, L.F. (1979) Nutritional value of brown midrib corn silage for lactating dairy cows. J. Dairy Sci. 52:788–792.

Kim, H., Ralph, J., Lu, F., Pilate, G., Leplé, J.C., Pollet, B. and Lapierre, C. (2002) Identification of the structure and origin of thioacidolysis marker compounds for cinnamyl alcohol dehydroge-nase deficiency in angiosperms. J. Biol. Chem. 277:47412–47419.

Krakowsky, M.D., Lee, M., Woodman-Clikeman, W.L., Long, M.J. and Sharopova, N. (2004) QTL mapping of resistance to stalk tunneling by the European corn borer in RILs of maize population B73 x De811. Crop Sci. 44:274–282.

Krakowsky, M.D., Lee, M. and Coors, J.G. (2005) Quantitative trait loci for cell-wall components in recombinant inbred lines of maize (*Zea mays* L.) 1: Stalk tissue. Theor. Appl. Genet. 111:337–346.

Kuc, J. and Nelson, O.E. (1964) The abnormal lignins produced by the brown midrib mutants of maize. 1. The brown-midrib-1 mutant. Arch. Biochem. Biophys. 105:103–113.

Lauer, J.G., Coors, J.G. and Flannery, P.J. (2001) Forage yield and quality of corn cultivars developed in different eras. Crop Sci. 41:1441–1455.

Li, L.G., Cheng, X.F., Leshkevich, J., Umezawa, T., Harding, S.A. and Chiang, V.L. (2001) The last step of syringyl monolignol biosynthesis in angiosperms is regulated by a novel gene encoding sinapyl alcohol dehydrogenase. Plant Cell 13:1567–1585.

Lübberstedt, T., Melchinger, A.E., Klein, D., Degenhardt, H. and Paul, C. (1997) QTL mapping in testcrosses of European flint lines of maize: II. Comparison of different testers for forage quality traits. Crop Sci. 37:1913–1922.

Lübberstedt, T., Zein, I., Andersen, J., Wenzel, G., Krutzfeldt, B., Eder, J., Ouzunova, M. and Chun, S (2005) Development and application of functional markers in maize. Euphytica 146:101–108.

Lusk, S.W., Karau, P.K., Balogu, D.O. and Gourley L.M. (1984) Brown midrib sorghum or corn silage for milk production. J. Dairy Sci. 67:1739–1744.

MacAdam, J.W. and Grabber J.H. (2002) Relationship of growth cessation with the formation of diferulate cross-links and p-coumaroylated lignins in tall fescue leaf blades. Planta 215:785–793.

Mahanta, S.K. and Pachauri, V.C. (2005) Nutritional evaluation of two promising varieties of forage sorghum in sheep fed as silage. Asian-Aust. J. Anim. Sci. 18:1715–1720.

Mahesh, V., Million-Rousseau, R., Ullmann, P., Chabrillange, N., Bustamante, J., Mondolot, L., Morant, M., Noirot, M., Hamon, S., de Kochko, A., Werck-Reichhart, D. and Campa, C. (2007) Functional characterization of two p-coumaroyl ester 3′-hydroxylase genes from coffee tree: evidence of a candidate for chlorogenic acid biosynthesis. Plant Mol. Biol. 64:145–159.

Méchin V., Argillier, O., Barrière Y. and Menanteau V. (1998) Genetic variation in stems of normal and brown-midrib3 maize inbred lines. Towards similarity for in vitro digestibility and cell-wall composition. Maydica 43:205–210.

Méchin, V., Argillier, O., Barrière, Y., Mila, I., Polet B. and Lapierre C. (2000) Relationships of cell-wall composition to in vitro cell-wall digestibility of maize inbred line stems. J. Sci. Food Agric. 80:574–580.

Méchin, V., Argillier, O., Hébert Y., Guingo, E., Moreau, L., Charcosset, A. and Barrière Y. (2001) QTL mapping and genetic analysis of cell wall digestibility and lignification in silage maize. Crop Sci. 41:690–697.

Minson, D.J. and Wilson, J.R. (1994) Prediction of intake as an element of forage quality. In: Fahey G.C. (Ed.), Forage quality, evaluation and utilisation, American Society of Agronomy, Inc., Crop Science Society of America, Inc., Soil Science Society of America, Inc., Madison, WI, pp. 533–563.

Mitchell, R.A.C. and Shewry P.R. (2007) A novel bioinformatics approach identifies candidate genes for the synthesis and feruloylation of arabinoxylan. Plant Physiol. 144:43–53.

Moreira, V.R., Santos, H.S., Satter, L.D. and Sampaio, I.B.M. (2003) Feeding high forage diets to lactating dairy cows. Arquivo Brasileiro de Medicina Veterinaria e Zootecnia 55:197–202.

Nair, R.B., Bastress, K.L., Ruegger, M.O., Denault, J.W. and Chapple, C. (2004) The *Arabidopsis thaliana* reduced epidermal fluorescence1 gene encodes an aldehyde dehydrogenase involved in ferulic acid and sinapic acid biosynthesis. Plant Cell 16:544–554.

Oba, M. and Allen, M.S. (1999) Evaluation of the importance of the digestibility of neutral detergent fiber from forage. Effects on dry matter intake and milk yield of dairy cows. J. Dairy Sci. 82:589–596.

Oba, M. and Allen, M.S. (2000) Effect of brown midrib 3 mutation in corn silage on productivity of dairy cows fed two concentrations of dietary neutral detergent fiber, 1. Feeding behavior and nutrient utilization. J. Dairy Sci. 83:1333–1341.

Oliver, A.L., Grant, R.J., Pedersen, J.F. and O'Rear, J. (2004) Comparison of brown-midrib-6 and -18 forage sorghum with conventional sorghum and corn silage in diets of lactating dairy cows. J. Dairy Sci. 87:637–644.

Oliver, A.L., Pedersen, J.F., Grant, R.J. and Klopfenstein, T.J. (2005a). Comparative effects of the sorghum *bmr*-6 and *bmr*-12 genes. I. Forage sorghum yield and quality. Crop Sci. 45:2234–2239.

Oliver, A.L., Pedersen, J.F., Grant, R.J., Klopfenstein, T.J. and Jose, H.D. (2005b). Comparative effects of the sorghum *bmr*-6 and *bmr*-12 genes. I. Grain Yield, stover yield, and stover quality in grain sorghum. Crop Sci. 45:2240–2245.

Orskov, E.R., Tait, G.A.G., Reid, G.W. and Flachowski, G. (1988) Effect of straw quality and ammonia treatment on voluntary intake, milk yield and degradation characteristics of faecal fibre. Anim. Prod. 46:23–27.

Pedersen, J.F., Funnell, D.L., Toy, J.J., Oliver, A.L. and Grant, R.J. (2006a). Registration of Atlas bmr-12 forage sorghum. Crop Sci. 46:478.

Pedersen, J.F., Funnell, D.L., Toy, J.J., Oliver, A.L. and Grant, R.J. (2006b). Registration of seven forage sorghum genetic stocks near-isogenic for the brown midrib genes *bmr-6* and *bmr-12*. Crop Sci. 46:490–491.

Pedersen, J.F., Funnell, D.L., Toy, J.J., Oliver, A.L. and Grant, R.J. (2006c). Registration of twelve forage sorghum genetic stocks near-isogenic for the brown midrib genes *bmr-6* and *bmr-12*. Crop Sci. 46:491–492.

Pichon, M., Courbou, I., Beckert, M., Boudet, A.M. and Grima-Pettenati, J. (1998) Cloning and characterization of two maize cDNAs encoding cinnamoyl-CoA reductase (CCR) and differential expression of the corresponding genes. Plant Mol. Biol. 38:671–676.

Pichon, M., Deswarte, C., Gerentes, D., Guillaumie, S, Lapierre, C., Toppan, A., Barrière, Y. and Goffner, D. (2006) Variation in lignin and cell wall digestibility traits in caffeic acid O-methyltransferase down-regulated maize half-sib progenies in field experiments. Mol. Breed. 18:253–261.

Piquemal, J., Chamayou, S., Nadaud, I., Beckert, M., Barrière, Y., Mila, I., Lapierre, C., Rigau, J., Puigdomenech P., Jauneau A., Digonnet, C., Boudet, A.M., Goffner, D. and Pichon M. (2002) Down-regulation of caffeic acid O-methyltransferase in maize revisited using a transgenic approach. Plant Physiol. 130:1675–1685.

Plénet, D. and Cruz, P. (1997) Diagnosis of the nitrogen status in crops. Maize and Sorghum, Chap. 5. In: G. Lemaire (Ed.), Springer-Verlag, Berlin, Heidelberg, pp. 93–106.

Porter, K.S., Axtell, J.D., Lechtenberg, V.L. and Colenbrander, V.F. (1978) Phenotype, fiber composition, and in vitro dry matter disappearance of chemically induced brown midrib (bmr) mutants of sorghum. Crop Sci. 18:205–208.

Ralph, J., Guillaumie, S., Grabber, J.H., Lapierre, C. and Barrière, Y. (2004) Genetic and molecular basis of grass cell wall biosynthesis and degradability. III. Towards a forage grass ideotype. C. R. Biol. 327:467–479.

Ranocha, P., Chabannes, M., Chamayou, S., Danoun, S., Jauneau, A., Boudet, A.M. and Goffner, D. (2002) Laccase down-regulation causes alterations in phenolic metabolism and cell wall structure in poplar. Plant Physiol. 129:145–155.

Reddy, M.S., Chen, F., Shadle, G., Jackson, L., Aljoe, H. and Dixon, R.A. (2005) Targeted down-regulation of cytochrome P450 enzymes for forage quality improvement in alfalfa (*Medicago sativa* L.). Proc. Natl. Acad. Sci. USA. 102:16573–16578.

Reid, G.W., Orskov, E.R. and Kay, M. (1988) A note on the effect of variety, type of straw and ammonia treatment an digestibility and on growth rate on steers. Anim. Prod. 47:157–160.

Rohde, A., Morreel, K., Ralph, J., Goeminne, G., Hostyn, V., De Rycke, R., Kushnir, S., Van Doorsselaere, J., Joseleau, J.P., Vuylsteke, M., Van Driessche, G., Van Beeumen, J, Messens, E. and Boerjan W. (2004) Molecular phenotyping of the pal1 and pal2 mutants of *Arabidopsis thaliana* revealed far-reaching consequences on phenylpropanoid, amino acid, and carbohydrate metabolism. Plant Cell 16:2749–2771.

Rook, J.A., Muller, L.D. and Shank, D.B. (1977) Intake and digestibility of brown midrib corn silage by lactating dairy cows. J. Dairy Sci. 60:1894–1904.

Roussel, V., Gibelin, C., Fontaine, A.S. and Barrière, Y. (2002) Genetic analysis in recombinant inbred lines of early dent forage maize. II – QTL mapping for cell wall constituents and cell wall digestibility from per se value and top cross experiments. Maydica 47:9–20.

Russell, W.A. (1984) Agronomic performance of maize cultivars representing different eras of breeding. Maydica 29:375–390.

Schiere, J.B., Joshi, A.L., Seetharam, A., Oosting, S.J., Goodchild, A.V., Deinum, B. and Van Keulen, H., (2004) Grain and straw for whole plant value. Implications for crop management and genetic improvement strategies. Explor. Agric. 40:277–294.

Schoch, G., Goepfert, S., Morant, M., Hehn, A., Meyer, D., Ullmann, P. and Werck-Reichhart, D. (2001) CYP98A3 from *Arabidopsis thaliana* is a 3′-hydroxylase of phenolic esters, a missing link in the phenylpropanoid pathway. J. Biol. Chem. 276:36566–36574.

Shadle, G., Chen, F., Srinivasa Reddy, M.S., Jackson, L., Nakashima, J., Dixon, R.A. (2007) Down-regulation of hydroxycinnamoyl CoA: Shikimate hydroxycinnamoyl transferase in transgenic alfalfa affects lignification, development and forage quality. Phytochem. 68:1521–1529.

Skibbe, D., Liu, F., Wen, T., Yandeau, M., Cui, X., Cao, J., Simmons, C. and Schnable, P. (2002) Characterization of the aldehyde dehydrogenase gene families of *Zea mays* and *Arabidopsis*. Plant Mol. Biol. 48:751–764.

Sommerfeldt, J.L., Schingoethe, D.J. and Muller, L.D. (1979) Brown midrib corn silage for lactating dairy cows. J. Dairy Sci. 62:1611–1618.

Stallings, C.C., Donaldson, B.M., Thomas, J.W. and Rossman, E.C. (1982) *In vivo* evaluation of brown-midrib corn silage by sheep and lactating dairy cows. J. Dairy Sci. 65:1945–1949.

Struik, P.C. (1983) Physiology of forage maize (*Zea mays* L.) in relation to its productivity. Doctoral thesis, Wageningen, the Netherlands, 97 p.

Tamagone, L., Merida, A., Parr, A., Mackay, S., Culianez-Marcia, F.A., Roberts, K. and Martin, C. (1998) The AmMYB308 and AmMYB330 transcription factors from Antirrhinum regulate phenylpropanoid and lignin biosynthesis in transgenic tobacco. Plant Cell 10:135–154.

Taylor, C.C. and Allen, M.S. (2005) Corn grain endosperm type and brown midrib 3 corn silage. Feeding behaviour and milk yield of lactating dairy cows. J. Dairy Sci. 88:1425–1433.

Tilley, J.M.A. and Terry, R.A. (1963) A two stage technique for the in vitro digestion of forage crops. J. Br. Grassland Soc. 18:104–111.

Tine, M.A., McLeod, K.R., Erdman, R.A. and Baldwin R.L. (2000) Effects of brown midrib corn silage on the energy balance of dairy cattle. J. Dairy Sci. 84:885–895.

Tingle, J.N. and Dawley, W.K. (1974) Yield and nutritive value of whole plant cereals at a silage stage. Can. J. Plant Sci. 54:621–624.

Troyer, A.F. (1999) Background of US hybrid corn. Crop Sci. 39:601–626.

Troyer, A.F. (2002) Germplasm ownership: related corn inbred. Crop Sci. 42:3–11.

Vadiveloo, J. (1992) Varietal differences in the chemical composition and in vitro digestibility of rice straw. J. Agric. Sci. Camb. 119:27–33.

Vignols, F., Rigau, J., Torres, M.A., Capellades, M. and Puigdomenech, P. (1995) The *brown midrib 3* (*bm3*) mutation in maize occurs in the gene encoding caffeic acid *O*-methyltransferase. Plant Cell 7:407–416.

Watanabe, H. and Kasuga, S. (2000) Effect of brown midrib and water soluble matter content on digestibility of forage sorghum (Sorghum bicolor Moench, Sorghum sudanense Stapf) foliage. Grassland Sci. 45:397–403.

Weller, R.F. and Phipps, R.H. (1986) The feeding value of normal and brown midrib-3 maize silage. J. Agric. Sci. 106:31–35.

Wilson, W.A., Harrington, S.E., Woodman, W.L., Lee, M., Sorrells, M.E. and McCouch, S. (1999) Inferences on the genome structure of progenitor maize through comparative analysis of rice, maize and the domesticated panicoids. Genetics 153:453–473.

Participatory Plant Breeding in Cereals

S. Ceccarelli and S. Grando

Abstract It is widely recognized that conventional plant breeding has been more beneficial to farmers in high potential environments or those who could profitably modify their environment to suit new cultivars than to the poorest farmers who could not afford to modify their environment through the application of additional inputs and could not risk the replacement of their traditional, well-known, and reliable varieties. As a consequence, low yields, crop failures, malnutrition, famine, and eventually poverty still affect a large proportion of humanity. Participatory plant breeding (PPB) is seen by several scientists as a way to overcome the limitations of conventional breeding by offering farmers the possibility to decide which varieties better suit their needs and conditions without exposing the household to any risk during the selection progress. PPB exploits the potential gains of breeding for specific adaptation through decentralized selection, defined as selection in the target environment, and is the ultimate conceptual consequence of a positive interpretation of genotype × environment interactions. The chapter describes a model of PPB in which genetic variability is generated by breeders, selection is conducted jointly by breeders, farmers, and extension specialists in a number of target environments, and the best selections are used in further cycles of recombination and selection. Therefore, from a scientific viewpoint, the process is similar to a conventional breeding program with three main differences, namely (a) testing and selection take place on-farm rather than on-station, (b) key decisions are made jointly by farmers and breeder, and (c) the process can be independently implemented in a large number of locations. Farmers handle the first phases of seed multiplication of promising breeding material in village-based seed production systems. The model has the following advantages: the varieties reach the release phase earlier than in conventional breeding, the release and seed multiplication concentrate on varieties known to be acceptable by farmers, biodiversity increases because different varieties are selected in different locations, the varieties fit the agronomic management that farmers are familiar with and can afford, and, therefore, the varieties can be beneficial to poor farmers. These advantages are particularly relevant to developing

S. Ceccarelli(✉)
The International Center for Agricultural Research in the Dry Areas (ICARDA), e-mail: s.ceccarelli@cgiar.org

M.J. Carena (ed.), *Cereals*,
DOI: 10.1007/978-0-387-72297-9, © Springer Science + Business Media, LLC 2009

countries where large investments in plant breeding have not resulted in production increases, especially in marginal environments. In addition to the economical benefits, participatory research has a number of psychological, moral, and ethical benefits which are the consequence of a progressive empowerment of the farmers' communities; these benefits affect sectors of their life beyond the agricultural aspects. In conclusion, PPB, as a case of demand-driven research, gives voice to farmers, including those who have been traditionally the most marginalized such as the women, and elevates local knowledge to the role of science.

1 Introduction

In recent years there has been increasing interest toward participatory research, in general, and toward participatory plant breeding (PPB), in particular. Following the early work of Rhoades and Booth (1982), scientists have become increasingly aware that users' participation in technology development may in fact increase the probability of success for the technology.

The interest is partly associated with the perception that the impact of agricultural research, including plant breeding, particularly in developing countries and for marginal environments and poor farmers has been below expectations. In fact about 2 billion people still lack reliable access to safe, nutritious food, and 800 million of them are chronically malnourished (Reynolds and Borlaug, 2006).

Three common characteristics of most agricultural research which might help to explain its limited impact in marginal areas are as follows:

1. The research agenda is usually decided unilaterally by the scientists and is not discussed with the users;
2. Agricultural research is typically organized in compartments, that is, disciplines and/or commodities, and seldom uses an integrated approach; this contrasts with the integration existing at farm level;
3. There is a disproportion between the large number of technologies generated by the agricultural scientists and the relatively small number of them actually adopted and used by the farmers.

When one looks at these characteristics as applied to plant-breeding programs, most scientists would agree that

1. Plant breeding has not been very successful in marginal environments and for poor farmers;
2. It still takes a long time (about 15 years) to release a new variety as reported in the conclusions of Interdrought-II (2005) "While basic research in plant biotechnology research towards the genetic improvement of crop productivity in water-limited conditions has expanded in recent years, the collaboration with plant breeding has been insufficient (with the exception perhaps of the private sector). This lack of collaboration hinders the delivery of biotechnology-based solutions to the end-user in the field, i.e. the farmer. There is an exponential growth of information in

genomics with a proportionally minute rate of application of this information to effective problem-solving infarming under water-limited conditions.
3. Many varieties are officially released, but few are adopted by farmers; by contrast, farmers often grow varieties which were not officially released;
4. Even when new varieties are acceptable to farmers, their seed is either not available or too expensive;
5. There is a widespread perception of a decrease of biodiversity associated with conventional plant breeding.

Participatory research, in general, defined as that type of research in which users are involved in the design – and not merely in the final testing – of a new technology, is now seen by many as a way to address these problems. PPB, in particular, defined as that type of plant breeding in which farmers, as well as other partners, such as extension staff, seed producers, traders, and NGOs, participate in the development of a new variety, is expected to produce varieties which are targeted (focused on the right farmers), relevant (responding to real needs, concerns, and preferences), and appropriate (able to produce results that can be adopted) (Bellon, 2006).

The objective of this chapter is to illustrate some of the characteristics of PPB using examples from projects implemented by the International Center for Agricultural Research in the Dry Areas (ICARDA) in a number of countries.

2 Genotype × Environment Interactions and Breeding Strategies

Plant breeding is a complex process, and in the majority of cases (the only notable exception being the breeding programs in Australia), only a small fraction of it takes place in farmers' fields; usually, most of the process takes place in one, or more often in a number of research stations, and all the decisions are made by the breeders and collaborating scientists (pathologist, entomologist, quality specialists, etc.). One of the main consequences is that a large amount of breeding material is discarded before knowing whether it could have been useful in the real conditions of farmers' fields, and the one which is selected is likely to perform well in environments similar to the research stations, but not in environments which are very different. This is because of genotype × environment (GE) interactions which are one of the major factors limiting the efficiency of breeding programs when they cause a change of ranking between genotypes in different environments (crossover interaction). An example of crossover GE interactions between research stations and farmers' fields is given in Fig. 1. In both cases there was much more similarity between research stations than between farmers' fields, and low or negative correlations between research stations and most of the farmers' fields.

In general, when different lines or cultivars of a given crop are evaluated in a sufficiently wide range of environments, GE interactions of crossover type seem to be very common (Ceccarelli et al., 2001). We have argued (Ceccarelli, 1989) that for crops grown in environments poorly represented by the research stations this

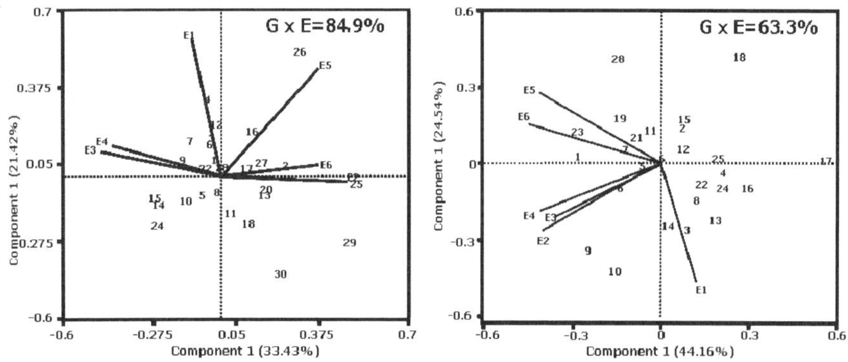

Fig. 1 Biplots of 30 barley genotypes grown in six locations in Morocco (*left*) including two research stations (E3 and E4) and four farmers' fields (E1, E2, E5, and E6) and of 25 barley genotypes in six locations in Tunisia (*right*) including two research stations (E5 and E6) and four farmers' fields (E1, E2, E3, and E4)

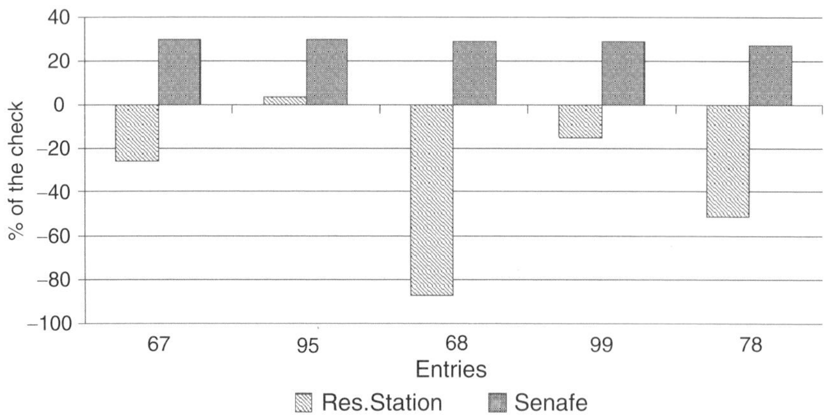

Fig. 2 Yield (in percent of the local check) of five barley lines in a farmer's field in Senafe (Eritrea) and in the research station at Halale (40 km south of Asmara)

often results in useful breeding materials being discarded. An example of the danger of discarding useful breeding material on station is shown in Fig. 2 where the five highest-yielding barley lines in a farmer field in Senafe (Eritrea), with yield advantages over the local check of between 27% and 30%, when tested on station showed a yield disadvantage of between 15% and 87% except entry 95 which had a yield advantage of only 4%.

When GE interactions are present the plant breeder can ignore them, avoid them, or exploit them (Eisemann et al., 1990). When GE interactions are significantly large, it is not possible to ignore them, and the two remaining strategies are (1) to avoid them by selecting material that is broadly adapted to the entire range of target environments or (2) to exploit them by selecting a range of material, each adapted to

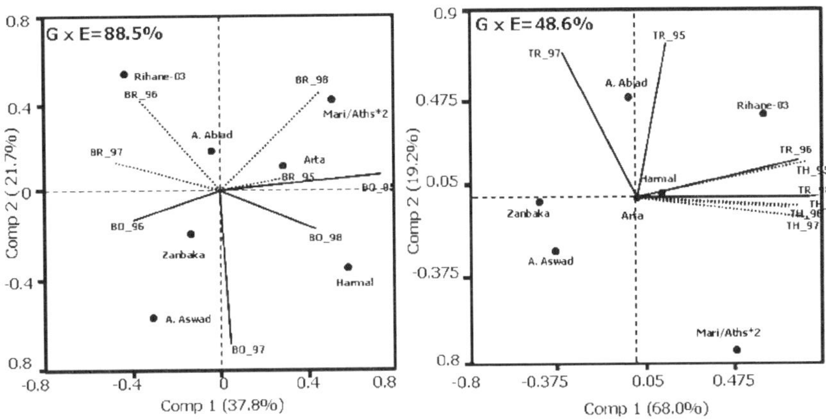

Fig. 3 Biplots of grain yield of seven barley cultivars grown for 4 years (1995–1998) in two dry locations, Bouider (BO) and Breda (BR) with a grand mean of 1.3 t/ha (*left*) and in two locations, Tel Hadya (TH) and Terbol (TR) with a grand mean of 3.5 t/ha (*right*)

a specific environment (Ceccarelli, 1989). The choice is based on a separate analysis of the two components of GE interactions, namely genotype × years (GY) and genotype × locations (GL), the first of which is largely unpredictable, while the second, if repeatable over time, identifies distinct target environments (Annicchiarico et al., 2005, 2006).

Selection for specific adaptation to each of the target environments is particularly important in breeding crops predominantly grown in unfavorable conditions, because unfavorable environments tend to be more different from each other than favorable environments (Ceccarelli and Grando, 1997). An example is shown in Fig. 3 where the total GE in the case of the two dry locations (*left*) was nearly 90%, while in the case of the two high-rainfall locations was less than 50%.

Selecting for specific adaptation has the advantage of adapting cultivars to the physical environment where they are meant to be cultivated, and, hence, is more sustainable than other strategies which rely on modifying the environment to fit new cultivars adapted to more favorable conditions (Ceccarelli and Grando, 2002). Selection theory shows that selection for specific adaptation is more efficient because it exploits the larger heritabilities within each specific target environment.

The similarity between research stations observed in Fig. 1 and between high-rainfall locations and years observed in Fig. 3 are likely to be also associated with the larger use of inputs (fertilizers, weed control, etc.) common to both research stations and high-rainfall areas, which tend to smooth out differences between locations and years.

Selection for specific adaptation is based on direct selection in the target environment, which has been also defined as decentralized selection (Falconer, 1981; Simmonds, 1984, 1991). These concepts started to be adopted also in relation with organic agriculture (Murphy et al., 2007).

The most serious challenge to decentralized selection for unfavorable environments is the large number of potential target environments. Moreover, the number of target environments is often increased by different uses of the crop (cash vs local consumption), different access to inputs, different market opportunities, etc. Clearly, selection for specific adaptation to unfavorable conditions targets a larger sample of environments than selection for favorable environments. Consequently, the number of selection sites will need to be larger.

The participation of farmers in the very early stages of selection offers a solution to the problem of fitting the crop to a multitude of both target environments and users' preferences (Ceccarelli, 1996).

3 Defining Decentralized PPB

Although plant breeding programs differ from each other depending on the crop, on the facilities, and on the breeder, they all have in common some major stages that Schnell (1982) has defined as "generation of variability," "selection," and "testing of experimental cultivars" (Fig. 4, *left*). To illustrate the process we will use as an example a self-pollinated crop and the more common breeding practices. The generation of variability is the shortest stage, consisting of the process of making crosses (or, less frequently, inducing mutations) and producing segregating populations, and takes place in research stations. The second stage is longer and consists, first, of the evaluation of the breeding value of the different segregating populations (by "cross-evaluation" or "selection between crosses"), and then in the selection

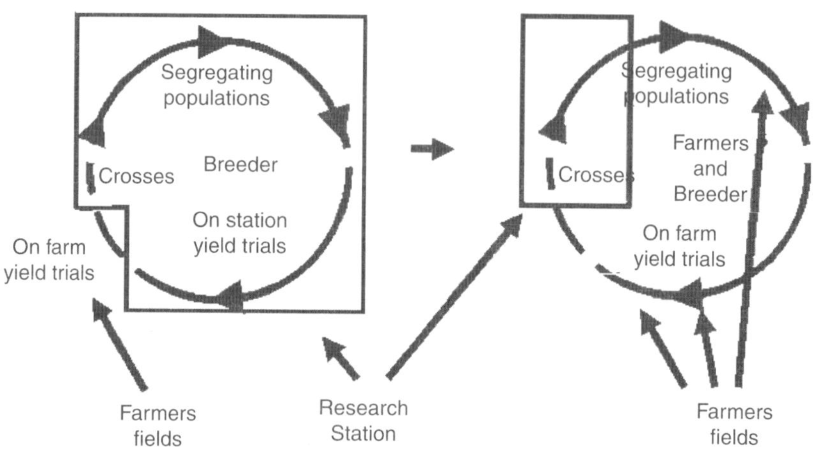

Fig. 4 Conventional plant breeding is a cyclic process that takes place largely within one or more research stations (*left*) with the breeder making all decisions; decentralized-participatory plant breeding is the same process, but takes place mostly in farmers' fields (*right*) and the decisions are made jointly by farmers and breeders

of the best plants within the superior populations, or in various combinations of the two while reducing heterozygosity. The second stage, like the first, usually takes place in research stations (although there are exceptions), and in some crops it can be shortened by the use of techniques such as single seed descent (SSD) and doubled haploids (DH). During the second stage, the breeding material is exposed to relevant biotic and abiotic stresses, often on more than one research station. The end product of the second stage is usually a population of several thousand pure lines even in those situations where uniformity is not a farmer's necessity or requirement.

The third stage is also long, consisting in the comparison of yield (usually of grain in those crops where the grain is the main commercial product) between the breeding lines produced during the second stage. This phase is usually subdivided into two substages. The first takes place on one or more research stations and the trials are referred to as multienvironment trials (MET). The second, when the number of breeding lines has been reduced to between 10 and 20, takes place in farmers' fields and the trials are referred to as on-farm trials even though they also are typically MET. In some exceptional cases, such as in most of the breeding programs in Australia, yield testing takes place entirely in farmers' fields, and therefore is fully decentralized.

Plant breeding is a cyclic process (Fig. 4); each year (or cropping season) a new cycle begins with new crosses, which are being made using largely as parents lines derived from previous cycles. Therefore, each year, breeding materials belonging to the three stages described earlier, and to different steps within each stage, are grown simultaneously. This implies a considerable investment not only in land to grow the parental material, the various generations of segregating populations, and the various levels of yield testing, each representing a different breeding cycle (amounting at several tens of thousand plots) but also in people, and in facilities to handle the considerable amount of seed and of data that the process generates. One important aspect of the process is that it is cyclic. This implies that the breeders accumulate a considerable amount of knowledge about the germplasm during the years. If this aspect of the process is not maintained in a PPB program, it is very difficult to talk about the process as "plant breeding" and is also very difficult to have farmer empowerment. In fact, this is strictly associated with the increasing farmers' knowledge which in turn is associated with the increasing farmers' familiarity about the process and the genetic material.

A decentralized PPB program (Fig. 4, *right*) is exactly the same process as described in the previous paragraph with three differences: (1) most of the process takes place in farmers' fields, (2) the decisions are taken jointly by the farmers and the breeder, and (3) the process can be implemented in a number of locations involving a large number of farmers with different breeding materials.

There is a considerable amount of debate among scientists about defining PPB; as many of those scientists are not plant breeders, the debate is often on the participatory rather than the breeding side of the definition. Two terms has been widely used, namely participatory variety selection (PVS) and PPB. In PVS farmers participate at the very end of the cyclic process described earlier when the number of choices and the genetic variability are limited. In PPB farmers participate as early as it is feasible, and in practice this can be achieved in a multitude of ways as long as, as mentioned earlier, the process is cyclic. The actual methods can vary with the crop, and for the same crop they may vary with the type of agriculture (subsistence or commercial) so that different types of farmers within the same country and growing the same crop for different purpose may require a different method. One of the main advantages of PPB is its flexibility which makes it adaptable to a multitude of requirements.

In the following sections we describe a model of PPB that can be applied to self-pollinated crops. The method is based on three main concepts which can be generalized to any PPB program.

1. The trials are grown in farmers' fields using the farmer's agronomic practices (to avoid GE interactions between research stations and farmers' fields).
2. Selection is conducted jointly by breeders and farmers in farmers' fields, so that farmers participate in all key decisions.
3. The traditional linear sequence scientist➔extension➔farmers is replaced by a team approach with scientists, extension staff, and farmers participating in all major steps of variety development.

The breeding method that the model assumes is a bulk-pedigree method in which selection between populations (cross evaluation) is conducted in the field together with farmers and selection within the superior population, when necessary, is conducted on station (Ceccarelli and Grando, 2005).

4 A Model of Decentralized PPB for Self-Pollinated Crops

4.1 The Model

The method of plant breeding we use in a number of countries has been described in detail by Ceccarelli and Grando (2005) and by Mangione et al. (2006) and more recently by Ceccarelli and Grando (2007) and Ceccarelli et al. (2007); the crosses are done on station, where we also grow the F1 and the F2, while in the farmers' fields the bulks are yield tested over a period of 4 years (Fig. 5).

The activities in farmers' fields begin with the yield testing of F3 bulks in trials called Farmers Initial Trials (FIT), which are unreplicated trials with systematic checks or partially replicated trials. The number of entries varies from about 50 in Egypt, to 75 in Eritrea Iran and Algeria, to 165 in Jordan and Syria, and the total number of plots varies from 60 in Egypt, to 100 in Eritrea Iran and Algeria, and to 200 in Jordan and Syria. Plot size varies from 2 m^2 to 12 m^2.

The bulks selected from the FIT with the process described in the next section are yield tested as F4 bulks for a second year in the Farmer Advanced Trials (FAT) with a number of entries and checks that varies from village to village and from year

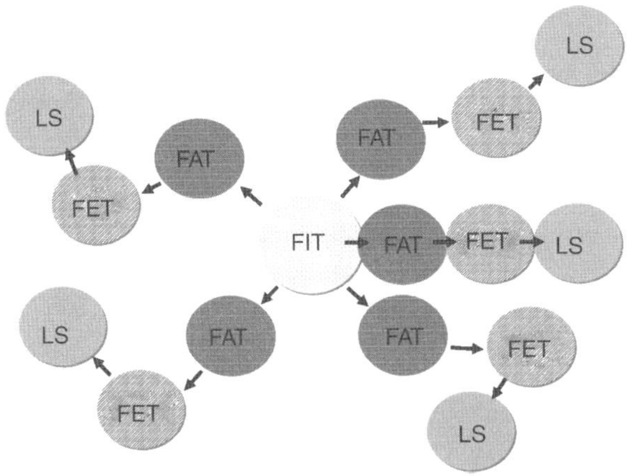

Fig. 5 A model of participatory plant breeding in one village: from the Farmer Initial Yield Trial (*FIT*), grown by one farmer, participatory selection identifies the lines to be grown in the Farmers Advanced Yield Trials (*FAT*) by more farmers (five in the figure). The process is repeated to identify lines to be grown in Farmer Elite Trials (*FET*) and in the initial adoption stage (*LS* or Large-Scale Trials). The model takes 4 years for the full implementation

to year. The plot size in the FAT is larger (10–45 m^2 depending on the country), and the number of FAT in each village depends on how many farmers are willing to grow this type of trial. In each village, the FAT contains the same entries. Each farmer decides the rotation, the seed rate, the soil type, the amount, and the time of application of fertilizer. Therefore, the FAT are planted in a variety of soil types and of agronomic managements. During selection, farmers exchange information about the agronomic management of the trials and rely greatly on this information before deciding which entries to select. Therefore, the breeding materials start to be characterized for their responses to environmental or agronomic factors at an early stage of the selection process.

The F4 bulks selected from the FAT are tested as F5 bulks in the Farmer Elite Trials (FET), with a plot size twice as large as the FAT, and after one more cycle of selection, a number of bulks (usually less than five) are planted by the farmers on large-scale (LS) unreplicated plots (few thousand m^2) as the first step in the adoption process.

The PPB trials (FIT, FAT, and FET) are in all respects like the MET in a conventional breeding program as described earlier. Even when the MET are conducted in farmers' fields, like in the breeding programs in Australia, there are still at least two major differences between the MET and the PPB trials. The first is that MET are established with the primary objective of sampling target physical environments, while the PPB trials are meant to sample both physical and socio-economic environments including different types of users. The second is that MET data are usually analyzed to estimate or predict the genotypic value of each line across all locations, while in PPB trials the emphasis is on estimating or predicting the genotypic value of each line over time in a given location.

4.2 Farmers' Selection and Data Collection

At the time of selection, farmers are provided with field books to register both qualitative and quantitative observations. Farmers' preferences are usually recorded from 0 (discarded) to 4 (most preferred plots) by between 10 and 30 farmers including (in some countries) women, occasionally assisted by scientists (or literate farmers) to record their scores. Breeders collect quantitative data on a number of traits indicated by farmers as important selection criteria (such as growth vigor, plant height, spike length, grain size, tillering, grain yield, biomass yield, harvest index, resistance to lodging and to diseases and pests, and cold damage), as usually done in the MET in a conventional breeding program.

The data are processed (see Sect. 4.3) and the final decision of which bulks to retain for the following season is made jointly by breeders and farmers in a special meeting and is based on both quantitative data and visual scores.

In parallel to the model shown in Fig. 5, and in those countries where varieties of self-pollinated crops can be released only if genetically uniform, pure line selection within selected bulks is conducted on station. The head rows are promoted to a screening nursery only if farmers select the corresponding bulks. The process is repeated until there is enough seed to include the lines (as F7) in the yield-testing phase (Ceccarelli and Grando, 2005). Therefore, when the model is fully implemented, the breeding material which is yield tested in the FIT, FAT, and FET includes new bulks as well as pure lines extracted from the best bulks of the previous cycle. If in a given country the requirements for the genetic uniformity of the varieties to be released are very strict, only the pure lines will be considered as candidates for release.

4.3 Experimental Designs and Statistical Analysis

An experimental design, which has proven to be suitable in the first stage where there is one host farmer in each location, is the unreplicated design with systematic checks every ten or every five entries arranged in rows and columns or a partially replicated design in which about 20–25% of the entries are replicated twice.

In the second and third level, the trials can be designed as α-lattices with two replications or as randomized complete blocks with farmers as replicates, or as standard replicated trials.

The data are subjected to different types of analysis, some of which where developed at ICARDA, such as the spatial analysis of unreplicated or replicated trials (Singh et al., 2003). The environmentally standardized best lineal unbiased predictors (BLUPs) obtained from the analysis are then used to analyze GE interaction using the GGE G=genotypic main effect plus GE=genotype×environment interaction biplot software (Yan et al., 2000).

Therefore, the PPB trials generate the same quantity and quality of data generated by the MET in a conventional breeding program with the additional information on farmers' preferences usually not available in the MET. As a consequence, varieties produced by PPB are eligible to be submitted to the process of officially variety release that in several countries, including many in the developing world, is the legal prerequisite for the commercial seed production.

4.4 *Time to Variety Release*

In a typical breeding program of a self-pollinated crop and following a classical pedigree method, it takes normally about 15 years to release a variety. With the method described in the previous section the time is reduced by half. However, the comparison is biased because of the difference in the genetic structure of the material being released, that is, pure lines in one case and populations in the second.

If populations are not acceptable by the variety release authorities, and the model includes pure line selection within the superior bulks, it can be shown that the time to variety release in the PPB program is still 3–4 years shorter than the conventional program based on the pedigree method, and again the comparison is biased because the conventional breeding program does not generate the information on farmers' preferences which is one of the main characteristics of a PPB program.

The method is, therefore, very flexible because it can generate populations, pure lines, and eventually mixtures of pure lines. Similarly, when applied to cross-pollinated crops, PPB can be used to produce hybrids, populations, and synthetics.

4.5 *Effect on Biodiversity*

One of the main benefits expected from PPB is an increase in crop biodiversity as a consequence of the joint effect of decentralized selection and of the farmers' participation. The effect on biodiversity is illustrated using the data of the 2001–2004 breeding cycle in Syria (Table 1). As indicated earlier, in each village the starting point of the breeding cycle in farmers' fields are the initial yield trials with

Table 1 Flow of germplasm, selection pressure, number of farmers participating in the selection and number of lines in initial adoption in one cycle of participatory plant breeding on barley in Syria

	FIT	FAT	FET	LS
Entries tested per village	165	17.3	7	3
Trials per village	1	3.2	3.4	2.8
Entries selected per village	17	8	3.5	1–2
Farmers selecting	9–10	8–9	8–9	8–9
No. of different entries	412	238	51	19

FIT Farmer Initial Trials, *FAT* Farmer Advanced Trials, *FET* Farmer Elite Trials, *LS* Large-Scale Trials

165 genetically different entries; the number of entries tested in the subsequent trials decreases to about 17 in the FAT, to 7 in the FET, and to 3 in the LS. The number of trials per village varies from one in the case of the FIT to about three in the case of the other trials. The number of lines selected by between eight and ten farmers per village was on average 17, 8, 3.5, and between 1 and 2.

Because different germplasm is tested in different villages, the total number of genetically different entries tested in the various trials was 412 in the FIT, 238 in the FAT, 51 in the FET, and 19 in the LS. In the case of Syria, the number of different entries at the end of a breeding cycle in farmers' fields is higher than the number of lines the Syrian National Program tests at the beginning of its on-farm testing which usually ends with one or two recommended varieties across the country.

5 Variety Release and Seed Production

The potential advantages of PPB, such as the speed with which new varieties reach the farmers, the increased adoption rate, and the increased biodiversity within the crop due to the selection of different varieties in different areas will not be achieved if the seed of the new varieties does not become available in sufficient amounts to the entire farmer community. In many countries this is associated with, and depends on, the official recognition of the new varieties. This process, called variety release, is usually the responsibility of a committee (the variety release committee) nominated by the Minister of Agriculture, which decides whether to release varieties based on a scientific report on the performance, agronomic characteristics, resistance to pests and diseases, and quality characteristics of the new variety. The process suffers from several drawbacks: (1) it takes a long time, (2) testing sites are poorly chosen, (3) the trial management is often not representative, (4) the trial analysis is biased against poor environments, (5) traits important to the farmers are not included, (6) farmers' opinion is not considered, (7) there is often lack of transparency in sharing the information, and (8) the trials are often conducted using the same methodologies for very many years.

As a consequence, there are several cases of varieties released which have never been grown by any farmer and also of varieties grown by farmers without being released. In the former case, the considerable investment made in developing the new variety and in producing its seed has no benefits.

One of the most important advantages of PPB is associated with reversing the delivery phase of a plant breeding program (Fig. 6). In a conventional breeding program, the most promising lines are released as varieties, their seed is produced under controlled conditions (certified seed) and only then do farmers decide whether to adopt them or not; therefore, the entire process is supply-driven. As a consequence, in many developing countries the process results in many varieties being released and only a small fraction being adopted. With PPB, it is the initial farmers' adoption which drives the decision of which variety to release, and, therefore, the process is demand-driven. Adoption rates are expected to be higher,

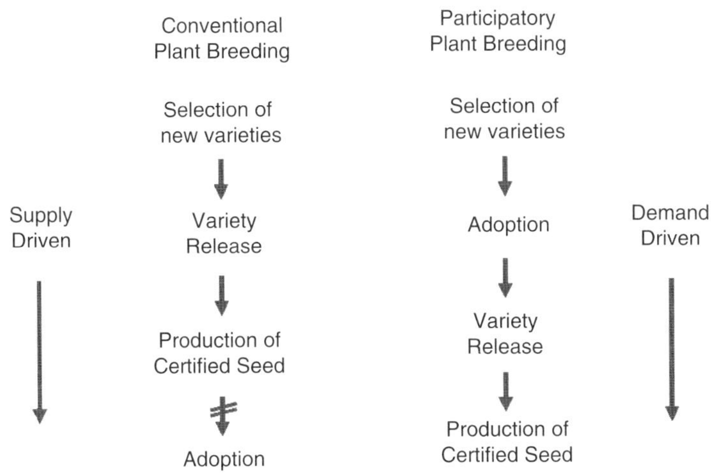

Fig. 6 In conventional plant breeding new varieties are released before knowing whether the farmers like them or not and the process is typically supply driven. In participatory plant breeding the delivery phase is turned upside down because the process is driven by the initial adoption by farmers at the end of a full cycle of selection and is, therefore, demand driven

and risks are minimized, as an intimate knowledge of varietal performance is gained by farmers as part of the selection process. Last but not least, the institutional investment in seed production is nearly always paid off by farmers' adoption.

The implementation of a PPB program not only implies a change in the process of variety release but also assumes changes in the seed sector. Conventional plant breeding and the formal seed sector have been successful in providing seeds of improved varieties of some important staple or cash crops to farmers in favorable areas of developing countries. However, the policy, regulatory, technical, and institutional environment under which these institutions operate limits their ability to serve the diverse needs of the small-scale farmers in marginal environments and remote regions.

The model we are implementing (Fig. 7) is based on the integration between the informal and the formal seed systems. During the selection and testing phase (the PPB trials described in Fig. 5) the seed required, which varies from 50 kg to 100 kg for each variety while the number of varieties in each village varies between 15 and 30, is produced in the village and is cleaned and treated with locally manufactured equipment. These are small seed cleaners which are able to process about 400 kg of seed per h. After the FET, the first initial adoption usually takes place, seed requirement goes up to few tons per farmer, and the number of varieties is reduced to two to three in each village. At this stage, seed production is still handled at village level, using locally manufactured larger equipment capable of cleaning and treating 1 t/h of seed. In this phase the staff of the Seed Organization starts supervising the LS village-based seed production. At the same time, the procedure for variety release can be initiated, and if the initial adoption if followed by a wider demand for seed, the variety is released, and the formal seed system can initiate LS

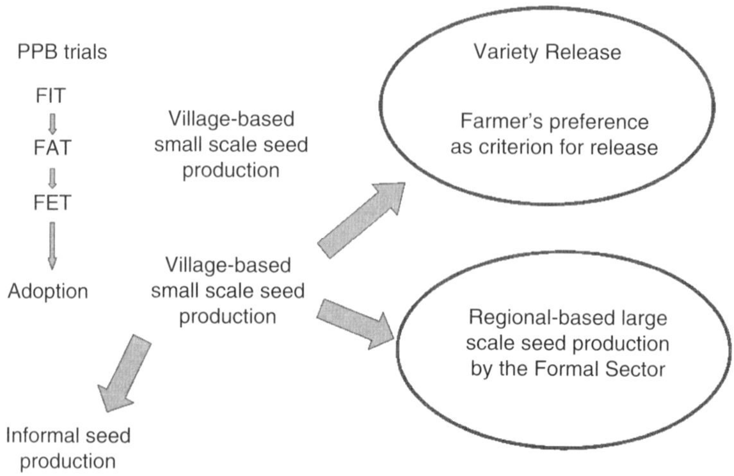

Fig. 7 Linking participatory plant breeding and variety release, with informal and formal seed production

Table 2 Countries where the participatory breeding program is implemented and program details

Country	Crop(s)	Locations	Trials	Plots
Syria	Barley	24	176	10,020
	Wheat	6	42	710
Jordan	Barley, wheat, chickpea	9	21	2,798
Egypt	Barley	6	20	460
Eritrea	Barley, wheat, hanfetse, chickpea, lentil, faba bean	7	36	1,475
Iran	Barley and bread wheat	5	3	100
Algeria	Barley	5	5	500
	Durum wheat	2	2	200

regional seed production using the few tons of seeds produced in the villages as a starting point.

In those countries where most of the seed used is produced by the informal seed system, the model can provide the informal system with quality seed of improved varieties.

6 Impact of PPB

By 2007, the model shown in Fig. 5 was fully implemented in Syria, Jordan, Egypt, and Eritrea, was in its second year in Algeria and started in Iran (Table 2). PPB programs based on the methodology described above have also been implemented in Tunisia and Morocco (Ceccarelli et al., 2001), and Yemen. These PPB projects had four main types of impact.

Table 3 Number of varieties selected and adopted by farmers in the participatory plant breeding (PPB) programs in five countries

Country	Crop(s)	Varieties
Syria	Barley	19
Jordan	Barley	1 (submitted)
Egypt	Barley	5
Eritrea	Barley	3
Yemen	Barley	2
	Lentil	2

Table 4 Varieties adopted from the participatory plant breeding (PPB) program by farmers in Syria in various rainfall zones

Pedigree	Name	Location	Rainfall*
H.spont.41-1/Tadmor	Raqqa-1	Bylounan	212.4
Arta//H.spont.41-5/Tadmor	Raqqa-2	Bylounan	212.4
Zanbaka/JLB37-064	Karim	Bylounan	212.4
Tadmor/3/Moroc9-75/ArabiAswad//H.spont.41-4	Akram	Bylounan	212.4
Mo.B1337/WI2291//Moroc9–75/3/SLB31–24	Suran-1	Suran	383.7
ChiCm/An57//Albert/3/Alger/Ceres.362-1-1/4/Arta	Suran-2	Suran	383.7
ER/Apm//Lignee131/3/Lignee131/ArabiAbiad/4/ Arta	Suran-3	Suran	383.7
Hml-02/5/..Alger/Ceres362-1-1/4/Hml	Nawair-1	Suran	383.7
Hml-02/5/..Giza 134-2L/6/Tadmor	Nawair-2	Suran	383.7
SLB03-10/Zanbaka	Yazem	J. Aswad	226.4
Tadmor//Roho/Mazurka/3/Tadmor	Salam	J. Aswad	226.4
ArabiAswad/WI2269/3/ArabiAbiad/WI2291// Tadmor/4/Akrash//WI2291/WI2269	Ethiad	J. Aswad	226.4

*Annual rainfall in mm in the period 2000–2005

1. Variety development: A number of varieties have been already adopted by farmers even though the program is relatively young in breeding terms (Table 3). In Syria, adoption is taking place for the first time in low-rainfall areas (<250 mm annual rainfall) (Table 4).
2. Institutional: In several countries, the interest of policymakers and scientists in PPB as an approach which is expected to generate quicker and more relevant results, has considerably increased.
3. Farmers' skills and empowerment: The cyclic nature of the PPB programs has considerably enriched farmers' knowledge, improved their negotiation capability, and enhanced their dignity (Soleri et al., 2002).
4. Enhancement of biodiversity: Different varieties have been selected in different areas within the same country, in response to different environmental constraints and users' needs. In Syria, where this type of impact has been measured more carefully, the number of varieties selected after three cycles of selection is four to five times higher than the number of varieties entering the on-farm trials in the conventional breeding program.

An economic analysis of the PPB barley-breeding program in Syria shows that PPB increases the benefits to resource poor farmers. The total estimated discounted research-induced benefits to Syrian agriculture were estimated at US $21.9 million for conventional breeding and US $42.7–113.9 million for three different PPB approaches (Lilja and Aw-Hasaan, 2002).

Using case studies on different crops, Ashby and Lilja (2004) have shown that

1. The use of participatory approaches improves the acceptability of varieties to disadvantaged farmers by including their preferences as criteria for developing, testing, and releasing new varieties. A survey conducted on over 150 PPB projects showed that (a) PPB improved program's effectiveness in targeting the poor; (b) by consulting women and involving them in varietal evaluation, there was a better acceptability and faster adoption of the varieties; and (c) involvement of women farmers in the development of maize seed systems in China resulted in a broadened national maize genetic base, in improved maize yields, and in strengthened women's organizations.
2. PPB improves research efficiency. A case study conducted using the PPB program in Syria (Ceccarelli et al., 2000, 2003) found that farmers' selections are as high yielding as breeders' selections. Another study found that by introducing farmer participation at the design stage, a 3-year reduction was achieved in the time taken from initial crosses to release. In another example, breeders concluded that it was faster, less expensive, and more reliable to involve farmers directly in the identification of promising accessions for use in the breeding program. Efficiency gains depend also on the extent to which farmer involvement enables the breeding program to minimize its investment in the development of varieties which, after release, turn out to be of little if any interest to farmers.
3. PPB accelerates adoption. The incorporation of participatory approaches consistently enables breeding programs to "break through" adoption bottlenecks caused by low levels of acceptability of new varieties by poor farmers. In addition to the examples given in Tables 3 and 4, other examples are Ethiopia, where out of over 122 varieties of cereals, legumes, and vegetables which had been released, only 12 were adopted by farmers; Brazil, where after years of nonadoption, the implementation of PPB led to the adoption of several clones of cassava which were both resistant to root rot and highly acceptable to farmers; and Ghana, where maize breeders had released several modern varieties (MVs) which had poor acceptability and poor adoption, while with farmers' participation the overall adoption of MVs increased to over two thirds.

Finally, there is increasing evidence that one of the most widespread impacts of PPB, and possibly of participatory research in general, is of a psychological and ethical nature; when farmers are asked which benefits they believe they receive from PPB, they refer that their quality of life has improved, that they feel happier as a consequence of changing their role from passive receivers to active protagonists, that their opinion is valued, and that, as an Eritrean farmer said, they have taken science back into their own hands.

7 Conclusions

The results presented in this chapter indicate that is possible to organize a plant breeding program in a way that addresses not only those plant characteristics that maximize yield and stability over time in a given physical environment but also the preferences of the users, by developing varieties which are specifically adapted to different physical and socioeconomic environments. Such an objective can be achieved by using a decentralized participatory approach, which needs to be extended also to seed production aspects. A breeding program organized according to these principles will have the advantages of producing environmentally friendly varieties and of maintaining or even enhancing biodiversity.

The main objections to PPB are usually that (1) plant breeding is "plant breeder's business", and if plant breeders do their job properly there should not be the need for PPB, (2) it is not possible for seed companies to cope with the multitude of varieties generated by PPB, and (3) varieties bred through PPB do not meet the requirements for official variety release.

With regards to the first objection, circumstantial evidence suggests that while plant breeding has been a success story in climatically, agronomically, and economically favorable areas, and in areas where the agronomic environment could be modified to create near-optimum growing conditions, it has been much less successful in less-favorable areas. In those areas where it has been successful, plant breeding has raised both environmental concerns due to high levels of chemical inputs required by MVs and biodiversity concerns because of the narrowing of the genetic basis of agricultural crops. More recently, there is a widespread concern about the use of the improperly called genetically modified organisms (GMOs) which, regardless of other considerations, represent yet another type of top–down technology. For these reasons, it may be useful to explore alternative avenues of plant breeding where the same science can be used in a different way.

The objection that seed companies have difficulties in coping with several varieties assumes implicitly the need to breed taking into account the requirements of the seed companies rather that the interest of the farmers, the consumers, and the society at large. It also ignores that in the case of the major food crops and in developing countries, farmers and not seed companies are the main suppliers of seed with over 90% of the seed which is currently planted; PPB can introduce new varieties directly into the most efficient seed system currently operating.

Against the third objection, the chapter has shown that it is possible to organize a PPB in such a way that it generates the same quantity of information of the same (or even better) quality than a conventional breeding program. In addition to the usual data set on agronomic characteristics, a PPB also generates information on farmers preferences (which is missing in the data set generated in a conventional breeding program), and, therefore, it makes the process of variety release more efficient and effective.

The third objection usually addresses also the genetic structure of the varieties produced by PPB. It assumes that varieties produced by PPB are inevitably genetically heterogeneous, unstable and not distinct, and, therefore, not suited for release. On this issue there are three points to make. First, the majority of cultivars still grown in marginal environments are genetically heterogeneous, and, in several cases, their seed is multiplied officially by the same authorities which deny the right of populations to be released; second, it is disputable how wise it is to replace them with genetically uniform material, and it has been recently shown (Di Falco and Chavas, 2006) that crop genetic diversity can increase farm productivity and can reduce the risk of crop failure; third, we have shown that PPB, like conventional plant breeding, is flexible and can be used to produce varieties with different genetic structure including pure lines and hybrids.

Therefore, the most frequent objections to PPB are unfounded; they ignore the fact that farmers have domesticated the crops that feed the world, and that they have continued to modify these crops for millennia. In this process they have planted, harvested, exchanged seed, introduced new crops and new varieties, and fed themselves and others, and, in doing so, they have accumulated a wealth of knowledge that modern science tends to ignore. PPB is one way of recognizing farmers' science and to merge it with modern science.

Acknowledgments The authors thank the several hundreds farmers who make their knowledge freely available, and the several researchers extension staff and NGOs who made this work possible, and several donors who have supported PPB at ICARDA. These include the OPEC Fund for International Development, the Governments of Italy and Denmark, der Bundesminister für Wirtschaftliche Zusammenarbeit (BMZ, Germany), the International Development Research Centre (IDRC, Canada), the System Wide Program on Participatory Research and Gender Analysis (SWP PRGA), and the Water and Food Challenge Program of the CGIAR.

References

Annicchiarico, P., Bellah, F. and Chiari, T. (2005) Defining Subregions and Estimating Benefits for a Specific-Adaptation Strategy by Breeding Programs: A Case study. *Crop Science* 45, 1741–1749.

Annicchiarico, P., Bellah, F. and Chiari, T. (2006) Repeatable Genotype x Location Interaction and its Exploitation by Conventional and GIS-Based Cultivar Recommendation for Durum Wheat in Algeria. *European Journal of Agronomy* 24, 70–81.

Ashby, J. A. and Lilja, N. (2004) Participatory Research: Does it Work? Evidence from Participatory Plant Breeding. In *New Directions for a Diverse Planet*. Proceedings of the 4th International Crop Science Congress. Brisbane, Australia, 26 September–1 October 2004. Available at http://www.cropscience.org.au

Bellon, M. R. (2006) Crop Research to Benefit Poor Farmers in Marginal Areas of the Developing World: A Review of Technical Challenges and Tools. CAB Reviews: Perspectives in Agriculture, Veterinary Science, Nutrition and Natural Resources. Online ISSN 1749-8848.

Ceccarelli, S. (1989) Wide Adaptation. How Wide? *Euphytica* 40, 197–205.

Ceccarelli, S. (1996) Positive Interpretation of Genotype by Environment Interactions in Relation to Sustainability and Biodiversity. In: M. Cooper and G. L. Hammers (Eds.), *Plant Adaptation and Crop Improvement*. CAB International, Wallingford, UK, ICRISAT, Andra Pradesh, India, IRRI, Manila, Philippines. pp. 467–486.

Ceccarelli, S. and Grando, S. (1997) Increasing the Efficiency of Breeding Through Farmer Participation. In: *Ethics and Equity in Conservation and Use of Genetic Resources for Sustainable Food Security*. Proceedings of a Workshop to Develop Guidelines for the CGIAR, April 21–25. Foz de Iguacu, Brazil, IPGRI; Rome, Italy, IPGRI. pp. 116–121.

Ceccarelli, S. and Grando, S. (2002) Plant Breeding with Farmers Requires Testing the Assumptions of Conventional Plant Breeding: Lessons from the ICARDA Barley Program. In: D. A. Cleveland David and D. Soleri (Eds.), *Farmers, Scientists and Plant Breeding: Integrating Knowledge and Practice*. Wallingford, Oxon, UK: CAB I Publishing International. pp. 297–332.

Ceccarelli, S. and Grando, S. (2005) Decentralized-Participatory Plant Breeding (pp. 145–156). In: R. Tuberosa, R. L. Phillips and M. Gale (Eds.), *In the Wake of the Double Helix: From the Green Revolution to the Gene Revolution*. Avenue Media, Bologna, Italy. pp. 145–156.

Ceccarelli, S. and Grando, S. (2007) Decentralized-Participatory Plant Breeding: An Example of Demand Driven Research. *Euphytica* 155, 349–360.

Ceccarelli, S., Grando, S., Tutwiler, R., Baha, J., Martini, A. M., Salahieh, H., Goodchild, A. and Michael, M. (2000) A Methodological Study on Participatory Barley Breeding. I. Selection Phase. *Euphytica* 111, 91–104.

Ceccarelli, S., Grando, S., Amri, A., Asaad, F. A., Benbelkacem, A., Harrabi, M., Maatougui, M., Mekni, M. S., Mimoun, H., El Einen, R. A., El Felah, M., El Sayed, A. F., Shreidi, A. S. and Yahyaoui, A. (2001) Decentralized and Participatory Plant Breeding for Marginal Environments. In: D. Cooper, T. Hodgink and C. Spillane (Eds.), *Broadening the Genetic Base of Crop Production*. CAB International. pp. 115–135.

Ceccarelli, S., Grando, S., Singh, M., Michael, M., Shikho, A., Al Issa, M., Al Saleh, A., Kaleonjy, G., Al Ghanem, S. M., Al Hasan, A. L., Dalla, H., Basha, S. and Basha, T. (2003) A Methodological Study on Participatory Barley Breeding. II. Response to Selection. *Euphytica* 133, 185–200.

Ceccarelli, S., Grando, S. and Baum, M. (2007) Participatory Plant Breeding in Water-Limited Environment. *Experimental Agriculture* 43, 1–25.

Di Falco, S. and Chavas, J. P. (2006) Crop Genetic Diversity, Farm Productivity and the Management of Environmental Risk in Rainfed Agriculture. *European Review of Agricultural Economics* 33, 289–314.

Eisemann, R. L., Cooper, M. and Woodruff, D. R. (1990) Beyond the Analytical Methodology: Better Interpretation and Exploitation of Genotype-by-Environment Interaction in Breeding. In: M. S. Kang (Ed.), *Genotype-by-Environment Interaction and Plant Breeding*. Department of Agron. Louisiana State Agric Experiment Stn, Baton Rouge, LA. pp. 108–117.

Falconer, D. S. (1981) Introduction to Quantitative Genetics (2nd Edn.). Longman Group Ltd., London.

Interdrought-II. The 2nd International conference on Integrated Approaches to Sustain and Improve Plant Production Under Drought Stress; Rome, Italy, September 24-28, 2005. Conference Conclusions and Recommendations (http://www.plantstress.com/ID2/ID2-Report.pdf, accessed 18 December, 2008).

Lilja, N. and Aw-Hasaan, A. (2002) Benefits and costs of participatory barley breeding in Syria. A Background Paper to a Poster Presented at the 25th International Conference of IAAE, Durban, South Africa, August 16–22 2003.

Mangione, D., Senni, S., Puccioni, M., Grando, S. and Ceccarelli, S. (2006) The Cost of Participatory Barley Breeding. *Euphytica* 150 (3), 289–306.

Murphy, K. M., Campbell, K. G., Lyon, S. R. and Jones, S. S. (2007) Evidence of Varietal Adaptation to Organic Farming Systems. *Field Crops Research* 102, 172–177.

Reynolds, M. P. and Borlaug, N. E. (2006) Applying Innovations and New Technologies for International Collaborative Wheat Improvement. *Journal of Agricultural Science* 144, 95–110.

Rhoades, R. and Booth, R. (1982) Farmer-Back-To-Farmer: A Model for Generating Acceptable Agricultural Technology. *Agricultural Administration* 11, 127–137.

Schnell, F. W. (1982) A Synoptic sSudy of the Methods and Categories of Plant Breeding. *Zeitshrift für Pflanzenzüchtung* 89, 1–18.

Simmonds, N. W. (1984) Decentralized Selection. *Sugar Cane* 6, 8–10.

Simmonds, N. W. (1991) Selection for Local Adaptation in a Plant Breeding Programme. *Theoretical and Applied Genetics* 82, 363–367.

Singh, M., Malhotra, R. S., Ceccarelli, S., Sarker, A., Grando, S. and Erskine, W. (2003) Spatial Variability Models to Improve Dryland Field Trials. *Experimental Agriculture* 39, 1–10.

Soleri, D., Cleveland, D. A., Smith, S. E., Ceccarelli, S., Grando, S., Rana, R. B., Rijal, D. and Labrada, H. R. (2002) Understanding Farmers' Knowledge as the Basis for Collaboration with Plant Breeders: Methodological Development and Examples from Ongoing Research in Mexico, Syria, Cuba, and Nepal. In: D. A. Cleveland David and D. Soleri (Eds.), *Farmers, Scientists and Plant Breeding: Integrating Knowledge and Practice*. Wallingford, Oxon, UK: CAB I Publishing International. pp. 19–60.

Yan, W., Hunt, L. A., Qinglai S. and Szlavnics, Z. (2000) Cultivar Evaluation and Mega-Environment Investigation Based on the GGE Biplot. *Crop Science* 40, 597–605.

Index

I

M

Q

R

S

T

U

V

W

X

Z

Printed in the United States of America